Workshop on the Tau/Charm Factory

Workshop on the Tau/Charm Factory

Argonne National Laboratory
Argonne, IL

June 21–23, 1995

Edited by
José Repond

AIP CONFERENCE PROCEEDINGS 349

American Institute of Physics   Woodbury, New York

Authorization to photocopy items for internal or personal use, beyond the free copying permitted under the 1978 U.S. Copyright Law (see statement below), is granted by the American Institute of Physics for users registered with the Copyright Clearance Center (CCC) Transactional Reporting Service, provided that the base fee of $6.00 per copy is paid directly to CCC, 222 Rosewood Drive, Danvers, MA 01923. For those organizations that have been granted a photocopy license by CCC, a separate system of payment has been arranged. The fee code for users of the Transactional Reporting Service is: 1-56396-523-2/ 96 /$6.00.

© 1996 American Institute of Physics.

Individual readers of this volume and nonprofit libraries, acting for them, are permitted to make fair use of the material in it, such as copying an article for use in teaching or research. Permission is granted to quote from this volume in scientific work with the customary acknowledgment of the source. To reprint a figure, table, or other excerpt requires the consent of one of the original authors and notification to AIP. Republication or systematic or multiple reproduction of any material in this volume is permitted only under license from AIP. Address inquiries to Office of Rights and Permissions, 500 Sunnyside Boulevard, Woodbury, NY 11797-2999; phone 516-576-2268; fax: 516-576-2499; e-mail: rights@aip.org.

L.C. Catalog Card No. 95–81467
ISBN 1-56396-523-2
DOE CONF- 9506186

Printed in the United States of America.

CONTENTS

Preface ix
    J. Repond

INTRODUCTION AND OVERVIEW

Welcome to the Workshop on the Tau/Charm Factory 5
    A. Schriescheim
Welcome to the Tau/Charm Factory Workshop............................ 7
    Z. Zheng
Overview of the Tau/Charm Factory.................................... 11
    J. Kirkby

TAU LEPTON PHYSICS

Importance of Precision Measurements in the Tau Sector................ 45
    A. Pich
Review of Tau Physics at CLEO II with Prospects for the
B Factories.. 62
    C. G. White for the CLEO Collaboration
Measurements of Properties of the τ Lepton at LEP 72
    S. R. Wasserbaech for the ALEPH Collaboration
Prospects for High Precision Measurements at the τcF Including
Current Achievements of BES 89
    T. Huang
Test of T and CP Violation in Leptonic Decay of $\tau^{\pm}$......... 104
    Y. S. Tsai
Addendum to the Test of CP Violation in Tau Decay 113
    Y. S. Tsai
Decay Rates, Structure Functions and New Physics Effects in
Hadronic Tau Decays... 119
    M. Finkemeier and E. Mirkes

ACCELERATOR DESIGNS

Tau/Charm Factory Collider Design at BEPC........................... 139
    Y. Z. Wu, N. Huang, L. H. Jin, and D. Wang
JINR Tau/Charm Factory: Status and Perspectives 152
    E. Perelstein
Argonne Tau/Charm Factory Collider Design Study 160
    L. C. Teng, E. A. Crosbie, J. Norem, and J. Repond
B Factory Collider Designs and Future Plans 172
    M. S. Zisman

Round Table Discussion of Collider Designs 196
  S. Kurokawa, E. Perelstein, L. Teng, M. Tigner, Y. Z. Wu, and M. Zisman

DETECTOR STUDIES

Monte Carlo Simulation of the Tau/Charm Factory at IHEP 203
  Y. Z. Huang, H. M. Liu, W. J. Xiong, J. L. Hu, B. S. Cheng, D. H. Zhang,
  S. Jin, S. M. Chen, X. L. Fan, A. M. Ma, and S. Z. Ye
A Fast Time-of-Flight Detector also Used as a Tracker 212
  T.-Y. Chen, M. He, N.-J. Zhang, and X.-Y. Zhang
Program of the Beijing TcF Feasibility Study 222
  S.-H. Wang

CHARMONIUM PHYSICS AND HADRONIC SPECTROSCOPY

Charmonium Theory 233
  A. X. El-Khadra
Fermilab E760 and E835: Charmonium Formation in $\bar{p}p$ Annihilation 251
  C. M. Ginsburg for the E760/E835 Collaboration
Electroweak Radiative Corrections and Measurements of R_{had} 270
  M. L. Swartz
Theoretical Predictions for Exotic Hadrons 285
  T. Barnes
Status of Hadron Spectroscopy at LEAR 307
  C. A. Meyer
Precision Charmonium Physics 319
  W. Toki

CHARM PHYSICS: TESTS OF THE STANDARD MODEL

Quo Vadis, Fascinum? 331
  I. I. Bigi
Charm Decay in Fixed Target: Present and Future 345
  J. Wiss
Charm Physics at CLEO and Future Prospects at a B-Factory 375
  D. Fujino
Prospects for High Precision Measurements of Charmed Hadron
Properties at the τCF 391
  P. Roudeau

CHARM PHYSICS: BEYOND THE STANDARD MODEL

Potential for Discoveries in Charm Meson Physics 409
  G. Burdman

High-Impact Charm Physics at the Turn of the Millennium................. 425
    D. M. Kaplan
An Overview of $D^0\overline{D}^0$ Mixing Search Techniques: Current Status
and Future Prospects... 447
    T. Liu
Rare D Decays, $D^0\overline{D}^0$ Mixing, and CP Violation at a Tau/Charm
Factory... 480
    G. Gladding

CONCLUSION

Tau/Charm Workshop Summary.. 491
    F. J. Gilman

APPENDICES

Program Advisory Committee 509
Local Organizing Committee 509
Scientific Program 510
List of Participants... 515
Author Index.. 517

PREFACE

The workshop on the Tau/Charm factory took place over the period June 21–23, 1995 at Argonne National Laboratory, Argonne, IL. The meeting was co-organized by the Institute of High Energy Physics in Beijing, China and Argonne National Laboratory. About 60 physicists from China, Europe, Japan, Russia, and the U. S. attended the workshop.

The meeting had a heavy schedule: more than 30 presentations in less than three days. The high quality of the presentations and the mixture of theoretical, experimental, and accelerator talks stimulated lively discussions culminating in a spirited round table discussion led by Maury Tigner.

The workshop banquet took place at Emilio's in Naperville where those excellently prepared Tapas quickly made us resent the limited capacity of our stomachs. And this despite the abundant flow of Rioja...

We would like to thank all the members of the Program Advisory Committee and the Local Organizing Committee who helped make this meeting a success. Special thanks go to Mary Burke and Sandra Klepec. Without their long hours of preparation and their professional organizational skills, the workshop would not have been nearly as successful as it was. Financial support was generously provided by Frank Y. Fradin, Associate Laboratory Director and Lawrence E. Price, Director of the High Energy Physics Division at Argonne.

José Repond
Editor

INTRODUCTION AND OVERVIEW

Welcome to the Workshop on the Tau/charm Factory

Alan Schriesheim

Argonne National Laboratory
Argonne, IL, U.S.A.

Good morning all. And welcome to Argonne.

Almost 50 years ago, this laboratory began as a center for research in basic nuclear science. In the following years, Argonne grew from a small group of pioneering physicists with one research objective to the facility you are visiting today... almost 5,000 people in over 100 buildings pursuing more than 200 research projects.

Through all that growth over those years, we remembered our roots and always fostered active programs in basic nuclear and elementary particle physics. The high energy physics program at Argonne has produced outstanding science, as you well know, and its facilities have benefitted the other programs of the laboratory. For example, part of the ZGS particle accelerator is now the IPNS, a marvelous tool for solid state science.

So it is especially pleasing for me today to continue Argonne's presence in the field by welcoming leaders of the international High Energy Physics community.

If my colleagues will permit me, I want to issue a special welcome to the distinguished scientists from China. I returned only a few weeks ago from you country ... a visit that was as enlightening as it was enjoyable. I hope we prove as gracious a host to you as your colleagues in China were to me.

The workshop opening today focuses on an important opportunity for broadened knowledge ... the Tau/charm Factory. An electron-positron collider optimized for producing large rates of tau leptons and charmed quarks would, we believe, make enormous advances in basic knowledge, including perhaps the understanding of the matter/antimatter asymmetry in the universe.

With such an important agenda before you, the best contribution I can make is to offer my very best wishes... and let you get on with your work.

Thank you.

Welcome to the Tau/charm Factory Workshop

Zhipeng Zheng

Institute of High Energy Physics
Chinese Academy of Sciences
Beijing 100039, China

Dear Ladies and Gentlemen, Colleagues and Friends:

It is my pleasure to be here and to welcome all of you to this workshop on the τ-charm factory. This very important meeting follows previous meetings on the same topic which were held at SLAC in 1989, at Seville and JINR in 1991, at Marbella and Dubna in 1993, at SLAC in 1994, and in France in 1995.

As you know, the purpose of this workshop is to motivate the construction of a τ-charm factory and explore possible technical options for both the collider and the detector. The following topics will be discussed at this workshop:

1) The physics potential of a τ-charm factory,

2) Progress in collider and detector designs,

3) Worldwide interest in a τ-charm factory.

In the different sessions of the workshop we will learn about and exchange new ideas and experiences related to the physics, the machine and the detector of a τ-charm factory. We will explore the possibilities of international collaboration in different stages of the τ-charm factory project: the initial feasibility study, the R&D phase, and the construction phase.

I would like to take this opportunity to inform you about our efforts in promoting the construction of a τ-charm factory in China.

First of all, I would like to emphasize the importance of last year's workshop held at SLAC. At that workshop it was recognized that a τ-charm factory was necessary since several physics topics, such as the search for glueballs, CP violation in the lepton sector, and physics of the Charmonium system, are uniquely accessible to the τ-charm factory and cannot be pursued with the same sensitivity at the B-factories. In addition, the τ-charm factory has the great advantage of high precision and low systematic errors in exploring the physics of the τ lepton and the charm quark. These conclusions were persuasive for the Chinese scientific community and the leadership responsible for defining the policy for the future development of science and technology in China.

I would like to take this opportunity to express particular gratitude towards Prof. T.D. Lee, who has constantly and strongly supported the development of China's High Energy Physics program. During his visit in China last October, he greatly advanced the Chinese τ-charm factory project by persuading the Chinese leadership to approve a budget of 5 million Chinese Yuan to be used for a feasibility study. Despite its moderate financial value, it bears a great significance in showing the support of the Chinese government, similar to a green traffic light saying: "go ahead".

Last February, the feasibility study of the τ-charm factory was also approved by a committee consisting of High Energy Physicists, leading to the formal and official approval by the Chinese Academy of Sciences.

Soon afterwards, a team consisting of about one hundred physicists and engineers was organized. They are mainly from IHEP, but also come from Beijing University, Qinghua University, the University of Chinese Science and Technology, Shandong University, Nanjing University and the Institute of Theoretical Physics. The team consists not only of experienced physicists, but also includes many younger scientists. All show great interest and enthusiasm toward the τ-charm factory. Without exaggeration we can claim that we have attracted the best part of the Chinese High Energy Physics community.

The topics and goals of our feasibility study are:

1) Define and address the critical issues, such as how to meet the challenge of its physics goals, how to achieve a luminosity of $10^{33}/\text{cm}^2/\text{s}$, how to handle the expected high data acquisition rate, and how to achieve the required detector performance.

2) Complete a detailed conceptual design including both the collider and detector.

Our strategy is the following:

1) Use of the existing knowledge and experience of the international community, by studying the information presented in the previous τ-charm factory workshops, and by encouraging international collaboration and free exchange of information.

2) Study the new technologies employed in the construction of new accelerators, such as the B-factories and LHC.

3) Use our own facilities and the experience gained with BEPC and BES.

4) Study possible upgrades of the BEPC linac for future use as the injector for the τ-charm factory.

We foresee about one and a half years for the feasibility study, to be followed by a two year R&D program. For the latter, we will request a budget of 40 million Chinese yuan. According to our plans, a detailed engineering design of both the collider and the detector will be complete by the end of the century.

We plan to hold the next workshop on the τ-charm factory in Bejing at the beginning of 1996. The meeting will be organized jointly by IHEP and CCAST (Center of Chinese Advanced Science and Technology, directed by Prof. T.D. Lee). Selected topics related to the physics, the collider and the detector of the τ-charm factory will be discussed. Colleagues from abroad are warmly welcome to attend and contribute to the workshop.

Dear colleagues and friends, the international High Energy Physics community is now facing a severe challenge. Advancement of the field requires large investments for which neither governments nor society yet recognizes the need. This contradiction makes it difficult to propose the construction of new accelerators. In my view, it is our responsibility to find a solution to this problem, a solution which must include worldwide cooperation.

International collaboration was one of the major factors contributing to the success of BEPC. Now, in its turn, the R&D program of the Chinese τ-charm factory needs to attract a broad international team. In this regard, we are glad to be able to report some success:

1) At the 15th Meeting of the PRC/US Joint Committee for Cooperation on High Energy Physics, both sides expressed strong interest in the τ-charm factory project. Since then, Chinese and American physicists are exploring the possibility and the mode of cooperation.

2) Prof. Yuji Sugawara, the director of KEK, showed active support for the τ-charm factory to be constructed in China. Prof. Shin-Ichi Kurokawa and Prof. Shu-Hong Wang of IHEP plan to set up a joint research group including members of the two institutes and centered on the τ-charm factory project.

3) Prof. Maury Tigner was invited by IHEP to serve as adviser for the τ-charm factory design. He worked at IHEP for two months this year and helped us with both the τ-charm factory project and with BEPC machine studies. We are looking forward to his next visit starting in September 1995.

I am confident that this meeting will play a historic role in expanding our understanding of the physics potential of the τ-charm and in promoting international collaboration on the τ-charm factory. To conclude my remarks, I would like thank Prof. José Repond and the Local Organizing Committee for the excellent work done in preparation of this workshop. I would also like to thank Prof. Alan Schriesheim and Prof. Lawrence E.Price for their support in organizing this workshop at Argonne.

And I wish the workshop full success

Overview of the Tau-Charm Factory

Jasper Kirkby

CERN, Geneva, Switzerland

Abstract. We present a brief overview of the physics interest in the Tau-Charm Factory and the designs of the accelerator and detector.

1 Introduction

There is an almost universal conviction among physicists that the Standard Model represents an incomplete picture of Nature—that 'New Physics' has to exist. There is, however, no agreement on even the general character of the New Physics and, moreover, data have yet to reveal any signs of it. We are thus in crucial need of further experimentation to uncover such evidence. The traditional experimental approach has been to go to higher energies to search for new particles and interactions. However, more recently a new approach has emerged which can answer different sorts of questions than those addressed by the high energy machines; it is simply to gather tens or hundreds of times more data at lower energies than has been previously available. This profusion of data, recorded in a new generation of advanced detectors, offers a study of the presently-known particles with unprecedented precision and it may solve some of the most profound questions facing particle physics.

Among the most fundamental puzzles at present is the replication of quark/lepton families or flavours—the origin of which is completely obscure—and the associated questions of the origins of the CKM matrix and CP violation for quarks, and of their apparent absence for leptons. A detailed study of the heavy flavour states like the τ lepton, charm hadrons and beauty hadrons holds the highest promise of shedding light on these questions. There are three basic reasons for this: i) heavy-flavoured particles possess the highest sensitivity to New Physics at high mass scales, ii) they can be treated with more reliable theoretical tools and iii) their decays have so far only been superficially explored in comparison with the lighter-flavoured particles.

The particles to be studied are generated by a new generation of intense low-energy accelerators known as particle factories. The primary requirements of a particle factory are to produce specific particles in copious quantities and with low backgrounds. These characteristics are well suited to high-luminosity e^+e^- colliders. In general the optimum machine energy to study a particular particle corresponds to the region near its pair-production threshold, which can provide the highest cross-sections, lowest backgrounds and other favourable ex-

© 1996 American Institute of Physics

perimental conditions. Other important energies at e$^+$e$^-$ colliders correspond to production of the narrow vector resonances (ϕ, J/ψ, ψ', Υ, Υ', Z^0, ...) which, in addition to being interesting particles in their own right, constitute high-rate secondary sources of lighter particles. The energies of interest for e$^+$e$^-$ collider particle factories are summarised in Table 1, and the production cross sections and typical event rates for these machines are indicated in Fig. 1.

Table 1: Summary of the energies of interest for e$^+$e$^-$ collider particle factories.

E_{cm} (GeV)	Particle resonance	Constituent threshold	Accelerator
1	ϕ	s$\bar{\text{s}}$	ϕ Factory
3–5.6	J/ψ, ψ'...	$\tau\bar{\tau}$, c$\bar{\text{c}}$	τc Factory
9–11	Υ(1S), Υ(2S)...	b$\bar{\text{b}}$	B Factory
91	Z^0	Z^0	Z Factory
350	θ	t$\bar{\text{t}}$	T Factory

The essential experimental tools for exploring the heavy-flavoured particles are a τ-charm Factory (τcF) and a beauty Factory. These machines address similar basic questions, but in complementary ways: the beauty Factory is optimised for beauty particles and CP violation in B decays; and the τ-charm Factory is optimised for the τ lepton, charm particles, and the spectroscopy of hidden charm states and light flavour hadrons. At the τcF, CP violation will be explored in the charm (D meson), strange (Λ and Ξ hyperons) and lepton (τ) sectors. Whereas a beauty Factory also generates large τ and charm samples—with statistics comparable to those of a τ-charm Factory (Table 2)—the key element of most future precision measurements will be how well the systematic errors can be controlled. Here the unique experimental environment of the τ-charm Factory is likely to prove an important advantage, as we will now describe.

2 Experimental environment

The τ-charm Factory operates primarily in the total energy range 3–5.6 GeV, with a peak luminosity of 10^{33} cm^{-2}s^{-1} at 4 GeV. This region is rich with resonances and particle thresholds (Fig. 2). The cross-sections for τ and charm production in e$^+$e$^-$ collisions are higher in this region than at any other energy. This is due to the overall E_{cm}^{-2} dependence of the cross-sections (as seen in Fig. 1) and also to the presence of charm resonances such as the ψ''(3.77), which has a peak cross section of 5 nb (over a continuum background of 13 nb) and decays with almost equal probability to pure D$^0\bar{\text{D}}^0$ or D$^+$D$^-$ final states. Furthermore, the various τ and charm signals can be turned on or off by adjusting the beam energy above or below each particular threshold. This

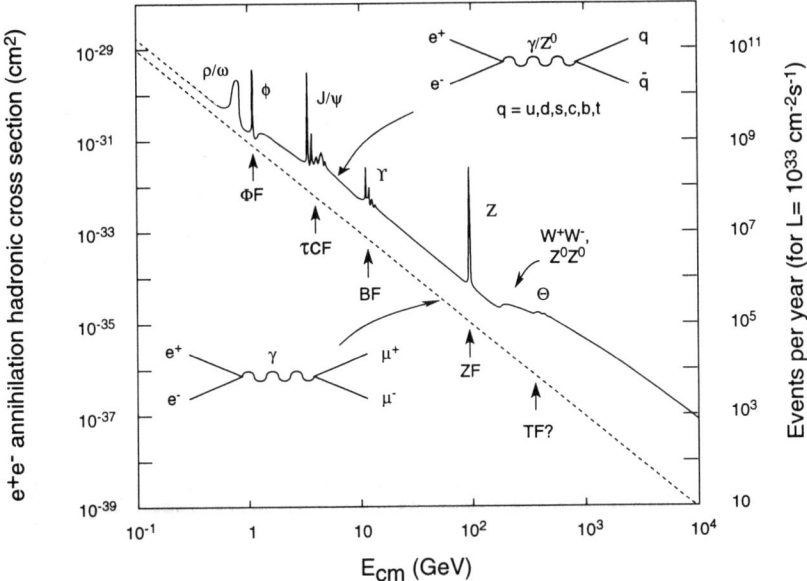

Figure 1: The e⁺e⁻ annihilation cross-section in the energy range 1 GeV <E_{cm}< 10 TeV. The energies of interest for particle factories are indicated, together with the typical events rates.

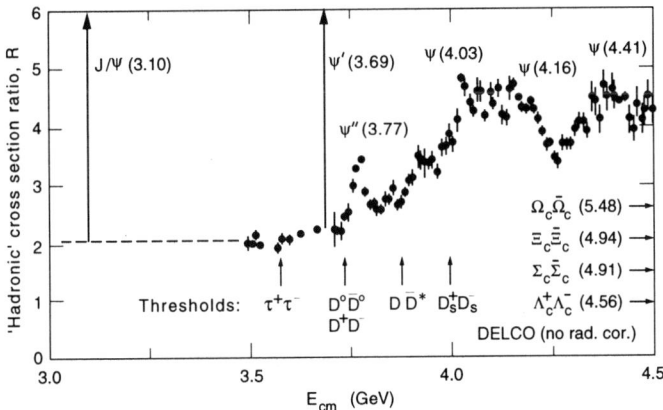

Figure 2: The hadronic cross section ratio, R, in the τ-charm threshold region. The ratio $R = \sigma(e^+e^- \to \text{'hadrons'}) / \sigma(e^+e^- \to \mu^+\mu^-)$, where 'hadrons' include both $q\bar{q}$ and $\tau^+\tau^-$ events. The data are from DELCO at SPEAR.

Table 2: Estimated statistics per year at the τ-charm Factory, at the indicated energies and integrated luminosities. (An integrated luminosity of 10 fb^{-1} per year corresponds to $L_{peak} = 10^{33}$ cm^{-2}s^{-1}.)

Event type		Events/year	Energy [GeV]	Integrated Luminosity [fb^{-1}]
$D^0\bar{D}^0$		2.9×10^7	$\psi''(3.77)$	10
D^+D^-		2.1×10^7	$\psi''(3.77)$	10
$D_s^+D_s^-/D_s^\pm D_s^{*\mp}$		0.9×10^7	4.14	10
$\Lambda_c^+\Lambda_c^-$		0.3×10^7	4.8?	3
$\Sigma_c\bar{\Sigma}_c$		0.1×10^7	5.2?	2
$\Xi_c\bar{\Xi}_c$		0.3×10^6	5.2?	2
$\Omega_c\bar{\Omega}_c$		0.3×10^5	5.6?	1
$\tau^+\tau^-$		0.5×10^7	3.56	10
″		2.4×10^7	3.67	10
″		3.5×10^7	4.25	10
″	(monochromator optics)	10.0×10^7	$\psi'(3.69)$	7
J/ψ	(standard optics)	1.3×10^{10}	$J/\psi(3.10)$	6
″	(monochromator optics)	8.0×10^{10}	$J/\psi(3.10)$	4
ψ'	(standard optics)	0.4×10^{10}	$\psi'(3.69)$	10
″	(monochromator optics)	1.4×10^{10}	$\psi'(3.69)$	7

ability to measure *experimentally* all backgrounds in τ and charm studies is unique to the τcF; at other machines the backgrounds are generally estimated by Monte Carlo simulations.

Operating near threshold also completely excludes background contributions from higher-mass particles such as b quarks. In addition, the heavy-flavoured particles appear in simple particle–anti-particle final states. As a consequence, each of the heavy-flavoured particles (D^0, $D^\pm$, $D_s^\pm$, $\Lambda_c^\pm$, Σ_c, Ξ_c, Ω_c, etc.) can be tagged simply by observing the decay of its partner. Tagging of charm particles is unique to the threshold region and has important advantages: measurement of the four-vector and identification of the recoiling particle without pre-selection of its decay mode, reduced biases, absence of additional particles, reduction of backgrounds, and exact flux normalisation (which allows measurement of *absolute* branching ratios).

Tau leptons can also be singly tagged at the τcF. Below $c\bar{c}$ and $b\bar{b}$ thresholds, the $\tau^+\tau^-$ events can be cleanly isolated with simple selection criteria on only one of the two τ's. For example, the requirement e + E_{miss} (which is satisfied by *one* of the two τ's in an event decaying via $\tau^- \rightarrow e^-\bar{\nu}_e\nu_\tau$) is expected to select $\tau^+\tau^-$ events with 24% efficiency and <0.1% background.

An additional feature of the threshold region is the small Lorentz boost of the particles under study. This has a number of advantages, including monochromatic spectra for two-body decays, decreased overlap of the secondary particles, easier (i.e. cleaner) $\pi/K/p$ separation and increased sensitivity for tests of CP violation in τ decays (Section 3.2).

Finally, achieving small systematic errors requires a precise knowledge of the detector performance: resolutions, efficiencies, particle mis-identifications, etc. Here the τcF has the important advantage of high-rate calibration sources —the J/ψ and ψ'—which provide numerous exclusive physics channels that can be used to calibrate and monitor all aspects of the detector. Examples of some J/ψ decays channels (1.3 kHz total event rate) that can be used to calibrate the particle identification are as follows: $\pi^\pm/\pi^0/\gamma$ ($J/\psi \to \rho\pi$, 20 Hz), $K^\pm$ (K^*K, 7 Hz), K_L^0 ($K_L^0 K_S^0$, 0.1 Hz), p ($p\bar{p}/p\bar{p}\pi^0/p\bar{p}2\pi^\pm$, 13 Hz), n ($np\pi^\pm$, 3 Hz), $e^\pm$ ($e^+e^-/e^+e^-\gamma$, 80 Hz) and $\mu^\pm$ ($\mu^+\mu^-/\mu^+\mu^-\gamma$, 80 Hz).

3 Physics interest

3.1 Introduction

Since the initial idea for a τ-charm Factory [1] in 1987 and the first machine design [2, 3], numerous workshops [4, 5, 6, 7, 8, 9, 10, 11] have been held to explore the physics potential and develop the designs of the accelerator and detector. In parallel—and in part stimulated by the interest in the τcF—a series of workshops dedicated to the τ lepton [12, 13, 14] has been established to review the current experimental and theoretical developments, and to discuss the future experimental facilities.

In the coming years, substantial experimental progress is expected in τ and charm physics at BEPC, LEP, CESR and B Factories, and in fixed-target experiments. Detailed studies have been made at the more-recent workshops [8, 9, 10, 11, 15, 16, 17] of the expected progress in τ-charm physics at these accelerators during the next 10 years. The conclusions are as follows. Other machines are competitive for some τ-charm measurements—especially those that are not expected to be limited by backgrounds and systematic errors. However, there are important experiments that are accessible only at a τcF, or else cannot be done elsewhere with the necessary precision. The τcF is the unique machine for performing a *comprehensive* precision study of τ and charm decays.

An indication of the scope of the physics programme can be seen in Table 3, which lists some of the physics that will be accessible as the luminosity of the collider progresses. Among the highlights are:

- τ and ν_τ masses.

Table 3: Physics-reach versus luminosity of the τ-charm Factory.

L_{peak} [cm^{-2}s^{-1}]	Physics-reach per 1 year's data
10^{32}	▷ glueballs, hybrid exotics and hybrid charmonium ▷ excited ψ and D states ▷ semi-leptonic D decays to $\mathcal{O}(1\%)$ precision ▷ τ decay Br's to $\mathcal{O}(0.1\%)$ precision ▷ $\Lambda_c^\pm$, Σ_c, Ξ_c, Ω_c, etc. decays to $\mathcal{O}(5\%)$ precision ▷ V-A structure in τ decays comparable to precision in μ decays ▷ doubly Cabibbo suppressed D^0, $D^\pm$, $D_s^\pm$ decays to $\mathcal{O}(3\%)$ precn. ▷ pure leptonic D decays, f_D and f_{D_s} to $\mathcal{O}(2\%)$ precision
10^{33}	▷ $\tau \to eX$ limit $\simeq 10^{-5}$; constraints on ν_τ masses below 1 MeV/c^2 ▷ $D^0\bar{D}^0$ mixing at 10^{-5}, within SM level ▷ rare $\tau/D/J$-ψ decays (LFV, FCNC, etc.) to limits $\simeq 10^{-7}$-10^{-8} ▷ direct ν_τ mass limit $\simeq$ 1 MeV/c^2 ▷ CP violation in D decays at SM level ▷ CP violation in τ decays at milli-weak level (10^{-3}) ▷ CP violation in Λ, Ξ decays at SM level ▷ ?? ▷ ??
10^{34}	▷ ??

- CP violation in the lepton (τ) and charm (D meson) sectors.

- Structure of the weak current in τ decays.

- Pure leptonic D decays; precision measurements of f_D, f_{D_s}, V_{cs} and V_{cd}.

- $D^0\bar{D}^0$ mixing.

- Gluonic spectroscopy; glueballs and hybrid gluonium.

These experiments and others are summarised in the remainder of this section.

3.2 Tau physics summary

There is no known difference between like-charge leptons (or quarks) of different families, except the value of their mass. In the Standard Model, masses are proportional to couplings to the Higgs particle. Even the discovery of "the Higgs" will not solve the problem of what makes the couplings different. Is there something other than mass that distinguishes like-charge objects from each other? For e's and μ's we know the answer to be negative to a high degree of precision. However, in comparison, our experimental knowledge of τ's is superficial; and we know even less about ν_τ's.

The τ and ν_τ leptons may offer the best *a priori* prospects to find the underlying reason for the existence of three families. The τ is a third generation lepton; its sensitivity to new physics is greatly increased by its relatively high mass; and it is the only lepton with a wide variety of decay channels—both leptonic and hadronic—of which many can be calculated with high precision in the Standard Model.

Some of the main τ experiments [8, 18, 19] are as follows:

- A precise measurement will be made of the $\tau^+\tau^-$ production cross-section near threshold (expected accuracy $\simeq 0.1$ %). The τ mass will be determined with a precision better than 100 keV (~10 keV with monochromator optics). The τ anomalous magnetic moment will be measured with a sensitivity at the level of the first QED contribution ($\alpha/2\pi$).

- The τcF will substantially improve on the ν_τ-mass limits, through accurate measurements of the end-points of the hadronic invariant-mass distributions in high-multiplicity τ decays like $\tau^- \to 5\pi^\pm \nu_\tau$, $3\pi^\pm 2\pi^0 \nu_\tau$ and $K^-K^+\pi^-\nu_\tau$. The estimated sensitivity (95% CL) is $\simeq 1$ MeV, to be compared with the present limit of 24 MeV.

- A sensitive search will be made for CP violation in τ decays. The Standard Model makes the *ad hoc* assumption of no CP violation in the lepton sector, but this has been barely tested experimentally. Tests of CP violation in τ decays have recently been proposed involving a search for finite CP-odd observables in the final state in pure leptonic decays $e^-\bar{\nu}_e\nu_\tau/\mu^-\bar{\nu}_\mu\nu_\tau$ [20, 21] (which avoids any uncertainty in assigning a possible CP violation to the leptonic or hadronic vertices). Since the initial e^+e^- state is CP-even, any CP-odd final state would be evidence of CP violation. Possible observables are $< \hat{p}_{in}.(\hat{p}_e \times \hat{p}_\mu) >$ and $< \vec{\sigma}_\tau.(\hat{p}_e \times \hat{p}_\mu) >$, where $\hat{p}_{in}$ is the unit vector parallel to the incident e^- momentum, $\hat{p}_e$ and $\hat{p}_\mu$ are the unit vectors along the decay lepton directions, and $\vec{\sigma}_\tau$ is the direction of spin of the τ (requiring longitudinal polarization of one or both beams). If CP violation occurs at a level similar to the quark system (i.e. at the milli-weak level, 10^{-3}) then it should be observable at the τcF.

Non-zero τ electric dipole moment, d_τ, would be an unambiguous signal of T (CP) violation [22]. The present limit on the τ electric dipole moment could be substantially improved by studying T-odd triple correlations of the final $\tau^+\tau^-$ decay products. Moreover, both the real and the imaginary parts of the electric dipole form factor could be tested. It has been shown recently [23] that the T-odd correlations are significantly enhanced when the e^+ and/or e^- beams are longitudinally polarized. The τcF with longitudinally polarized beams could reach sensitivities of 1×10^{-19} e cm for $\Delta(\mathrm{Re}d_\tau)$ and 6×10^{-15} e cm for $\Delta(\mathrm{Im}d_\tau)$.

- The leptonic (V-A) τ-decay parameters will be precisely measured, constraining the underlying dynamics with a precision comparable to, or even better than, that of the μ. The expected accuracies are better than 0.3% for ρ, 0.003 for $|\eta_\mu|$, 1.6% for ξ, 2.8% for δ and 15% for ξ'_μ.

- A systematic programme will be carried out to measure the branching ratios of all τ decay channels and search for signs of discrepancies with the theoretical expectations. The τcF has notable strengths for decays involving K's and multiple γ's, which are poorly-known at present. In a one-year data sample, the expected precisions on the 1-prong branching ratios are 0.15% (e, μ), 0.2% (π) and 0.8% (K), whereas present errors are 1% - 6%. These precise measurements will allow a test of the universality of the leptonic charged currents at the 0.2–0.3% level. In addition present experimental controversies, such as the so-called '1-prong problem', should be definitively settled.

- A detailed analysis will be made of each τ hadronic decay channel, allowing many interesting tests of QCD to be made. This analysis will provide, among other results, a precise determination of the vector- and axial-vector-current spectral functions, both in the Cabibbo-allowed and Cabibbo-suppressed channels.

- Rare and forbidden τ decays will be explored with a factor $\gtrsim 100$ improvement in sensitivity beyond present experiments. The τcF is especially suited to study two-body decays like $\tau^- \to l^- X$ (l = e, μ; X = Majoron, familon, flavon,...), which lead to the distinctive signature of monochromatic leptons. The expected branching-ratio sensitivity for this decay is better than 10^{-5}, to be compared with the present limits of 1-2%. Second-class-current decays like $\tau^- \to \pi^- \eta \nu_\tau$ will be cleanly measured at the expected Standard Model level of about 10^{-5}.

In summary the τcF provides a unique and ideal experimental environment for precision τ studies: unsurpassed statistics, complete absence of backgrounds from heavy-flavoured hadrons, production near rest, and the possibility of longitudinally-polarized beams. This will allow the τ to be probed with a precision between one and two orders of magnitude beyond current experiments, potentially leading to fundamental discoveries on the origin of the three families of leptons and quarks.

3.3 Charm physics summary

The *qualitative* pattern in the observed lifetimes of the charm hadrons—D mesons, $\Lambda_c^\pm$, Ξ_c and Ω_c baryons—can be understood in a natural way due to the interplay of various non-spectator effects. Furthermore, phenomenological models have been able to describe a large body of exclusive D meson decays without major conflict with the data. However, despite this qualitative success, there are compelling reasons for raising charm studies to a higher level of precision. These reasons can be distilled into three main points:

1. In the Standard Model with three families and a beauty lifetime ~ 1 ps, the two CKM parameters that are relevant for weak charm transitions, namely V_{cs} and V_{cd}, are tightly constrained by the unitarity of the CKM matrix. Furthermore, second generation theoretical technologies have emerged, allowing a more rigorous treatment of charm decays: QCD sum rules, heavy quark expansions and QCD lattice simulations. Charm decays then constitute a unique laboratory to probe QCD *quantitatively* at the interface between the perturbative and non-perturbative regimes.

2. The theoretical tools that are needed to fully exploit the discovery potential of beauty decays must first to be tested and calibrated with charm decays. The same is true for the experimental tools; improved measurements of charm absolute Br's, etc. at a τcF are required for analysing beauty decays.

3. There is the potential for fundamental surprises in areas such as D^0-$\bar{D}^0$ mixing, CP violation in D decays, or rare D (or J/ψ) decays.

Some of the main experiments [8, 24, 25] are as follows:

- The τcF will precisely determine f_D and f_{D_s} by measuring each of the pure leptonic branching ratios, Br(D$\rightarrow \mu\nu$) and Br(D$_s \rightarrow \mu\nu, \tau\nu$), to $\sim 2\%$ accuracy in a one-year data sample. On the one hand this will provide an important test of QCD and $SU(3)$ symmetry and, on the other hand, it is an essential stepping stone in predicting f_B—a quantity crucial in interpreting B^0-$\bar{B}^0$ mixing and CP violation in B decays. A measurement of pure leptonic charm decays at this level of precision can only be made at a τcF, which has the important advantages of tagged $D^\pm$/ $D_s^\pm$ and a kinematic constraint on the missing ν('s).

- Similarly the study of inclusive and exclusive semileptonic charm decays will test our understanding of QCD and calibrate the methods used in extracting the CKM parameters V_{cb} and V_{ub} from semileptonic B decays. It should be noted that *inclusive* semileptonic charm decays can be studied only near threshold. Measurements of the semileptonic branching ratios for $D_s^\pm$, $\Lambda_c^\pm$ and Ξ_c are especially needed, and are difficult to make at other machines. In particular for the baryons, the semi-leptonic branching ratios will provide information on the underlying dynamics that is independent of and thus in addition to that inferred from lifetime measurements.

The pure- and semi-leptonic decays will provide a precise determination of V_{cs} and V_{cd}. Although unitarity tightly constrains these values (uncertainties of 0.1% for V_{cs} and 1% for V_{cd}), this has been poorly checked experimentally: the direct measurements have an accuracy of only 10–20%. At a τcF the D semileptonic branching ratios can be measured to an accuracy better than 1%, whereas present errors are 6% for $D^0 \to K^- e^+ \nu_e$ and 50% for $D^0 \to \pi^- e^+ \nu_e$. Overall, V_{cd}/V_{cs} can be determined to $\sim 1\%$ at a τcF, which is comparable to the present precision of $\theta_{Cabibbo}$.

- Although absolute branching ratios have been obtained for D^0 and $D^\pm$ mesons with an accuracy of 10–15%, there exist no useful direct measurements of the absolute branching ratios for any other charm hadrons: $D_s^\pm$, $\Lambda_c^\pm$, Σ_c, Ξ_c, Ω_c, etc. A τcF provides the only way—through the unique capability to tag each of the various charm hadrons—to improve the precision of the absolute branching ratios for $D^0/D^\pm$ mesons to the per cent level, and to establish absolute branching ratios at 5% precision for $D_s^\pm$, $\Lambda_c^\pm$, Σ_c, Ξ_c and Ω_c.

A comprehensive analysis will be made of not only Cabibbo-allowed decay modes but also of once- and twice-Cabibbo-suppressed modes of D^0, $D^\pm$ and $D_s^\pm$; likewise for Cabibbo-allowed and once-suppressed decay modes of $\Lambda_c^\pm$. Such data would also allow us to refine predictions on CP-violating asymmetries in charm decays.

- Within the Standard Model, it is expected that D^0-$\bar{D}^0$ mixing occurs via long-range interactions at the level of 10^{-4} and that direct CP violation will give rise to asymmetries of the order of 10^{-3} in certain Cabibbo-suppressed D decay channels [$\mathcal{O}(0.5\%)$ Br]. At these levels, the τcF will be able to observe D^0-$\bar{D}^0$ mixing (the sensitivity is 2×10^{-5} in a one-year data sample) and to just reach the sensitivity where CP violation in D decays may be observed. This opens up the exciting possibility of exploring the mechanism for CP violation in a system that is complementary to the B or K systems and, moreover, especially sensitive to sources beyond the Standard Model.

It may also be possible to observe direct CP violation by measuring differences in the decay parameters of hyperons and anti-hyperons (Λ and Ξ) produced in J/ψ decays [26]. Since the expected differences are small [e.g. $\mathcal{O}(10^{-3})$ in the polarization asymmetry parameter B'_Ξ], very large statistics ($> 10^{11}$ J/ψ decays) are required. These may be reached in one-year's operation of the τcF collider with monochromator optics.

- Searches for rare decays of charm and charmonium can be made with improvements in Br sensitivity of 2–3 orders of magnitude compared with present experiments. These processes include lepton-flavour-violating decays (such as $D^0 \to e^{\pm}\mu^{\mp}/Xe^{\pm}\mu^{\mp}$), which are completely forbidden for massless ν's, and flavour changing neutral-current decays [such as $D^0 \to l^+l^-/Xl^+l^-$ (l=e,μ), $X\nu\bar{\nu}$ and $X\gamma$], which may occur in the Standard Model but are highly suppressed by the GIM mechanism. Searches for rare decays of D mesons are complementary to those of K or B mesons since the New Physics may be flavour-dependent, i.e. different for up-like and down-like quarks.

The τcF will measure radiative decays like D $\to K^{\star}\gamma$ (Br $\sim 10^{-5}$). These decays—which are thought to be dominated by long-range interactions—will calibrate the non-penguin contributions in the corresponding B decays, e.g. B $\to \rho\gamma/K^{\star}\gamma$, thereby allowing precision determination of the CKM matrix elements V_{td} and V_{ts}. This again illustrates the importance of precision charm measurements for extracting B physics.

The high statistics and narrow width of the J/ψ may provide the first opportunity to measure weak decays of a vector meson, such as J/$\psi \to D_s e \nu_e$ (Br $\simeq 10^{-8}$), or the C-violating decay J/$\psi \to \phi\phi$ (Br $\simeq 10^{-8}$). The J/ψ can be tagged via $\psi' \to \pi^+\pi^-$ J/ψ to allow searches for rare processes involving J/$\psi \to$ 'nothing'. Axions or other evasive neutrals can be produced in J/$\psi \to \gamma +$ 'nothing'; the expected Br sensitivity is $\simeq 10^{-8}$.

In summary the motivation for dedicated precision studies of heavy flavour hadron decays is based on three goals:

1. To probe the strong interactions in a novel environment.

2. To extract fundamental parameters of the Standard Model—such as the CKM parameters—in the most reliable way.

3. To search for signs of 'New Physics' via CP violation, mixing and rare decays.

Viewed in this framework, the physics at a τ-charm Factory and at a beauty Factory should be seen as complementing each other rather than competing against each other. It is the detailed and comprehensive analysis of charm decays at a τ-charm Factory that will prove essential for gauging the impact of non-perturbative dynamics on beauty decays; once the non-perturbative corrections are understood in charm decays, they can be scaled to beauty decays with confidence.

3.4 Charmonium physics summary

The J/ψ has revolutionised our understanding of light hadrons. Historically, as more J/ψ data have been accumulated, increasingly rich structures have appeared in the final state hadrons, and new states been discovered. This is an area of physics which is still limited by statistics and where a τcF—which will increase the existing world data on J/ψ decays by *three orders of magnitude* (Table 2)—will have a major impact. Some of the main experiments [8, 17, 27, 28] are as follows:

- The decays J/$\psi \to \gamma X$ have proven to be among the most productive in strong interaction studies. These have revealed the $\iota(1440)$ and $\theta(1700)$, which are candidates for glueballs and whose dynamical structure is still controversial. The τcF should definitively settle these controversies.

- The J/ψ is a unique filter of the internal structure of light hadrons. Exploiting the ideal mixing within the ω/ϕ enables the flavour content of associated mesons, M, to be determined via J/$\psi \to$M+ω/ϕ. Improved data for M= ι and θ are especially needed. In addition, J/$\psi \to \gamma A \to \gamma(\gamma\rho : \gamma\omega : \gamma\phi)$ filters the flavour content of all $C = +$ states; this requires τcF statistics since only the channel J/$\psi \to \gamma\gamma\rho$ has been measurable so far. These channels, together with J/$\psi \to$A+*hadrons*, are particularly interesting in the region $m_A > 2$ GeV. The meson spectrum in the 2–3 GeV mass range is expected to be rich, but has been barely explored in J/ψ decays. Presently there are hints of pseudoscalar states which may be the sign of new physics (hybrid states are predicted to occur in this region).

- Experimentation at the ψ' will, via $\gamma\chi_c$ decays, yield 10^8 χ_c's per month—equivalent to about ten times the present world sample of J/ψ. The χ_c offers special opportunities. In particular, the spin, J, can be fixed at 0, 1 or 2 and thereby the partial wave analysis of hadronic decays can be constrained. Examples include $\chi_{c0} \to f_0(975)$+M which enhances the scalar hadrons, M; and $\chi_{c1} \to \pi$+H which is sensitive to the hybrid exotic sector H ($J^{PC} = 1^{-+}$). These complement the $\bar{p}p$ experiments at LEAR, providing greater statistics and the added bonus of constrained partial waves. The data on χ_c decays is presently notable for its almost complete absence. For such a potentially rich source of unique information on the structure and dynamics of light hadrons, this is a remarkable gap in our knowledge. With the statistics of the τcF, the physics output from χ_c decays can exceed that already flowing from the J/ψ.

- The Particle Data Group recognises four ψ states above 4 GeV. Nothing is known about their hadronic decays nor even if they are truly resonant states. The relative fractions of D, D^* and D^{**} in their decays will provide important information and tests of Heavy Quark Effective Theory, and can reveal the internal structure of these higher mass ψ states.

- In lattice QCD and related models, hybrid charmonium is predicted to exist within 300 MeV of charm threshold. These hybrids include vector states which can be formed directly in e^+e^- annihilation. Such states have been actively sought in the light quark sector during the last decade, and tantalising hints are emerging. Theory suggests that these states may be more clearly identified in the heavy quark sector and that charm is the optimal flavour (light enough that the states are produced in e^+e^- annihilation with leptonic widths $\mathcal{O}(0.1\text{ keV})$ and heavy enough that the conventional quarkonium potential states are well understood and that extra states can be readily identified). A fine-grained energy scan to search for these states can be carried out at a τcF over a period of about one month; this involves measuring the hadronic cross section above charm threshold in 1 MeV steps to an accuracy of 1%. (The measurements in Fig. 2 have a statistical precision of about 5% and are in 10 MeV steps; DELCO took 12 weeks to record these data at SPEAR.) The 1^- entry channel provides a well-understood, clean production mechanism with an optimal signature for hybrids. If hybrids states exist then it is likely that the τcF will be required for their discovery.

- A tag on a final-state J/ψ at any operating energy of the τcF above open charm threshold may reveal the existence of further exotic states. An example is $J/\psi + \pi$, which would reveal isovector charmonium, such as $D\bar{D}$ or $D^\star\bar{D}^\star$ molecules. Sighting the J/ψ in association with an η could reveal the existence of ψ^* and complement the direct search $e^+e^- \to \psi^*$, thereby clarifying the dynamical structure of the ψ^*.

- There is strong additional physics interest for an energy scan beyond the searches for new charmonium and charm states (and indeed for identifying optimised operating energies for charm studies). The interpretation of precision electroweak data requires a precise knowledge of the fine structure constant evaluated at the Z pole, $\alpha(m_Z^2)$. Recent calculations [29, 30, 31, 32] indicate that the dominant source of error in the theoretical calculation of the radiative corrections to $\alpha(m_Z^2)$ is due to the 15–20% experimental uncertainties in the cross-section $\sigma(e^+e^-\to \text{hadrons})$ in the low-energy region $1 \lesssim E_{cm} \lesssim 5$ GeV. This uncertainty is beginning to limit the determination of $\sin^2\theta_W$, $(g-2)_\mu$ and other precision electroweak parameters. The τcF will make a precision $[\mathcal{O}(1\%)]$ measurement of the cross-section $\sigma(e^+e^-\to \text{hadrons})$ over the full energy range $1 \lesssim E_{cm} \lesssim 5.6$ GeV, thereby allowing precise determination of $\alpha(m_Z^2)$.

In summary the τcF offers a new and unique window on the dynamics and structure of hadrons made of charm or of light flavours, and promises to elucidate the dynamical role of the gluonic degrees of freedom in the nonperturbative sector of QCD. New states of hadronic matter are predicted to exist in the energy range directly probed by the τcF. The high statistics of the τcF, together with a high-resolution detector, will enable important discoveries to be made about hadrons, and potentially fundamental discoveries on the role of gluons in the hadronic spectrum.

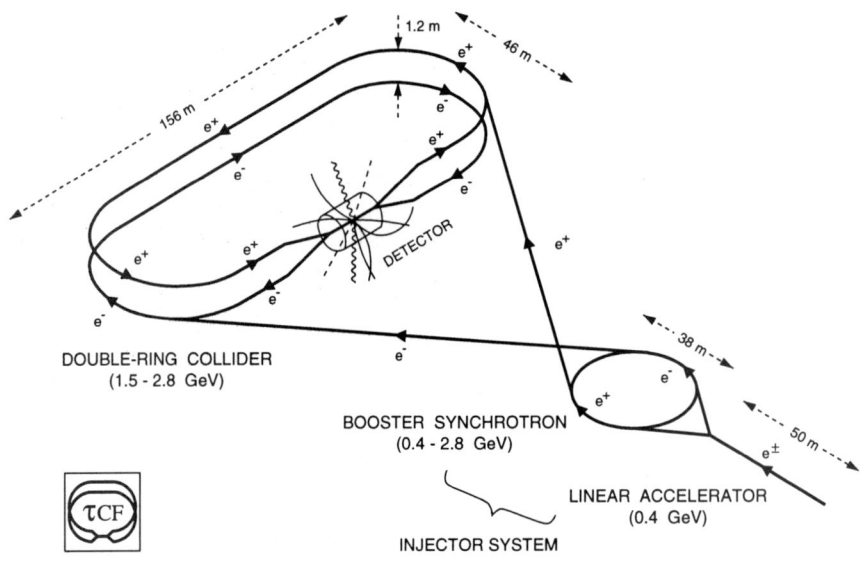

Figure 3: Layout of the τcF accelerator. Representative dimensions are indicated.

4 Accelerator design

The τcF accelerators comprise an injector system and a double-ring collider (Fig. 3). In the absence of re-usable accelerators, the most cost-effective injector chain involves a linear accelerator, providing intense 0.4 GeV e^-/e^+ beams, followed by a booster cyclotron operating between 0.4 GeV and 2.8 GeV. It is important that the injector energy reaches the maximum energy of the collider in order to allow rapid "topping-up" of the beam, without need to ramp the energy of the collider.

The primary energy range of the τcF collider extends from the $J/\psi(3.10)$

to 5.6 GeV total energy, which covers the charm baryon thresholds up to $\Omega_c\bar{\Omega}_c$ production. The collider is also designed to operate at energies below the J/ψ—with a luminosity decreasing as E^2, assuming a constant emittance controlled by wigglers—in order to allow measurement of the hadronic cross-section down to $E_{cm} \sim$ 1 GeV. The peak luminosity is 10^{33} cm^{-2}s^{-1} at 4 GeV total energy, close to the $\tau^+\tau^-$ and open-charm thresholds. The improvement in luminosity relative to previous machines at this energy (Table 4) is due to a higher stored current and tighter focusing at the interaction point. The increase in luminosity relative to the current machine at these energies, the Beijing Electron Positron Collider (BEPC) [33], is about two orders of magnitude.

Table 4: Comparison of e$^+$e$^-$ colliders at $E_{beam} \simeq$ 2.0 GeV. Recall that the luminosity, assuming optimum coupling, is $L = \left(I\gamma\xi_y/2er_e\beta_y^*\right)\left(1+\sigma_y^*/\sigma_x^*\right)$, where γ is the Lorentz factor, e is the electron charge, r_e is the classical radius of the electron, σ_x^*/σ_y^* are the r.m.s. radial/vertical beam sizes at the interaction point, and the other quantities are defined in the Table below. With standard optics, the beams are flat $(\sigma_y^* \ll \sigma_x^*)$ and the luminosity $L \propto (I\xi_y/\beta_y^*)E_{beam}$.

			SPEAR	BEPC	τcF
Number of bunches	k_b		1	1	32
Effective bunch spacing	S_b	[m]	234	240	11.2
Beam current	I	[mA]	10	50	564
Vert. β-function at IP	β_y^*	[m]	0.08	0.10	0.01
Beam-beam tune shift	ξ_y		0.025	0.04	0.04
Luminosity	L	[cm^{-2}s^{-1}]	1.0×10^{30}	1.0×10^{31}	1.0×10^{33}

The design of the τcF accelerator has been developed at several laboratories, including CERN [2, 3, 34], SLAC [35], LAL-Orsay [36], CERN/Spain [37], INP-Novosibirsk /JINR-Dubna [38], ITEP-Moscow [39], IHEP-Beijing [40] and Argonne National Laboratory [41]. All studies have confirmed that the τcF collider can achieve its design luminosity with a relatively "conservative" design that assumes previously-achieved performances for the machine parameters: beam-beam tune shift of 0.04 with head-on crossings, realistic beam current (0.6 A), low heat load on the vacuum chambers (2 kW/m), etc.

As indicated qualitatively in Table 3, certain experiments will benefit from increased performance of the τcF collider, beyond the basic machine with a luminosity of 10^{33} cm^{-2}s^{-1}. The concept of a flexible lattice was therefore discussed at the Marbella workshop [42, 43] and subsequently successfully developed by the LAL Orsay-Dubna collaboration [44]. The phases of operation of this machine are as follows (refer to the parameter list in Table 5):

Table 5: Parameter list for the τcF collider design from the LAL Orsay-Dubna study [44].

		Standard Scheme	Monochrom. Scheme	Cross. angle Scheme	
Beam energy	E	2.0	2.0	2.0	GeV
Luminosity	L	1.0×10^{33}	1.0×10^{33}	3.6×10^{33}	cm^{-2}s^{-1}
CM energy resolution	σ_w	1.8	0.14	1.6	MeV
Circumference	C	359	359	359	m
Natural emittance	ϵ_0	378	17	225	nm
Vertical emittance	ϵ_y	19	2	5	nm
Coupling factor	κ	0.05	(wigglers)	0.05	
Damping partition nos.	J_x	0.58	2	0.66	
	J_y	1	1	1	
	J_ϵ	2.42	1	2.34	
Bending radius in arc	ρ	10.5	10.5	10.5	m
Damping times	τ_x	39	17	37	ms
	τ_y	23	34	24	ms
	τ_s	9	32	10	ms
Momentum compaction	α	1.67×10^{-2}	8.43×10^{-3}	1.67×10^{-2}	
Fractional energy spread	σ_E	6.23×10^{-4}	7.31×10^{-4}	5.59×10^{-4}	
Total current per beam	I	0.564	0.537	2.0	A
Particles per bunch	N_b	1.32×10^{11}	1.26×10^{11}	0.78×10^{11}	
Number of bunches	k_b	32	32	192	
RF voltage	V_{RF}	8	5	7	MV
RF frequency	f_{RF}	481	481	481	MHz
Harmonic number	q	576	576	576	
Rad. energy loss/turn	U_0	211	142	196	keV
RMS bunch length	σ_s	7.43	8.01	7.13	mm
Bunch spacing	S_b	11.2	11.2	1.9	m
Reqd. long. impedance	$|Z/n|$	0.24	0.19	0.31	Ω
β-functions at IP	β_x^*	0.20	0.01	0.50	m
	β_y^*	0.01	0.15	0.01	m
Vertical dispersion at IP	D_y^*	0	0.36	0	m
Beam-beam parameters	ξ_x	0.04	0.04	0.04	
	ξ_y	0.04	0.03	0.04	

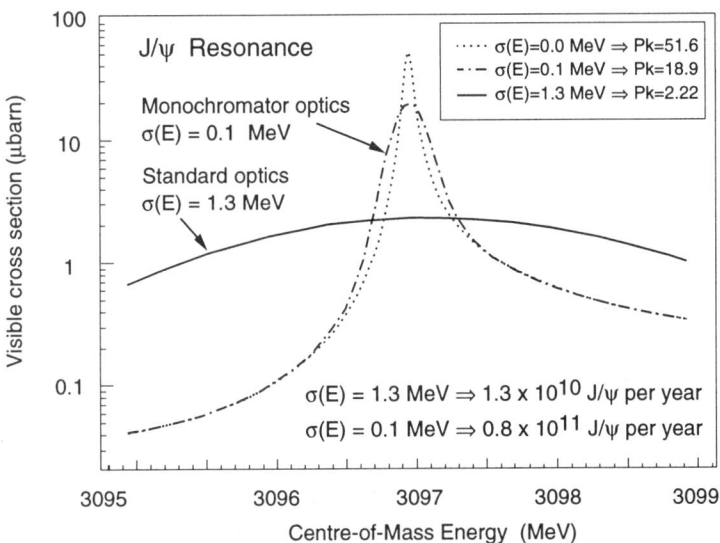

Figure 4: The visible cross section at the J/ψ resonance for several values of the beam collision energy spread σ(E).

- First-stage operation of the collider at 10^{33} cm^{-2}s^{-1} with a conservative design.

- Subsequent operation of the collider with several upgrade options:

  - **Increased luminosity**, 3.6×10^{33} cm^{-2}s^{-1}, with more bunches and a finite crossing angle. This would involve re-building the experimental insertion but leaving the rest of the collider unmodified.

  - **Monochromator optics** to reduce the collision energy spread from 1.8 MeV to 0.14 MeV (at 2 GeV beam energy) [45, 46, 47, 48, 49]. This has several important physics consequences: increase of the J/ψ rate (Fig. 4), increase of the ψ' rate and maximum $\tau^+\tau^-$ rate (Table 2), precise measurement of the $\tau^+\tau^-$ cross-section at threshold, operation closer to τ threshold (narrower monochromatic peaks for two-body decays), and tighter beam-energy kinematic constraints (which, for charm physics, results in narrower mass peaks and smaller backgrounds).

  - **Longitudinal beam polarization.** This would involve longitudinal polarization of either a single beam (e$^-$) or both beams. The former can be achieved relatively simply using a SLAC-type polarized electron source, together with spin rotators installed in the collider. The latter option would require an additional small storage ring in the injector chain to polarize the beams before they are injected into the collider [46]. Longitudinal beam polarization significantly increases the experimental sensitivity for studies of CP violation in τ decays and hyperon decays.

5 Detector

5.1 Design considerations

Experiments at the τ-charm Factory are characterised in two ways: i) high-precision measurements, and ii) searches for rare processes. The basic requirement for both is high statistics but, in order to fully exploit the very large data samples generated at the τcF, backgrounds and systematic errors must be small and well understood. As described previously, the *physics* backgrounds at the τcF are small and experimentally measurable. However, special care must be taken to ensure that the *detector-induced* backgrounds and systematic errors are maintained at a comparably low level and are accurately monitored. These may arise from several sources, such as:

- Uncertainties in geometrical acceptance; blind regions in the detector.
- Uncertainties in detector efficiencies.
- Uncertainties in detector resolutions, and non-Gaussian tails.
- Detector mis-measurements, such as:
  - Mis-tracked particles.
  - Overlapping particles.
  - Fake photons ("splitoffs").
  - Particle mis-identification.
- Luminosity measurement errors.

These considerations have shaped the underlying principles in the detector design:

- *Uniform and efficient sub-detectors covering the full solid angle,* for reduced losses, reduced background feed-ins, and reduced uncertainties in efficiency corrections.
- *Highly granular detectors,* for improved precision, improved recognition of secondary interactions and reduced particle overlaps.

- *Redundant measurements,* for cross-checks and improvement of the overall performance.

- *Frequent calibration and monitoring* (at the J/ψ and ψ'), for precise knowledge of the detector performance.

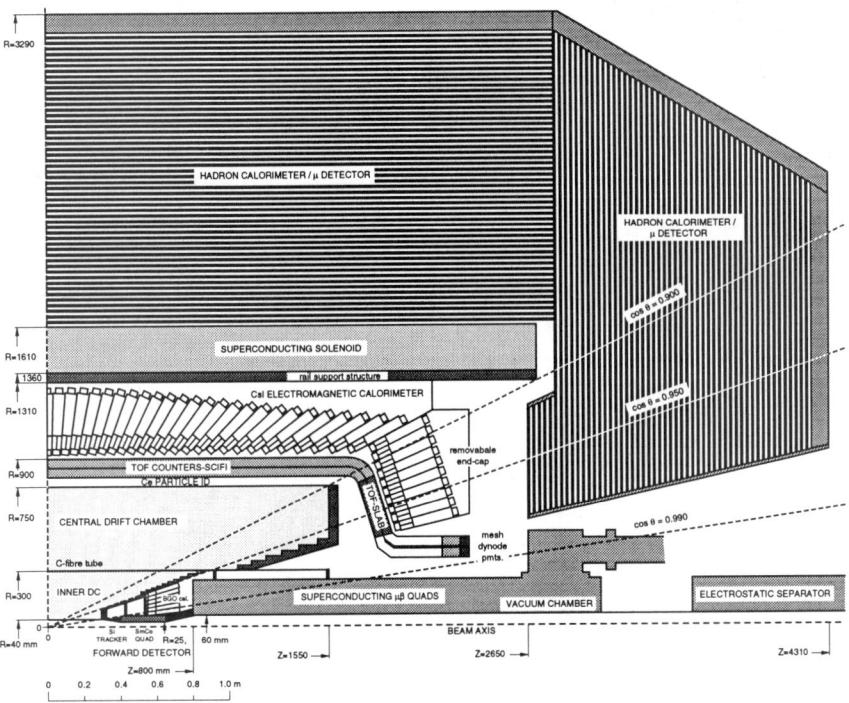

Figure 5: The τcF detector, EXACT; rz view.

5.2 Detector design

Guided by these design considerations, a detector for the τ-charm Factory has been developed over the course of several workshops [4, 5], and further refined[1] at the Marbella meeting [8]. The detector, which is shown in Figs. 5 and 6, covers all the foreseen physics and can be built with present technologies. The main features are summarised below.

---

[1]Among the refinements is a detector name: EXACT (Experimental Apparatus for Charm and Tau).

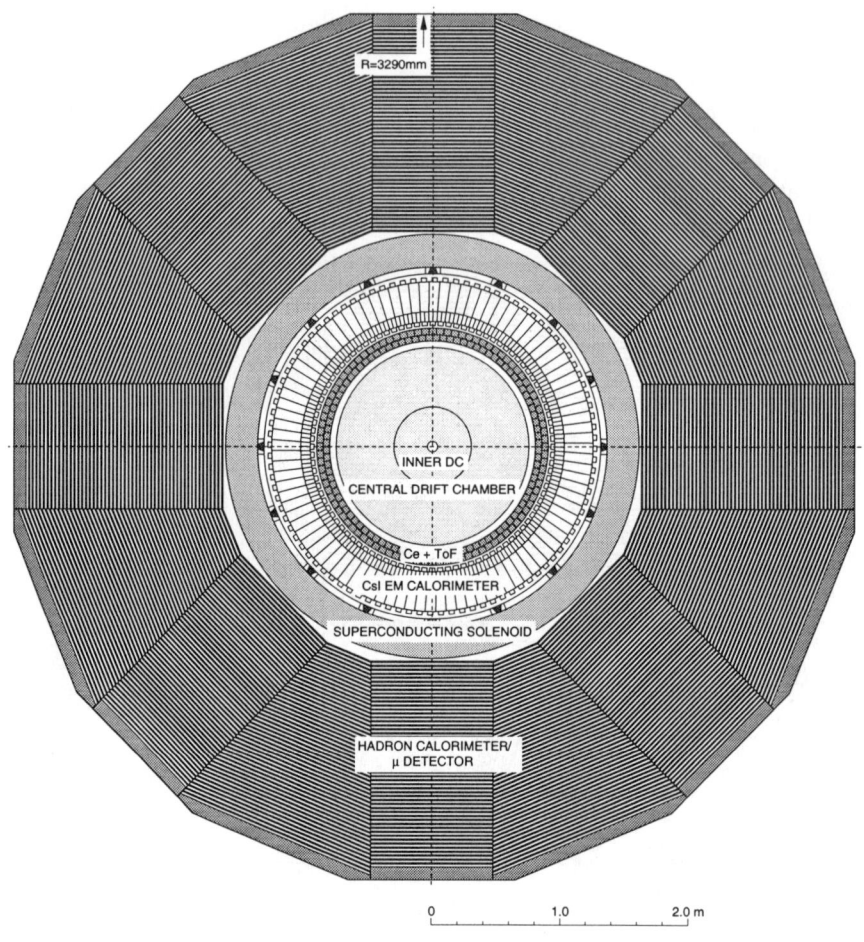

Figure 6: The τcF detector, EXACT; rϕ view.

5.2.1 Central drift chamber

The primary functions are to measure precisely the directions and momenta of the charged particles emerging from the interaction point. Other important functions include vertex measurement, provision of charged-particle triggering at Level 2, dE/dx measurement for particle identification, and the measurement of 'kinks' that could signal either fake μ's from π/K decays or large scatters from direct wire collisions. The central drift chamber is optimised for the measurement of low-momentum particles: reduced entrance material (0.2% X_0), reduced end-plate material, Al field wires and a He-based gas. The analysing magnetic field (B = 1–1.2 T) is provided by a superconducting solenoid (whose dimensions and performance are identical to the Topaz solenoid at KEK).

5.2.2 Time-of-flight counters

The functions of the time-of-flight counters are to identify the particles, to provide a fast (Level 1) charged particle trigger and to tag the beam crossing in which the event occurred. The counters are constructed from either solid scintillator or 2mm-diameter plastic scintillating fibres. Tests at KEK and elsewhere have indicated that a timing resolution $\sigma_t \sim 100$ ps can be achieved using 4 cm-thick scintillator and mesh-dynode phototubes placed directly in the magnetic field.

5.2.3 Cerenkov detector

The function of the Cerenkov detector is to provide additional particle identification, both to improve the overall performance and also to provide redundancy for precision calibration of the particle identification detectors. A Cerenkov detector has potentially excellent performance in the energy range of interest to the τcF, namely up to a kinematic limit of $\sim$1–1.5 GeV/c. This detector is especially useful in strengthening the K/π separation near 1 GeV/c (for $D^0\bar{D}^0$ mixing) and in improving the e/π separation at the same energy (for limits on the rare decay $\tau \to eX$). A Cerenkov detector would also substantially improve the μ/π separation in the difficult region at low energy, below the range of the outer μ detector (< 0.5 GeV/c). The candidate devices are a fast RICH, a DIRC or an aerogel counter. In the case of aerogel, impressive performances—which are suitable for the τcF—are now being achieved: signal pulse heights of 15 p.e. and a background rejection $\sim 3 \times 10^{-2}$, using aerogels of refractive index $\sim$ 1.01–1.02 and thickness 12 cm.

5.2.4 Forward detectors

The functions of the forward detectors are to provide precision tracking and electromagnetic calorimetry in the region $0.95 < |\cos\theta| < 0.99$, to complete the hermeticity of the τcF detector, and to provide a precision (0.1%) luminosity measurement. Each of the two forward trackers comprises a system of double-sided Si μ-strip detectors followed by a BGO calorimeter. BGO is chosen for its high density and small Molière radius, along with crystal calorimeter performance.

5.2.5 Electromagnetic calorimeter

The primary functions of the electromagnetic calorimeter are to provide precision measurements of the directions and energies of γ's emerging from the interaction point. Other important functions include e/π separation, detector hermeticity (absence of projective cracks, low γ energy threshold and precision γ energy measurement), provision of a neutral energy trigger, and μ identification, especially in the range below 0.5 GeV/c. The design involves a CsI(Tl) crystal electromagnetic calorimeter to achieve high resolution measurements of γ's, and high efficiency down to low energies. Novel features are longitudinal segmentation of the calorimeter towers, with sub-division of the front section, and redundant photodiode readout. These should lead to improvements in the performance for rejecting fake photons, in the e/π separation and in the overall reliability.

5.2.6 Hadron calorimeter/μ detector

The functions of the outer hadron calorimeter/μ detector are to complete the hermeticity of the detector for neutral hadrons (K_L^0/n), to identify μ's by precise range-energy measurements and to provide a flux return for the solenoidal magnetic field. In order to preserve hermeticity, the boundaries between the hadron calorimeter modules do not project to the interaction point (Fig. 6). The barrel modules are designed to be individually removable in order to allow the future installation of a μ polarimeter for experiments involving the μ polarization in τ decays (measurement of ξ'_μ and tests of CP violation). The hadron calorimeter has a fine sampling (1 cm-thick Fe plates) in order to provide a good efficiency for tagging K_L^0/n interactions and to provide sufficient $\pi - \mu$ range discrimination (the difference in range between a μ and a non-interacting π is ~3.3 cm Fe, over the full energy range). Suitable choices for the active layers are drift chambers or resistive plate chambers.

5.2.7 Machine-detector backgrounds

Synchrotron radiation from the vertical bending magnets and quadrupoles in the interaction region has sufficiently low critical energies ($\simeq 0.35$ keV) that simple masking and absorption in the beam-pipe prevent significant detector backgrounds. Protection of the detector from bremsstrahlung and from beam particles scattered by residual gas in the vacuum system requires adjustable masks upstream of the electrostatic separators, and between the separators and the $\mu\beta$ quadrupoles. Simulations indicate that the backgrounds will be mild: a 6 kHz rate of electrons with $p_\perp \geq 100$ MeV/c strike the beam-pipe within 80 cm of the interaction point, and most of these exit the ends of the detector at small radii. The flux of hard photons is comparable.

Protection of the interaction region during injection is helped by having each of the injection points as far from the interaction region as possible—about 3/4 turn. A system of adjustable collimators in the arcs protect the interaction region during injection and serve as an aperture limit in both planes during physics running. Finally, as regards radiation damage to the detector during normal machine running, the radiation doses are well-within safe limits.

5.2.8 Triggering and computing

The Level 1 trigger uses loose, redundant triggers and has an estimated output rate of a few 10's kHz. The Level 2 trigger refines the requirements and is expected to accept a few 100 Hz of beam-related background events in addition to the desired physics events. The Level 3 trigger accepts data from the event builder and performs full event reconstruction. This rejects all the remaining background and allows the possibility of software selection before writing events to tape. The computational power of the Level 3 filter must be 2×10^4 MIPS to process a 2 kHz input event rate (J/ψ).

The data storage requirements are considerable. Assuming that the τcF operates each year (10^7 s) with 10% on-resonance (2 kHz event rate) and the remaining 90% off-resonance (100 Hz event rate), the number of events recorded per year is about 3×10^9. Twice this number of events will be generated each year by Monte Carlo simulations. Assuming both the raw and DST data are retained for real events, while only the DST is kept for simulated events, the raw data will be accumulated at the rate of 60 Terabytes per year and the DST data at 9 Terabytes per year.

As regards the off-line computing, there are three main components: i) generation of Monte Carlo data, ii) re-processing of raw data and iii) selection and physics analysis of the DST events. Together these require an off-line processing power of $\sim 5 \times 10^4$ MIPS. The general model for physics analysis involves the extraction of a compressed subset of the on-line DST library located at the host laboratory. This work would be carried out on clusters of workstations with high-capacity disk servers, located both at the host laboratory and in the collaborating institutes. A high-bandwidth network connection with the host laboratory is an important requirement.

5.3 Detector performance

A summary of the expected performance of the τcF detector is presented in Table 6, including a comparison with previous detectors at these energies. With careful design, and with the benefit of the experience of present experiments, it appears likely that a detector can be built that can maintain very low systematic errors and take full advantage of the large increase in statistics expected at the τ-charm Factory. The τcF detector represents a dramatic improvement in performance compared with previous detectors at these energies: Mark I/II/III, LGW, DELCO, Pluto, DASP, Crystal Ball, BES, etc.

Table 6: Comparison of the performances of the τcF detector EXACT with previous detectors at these energies. The symbol '$\oplus$' denotes addition in quadrature.

	Mark III (SPEAR) / BES (BEPC)	EXACT (τcF)
Charged particles:		
Momentum res.: σ_p/p(GeV/c)	1.5%$p \oplus$ 1.5%/β [MkIII] 0.7%$p \oplus$ 1.3%/β [BES]	0.4%$p \oplus$ 0.4%/β
Angular resolution: σ_ϕ (mr)	2 $\oplus$ 2/pβ	0.5 $\oplus$ 1.1/pβ
p^π_{min}(MeV/c) for efficient tracking	80	50
Ω(barrel) ($\times 4\pi$ sr)	70%	90%
Photons:		
Energy resolution: σ_E/E(GeV)	17%/$\sqrt{E}$	2%/$E^{\frac{1}{4}} \oplus$ 1%
Angular resolution: $\sigma_{\theta,\phi}$ (mr)	10 [MkIII]; 5 [BES]	1.7 + 2/$\sqrt{E}$ (θ=90°)
2γ angular separation: $\Delta\theta_{2\gamma}$ (mr)	20	50
E^γ_{min}(MeV) for efficient detection	100	10
Particle identification:		
$\pi \to$ K separation	3σ at 0.7 GeV/c	3σ at 1.0 GeV/c (10^{-4} inc. Čе.)
π/K $\to$ e separation	4% at 0.5 GeV/c	0.1% (10^{-5} inc. Čе.)
π/K $\to \mu$ separation	5% at 1.0 GeV/c	1.5%/p + (1-4)%
K^0_L detection efficiency	60%	95%
n mean detection efficiency	-	50%
ν 'detection': $p^\perp_{min}$(MeV/c)	-	100

6 Cost estimates

Table 7 [50] compares the estimated materials cost and personnel requirements for constructing the τcF accelerators on a) a green-field site and b) a site with an existing injector (in this case, the CERN ISR site). Compared with a green-field site, the materials costs at an existing laboratory are reduced by a factor 3 and the staff requirements halved. These substantial savings assume the presence of a suitable injector and general laboratory infrastructure. Part of the costs could be further offset by the use of previous R&D results and components that have been developed for other projects, such as the B Factories and LHC (which has similar requirements for its 400 MHz superconducting RF cavities.) With an estimated construction cost of 70 MCHF for the τcF detector (Table 8), the total construction cost of the τ-charm Factory at an existing laboratory is about 170 MCHF.

Table 7: Comparison of the estimated construction costs and personnel requirements for the τcF accelerator on a green-field site and on a site with an existing injector (the CERN ISR site was used in this study [50]).

Item	Green-Field Site [MCHF]	Site with injector [MCHF]
Site & infrastructure:		
Buildings & civil engineering	76	9
Experimental zone infrastructure	12	8
Site equipment (electrical plant, etc.)	37	0
Lab. computing & networks	15	0
Subtotal:	140	17
Accelerators:		
Collider	87	79
Booster	23	0
Linac	17	0
Transfer lines	8	6
Technical equipment	7	0
Recurrent cost (lab. electricity, etc.)	7	0
Subtotal:	149	85
Total (material):	**289**	**102**
Personnel:		
Total (personnel):	660 man-y	315 man-y

Table 8: Preliminary construction cost estimate for the τcF detector, EXACT.

Item	Cost [CHF]	Cost [%]
Inner & central drift chamber	3.8	6.0
Superconducting solenoid	7.0	11.1
Cerenkov	2.3	3.7
Time-of-flight	1.2	1.9
Electromagnetic calorimeter [CsI(Tl)]	30.7	48.7
Forward BGO calorimeter	1.2	1.9
Forward Si tracker	1.6	2.5
Hadron calorimeter/μ detector	7.3	11.6
Trigger & data acquisition	7.9	12.6
Subtotal:	63.0	100.0
Contingency (10%):	6.3	
Total:	**69.3**	

7 Conclusion

The τ-charm Factory is an essential machine for exploring the heavy-flavoured particles. It provides a unique experimental environment for precision studies of tau, charm and light quark/gluon spectroscopy. Many important τ-charm experiments are accessible only at a τcF, or else cannot be done with the necessary precision elsewhere—at B Factories or in fixed-target experiments.

The physics of the τ-charm Factory and the B Factory are complementary. Both machines are needed to fully exploit the potential of the heavy-flavoured particles to consolidate and extend the Standard Model. It is the detailed and comprehensive analysis of charm decays at a τcF that will prove essential for gauging the impact of non-perturbative dynamics on beauty decays.

The required performances for the τcF machine and detector can be achieved using present technologies. For the future, the τcF collider has several upgrade options that significantly extend the physics capabilities: increased luminosity, monochromatisation of the collision energy and longitudinal polarization of the beams.

Together with a high-resolution detector, the τ-charm Factory constitutes a superb experimental tool. It offers the tantalising possibility of shedding light on the reason for the replication of the quark/lepton families and on the associated questions of the origins of the CKM matrix and CP violation for quarks, and of their apparent absence for leptons.

Acknowledgements

I would like to thank many colleagues for interesting discussions on the τ-charm Factory. In particular, for the present paper, I would like to acknowledge the contributions of I.I. Bigi, F. Close, J. Jowett, R. Landua, J. Le Duff, A. Pich, P. Roudeau and N. Wermes. Finally I would like to thank Jose Repond for having organised such an enjoyable and stimulating meeting.

References

[1] J. Kirkby, *A τ-charm Factory at CERN*, CERN-EP/87-210 (1987), and Proc. International School of Physics with Low-Energy Antiprotons; Spectroscopy of Light and Heavy Quarks, Erice, Sicily, 1987 (Plenum Press, New York, 1989) 401.

[2] J.M. Jowett, *Initial design of a τ-charm Factory at CERN*, CERN LEP-TH/87-56 (1987).

[3] J.M. Jowett, *The τ-charm Factory storage ring*, CERN-LEP/88-22 (1988), and Proc. 1st European Particle Accelerator Conference, Rome, 1988 (World Scientific, Singapore, 1989) 368.

[4] Proc. *Tau-Charm Factory Workshop*, SLAC, California, USA, 23-27 May 1989, ed. L.V. Beers, SLAC-Report-343 (1989).

[5] Proc. *Meeting on the Tau-Charm Factory Detector and Machine*, Sevilla, Spain, 29 April-2 May 1991, eds. J. Kirkby and J.M. Quesada (Univ. Sevilla, 1992).

[6] Proc. *Workshop on JINR c-tau Factory*, JINR, Dubna, Russia, 29-31 May 1991, eds. V.A. Bednyakov and G.A. Chelkov (Dubna, 1992).

[7] Proc. *Second Workshop on JINR Tau-Charm Factory*, JINR, Dubna, Russia, 27-29 April 1993, eds. V.S. Alexandrov and E.A. Perelstein (Dubna, 1993).

[8] Proc. *Third Workshop on the Tau-Charm Factory*, Marbella, Spain, 1-5 June 1993, eds. J. Kirkby and R. Kirkby (Éditions Frontières, 1994).

[9] Proc. *The Tau-Charm Factory in the Era of B Factories and CESR*, SLAC, California, USA, 15-16 August 1994, ed. L.V. Beers and M. Perl, SLAC-451 (1994).

[10] Proc. *Journées sur les Projets de la Physique Hadronique*, Super-Besse, France, 12-14 January 1995, ed. A. Falvard (in preparation).

[11] Proc. *Workshop on the Tau/Charm Factory*, Argonne, Illinois, USA, 21-23 June 1995, ed. J. Repond (these proceedings).

[12] Proc. *Workshop on Tau Lepton Physics*, Orsay, France, 24-27 September 1990, eds. M. Davier and B. Jean-Marie (Éditions Frontières, 1991).

[13] Proc. *Second Workshop on Tau Lepton Physics*, The Ohio State University, U.S.A., 8-11 September 1992, ed. K.K. Gan (World Scientific, 1993).

[14] Proc. *Third Workshop on Tau Lepton Physics*, Montreux, Switzerland 19-22 September 1994, ed. L. Rolandi (North Holland, 1995).

[15] A. Pich, *Tau Physics Prospects at the Tau-Charm Factory and at Other Machines*, in [8], p. 51.

[16] P. Roudeau, *Charm Physics at the Tau/Charm Factory and at Other Machines*, in [8], p. 61.

[17] F.E. Close, *Hadrons and Glue at a Tau-Charm Factory*, in [8], p. 73.

[18] A. Pich, *Tau Physics*, in *Heavy Flavours*, eds. A.J. Buras and M. Lindner, Advanced Series on Directions in High Energy Physics, Vol. 10 (World Scientific, Singapore, 1992) p. 375; *Tau Physics and Tau-Charm Factories*, Nucl. Phys. B (Proc. Suppl.) 31 (1993) 213.

[19] A. Pich, *Perspectives on Tau-Charm Factory Physics*, in [8], p. 767.

[20] Y.S. Tsai, *Production of polarized τ pairs and tests of CP violation using polarized $e^{\pm}$ colliders near threshold*, Phys.Rev. D51 (1995) 3172. Also, SLAC-PUB-95-6916 (1995) and in [11].

[21] T.D. Lee, Proc. *International Workshop on B Physics: Physics Beyond the Standard Model at the B Factory*, Nagoya, Japan, 26–28 October 1994 (World Scientific, Singapore, 1994).

[22] W. Bernreuther and O. Nachtmann, *How to search for an electric dipole form factor of the τ lepton at a τ-charm Factory*, in [4], p. 545.

[23] B. Ananthanarayan and S.D. Rindani, *Measurement of the τ electric dipole moment using longitudinal polarization of e^+e^- beams*, Univ. Lausanne preprint PRL-TH-94/32 (1994).

[24] A. Le Yaouanc, L. Oliver, O. Pène, J.C. Raynal, and P. Roudeau, *Prospects for D Physics at a Tau-Charm Factory*, LPTHE Orsay 92/49 and LAL 92-53 (1992).

[25] I.I. Bigi, *Weak Decays of Charm Hadrons: The Next Lesson on QCD – and Possibly More!*, in [8], p. 239.

[26] E. González Romero and J.I. Illana, *CP Violation in Non-Leptonic Hyperon Decays*, in [8], p. 525.

[27] T. Barnes, *Charmonium Physics at a Tau-Charm Factory*, in [8], p. 411.

[28] R. Landua, *Signatures of non-$q\bar{q}$ states*, in [10].

[29] M.L. Swartz, *Re-evaluation of the Hadronic Contribution to $\alpha(M_Z^2)$*, SLAC-PUB-6710, submitted to Phys. Rev. D (November 1994).

[30] A.D. Martin and D. Zeppenfeld, Phys. Lett. B345 (1994) 558.

[31] S. Eidelman and F. Jegerlehner, *Hadronic contributions to $(g-2)$ of the leptons and to the effective fine structure constant $\alpha(M_Z^2)$*, PSI-PR-95-1 (January 1995).

[32] H. Burkhardt and B. Pietrzyk, *Update of the hadronic contribution to the QED vacuum polarization*, LAPP-EXP-95.05, submitted to Phys. Lett. B (June 1995).

[33] S.X. Fang, *The BEPC upgrade*, in [8], p. 5.

[34] J.M. Jowett, *Lattice and Interaction Region Design for Tau-Charm Factories*, Proc. *Joint US/CERN School on Particle Accelerators*, Benalmadena, Spain, 29 October–4 November 1992, (Springer-Verlag 1993).

[35] B. Barish et al., *Tau-Charm Factory Design*, SLAC-PUB-5180 (1990).

[36] J. Gonichon et al., *Preliminary Study of a High Luminosity e^+e^- Storage Ring at a C.M. Energy of 5 GeV*, Orsay Report, LAL/RT 90-02 (1990).

[37] J.M. Jowett, A. Zholents, C. Fernandez-Figueroa, M. Munoz, J.-M. Quesada, C. Willmott, *The Tau-Charm Factory*, Proc. *XVth International Conference on High Energy Accelerators*, Hamburg, 20-24 July 1992, Int. J. Mod. Phys. A (Proc. Suppl.) 2A (1993) 439.

[38] V.S. Alexandrov et al., *JINR Tau-Charm Factory Design Considerations*, Proc. *IEEE 1991 Particle Accelerator Conference*, San Francisco, USA, 6-9 May 1991, p. 195.
E.A. Perelstein et al., *JINR Tau-Charm Factory Study*, same volume as [37], p. 448.
E.A. Perelstein et al. *JINR Tau-Charm Factory Design Study*, in [8], p. 557.

[39] M.V. Danilov et al., *Conceptual design of Tau-Charm Factory at ITEP*, same volume as [37], p. 455.

[40] Y.Z. Wu, *TCF Collider Design at IHEP*, in [11].

[41] L.C. Teng, *Argonne TCF Collider Design*, in [11].

[42] J.M. Jowett et al., *Summary of the Optics Session*, in [8], p. 553.

[43] Y.I. Alexahin, *A Scheme of Monochromatic Tau-Charm Factory with Finite Crossing Angle*, in [8], p. 571.

[44] P. Beloshitsky et al., *Modern View of Tau-Charm Factory Design Principles*, Orsay Report, LAL/RT/94-05 (1994).

[45] Yu.I. Alexahin, A.N. Dubrovin and A.A. Zholents, *Proposal on a Tau-Charm Factory with Monochromatisation*, Proc. 2nd European Particle Accelerator Conf., Nice, 1990 (Éditions Frontiéres, 1990) 398.

[46] A. Zholents, *Polarized J/ψ Mesons at a Tau-Charm Factory with a Monochromator Scheme*, CERN SL/92-27 (1992).

[47] P. Beloshitsky, *A Magnet Lattice for a τ-charm Factory Suitable for both Standard Scheme and Monochromatisation Scheme*, Orsay Report, LAL/RT 92–09 (1992).

[48] A. Faus-Golfe and J. Le Duff, *A Versatile Lattice for a τ-charm Factory that includes a Monochromatisation Scheme*, Orsay Report, LAL/RT 92-01 (1992).

[49] A. Faus-Golfe and J. Le Duff, *A Versatile Lattice for a τ-charm Factory that includes a Monochromatisation Scheme (Low Emittance) and a Standard Scheme (High Emittance)*, Proc. Particle Accelerator Conference, Washington May 1993.

[50] J.L. Baldy et al., *Tau-Charm Factory Cost Estimate*, CERN ST/93-03 (1993), and in [8], p. 603.

TAU LEPTON PHYSICS

IMPORTANCE OF PRECISION MEASUREMENTS IN THE TAU SECTOR

A. Pich

Departament de Física Teòrica, IFIC, Universitat de València – CSIC
E-46100 Burjassot, València, Spain

Abstract. τ decays provide a powerful tool to test the structure of the weak currents and the universality of their couplings to the W boson. The constraints implied by present data and the possible improvements at the τcF are analyzed.

INTRODUCTION

The light quarks and leptons are by far the best known ones. Many experiments have analyzed in the past the properties of e, μ, ν_e, ν_μ, π, K, ... However, one naïvely expects the heavier fermions to be much more sensitive to New Physics, since they may couple more strongly to whatever dynamics is responsible for the fermion-mass generation. Obviously, new heavy-flavour facilities, such as the B and Tau-Charm Factories (τcF), are needed to match (at least) the precision attained for the light flavours.

Similarly to the bottom quark, the tau lepton is a third generation fermion, with a wide variety of decay channels into particles belonging to the first and second fermionic families. Therefore, one can expect that τ and b physics will provide some clues to the puzzle of the recurring generations of leptons and quarks. While the decays of the b-quark are ideally suited to look for quark mixing and CP-violating phenomena, the pure leptonic or semileptonic character of τ decays provides a much cleaner laboratory to test the structure of the weak currents and the universality of their couplings to the gauge bosons. Moreover, the tau is the only known lepton massive enough to decay into hadrons; its semileptonic decays are then an ideal tool for studying strong interaction effects in very clean conditions.

The last five years have witnessed a substantial change on our knowledge of the τ properties. The large (and clean) data samples collected by the most recent experiments have improved considerably the statistical accuracy and, moreover, have brought a new level of systematic understanding. The qualitative change of the τ data can be appreciated in Table 1, which compares the status of several τ measurements in 1990 [1,2] with the most recent world averages [3,4]. All experimental results obtained so far confirm the Standard Model (SM) scenario in which the τ is a sequential lepton, with its own quantum number and associated neutrino.

© 1996 American Institute of Physics

Table 1: Recent improvements in τ physics and expected precision at the τcF.

Parameter	1990 [1,2]	1995 [3,4]	τcF sensitivity
m_τ (MeV)	$1784.1^{+2.7}_{-3.6}$	1777.0 ± 0.3	0.1
m_{ν_τ} (MeV)	< 35 (a)	< 24 (a)	1–2
τ_τ (fs)	303 ± 8	291.6 ± 1.6	–
B_e (%)	17.7 ± 0.4	17.79 ± 0.09	0.02
B_μ (%)	17.8 ± 0.4	17.33 ± 0.09	0.02
$B(\pi^- \nu_\tau)$ (%)	11.0 ± 0.5	11.09 ± 0.15	0.01
$B(K^- \nu_\tau)$ (%)	0.68 ± 0.19	0.68 ± 0.04	0.003
$B(\pi^- \eta \nu_\tau)$	$< 9 \times 10^{-3}$ (a)	$< 3.4 \times 10^{-4}$ (a)	10^{-6}
$B(l^- G)$	$< 10^{-2}$	$< 2.7 \times 10^{-3}$ (a)	10^{-5}
$B(\mu^- \gamma)$	$< 5.5 \times 10^{-4}$ (b)	$< 4.2 \times 10^{-6}$ (b)	10^{-7}
$B(e^- e^+ e^-)$	$< 3.8 \times 10^{-5}$ (b)	$< 3.3 \times 10^{-6}$ (b)	10^{-7}
$\rho_{\tau \to \mu}$	0.84 ± 0.11	0.738 ± 0.038	0.002
$\eta_{\tau \to \mu}$	–	-0.14 ± 0.23	0.003
$\xi_{\tau \to \mu}$	–	1.23 ± 0.24	0.02
$(\xi \delta)_{\tau \to \mu}$	–	0.71 ± 0.15	0.02
$\xi'_{\tau \to \mu}$	–	–	0.15
h_{ν_τ}	–	-1.014 ± 0.027	0.003
a_τ^γ	< 0.1 (b)	< 0.01 (a)	0.001
d_τ^γ (e cm)	$< 6 \times 10^{-16}$ (b)	$< 5 \times 10^{-17}$ (a)	10^{-17}

(a) 95% CL ;   (b) 90% CL

The present experiments are soon going to reach their systematic limits. Further improvements in τ physics require then new high-precision facilities, to push the significance of the τ tests beyond the present few per cent level. The last column in Table 1 shows the sensitivities that could be achieved at the τcF. In some cases, a much better accuracy could be obtained with polarized beams or monochromatic optics.

In the following, I discuss several precision tests of the SM, using the present τ-decay data, and the expected improvements at the τcF. I will concentrate on the universality and Lorentz-structure of the charged leptonic currents. A discussion of other important topics in τ physics can be found in refs. [2,5–8].

CHARGED-CURRENT UNIVERSALITY

The leptonic decays $\tau^- \to e^- \bar{\nu}_e \nu_\tau, \mu^- \bar{\nu}_\mu \nu_\tau$ are theoretically understood at the level of the electroweak radiative corrections [9]. Within the SM,

$$\Gamma_{\tau \to l} \equiv \Gamma(\tau^- \to \nu_\tau l^- \bar{\nu}_l) = \frac{G_F^2 m_\tau^5}{192\pi^3} f(m_l^2/m_\tau^2) r_{EW}, \qquad (1)$$

where $f(x) = 1 - 8x + 8x^3 - x^4 - 12x^2 \log x$. The factor $r_{EW} = 0.9960$ takes into account radiative corrections not included in the Fermi coupling constant G_F, and the non-local structure of the W propagator [9].

Using the value of G_F measured in μ decay, Eq. (1) provides a relation [2] between the τ lifetime and the leptonic branching ratios $B_l \equiv B(\tau^- \to \nu_\tau l^- \bar{\nu}_l)$:

$$B_e = \frac{B_\mu}{0.972564 \pm 0.000010} = \frac{\tau_\tau}{(1.6321 \pm 0.0014) \times 10^{-12} \, s}. \qquad (2)$$

The errors reflect the present uncertainty of 0.3 MeV in the value of m_τ.

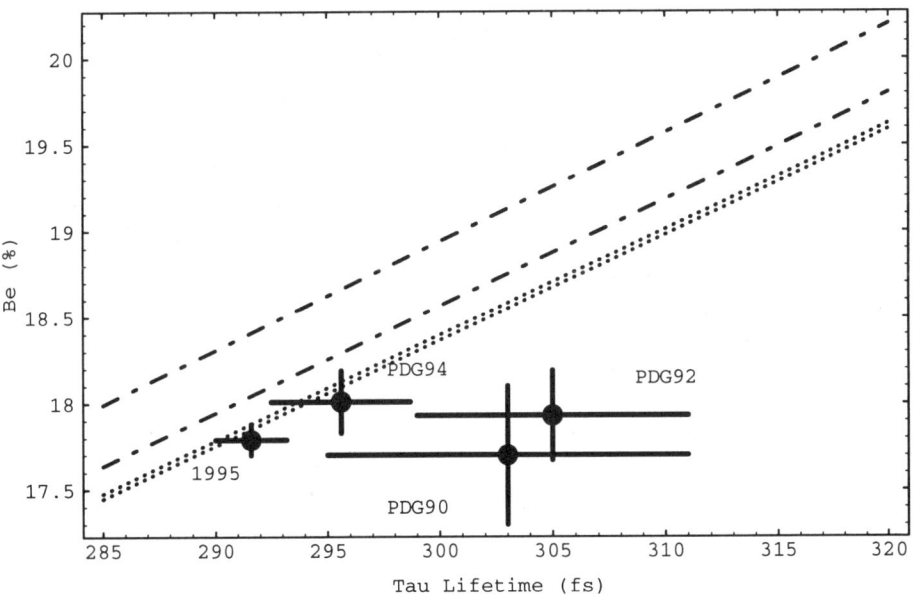

Figure 1: Relation between B_e and τ_τ. The narrow dotted band corresponds to the prediction in Eq. (2). The larger region between the two dot-dashed lines indicates the relation obtained with the old [10] value of m_τ. The experimental points show the present world averages [4], together with the values quoted by the Particle Data Group [1,3,10] since 1990.

Table 2: Present constraints on $|g_\mu/g_e|$.

	B_μ/B_e [4]	$R_{\pi \to e/\mu}$ [12]	$\sigma \cdot B_{W \to \mu/e}$ [13]
$|g_\mu/g_e|$	1.0008 ± 0.0036	1.0017 ± 0.0015	1.01 ± 0.04

Table 3: Present constraints on $|g_\tau/g_\mu|$.

	$B_e \tau_\mu/\tau_\tau$ [4]	$R_{\tau/\pi}$ [4]	$R_{\tau/K}$ [4]	$\sigma \cdot B_{W \to \tau/\mu}$ [13]
$|g_\tau/g_\mu|$	0.9979 ± 0.0037	1.006 ± 0.008	0.972 ± 0.029	0.99 ± 0.05

The predicted B_μ/B_e ratio is in perfect agreement with the measured value $B_\mu/B_e = 0.974 \pm 0.007$ [4]. As shown in Fig. 1, the relation between B_e and τ_τ is also well satisfied by the present data. Notice, that this relation is very sensitive to the value of the τ mass [$\Gamma_{\tau \to l} \propto m_\tau^5$]. The most recent measurements of τ_τ, B_e and m_τ have consistently moved the world averages in the correct direction, eliminating the previous ($\sim 2\sigma$) disagreement. The experimental precision (0.5%) is already approaching the level where a possible non-zero ν_τ mass could become relevant; the present bound [11] $m_{\nu_\tau} < 24$ MeV (95% CL) only guarantees that such effect is below 0.14%.

These measurements can be used to test the universality of the W couplings to the leptonic charged currents. The B_μ/B_e ratio constraints $|g_\mu/g_e|$, while the B_e/τ_τ relation provides information on $|g_\tau/g_\mu|$. The present results are shown in Tables 2 and 3, together with the values obtained from the ratio [12] $R_{\pi \to e/\mu} \equiv \Gamma(\pi^- \to e^- \bar{\nu}_e)/\Gamma(\pi^- \to \mu^- \bar{\nu}_\mu)$, and from the comparison of the $\sigma \cdot B$ partial production cross-sections for the various $W^- \to l^- \bar{\nu}_l$ decay modes at the p-$\bar{p}$ colliders [13].

The decay modes $\tau^- \to \nu_\tau \pi^-$ and $\tau^- \to \nu_\tau K^-$ can also be used to test universality through the ratios

$$R_{\tau/\pi} \equiv \frac{\Gamma(\tau^- \to \nu_\tau \pi^-)}{\Gamma(\pi^- \to \mu^- \bar{\nu}_\mu)} = \left|\frac{g_\tau}{g_\mu}\right|^2 \frac{m_\tau^3}{2 m_\pi m_\mu^2} \frac{(1 - m_\pi^2/m_\tau^2)^2}{(1 - m_\mu^2/m_\pi^2)^2} \left(1 + \delta R_{\tau/\pi}\right), \quad (3)$$

$$R_{\tau/K} \equiv \frac{\Gamma(\tau^- \to \nu_\tau K^-)}{\Gamma(K^- \to \mu^- \bar{\nu}_\mu)} = \left|\frac{g_\tau}{g_\mu}\right|^2 \frac{m_\tau^3}{2 m_K m_\mu^2} \frac{(1 - m_K^2/m_\tau^2)^2}{(1 - m_\mu^2/m_K^2)^2} \left(1 + \delta R_{\tau/K}\right), \quad (4)$$

where the dependence on the hadronic matrix elements (the so-called decay constants $f_{\pi,K}$) factors out. Owing to the different energy scales involved, the radiative corrections to the $\tau^- \to \nu_\tau \pi^-/K^-$ amplitudes are however not the same than the corresponding effects in $\pi^-/K^- \to \mu^- \bar{\nu}_\mu$. The size of the relative correction was first estimated by Marciano and Sirlin [14] to be $\delta R_{\tau/\pi} = (0.67 \pm 1.)\%$, where the 1% error amounts for the missing long-distance contributions to the tau decay rate. A recent evaluation of those long-distance corrections [15] quotes the more precise values:

$$\delta R_{\tau/\pi} = (0.16 \pm 0.14)\% , \qquad \delta R_{\tau/K} = (0.90 \pm 0.22)\% . \qquad (5)$$

Using these numbers, the measured [4] $\tau^- \to \pi^- \nu_\tau$ and $\tau^- \to K^- \nu_\tau$ decay rates imply the $|g_\tau/g_\mu|$ ratios given in Table 3. The inclusive sum of both decay modes, i.e. $\Gamma[\tau^- \to h^- \nu_\tau]$ with $h = \pi, K$, provides a slightly more accurate determination: $|g_\tau/g_\mu| = 1.004 \pm 0.007$.

The present data verifies the universality of the leptonic charged-current couplings to the 0.15% (e/μ) and 0.37% (τ/μ) level. The precision of the most recent τ-decay measurements is becoming competitive with the more accurate π-decay determination. It is important to realize the complementarity of the different universality tests. The pure leptonic decay modes probe the charged-current couplings of a transverse W. In contrast, the decays $\pi/K \to l\bar{\nu}$ and $\tau \to \nu_\tau \pi/K$ are only sensitive to the spin-0 piece of the charged current; thus, they could unveil the presence of possible scalar-exchange contributions with Yukawa-like couplings proportional to some power of the charged-lepton mass. One can easily imagine new-physics scenarios which would modify differently the two types of leptonic couplings [16]. For instance, in the usual two-Higgs doublet model, charged-scalar exchange generates a correction to the ratio B_μ/B_e, but $R_{\pi \to e/\mu}$ remains unaffected. Similarly, lepton mixing between the ν_τ and an hypothetical heavy neutrino would not modify the ratios B_μ/B_e and $R_{\pi \to e/\mu}$, but would certainly correct the relation between B_l and the τ lifetime.

At the τcF, the accurate measurement of the B_μ/B_e ratio would allow to test $|g_\mu/g_e|$ to the 0.05% level, compared to the present 0.36% precision (0.15% from $R_{\pi \to e/\mu}$). The final accuracy of the $|g_\tau/g_\mu|$ universality test will be limited by the knowledge of the τ lifetime. Assuming that the τ_τ measurement will be improved (at LEP or at the B Factory) by a factor of 2, i.e. $\delta \tau_\tau/\tau_\tau \sim 0.3\%$, $|g_\tau/g_\mu|$ would be tested with a 0.16% precision.

LORENTZ STRUCTURE OF THE CHARGED CURRENT

Let us consider the leptonic decays $l^- \to \nu_l l'^- \bar{\nu}_{l'}$, where the lepton pair (l, l') may be (μ, e), (τ, e), or (τ, μ). The most general, local, derivative-free, lepton-number conserving, four-lepton interaction Hamiltonian, consistent with locality and Lorentz invariance [17–23],

$$\mathcal{H} = 4\frac{G_{l'l}}{\sqrt{2}} \sum_{\epsilon,\omega=R,L}^{n=S,V,T} g_{l'_\epsilon l_\omega}^n \left[\overline{l'_\epsilon} \Gamma^n (\nu_{l'})_\sigma \right] \left[\overline{(\nu_l)_\lambda} \Gamma_n l_\omega \right] , \qquad (6)$$

contains ten complex coupling constants or, since a common phase is arbitrary, nineteen independent real parameters. $\epsilon, \omega, \sigma, \lambda$ are the chiralities (left-handed, right-handed) of the corresponding fermions, and n labels the type of interaction: scalar (I), vector (γ^μ), tensor ($\sigma^{\mu\nu}/\sqrt{2}$). For given n, ϵ, ω, the neutrino chiralities σ and λ are uniquely determined. Taking out a common factor $G_{l'l}$, which is determined by the total decay rate, the coupling constants $g_{l'_\epsilon l_\omega}^n$ are normalized to [21]

$$1 = \frac{1}{4}\left(|g_{RR}^S|^2 + |g_{RL}^S|^2 + |g_{LR}^S|^2 + |g_{LL}^S|^2\right) + 3\left(|g_{RL}^T|^2 + |g_{LR}^T|^2\right)$$
$$+ \left(|g_{RR}^V|^2 + |g_{RL}^V|^2 + |g_{LR}^V|^2 + |g_{LL}^V|^2\right). \quad (7)$$

In the SM, $g_{LL}^V = 1$ and all other $g_{\epsilon\omega}^n = 0$.

For an initial lepton-polarization $\mathcal{P}_l$, the final charged lepton distribution in the decaying lepton rest frame is usually parametrized in the form [18,19]

$$\frac{d^2\Gamma}{dx\, d\cos\theta} = \frac{m_l\omega^4}{2\pi^3}G_{l'l}^2\sqrt{x^2 - x_0^2}\left\{x(1-x) + \frac{2}{9}\rho\left(4x^2 - 3x - x_0^2\right) + \eta\, x_0(1-x)\right.$$
$$\left. - \frac{1}{3}\mathcal{P}_l\xi\sqrt{x^2 - x_0^2}\cos\theta\left[1 - x + \frac{2}{3}\delta\left(4x - 4 + \sqrt{1 - x_0^2}\right)\right]\right\}, \quad (8)$$

where θ is the angle between the l^- spin and the final charged-lepton momentum, $\omega \equiv (m_l^2 + m_{l'}^2)/2m_l$ is the maximum l'^- energy for massless neutrinos, $x \equiv E_{l'}/\omega$ is the reduced energy and $x_0 \equiv m_{l'}/\omega$. For unpolarized l's, the distribution is characterized by the so-called Michel [17] parameter ρ and the low-energy parameter η. Two more parameters, ξ and δ can be determined when the initial lepton polarization is known. If the polarization of the final charged lepton is also measured, 5 additional independent parameters [3] (ξ', ξ'', η'', α', β') appear.

The total decay rate is given by (neutrinos are assumed to be massless)

$$\Gamma = \frac{m_l^5 G_{l'l}^2}{192\pi^3}\left\{f\left(\frac{m_{l'}^2}{m_l^2}\right) + 4\eta\frac{m_{l'}}{m_l}g\left(\frac{m_{l'}^2}{m_l^2}\right)\right\}r_{\rm EW}, \quad (9)$$

where $g(z) = 1 + 9z - 9z^2 - z^3 + 6z(1+z)\ln z$, and the SM radiative corrections $r_{\rm EW}$ have been included[1].

Thus, the normalization $G_{e\mu}$ corresponds to the Fermi coupling G_F, measured in μ decay. The B_μ/B_e and $B_e\tau_\mu/\tau_\tau$ universality tests, discussed in the previous section, actually prove the ratios $|\hat{G}_{\mu\tau}/\hat{G}_{e\tau}|$ and $|\hat{G}_{e\tau}/\hat{G}_{e\mu}|$, respectively, where

$$\hat{G}_{l'l} \equiv G_{l'l}\sqrt{1 + 4\eta_{l\to l'}\frac{m_{l'}}{m_l}\frac{g(m_{l'}^2/m_l^2)}{f(m_{l'}^2/m_l^2)}}. \quad (10)$$

---

[1]Since we assume that the SM provides the dominant contribution to the decay rate, any additional higher-order correction beyond the effective four-fermion Hamiltonian (6) would be a subleading effect.

An important point, emphatically stressed by Fetscher and Gerber [22], concerns the extraction of $G_{e\mu}$, whose uncertainty is dominated by the uncertainty in $\eta_{\mu\to e}$.

In terms of the $g_{\epsilon\omega}^n$ couplings, the shape parameters in Eq. (8) are:

$$\rho - \frac{3}{4} = -\frac{3}{4}\left[|g_{LR}^V|^2 + |g_{RL}^V|^2 + 2|g_{LR}^T|^2 + 2|g_{RL}^T|^2 + \text{Re}(g_{LR}^S g_{LR}^{T*} + g_{RL}^S g_{RL}^{T*})\right],$$

$$\eta = \frac{1}{2}\text{Re}\left[g_{LL}^V g_{RR}^{S*} + g_{RR}^V g_{LL}^{S*} + g_{LR}^V\left(g_{RL}^{S*} + 6g_{RL}^{T*}\right) + g_{RL}^V\left(g_{LR}^{S*} + 6g_{LR}^{T*}\right)\right],$$

$$\xi - 1 = -\frac{1}{2}\Big[|g_{LR}^S|^2 + |g_{RR}^S|^2 + 4(-|g_{LR}^V|^2 + 2|g_{RL}^V|^2 + |g_{RR}^V|^2)$$
$$-4|g_{LR}^T|^2 + 16|g_{RL}^T|^2 - 8\text{Re}(g_{LR}^S g_{LR}^{T*} - g_{RL}^S g_{RL}^{T*})\Big], \quad (11)$$

$$(\xi\delta) - \frac{3}{4} = -\frac{3}{4}\Big[\frac{1}{2}(|g_{LR}^S|^2 + |g_{RR}^S|^2) + (|g_{LR}^V|^2 + |g_{RL}^V|^2 + 2|g_{RR}^V|^2)$$
$$+2(2|g_{LR}^T|^2 + |g_{RL}^T|^2) - \text{Re}(g_{LR}^S g_{LR}^{T*} - g_{RL}^S g_{RL}^{T*})\Big].$$

In the SM, $\rho = \delta = 3/4$, $\eta = \eta'' = \alpha' = \beta' = 0$ and $\xi = \xi' = \xi'' = 1$.

It is convenient to introduce [21] the probabilities $Q_{\epsilon\omega}$ for the decay of a ω-handed l^- into an ϵ-handed daughter lepton,

$$Q_{LL} = \frac{1}{4}|g_{LL}^S|^2 + |g_{LL}^V|^2 = \frac{1}{4}\left(-3 + \frac{16}{3}\rho - \frac{1}{3}\xi + \frac{16}{9}\xi\delta + \xi' + \xi''\right),$$

$$Q_{RR} = \frac{1}{4}|g_{RR}^S|^2 + |g_{RR}^V|^2 = \frac{1}{4}\left(-3 + \frac{16}{3}\rho + \frac{1}{3}\xi - \frac{16}{9}\xi\delta - \xi' + \xi''\right),$$

$$Q_{LR} = \frac{1}{4}|g_{LR}^S|^2 + |g_{LR}^V|^2 + 3|g_{LR}^T|^2 = \frac{1}{4}\left(5 - \frac{16}{3}\rho + \frac{1}{3}\xi - \frac{16}{9}\xi\delta + \xi' - \xi''\right), \quad (12)$$

$$Q_{RL} = \frac{1}{4}|g_{RL}^S|^2 + |g_{RL}^V|^2 + 3|g_{RL}^T|^2 = \frac{1}{4}\left(5 - \frac{16}{3}\rho - \frac{1}{3}\xi + \frac{16}{9}\xi\delta - \xi' - \xi''\right).$$

Upper bounds on any of these (positive-semidefinite) probabilities translate into corresponding limits for all couplings with the given chiralities.

For μ-decay, where precise measurements of the polarizations of both μ and e have been performed, there exist [21] upper bounds on Q_{RR}, Q_{LR} and Q_{RL}, and a lower bound on Q_{LL}. They imply corresponding upper bounds on the 8 couplings $|g_{RR}^n|$, $|g_{LR}^n|$ and $|g_{RL}^n|$. The measurements of the μ^- and the e^- do not allow us to determine $|g_{LL}^S|$ and $|g_{LL}^V|$ separately [21,24]. Nevertheless, since the helicity of the ν_μ in pion decay is experimentally known [25] to be -1, a lower limit on $|g_{LL}^V|$ is obtained [21] from the inverse muon decay $\nu_\mu e^- \to \mu^- \nu_e$. The present (90% CL) bounds [3,26] on the μ-decay couplings are given in Table 4. These limits show nicely that the bulk of the μ-decay transition amplitude is indeed of the predicted V−A type.

The experimental analysis of the τ-decay parameters is necessarily different from the one applied to the muon, because of the much shorter τ lifetime. The

Table 4: 90% CL experimental limits [3, 26] for the μ-decay $g^n_{e_e\mu_\omega}$ couplings.

$|g^S_{e_L\mu_L}| < 0.55$	$|g^V_{e_L\mu_L}| > 0.96$	–
$|g^S_{e_R\mu_R}| < 0.066$	$|g^V_{e_R\mu_R}| < 0.033$	–
$|g^S_{e_L\mu_R}| < 0.125$	$|g^V_{e_L\mu_R}| < 0.060$	$|g^T_{e_L\mu_R}| < 0.036$
$|g^S_{e_R\mu_L}| < 0.424$	$|g^V_{e_R\mu_L}| < 0.110$	$|g^T_{e_R\mu_L}| < 0.122$

Table 5: Experimental averages [3, 35–37] of the Michel parameters. The last column ($\tau \to l$) assumes identical couplings for $l = e, \mu$ (the quoted value for $\eta_{\tau \to l}$ is that obtained directly from measurements of the energy distribution). $\xi_{\mu \to e}$ refers to the product $\xi_{\mu \to e} \mathcal{P}_\mu$, where $\mathcal{P}_\mu \approx 1$ is the longitudinal polarization of the muon from pion decay.

	$\mu \to e$	$\tau \to \mu$	$\tau \to e$	$\tau \to l$
ρ	0.7518 ± 0.0026	0.738 ± 0.038	0.736 ± 0.028	0.733 ± 0.022
η	-0.007 ± 0.013	-0.14 ± 0.23	–	-0.01 ± 0.14
ξ	1.0027 ± 0.0085	1.23 ± 0.24	1.03 ± 0.25	1.06 ± 0.11
$\xi\delta$	0.7506 ± 0.0074	0.71 ± 0.15	1.11 ± 0.18	0.76 ± 0.09

measurement of the τ polarization and the parameters ξ and δ is still possible due to the fact that the spins of the $\tau^+\tau^-$ pair produced in e^+e^- annihilation are strongly correlated [27–34]. However, the polarization of the charged lepton emitted in the τ decay has never been measured. In principle, this could be done for the decay $\tau^- \to \mu^- \bar{\nu}_\mu \nu_\tau$ by stopping the muons and detecting their decay products [31]. The measurement of the inverse decay $\nu_\tau l^- \to \tau^- \nu_l$ looks far out of reach.

The present experimental status on the τ-decay Michel parameters is shown in Table 5 [23], which gives the world-averages of all published [3, 35–37] measurements. For comparison, the values measured in μ-decay [3] are also given. The improved accuracy of the most recent experimental analyses has brought an enhanced sensitivity to the different shape parameters, allowing the first measurements of $\eta_{\tau \to \mu}$ [35, 36], $\xi_{\tau \to e}$, $\xi_{\tau \to \mu}$, $(\xi\delta)_{\tau \to e}$ and $(\xi\delta)_{\tau \to \mu}$ [35]. (The ARGUS measurement [37] of $\xi_{\tau \to l}$ and $(\xi\delta)_{\tau \to l}$ assumes identical couplings for $l = e, \mu$. A measurement of $\sqrt{\xi_{\tau \to e}\xi_{\tau \to \mu}}$ was published previously [38]).

The determination of the τ-polarization parameters [35, 37] allows us to bound the total probability for the decay of a right-handed τ [31],

$$Q_{\tau_R} \equiv Q_{l'_R \tau_R} + Q_{l'_L \tau_R} = \frac{1}{2}\left[1 + \frac{\xi}{3} - \frac{16}{9}(\xi\delta)\right] . \qquad (13)$$

Table 6: 90% CL limits [23] for the τ_R-decay $g^n_{l_\epsilon \tau_R}$ couplings. The numbers with an asterisk use the measured value of $(\xi\delta)_e$.

$\tau \to \mu$	$\tau \to e$	$\tau \to l$
$|g^S_{\mu_R \tau_R}| < 1.05$	$|g^S_{e_R \tau_R}| < 0.75^*$	$|g^S_{l_R \tau_R}| < 0.74$
$|g^S_{\mu_L \tau_R}| < 1.05$	$|g^S_{e_L \tau_R}| < 0.75^*$	$|g^S_{l_L \tau_R}| < 0.74$
$|g^V_{\mu_R \tau_R}| < 0.53$	$|g^V_{e_R \tau_R}| < 0.38^*$	$|g^V_{l_R \tau_R}| < 0.37$
$|g^V_{\mu_L \tau_R}| < 0.53$	$|g^V_{e_L \tau_R}| < 0.38^*$	$|g^V_{l_L \tau_R}| < 0.37$
$|g^T_{\mu_L \tau_R}| < 0.30$	$|g^T_{e_L \tau_R}| < 0.22^*$	$|g^T_{l_L \tau_R}| < 0.21$

One finds (ignoring possible correlations among the measurements) [23]:

$$\begin{aligned} Q^{\tau\to\mu}_{\tau_R} &= 0.07 \pm 0.14 \; < \; 0.28 \quad (90\% \text{ CL}), \\ Q^{\tau\to e}_{\tau_R} &= -0.32 \pm 0.17 \; < \; 0.14 \quad (90\% \text{ CL}), \\ Q^{\tau\to l}_{\tau_R} &= 0.00 \pm 0.08 \; < \; 0.14 \quad (90\% \text{ CL}), \end{aligned} \qquad (14)$$

where the last value refers to the τ-decay into either $l = e$ or μ, assuming universal leptonic couplings. Since these probabilities are positive semidefinite quantities, they imply corresponding limits on all $|g^n_{l_R \tau_R}|$ and $|g^n_{l_L \tau_R}|$ couplings. The quoted 90% CL have been obtained adopting a Bayesian approach for one-sided limits [3]. Table 6 gives the implied bounds on the τ-decay couplings.

The central value of $Q^{\tau\to e}_{\tau_R}$ turns out to be negative at the 2σ level; i.e., there is only a 3% probability to have a positive value of $Q^{\tau\to e}_{\tau_R}$. Therefore, the limits on $|g^n_{e_R \tau_R}|$ and $|g^n_{e_L \tau_R}|$ should be taken with some caution, since the meaning of the assigned confidence level is not at all clear. The problem clearly comes from the measured value of $(\xi\delta)_e$. In order to get a positive probability Q_{τ_R}, one needs $(\xi - 1) > \frac{16}{3}[(\xi\delta) - \frac{3}{4}]$. Thus, $(\xi\delta)$ can only be made larger than $3/4$ at the expense of making ξ correspondingly much larger than one [23].

If lepton universality is assumed (i.e. $G_{l'l} = G_F$, $g^n_{l'_e l_\omega} = g^n_{e\omega}$), the leptonic decay ratios B_μ/B_e and $B_e \tau_\mu / \tau_\tau$ provide limits on the low-energy parameter η. The best sensitivity [39] comes from $\widehat{G}_{\mu\tau}$, where the term proportional to η is not suppressed by the small m_e/m_l factor. The measured B_μ/B_e ratio implies then [23]:

$$\eta_{\tau \to l} = 0.007 \pm 0.033 \, . \qquad (15)$$

This determination is more accurate that the one in Table 5, obtained from the shape of the energy distribution, and is comparable to the value measured in μ-decay: $\eta_{\mu \to e} = -0.007 \pm 0.013$ [40].

A non-zero value of η would show that there are at least two different couplings with opposite chiralities for the charged leptons. Since, we assume the

V−A coupling g_{LL}^V to be dominant, the second coupling would be [31] a Higgs-type coupling g_{RR}^S [$\eta \approx \text{Re}(g_{RR}^S)/2$, to first-order in new-physics contributions]. Thus, Eq. (15) puts the (90% CL) bound: $-0.09 < \text{Re}(g_{RR}^S) < 0.12$.

Model-Dependent Constraints

The general bounds in Table 6 look rather weak. The sensitivity of present experiments is not good enough to get interesting constraints from a completely general analysis of the four-fermion Hamiltonian. Nevertheless, stronger limits can be obtained within particular models, as shown in Tables 7, 8 and 9.

Table 7 assumes that there are no tensor couplings, i.e. $g_{\epsilon\omega}^T = 0$. This condition is satisfied in any model where the interactions are mediated by vector bosons and/or charged scalars [23]. In this case, the quantities $(1 - \frac{4}{3}\rho)$, $(1 - \frac{4}{3}\xi\delta)$ and $(1 - \frac{4}{3}\rho) + \frac{1}{2}(1 - \xi)$ reduce to sums of $|g_{l'_e l_\omega}^n|^2$, which are positive semidefinite; i.e., in the absence of tensor couplings, $\rho \leq \frac{3}{4}$, $\xi\delta \leq \frac{3}{4}$ and $(1 - \xi) > 2(\frac{4}{3}\rho - 1)$. This allows us to extract direct bounds on several couplings.

Table 7: 90% CL limits for the couplings $g_{\epsilon\omega}^n$, assuming that there are no tensor couplings [23]. The numbers with an asterisk use the measured value of $(\xi\delta)_e$.

	$\mu \to e$	$\tau \to \mu$	$\tau \to e$	$\tau \to l$
$|g_{LL}^S|$	< 0.55	≤ 2	≤ 2	≤ 2
$|g_{RR}^S|$	< 0.066	< 0.80	< 0.63*	< 0.62
$|g_{LR}^S|$	< 0.125	< 0.80	< 0.63*	< 0.62
$|g_{RL}^S|$	< 0.424	≤ 2	≤ 2	≤ 2
$|g_{LL}^V|$	> 0.96	≤ 1	≤ 1	≤ 1
$|g_{RR}^V|$	< 0.033	< 0.40	< 0.32*	< 0.31
$|g_{LR}^V|$	< 0.060	< 0.31	< 0.27	< 0.25
$|g_{RL}^V|$	< 0.047	< 0.23	< 0.27	< 0.18

If one only considers W-mediated interactions, but admitting the possibility that the W couples non-universally to leptons of any chirality, the stronger limits in Table 8 are obtained [23]. In this case, the $g_{l'_e l_\omega}^V$ constants factorize into the product of two leptonic W couplings, implying [41] additional relations among the couplings, such as $g_{LR}^V g_{RL}^V = g_{LL}^V g_{RR}^V$, which hold within any of the three channels, (μ,e), (τ,e), and (τ,μ). Moreover, there are additional equations relating different processes, such as [23] $g_{\mu_L \tau_L}^V g_{e_L \tau_R}^V = g_{\mu_L \tau_R}^V g_{e_L \tau_L}^V$. The normalization condition (7) provides lower bounds on the g_{LL}^V couplings.

Table 8: 90% CL limits on the $g^V_{\epsilon\omega}$ couplings, assuming that (non-standard) W-exchange is the only relevant interaction [23].

	$\mu \to e$	$\tau \to \mu$	$\tau \to e$
$|g^V_{LL}|$	> 0.997	> 0.95	> 0.96
$|g^V_{RR}|$	< 0.0028	< 0.019	< 0.013
$|g^V_{LR}|$	< 0.060	< 0.31	< 0.27
$|g^V_{RL}|$	< 0.047	< 0.060	< 0.047

For W-mediated interactions, the hadronic τ-decay modes can also be used to test the structure of the $\tau\nu_\tau W$ vertex, if one assumes that the W coupling to the light quarks is the SM one. The $\mathcal{P}_\tau$ dependent part of the decay amplitude is then proportional to the mean ν_τ helicity

$$h_{\nu_\tau} \equiv \frac{|g_R|^2 - |g_L|^2}{|g_R|^2 + |g_L|^2}, \tag{16}$$

which plays a role analogous to the leptonic-decay parameter ξ. The analysis of $\tau^+\tau^-$ decay correlations in leptonic–hadronic and hadronic–hadronic decay modes, using the π, ρ and a_1 hadronic final states, gives $h_{\nu_\tau} = -1.014 \pm 0.027$ [35,37,42]; this implies $|g_R/g_L|^2 = -0.007 \pm 0.013 < 0.018$ (90% CL). The sign of the ν_τ helicity can be determined [43] to be negative with the decay $\tau^- \nu_\tau a_1^-$, because there are two different amplitudes [corresponding to two different ways of forming the rho in $a_1^- \to (\rho\pi)^-$] and their interference contains information on the sign.[2]

Table 9: 90% CL limits for the $g^n_{\epsilon\omega}$ couplings, taking $g^S_{RR} = 0$, $g^S_{LL} = 0$, $g^V_{LR} = g^S_{LR} = 2g^T_{LR}$ and $g^V_{RL} = g^S_{RL} = 2g^T_{RL}$ [23].

	$\mu \to e$	$\tau \to \mu$	$\tau \to e$	$\tau \to l$
$|g^V_{LL}|$	> 0.998	> 0.95	> 0.96	> 0.97
$|g^V_{LR}|$	< 0.047	< 0.22	< 0.19	< 0.18
$|g^V_{RL}|$	< 0.033	< 0.16	< 0.19	< 0.13

Table 9 shows the constraints obtained under the assumption that the interaction is mediated by the SM W plus an additional neutral scalar [23].

---

[2]Once the h_{ν_τ} sign is fixed, the measurement of leptonic–hadronic correlations determines the signs of $\xi_{\tau \to e}$ and $\xi_{\tau \to \mu}$ to be positive. At the Z peak, the signs of $\xi_{\tau \to l}$ and h_{ν_τ} can be directly determined [44] from the sign of $\mathcal{P}_\tau$, which is fixed by combining the measurements of the polarization and left-right asymmetries.

The scalar contributions vanish for the LL and RR couplings and satisfy the relations $g^V_{LR} = g^S_{LR} = 2g^T_{LR}$, $g^V_{RL} = g^S_{RL} = 2g^T_{RL}$. This allows to express everything in terms of the vector couplings. The quantities $(1 - \frac{4}{3}\rho)$, $(1 - \frac{4}{3}\xi\delta)$ and $(1 - \frac{4}{3}\rho) + \frac{1}{2}(1 - \xi)$ are also positive semidefinite in this case. Moreover, $(1 - \frac{4}{3}\rho) = (1 - \frac{4}{3}\xi\delta)$.

EXPECTED SIGNALS IN MINIMAL NEW-PHYSICS SCENARIOS

All experimental results obtained so far are consistent with the SM. Clearly, the SM provides the dominant contributions to the τ-decay amplitudes. Future high-precision measurements of allowed τ-decay modes should then look for small deviations of the SM predictions and find out the possible source of any detected discrepancy.

In a first analysis, it seems natural to assume [23] that new-physics effects would be dominated by the exchange of a single intermediate boson, coupling to two leptonic currents. The new contribution could be originated by non-standard couplings of the usual W boson, or by the exchange of a new scalar or vector particle (intermediate tensor particles hardly appear in any reasonable model beyond the SM).

Table 10 [23] summarizes the expected effects of different new-physics scenarios on the measurable shape parameters. The four general cases studied correspond to adding a single intermediate boson-exchange, V^+, S^+, V^0, S^0 (charged/neutral, vector/scalar), to the SM contribution (a non-standard W would be a particular case of the SM + V^+ scenario). AS indicates that any sign is allowed.

Table 10: Theoretical constraints on the Michel parameters [23]

	SM + V^+	SM + S^+	SM + V^0	SM + S^0
$\rho - 3/4$	< 0	0	0	< 0
$\xi - 1$	AS	< 0	< 0	AS
$(\delta\xi) - 3/4$	< 0	< 0	< 0	< 0
η	0	AS	AS	AS

It is immediately apparent that $\rho \leq 3/4$ and $(\delta\xi) < 3/4$ in all cases studied. Thus one can have new physics and still ρ be equal to the SM value. In fact, any interaction consisting of an arbitrary combination of g^S_{ew}'s and $g^V_{\gamma\gamma}$'s yields this result [31]. On the other hand, $(\delta\xi)$ will be different from 3/4 in any of the cases above providing, in principle, a better opportunity for the detection of Physics Beyond the SM.

The above features are easy to understand by looking back at Eqs. (11) and recalling that the tensor couplings can only be generated by neutral scalar interactions (violating individual lepton flavours), in which case they are proportional to the scalar couplings. It is easy to see that having two such neutral scalars will not alter the situation. Indeed, to obtain $\rho > 3/4$ or $(\delta\xi) > 3/4$ one would need to get contributions from charged and neutral scalars simultaneously [23]. Moreover, $(\delta\xi) > 3/4$ can only happen through RL couplings and must be accompanied by $\xi > 1$.

The τcF offers an ideal experimental environment to perform this kind of analyses. The expected sensitivities to the different shape parameters, quoted in Table 1, would allow to prove the effective four-fermion Hamiltonian to a level where very interesting constraints on new-physics scenarios could be obtained. The numbers given in Table 1 are somehow conservative, since they only take into account the information obtained from correlated $\tau^+\tau^-$ events where both τ's decay into leptons [5,45]. Better precisions may be reached including the correlations of the leptonic decays with the hadronic ones [5,45].

DISCUSSION

The flavour structure of the SM is one of the main pending questions in our understanding of weak interactions. Although we do not know the reason of the observed family replication, we have learn experimentally that the number of SM fermion generations is just three (and no more). Therefore, we must study as precisely as possible the few existing flavours, to get some hints on the dynamics responsible for their observed structure. The construction of high-precision flavour factories is clearly needed.

Without any doubt, the τcF is the best available tool to explore the τ and ν_τ leptons and the charm quark. This facility combines the three ingredients required for making an accurate and exhaustive investigation of these particles: high statistics, low backgrounds and good control of systematic errors. The threshold region provides a series of unique features (low and measurable backgrounds free from heavy flavour contaminations, monochromatic particles from two-body decays, small radiative corrections, single tagging, high-rate calibration sources, ...) that create an ideal experimental environment for this physics.

Two basic properties make the τ particle an ideal laboratory for testing the SM: the τ is a lepton, which means clean physics, and moreover, it is heavy enough to produce a large variety of decay modes. In the previous sections I have discussed two particular topics, charged-current universality and Lorentz structure of the weak currents, which would greatly benefit from a high-precision experimental study of the τ lepton. There are, in addition, many other interesting subjects to be investigated.

The τcF could carry out a precise and exhaustive study of all exclusive τ decay channels, looking for signs of discrepancies with the theoretical expectations. The accurate measurement of the q^2 distribution of the final hadrons would allow a detailed analysis of the vector and axial-vector spectral functions and, therefore, a significant improvement of our knowledge of QCD. Rare and forbidden τ decays could be looked for, with a sensitivity better than 10^{-7} in some channels. The bound on the ν_τ mass could be pushed down to the 1–2 MeV level. The present knowledge of the τ electromagnetic moments could be improved by more than one order of magnitude. Last but not least, CP-violation in the lepton sector at the milli-weak (10^{-3}) level could be investigated (with longitudinal beam polarization).

In addition to the large improvement in our knowledge of the τ lepton, the τcF would also provide precious information on the c quark, through the detailed study of the D mesons and the J/Ψ and other charmonium states. A comprehensive set of precision measurements for τ, charm and light-hadron spectroscopy would be obtained, proving the SM to a much deeper level of sensitivity and exploring the frontiers of its possible extensions.

ACKNOWLEDGEMENTS

Many results discussed here have been obtained in collaboration with João P. Silva [23]. This work has been supported in part by CICYT (Spain), under grant No. AEN-93-0234.

REFERENCES

1. M. Aguilar-Benítez et al, *Review of Particle Properties*, Phys. Lett. **B239** (1990) 1.

2. A. Pich, *Tau Physics*, in *Heavy Flavours*, eds. A.J. Buras and M. Lindner, Advanced Series on Directions in High Energy Physics – Vol. 10 (World Scientific, Singapore, 1992), p. 375.

3. M. Aguilar-Benítez et al, *Review of Particle Properties*, Phys. Rev. **D50** (1994) 1173.

4. Proc. *Third Workshop on Tau Lepton Physics* (Montreux, 1994), ed. L. Rolandi, *Nucl. Phys. B (Proc. Suppl.)* **40** (1995).

5. A. Pich, *Perspectives on τ-Charm Factory Physics*, in Proc. *Third Workshop on the Tau-Charm Factory* (Marbella, 1993), eds. J. Kirkby and R. Kirkby (Editions Frontiéres, Gif-sur-Yvette, 1994), p. 767.

6. A. Pich, *Tau Physics Prospects at the τ-Charm Factory and at other Machines*, in Proc. *Third Workshop on the Tau-Charm Factory* (Marbella, 1993), eds. J. Kirkby and R. Kirkby (Editions Frontiéres, Gif-sur-Yvette, 1994), p. 51.

7. A. Pich, *QCD Predictions for the τ Hadronic Width: Determination of $\alpha_s(M_\tau^2)$*, in Proc. *QCD 94* (Montpellier, 1994), ed. S. Narison, *Nucl. Phys. B (Proc. Suppl.)* **39B,C** (1995) 326.

8. A. Pich, *QCD Tests from Tau Decay Data*, in Proc. *Tau-Charm Factory Workshop* (SLAC, California, 1989), ed. L.V. Beers, SLAC-Report-343 (1989), p. 416.

9. W.J. Marciano and A. Sirlin, *Phys. Rev. Lett.* **61** (1988) 1815.

10. M. Aguilar-Benítez *et al*, *Review of Particle Properties*, *Phys. Rev.* **D45** (1992) Part 2.

11. D. Buskulic *et al* (ALEPH), *Phys. Lett.* **B349** (1995) 585.

12. D.I. Britton *et al*, *Phys. Rev. Lett.* **68** (1992) 3000; G. Czapek *et al* *Phys. Rev. Lett.* **70** (1993) 17..

13. C. Albajar *et al* (UA1), *Z. Phys.* **C44** (1989) 15; J. Alitti *et al* (UA2), *Phys. Lett.* **B280** (1992) 137; F. Abe *et al* (CDF), *Phys. Rev. Lett.* **68** (1992) 3398; **69** (1992) 28.

14. W.J. Marciano and A. Sirlin, *Phys. Rev. Lett.* **71** (1993) 3629.

15. R. Decker and M. Finkemeier, *Nucl. Phys.* **B438** (1995) 17; *Nucl. Phys. B (Proc. Suppl.)* **40** (1995) 453.

16. W.J. Marciano, *Nucl. Phys. B (Proc. Suppl.)* **40** (1995) 3.

17. L. Michel, *Proc. Phys. Soc.* **A63** (1950) 514; 1371.

18. C. Bouchiat and L. Michel, *Phys. Rev.* **106** (1957) 170.

19. T. Kinoshita and A. Sirlin, *Phys. Rev.* **107** (1957) 593; **108** (1957) 844.

20. F. Scheck, *Leptons, Hadrons and Nuclei* (North-Holland, Amsterdam, 1983); *Phys. Rep.* **44** (1978) 187.

21. W. Fetscher, H.-J. Gerber and K.F. Johnson, *Phys. Lett.* **B173** (1986) 102.

22. W. Fetscher and H.-J. Gerber, *Precision Measurements in Muon and Tau Decays*, in *Precision Tests of the Standard Electroweak Model*, ed. P. Langacker, Advanced Series on Directions in High Energy Physics – Vol. 14 (World Scientific, Singapore, 1995), p. 657.

23. A. Pich and J.P. Silva, *Phys. Rev.* **D** (1995) in press [hep-ph/9505327].

24. C. Jarlskog, *Nucl. Phys.* **75** (1966) 659.

25. L.Ph. Roesch *et al*, *Helv. Phys. Acta* **55** (1982) 74;
    W. Fetscher, *Phys. Lett.* **140B** (1984) 117;
    A. Jodidio *et al*, *Phys. Rev.* **D34** (1986) 1967; *Phys. Rev.* **D37** (1988) 237.

26. B. Balke *et al*, *Phys. Rev.* **D37** (1988) 587;
    D. Geiregat *et al*, *Phys. Lett.* **B247** (1990) 131;
    S.R. Mishra *et al*, *Phys. Lett.* **B252** (1990) 170.

27. Y.S. Tsai, *Phys. Rev.* **D4** (1971) 2821;
    S. Kawasaki, T. Shirafuji and S.Y. Tsai, *Progr. Theor. Phys.* **49** (1973) 1656.

28. S.-Y. Pi and A.I. Sanda, *Ann. Phys. (NY)* **106** (1977) 171.

29. J.J. Gómez-Cadenas, *Beautiful τ Physics in the Charm Land*, in Proc. *Tau-Charm Factory Workshop* (SLAC, 1989), ed. L.V. Beers, SLAC-Report-343 (1989), p. 48.

30. C.A. Nelson, *Phys. Rev.* **D43** (1991) 1465; *Phys. Rev. Lett.* **62** (1989) 1347; *Phys. Rev.* **D40** (1989) 123 [Err: **D41** (1990) 2327];
    S. Goozovat and C.A. Nelson, *Phys. Rev.* **D44** (1991) 2818; *Phys. Lett.* **B267** (1991) 128 [Err: **B271** (1991) 468].

31. W. Fetscher, *Phys. Rev.* **D42** (1990) 1544.

32. J. Bernabéu, A. Pich and N. Rius, *Phys. Lett.* **B257** (1991) 219.

33. R. Alemany *et al*, *Nucl. Phys.* **B379** (1992) 3.

34. M. Davier *et al*, *Phys. Lett.* **B306** (1993) 411.

35. D. Buskulic *et al* (ALEPH), *Phys. Lett.* **B346** (1995) 379.

36. H. Albrecht *et al* (ARGUS), *Phys. Lett.* **B341** (1995) 441.

37. H. Albrecht *et al* (ARGUS), *Phys. Lett.* **B349** (1995) 576.

38. H. Albrecht *et al* (ARGUS), *Phys. Lett.* **B316** (1993) 608.

39. A. Stahl, *Phys. Lett.* **B324** (1994) 121.

40. H. Burkard *et al*, *Phys. Lett.* **160B** (1985) 343.

41. K. Mursula and F. Scheck, *Nucl. Phys.* **B253** (1985) 189;
    K. Mursula, M. Roos and F. Scheck, *Nucl. Phys.* **B219** (1983) 321.

42. H. Albrecht *et al* (ARGUS), *Phys. Lett.* **B337** (1994) 383;
    Z. Phys. **C58** (1993) 61.

43. H. Albrecht *et al* (ARGUS), *Phys. Lett.* **B250** (1990) 164.

44. D. Buskulic *et al* (ALEPH), *Phys. Lett.* **B321** (1994) 168.

45. A. Stahl, *The Lorentz Structure of the Charged Weak Current in τ Decays*, in Proc. *Third Workshop on the Tau-Charm Factory* (Marbella, 1993), eds. J. Kirkby and R. Kirkby (Editions Frontiéres, Gif-sur-Yvette, 1994), p. 175.

Review of Tau Physics at CLEO II with Prospects for the B Factories

Christopher G. White
Representing the CLEO Collaboration

Department of Physics
The Ohio State University
Columbus, Ohio 43210

Abstract. The CLEO II detector, located in Ithaca, New York, is an excellent tool for the study of the tau lepton. With a peak luminosity of about $3 \times 10^{32} cm^{-2} s^{-1}$, CESR (Cornell Electron-Positron Storage Ring) produces the highest luminosity of any e^+e^- collider. Having accumulated about 4 fb^{-1} of data (approximately 3.6 million τ pairs), CLEO II boasts the worlds largest sample of tau events. The CLEO II detector allows for a spectrum of measurements, including measurements of the tau mass, lifetime, a wide variety branching fractions, suppressed decays, study of rare decays, and a limit on the tau neutrino mass. As a consequence, CLEO has made a significant contribution to our understanding of the tau lepton. With the approval of the symmetric and asymmetric B factories, our knowledge will continue to increase.

Introduction

The τ lepton is a useful laboratory for studying the Standard Model of electroweak interactions[1]. Its properties, such as mass, lifetime, and leptonic branching fractions, are related to each other. If the τ is a sequential lepton, then its coupling to the W is the same as that of the μ and its lifetime is directly related to the μ lifetime (τ_μ). To lowest order[2] and neglecting the electron and neutrino masses, the Standard Model predicts:

$$\tau_\tau = \tau_\mu (m_\mu/m_\tau)^5 B(\tau^- \to e^- \nu_\tau \bar{\nu}_e). \tag{1}$$

The calculated τ lifetime (τ_τ) depends directly on experimental measurements of the μ mass (m_μ), μ lifetime (τ_μ), the τ mass (m_τ), and electronic branching ratio ($B(\tau^- \to e^- \nu_\tau \bar{\nu}_e)$). Using PDG94 average values for these quantities [3], the predicted lifetime becomes:

$$\tau_\tau = (1.63 \pm 0.002) \times 10^{-12} B(\tau^- \to e^- \nu_\tau \bar{\nu}_e)$$

$$= (2.92 \pm 0.03) \times 10^{-13} \text{ s}. \quad (2)$$

In addition, detailed measurements of tau decays provide information on Cabbibo suppression, second class currents, lepton flavor violation, and various theoretical models, such as the CVC (conserved vector current) hypothesis. Finally, a large sample of tau decays can be used to put a limit on the tau neutrino mass. In this paper we present measurements of the τ lifetime, the mass of the tau, and the electronic branching fraction. We also will present several branching fractions measured with the CLEO II detector, as well as our current limit on the tau neutrino mass.

The Data Set

The data sample was accumulated at the Cornell Electron-Positron Storage Ring (CESR). It corresponds to a total integrated luminosity of 4.0 fb^{-1}, with approximately two thirds of the data collected at the $\Upsilon(4S)$ ($\sqrt{s}$ =10.58 GeV), and the rest at energies slightly below the $\Upsilon(4S)$. This luminosity corresponds to the production of some 3.6x10^6 τ-pairs. Although this is the total current data set, only a sub-set has been used in the analyses presented here.

Detector

The CLEO-II detector[4], a cylindrical detector with symmetric end-caps, emphasizes precision charged particle tracking and high resolution electromagnetic calorimetry. The detector surrounds a 3.5 cm radius beryllium beampipe of thickness 0.5 mm. The inner wall of the pipe is coated with 25 μm of silver for synchrotron radiation shielding. The beampipe contributes 0.44% of a radiation length of material for a particle of normal incidence. Charged particle tracking is accomplished using information from three concentric chambers: a 6 layer straw tube chamber (PT), whose innermost layer is located 4.7 cm from the interaction point, a 10 layer high precision vertex wire chamber (VD), and a 51 layer (40 axial layers and 11 stereo layers) wire drift chamber (DR). The z position (coordinate along the beam axis) is determined using the stereo layers of the DR and cathode strips mounted on the inner and outer walls of both the VD and DR. All tracking layers reside inside a 1.5 Tesla superconducting coil. Electromagnetic calorimetry is accomplished using 7800 cesium-iodide (CsI) crystals. The crystals provide 95% solid angle coverage with energy and angular resolution, in the barrel region, of $(\sigma_E/E) = 0.35/E^{0.75} + 1.9 - 0.1E$ and $\sigma_\phi = 2.8/\sqrt{E} + 2.5$ mradians (E in GeV) respectively.

Electronic Branching Fraction

A precise measurement of the tau branching fraction into an electron, and a neutrino and anti-neutrino has been measured and published by CLEO [5]. The measurement is unique in that the technique actually measures B_e^2. This was accomplished by selecting events in which both taus decay electronically. Most systematic errors are therefore reduced by a factor of 2 in the final result. The measurement is expressed as follows:

$$B_e^2 = \frac{N \cdot (1 - f_{bck})}{\epsilon_t \cdot \epsilon_{acc} \cdot \epsilon_e^2 \cdot L\sigma_0(1 + \delta)} \qquad (3)$$

where N is the number of selected events, f_{bck} is the background fraction, ϵ_t, ϵ_{acc}, and ϵ_e^2 are the trigger, acceptance, and electron identification efficiencies, and $L\sigma_0(1 + \delta)$ represents the total number of tau pairs produced. The final result is:

$$B_e = 17.97 \pm 0.14 \pm 0.23\%. \qquad (4)$$

The first error is statistical, while the second is systematic.

Tau Mass

The tau mass has also been measured and published by CLEO[6]. This measurement uses a unique technique best employed at CESR energies. We consider two-body hadronic decays of the form $\tau \rightarrow h\nu_\tau$, where h is some hadronic system. We require both taus to decay in this fashion. Neglecting initial and final state radiation, the two taus must be back to back. Furthermore, the tau directions must lie on cones around the hadron directions. The half-angles $\theta_\pm$ are given by the relation:

$$P_\tau - P_h = P_\nu \rightarrow m_\tau^2 + m_{h\pm}^2 - 2E_\tau E_{h\pm} + 2p_\tau p_{h\pm} cos\theta_\pm = m_\nu^2. \qquad (5)$$

We take $E_\tau = E_{beam}$ under the assumption that there is no initial or final state radiation. $E_{h\pm}$ and $p_{h\pm}$ are the measured values for the observed hadronic systems with $m_{h/pm} = \sqrt{E_{h\pm}^2 - p_{h\pm}^2}$. The half angles $\theta_\pm$ can thus be calculated.

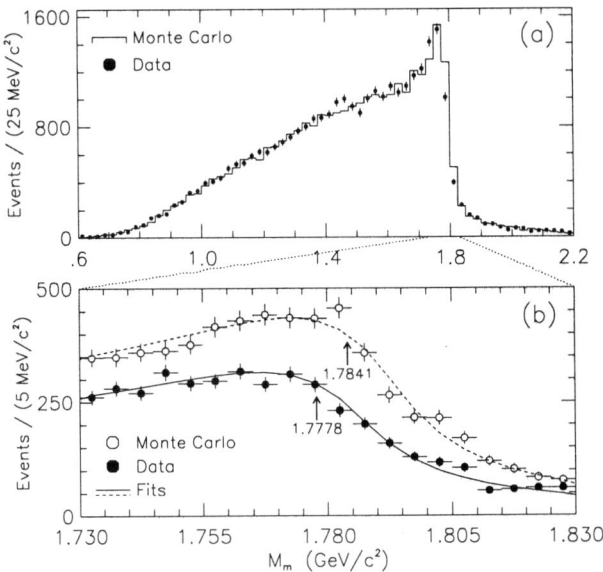

Figure 1: *(a) M_m distribution for data and Monte Carlo. (b) Expanded view of the edge. The Monte Carlo points have been vertically shifted to allow for a visual comparison with data.*

Since the taus are back to back, the true tau direction must be along the intersection of one cone with the parity inverse of the other. In general, there are two solutions to the intersection of two cones. If we adjust m_τ to smaller values, the half angles shrink. Eventually the two cones will intersect in a single ray such that the expression $\theta_+ + \theta_- + \theta = \pi$ (where θ is the angle between the the hadronic systems) is satisfied. This represents the minimum kinematically allowed tau mass for that event (M_m). The distribution of M_m is shown in figure 1 for data and Monte Carlo.

The edge of the distributions are fitted. By comparing the fitted edge in Monte Carlo to the known input tau mass, an offset is determined and applied to the fitted edge found in data. The tau mass is thus extracted. It should be noted that the fitted position of the edge follows the generated value of m_τ linearly over the entire region of relevance. The final result for the tau mass is $1777.8 \pm 0.7 \pm 1.7$ MeV/c². This technique assumes a zero tau neutrino mass.

By comparing the observed mass with measurements made at $\tau^+\tau^-$ threshold, which are independent of the tau neutrino mass, we derive a 95% confidence level upper limit of 75 MeV/c² for the tau neutrino mass.

Lifetime

CLEO is also undertaking a precision measurement of the tau lepton lifetime. The result presented here is a preliminary result based on an analysis from 1994. A more detailed measurement is due out in late 1995 with higher precision. The analysis looks at tau pairs where one or both taus decay into three charged particles, a 1 versus 3 topology, or 3 versus 3 topology. The displacement of the reconstructed decay vertices with respect to each other (in the 3 versus 3 case) or the beam center (1 versus 3 case) is used to measure the lifetime.

The tau proper flight distance, $c\tau$, is calculated using:

$$c\tau = \frac{1}{\gamma\beta}\frac{L_{xy}}{\sin\theta} = \frac{m_\tau}{p_\tau}\frac{L_{xy}}{\sin\theta} \qquad (6)$$

where γ, β, and p_τ, the magnitude of the τ's momentum, are calculated using the beam energy. L_{xy} is the flight path in the plane transverse to the z axis. We measure L_{xy} rather than L because the tracking resolution in the xy ($=r\phi$) plane is an order of magnitude better than the resolution in the z direction. This decay distance is converted to the full decay distance (L) using $\sin\theta$, with θ the polar angle. The τ polar angle is approximated using the vector momentum of the 3 charged tracks.

For each τ the most probable decay length in the transverse plane, L_{xy}, is determined via the equation:

$$L_{xy} = \frac{Xt_x\sigma_y^2 + Yt_y\sigma_x^2 - (Xt_y + Yt_x)\sigma_{xy}^2}{t_y^2\sigma_x^2 + t_x^2\sigma_y^2 - 2t_xt_y\sigma_{xy}^2} \qquad (7)$$

with :

$$X = X_v - X_b \text{ and } Y = Y_v - Y_b \qquad (8)$$

In the above equation, X_v and Y_v are the horizontal and vertical decay coordinates of the τ and X_b and Y_b are the horizontal and vertical production points of the τ. Here $\sigma_x^2(\sigma_y^2)$ is the variance of $X(Y)$ and σ_{xy}^2 is the correlation term for the X and Y vertex errors. Finally, t_x and t_y are direction cosines calculated from the momentum vector of the three charged tracks. Negative decay distances arise when the reconstructed vertex lies in the hemisphere opposite that of the 3-prong momentum vector. In the double vertex method we take 3 vs 3 τ-pairs, reconstruct both vertices, and compute the projection of the difference of these decay lengths along the estimated line of flight of the taus. The measurement is independent of the beam position. Hence systematic errors are limited to those associated with reconstruction of the vertices.

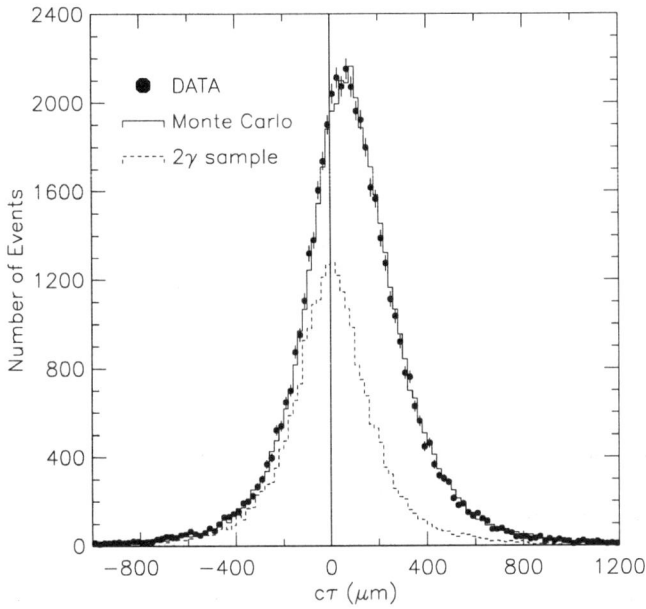

Figure 2: *Decay length distribution (in the τ rest frame) for both Monte Carlo and Data for 1 vs 3 events. The dash histogram is for a sample of 2-photon events with 4 charged tracks in the final state. This 2-photon sample corresponds to a zero lifetime control sample.*

The hadronic background in the 1 vs 3 sample is estimated using both data and Monte Carlo calculations to be 1.0%. From a Monte Carlo calculation we estimate the two photon background in the sample to be less than 0.2%, dominated by hadronic final states. The measured backgrounds in the 3 vs 3 sample are $4.7 \pm 0.9\%$ $q\bar{q}$, <0.1% $B\bar{B}$ and < 0.2 % two-photon events. These were determined from Monte Carlo calculations and data.

The measured $c\tau$ distributions for 1 vs 3 and 3 vs 3 events are shown in Fig. 1 and Fig. 2 respectively. Also displayed is a Monte Carlo calculation for this distribution (including contributions from the backgrounds) showing

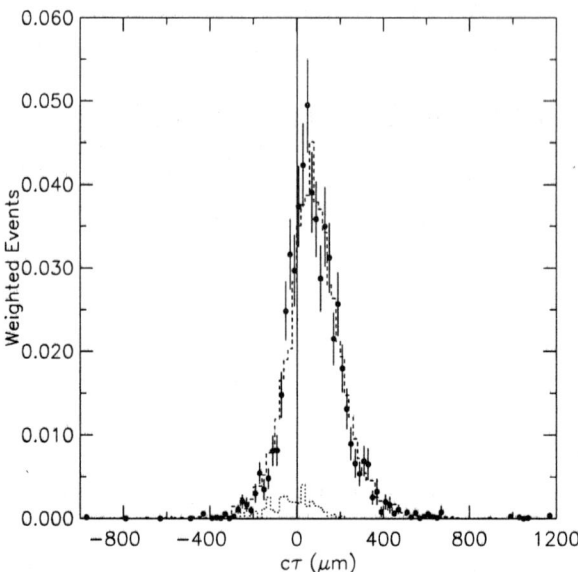

Figure 3: *Decay length distribution for Monte Carlo and Data for 3 vs 3 events in the τ rest frame. Data are indicated by points with error bars, simulation (signal plus background) by the dashed histogram, and background simulation by the dotted histogram.*

good agreement with the shape of the data distribution, and a zero lifetime sample of two photon events.

For the 1 vs 3 sample we measure $\tau_\tau = (2.91 \pm 0.03 \pm 0.07) \times 10^{-13}$ s, with the first error statistical and the second systematic. For the 3 vs 3 sample we measure $c\tau_\tau = 85.3 \pm 3.8 \pm 3.0$ μm corresponding to $\tau_\tau = (2.85 \pm 0.13 \pm 0.10) \times 10^{-13}$ s.

Branching Ratio Results

The large sample of tau pair events allow for many branching ratio measurements. CLEO has been very busy examining allowed, suppressed, rare, and forbidden decays of the tau lepton. A sampling of these measurements are shown below. Many more are under study.

$\tau^- \to h^-\pi^0\nu_\tau$	$25.87 \pm 0.12 \pm 0.42\%$	[7]
$\tau^- \to h^-2\pi^0\nu_\tau$	$8.92 \pm 0.15 \pm 0.38\%$	[8]
$\tau^- \to h^-3\pi^0\nu_\tau$	$1.07 \pm 0.07 \pm 0.12\%$	[8]
$\tau^- \to h^-4\pi^0\nu_\tau$	$0.16 \pm 0.04 \pm 0.05\%$	[8]
$\tau^- \to K^-\nu_\tau$	$0.66 \pm 0.07 \pm 0.09\%$	[9]
$\tau^- \to K^-\pi^0\nu_\tau$	$0.51 \pm 0.10 \pm 0.07\%$	[9]
$\tau^- \to 3\pi^-2\pi^+\nu_\tau$	$0.077 \pm 0.005 \pm 0.009\%$	[10]
$\tau^- \to 3\pi^\pm 2\pi^0\nu_\tau$	$0.48 \pm 0.04 \pm 0.04\%$	[11]
$\tau^- \to \pi^-\pi^0\omega\nu_\tau$	$0.39 \pm 0.04 \pm 0.04\%$	[11]
$\tau^- \to \pi^-\pi^0\eta\nu_\tau$	$0.17 \pm 0.02 \pm 0.02\%$	[12]
$\tau^- \to \pi^-\eta\nu_\tau$	$< 3.4 \times 10^{-4}$ @ 95%CL	[12]
$\tau^- \to \mu^-\gamma$	$< 4.2 \times 10^{-6}$ @ 90%CL	[13]

Tau Neutrino Mass Limit

A limit on the tau neutrino mass has been made using the decays $\tau^- \to 3\pi^-2\pi^+\nu_\tau$ and $\tau^- \to 3\pi^\pm 2\pi^0\nu_\tau$. The analysis employed a traditional one dimensional mass spectrum analysis with maximum sensitivity near the tau mass. CLEO is currently updating this analysis with a significantly larger data sample, and improved methods. A 2-dimensional analysis with a larger region of sensitivity to the tau neutrino mass should allow for a more stringent limit than is currently available. As a consequence, I will not detail this analysis since an improved version will be released shortly. The current CLEO limit is

$$m_{\nu_\tau} < 32.6 MeV/c^2 \text{ @ } 95\% CL [14].$$

CLEO conservatively expects to improve this limit to below 20 MeV/c^2.

Prospects for Future B Factories

CLEO II will collect a total integrated luminosity of about 5 fb^{-1}. Most of this has already been collected and is being analysed at this time. The limitation to precision measurements at CLEO currently is systematics. Most of the measurements presented here have larger systematic errors than statistical errors. Better understanding of systematics will be required to make good use of the even larger data sets to be associated with the symmetric and asymmetric B factories. The total number of tau pairs to be generated at these factories should increase the current world total by as much as 20 fold. Increased statistics will be useful in rare, and forbidden decay searches, and may allow for new or novel techniques in measuring fundamental tau parameters. One hopes to be able to trade off statistics for systematics in future detectors. There is still a lot of exciting physics to be done in the tau sector, and we expect the future to be profitable and productive.

Acknowledgements

We gratefully acknowledge the effort of the CESR staff in providing us with excellent luminosity and running conditions. This work was supported by the National Science Foundation, the U.S. Dept. of Energy, the Heisenberg Foundation, the SSC Fellowship program of TNRLC, Natural Sciences and Engineering Research Council of Canada, and the A.P. Sloan Foundation.

References

[1] S. L. Glashow, A Salam, S. Weinberg, Rev. Mod. Phys, **52**, (1980) 515.

[2] A summary of the corrections to the τ lifetime can be found in: W. J. Marciano, The Vancouver Meeting, in: Proc. Particles and Fields '91 (Vancouver, August 1991), Vol. 1, ed. D. Axen, D. Bryman, and M. Comyn (World Scientific Publishing Co. Pte. Ltd., Singapore, 1992) p. 461. The radiative corrections to the lifetime calculation are of the order of α and increase the average decay length by less than 1%.

[3] (Particle Data Group) L. Montanet et al., Phys. Rev. D **50**, 1173 (1994).

[4] Y. Kubota et al., Nucl Inst. Meth. **320** (1992) 66;

[5] D.S. Akerib et al., Phys. Rev. Lett. **69** (1992) 3610.

[6] R. Balest et al., Phys. Rev. D Rapid Comm **47** (1993) R3671.

[7] M. Artuso et al., Phys. Rev. Lett. **72** (1994) 3762.

[8] M. Procario et al., Phys. Rev. Lett. **70** (1993) 1207.

[9] M. Battle et al., Phys. Rev. Lett. **73** (1994) 1079.

[10] D. Gibaut et al., Phys. Rev. Lett. **73** (1994) 934.

[11] D. Bortoletto et al., Phys. Rev. Lett. **71** (1993) 1791.

[12] M. Artuso *et al.*, Phys. Rev. Lett. **69** (1992) 3278.

[13] A. Bean *et al.*, Phys. Rev. Lett. **70** (1993) 138.

[14] D. Cinabro *et al.*, Phys. Rev. Lett. **70** (1993) 3700.

Measurements of properties of the τ lepton at LEP

Steven R. Wasserbaech[1]

*Department of Physics, University of Washington,
P.O. Box 351560, Seattle, Washington 98195-1560*

Representing the ALEPH Collaboration

Abstract. Conditions are favorable for studies of the τ lepton in e^+e^- annihilation at the Z^0 resonance. The four LEP experiments have obtained a number of precise measurements of τ properties. The present status of this work is reviewed, and the ultimate precision of the LEP measurements is estimated.

INTRODUCTION

I begin this talk with a brief discussion of the conditions for τ lepton studies at LEP. The following sections cover the results that have been obtained in the categories of neutral-current and charged-current interactions of the τ. Finally, some attempts to measure the τ neutrino mass are described.

At LEP 1, $\tau^+\tau^-$ events may be selected with low background contamination and high efficiency. Moreover, the efficiency can be made roughly equal for all combinations of τ^+ and τ^- decay modes. These favorable circumstances result from the fact that the differences in the characteristics of $\tau^+\tau^-$ events and the various potential background reactions are accentuated when running on the Z^0 resonance. As an example, I consider the rejection of $e^+e^- \to q\bar{q}$ events, which represent a troublesome background in many τ studies at lower-energy machines. Although the cross section for $\tau^+\tau^-$ production at the Z^0 resonance is considerably smaller than that for hadronic final states,

$$\frac{\sigma(e^+e^- \to \tau^+\tau^-)}{\sigma(e^+e^- \to q\bar{q})} \sim 5\% \text{ at LEP 1,}$$

$$\sim 25\% \text{ at CESR, DORIS, PEP, PETRA,}$$

---

[1] E-mail: `wasser@u.washington.edu`

TABLE 1. $\tau^+\tau^-$ selection efficiencies and background contaminations in the LEP $\tau^+\tau^-$ cross section analyses.

Experiment	Efficiency	Background
ALEPH (1)	78%	1.6%
DELPHI (2)	53%	2.4%
L3 (3)	50%	2.9%
OPAL (4)	75%	2.4%

TABLE 2. Approximate number of produced $Z^0 \to \tau^+\tau^-$ events per experiment in each run period.

Run period	Events
1989–90	9K (scan)
1991	15K (scan)
1992	34K
1993	32K (scan)
1994	85K
1995	~100K? (scan)
Total	~275K

the high charged (and neutral) particle multiplicity in $q\bar{q}$ events clearly distinguishes such events from the $\tau^+\tau^-$ signal:

$$\langle N_{\text{charged}} \rangle \sim 21 \text{ for } q\bar{q} \text{ events at the } Z^0,$$
$$\sim 7 \text{ for } q\bar{q} \text{ events at the } \Upsilon(4S),$$

which are to be compared with $\langle N_{\text{charged}} \rangle = 2.6$ for $\tau^+\tau^-$ events.

One example of a $\tau^+\tau^-$ selection algorithm is described in (1). Briefly, the selection procedure consists of cuts on charged and neutral track multiplicities and jet opening angles, to reject $q\bar{q}$ backgrounds; a cut on acollinearity and the requirement of large visible energy or a large $p_\perp$ imbalance, to reject two-photon events; and cuts on the energies of the leading tracks, particle identification, and total energy, to reject Bhabha and dimuon events. The selection efficiencies and background contaminations applicable to the $\tau^+\tau^-$ cross section measurements from the four LEP experiments are compared in Table 1.

Table 2 shows the average numbers of $Z^0 \to \tau^+\tau^-$ events produced in each experiment for the run periods through 1994, along with estimates for 1995. The final LEP 1 total will be on the order of a quarter of a million produced $\tau^+\tau^-$ events per experiment. Most τ results presented until now represent analyses of the data samples obtained through 1992 or 1993. For 1989–1992 we have $N = 58K$, so $\sqrt{N_{\text{total}}/N} = 2.2$; for 1989–1992 we have $N = 90K$, so $\sqrt{N_{\text{total}}/N} = 1.7$. One may therefore conclude that the present statistical uncertainties on the various τ measurements will be reduced by a factor of roughly 2 by the end of the LEP 1 program. Through the remainder of this talk I give estimates of the final uncertainties that will be achieved at LEP for these measurements. A range of numbers is given in most cases; the limits correspond to the hypotheses that the systematic uncertainties will decrease as $1/\sqrt{N}$ or will remain at their present values. These estimates do not take into account the possible benefits of detector upgrades or innovations in the analysis techniques.

NEUTRAL-CURRENT INTERACTIONS

The Z^0 boson is a mixture of the W^3 and B gauge bosons, which are respectively associated with the SU(2) and U(1) factors of the gauge group of the standard electroweak model. Due to the different properties of left- and right-handed fermions under SU(2) transformations, the Z^0 is found to have unequal couplings to $f_L \bar{f}_R$ and $f_R \bar{f}_L$. This inequality leads to various observable asymmetries in Z^0 production and decay which are sensitive to the weak mixing angle and to physics beyond the Standard Model.

To further illustrate the origin of the asymmetries involving the τ, I consider the interaction of unpolarized e^+ and e^- beams at the peak of the Z^0 resonance. The four possible helicity combinations, $e_L^- e_R^+$, $e_R^- e_L^+$, $e_L^- e_L^+$, and $e_R^- e_R^+$, arise with equal luminosities. The Z^0 production cross section is denoted σ_L in the first case and σ_R in the second; the cross section is zero in the two remaining cases. We observe $\sigma_L > \sigma_R$. Since the spin of the Z^0 is determined by the spins of the incident e^+ and e^-, the difference in cross sections yields a net polarization of the Z^0's along the e^+ direction. This is illustrated in the top two sketches of Fig. 1. The degree of the Z^0 polarization is characterized by the asymmetry parameter $\mathcal{A}_e$.

The four lower sketches in Fig. 1 represent the possible configurations of the τ^+ and τ^- production directions and helicities, where the τ^+ and τ^- directions are here restricted to $\theta = 0$ or π for simplicity. The difference in the couplings of the Z^0 to $\tau_L^- \tau_R^+$ and $\tau_R^- \tau_L^+$ yields an asymmetry, characterized by $\mathcal{A}_\tau$, in the cross sections for the four final states shown.

From inspection of Fig. 1 we conclude that the numbers of τ^- produced in the forward and backward hemispheres are unequal; the forward-backward asymmetry is given by $(f_a + f_d) - (f_b + f_c) = \mathcal{A}_e \mathcal{A}_\tau$. (This result is reduced to $\frac{3}{4}\mathcal{A}_e \mathcal{A}_\tau$ when all θ values are considered.)

A mean τ^- polarization of $(f_b + f_d) - (f_a + f_c) = -\mathcal{A}_\tau$ is predicted from

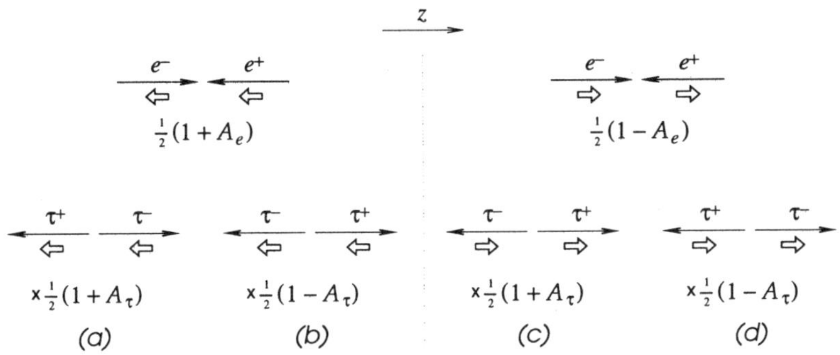

FIGURE 1. Asymmetries in $e^+e^- \to \tau^+\tau^-$. The incident e^- beam travels along the $+z$ direction. The left and right sketches at the top show the two possible helicity configurations of the incident e^+ and e^- that annihilate to form a Z^0. The fermion spins are indicated by the small open arrows. The four sketches at the bottom show the possible configurations for the outgoing τ^+ and τ^- in the case that the $\tau^+\tau^-$ production axis lies along the beam axis. The relative probabilities are $f_a = \frac{1}{4}(1+\mathcal{A}_e)(1+\mathcal{A}_\tau)$, $f_b = \frac{1}{4}(1+\mathcal{A}_e)(1-\mathcal{A}_\tau)$, $f_c = \frac{1}{4}(1-\mathcal{A}_e)(1+\mathcal{A}_\tau)$, and $f_d = \frac{1}{4}(1-\mathcal{A}_e)(1-\mathcal{A}_\tau)$.

Fig. 1. (This result also holds when integrating over θ.) Moreover, the polarization itself has a forward-backward asymmetry. The τ^- polarization at $\theta = 0$ is

$$P_\tau(\theta = 0) = \frac{f_d - f_a}{f_d + f_a} = \frac{-\mathcal{A}_e - \mathcal{A}_\tau}{1 + \mathcal{A}_e\mathcal{A}_\tau} \cong -(\mathcal{A}_e + \mathcal{A}_\tau), \qquad (1)$$

whereas at $\theta = \pi$ we have

$$P_\tau(\theta = \pi) = \frac{f_b - f_c}{f_b + f_c} = \frac{\mathcal{A}_e - \mathcal{A}_\tau}{1 - \mathcal{A}_e\mathcal{A}_\tau} \cong \mathcal{A}_e - \mathcal{A}_\tau. \qquad (2)$$

In the Standard Model, $\mathcal{A}_e = \mathcal{A}_\tau$, so the polarization is maximum in the forward direction and zero in the backward direction.

The asymmetry parameters $\mathcal{A}_\ell$ may be expressed in terms of the vector and axial vector coupling constants, g_{V_ℓ} and g_{A_ℓ}, of the Z^0 according to

$$\mathcal{A}_\ell = \frac{2g_{V_\ell}g_{A_\ell}}{g_{V_\ell}^2 + g_{A_\ell}^2}. \qquad (3)$$

Experiment reveals that $|g_{V_\ell}| \ll |g_{A_\ell}|$, so $\mathcal{A}_\ell \cong 2g_{V_\ell}/g_{A_\ell}$. Further information about the coupling constants is, of course, obtained from the total cross sections:

$$\sigma(e^+e^- \to \tau^+\tau^-) \propto g_{V_\tau}^2 + g_{A_\tau}^2. \qquad (4)$$

In the Standard Model (at tree level), $g_{A_\ell} = -\frac{1}{2}$ and $g_{V_\ell} = -\frac{1}{2} + 2\sin^2\theta_W$.

I now discuss the LEP measurements of the neutral-current interactions of the τ.

$\tau^+\tau^-$ Cross Section and Forward-Backward Asymmetry

From measurements of the $\tau^+\tau^-$ and other cross sections around the Z^0 peak, the LEP experiments have deduced the following Z^0 partial widths (5):

$$\Gamma_{ee} = 83.85 \pm 0.21 \,\text{MeV}, \tag{5}$$

$$\Gamma_{\mu\mu} = 83.95 \pm 0.30 \,\text{MeV}, \tag{6}$$

$$\Gamma_{\tau\tau} = 84.26 \pm 0.34 \,\text{MeV}. \tag{7}$$

These preliminary averages, compiled from results reported at Glasgow, were obtained from the 1990–1993 data samples. Under the assumption of lepton universality, we would expect $\Gamma_{ee} = \Gamma_{\mu\mu} = \Gamma_{\tau\tau} + 0.2\,\text{MeV}$, and the measurements are consistent with this expectation. The final LEP 1 uncertainty on $\Gamma_{\tau\tau}$ will be 0.2 to 0.3 MeV. Figure 2 shows measurements of $\sigma(e^+e^- \to \tau^+\tau^-)$ vs. $\sqrt{s}$ obtained by DELPHI.

Figure 3 shows the forward-backward asymmetry in $e^+e^- \to \tau^+\tau^-$ versus center of mass energy from L3. The peak forward-backward asymmetry for τ's is measured at LEP to be (5)

$$A_{\text{FB}}^{0,\tau} = 0.0228 \pm 0.0026. \tag{8}$$

The final uncertainty will be approximately 0.0015.

τ Polarization

All four experiments measure the τ polarization and the forward-backward polarization asymmetry. The ALEPH results for the τ decay modes $e\nu\bar{\nu}$, $\mu\nu\bar{\nu}$, $\pi\nu$, $\rho\nu$, and $a_1\nu$ (with the a_1 decaying into three charged pions) are depicted in Fig. 4. The analysis of the first three decay modes is based on the scaled energy x of the single observable final state particle. The polarization is measured in the $\rho\nu$ and $a_1\nu$ channels by means of optimized variables, generically denoted ω. For a τ^- decay, ω is defined to be

$$\omega = \frac{W_R(\xi) - W_L(\xi)}{W_R(\xi) + W_L(\xi)}, \tag{9}$$

where ξ represents the set of all decay observables for the channel in question and $W(\xi)$ is the probability density function in the ξ space for a particular τ^- helicity.

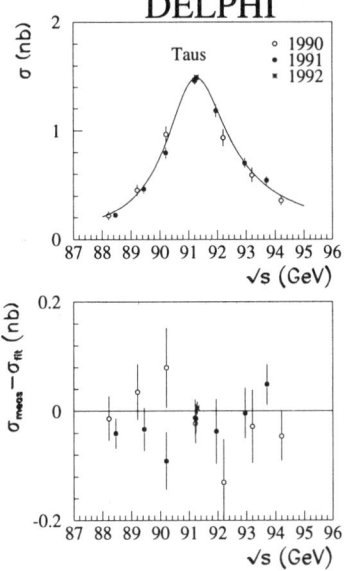

FIGURE 2. $\sigma(e^+e^- \to \tau^+\tau^-)$ vs. $\sqrt{s}$, as measured by DELPHI (6) from their 1990–1992 data. The lower plot shows the difference between the measured points and a fitted function.

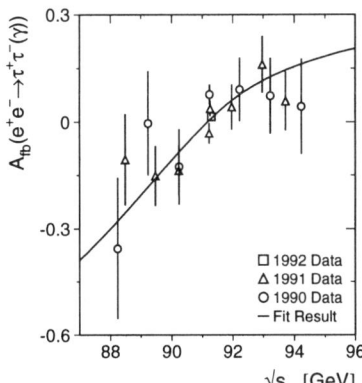

FIGURE 3. The forward-backward asymmetry A_{FB}^τ vs. $\sqrt{s}$, as measured by L3 (3) from their 1990–1992 data. The points with error bars are the measured values, and the curve represents the result of a fit to the combined cross section and forward-backward asymmetry data.

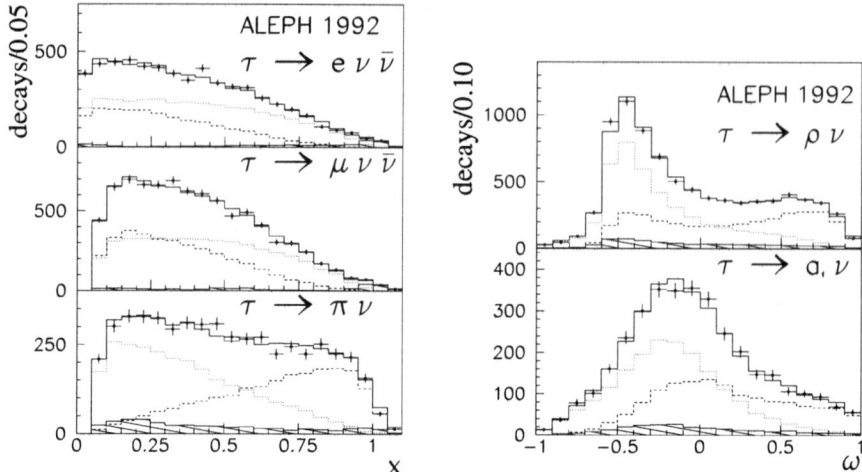

FIGURE 4. Distributions of $x = E/E_{\text{beam}}$ for $\tau \to e\nu\bar{\nu}$, $\mu\nu\bar{\nu}$, and $\pi\nu$; and of the polarization-sensitive variable ω for $\tau \to \rho\nu$ and $a_1\nu$ from ALEPH (7). The data are shown as points with error bars, the individual contributions from $\tau_L^+\tau_R^-$ and $\tau_R^+\tau_L^-$ by dashed and dotted histograms, respectively. The background is superimposed as a hatched histogram. The solid histogram shows the sum of all simulated contributions.

The asymmetry parameters $\mathcal{A}_e$ and $\mathcal{A}_\tau$ are extracted from the measured τ polarization as a function of θ. The preliminary LEP averages reported at Glasgow are

$$\mathcal{A}_e = 0.135 \pm 0.011, \tag{10}$$
$$\mathcal{A}_\tau = 0.143 \pm 0.010. \tag{11}$$

The final LEP 1 uncertainties will be 0.005 to 0.007 for $\mathcal{A}_e$ and 0.004 to 0.006 for $\mathcal{A}_\tau$.

Effective Z^0 Neutral-Current Couplings

The effective vector and axial vector couplings for e, μ, and τ are determined from the measurements of the leptonic partial widths of the Z^0, the forward-backward asymmetries, and the τ polarization. The results are summarized in Fig. 5. The coupling constants for the three lepton flavors are in agreement; no evidence for the violation of lepton universality is found. A fit to the data in which lepton universality is assumed yields

$$g_{V_\ell} = -0.0366 \pm 0.0013, \tag{12}$$
$$g_{A_\ell} = -0.50128 \pm 0.00054. \tag{13}$$

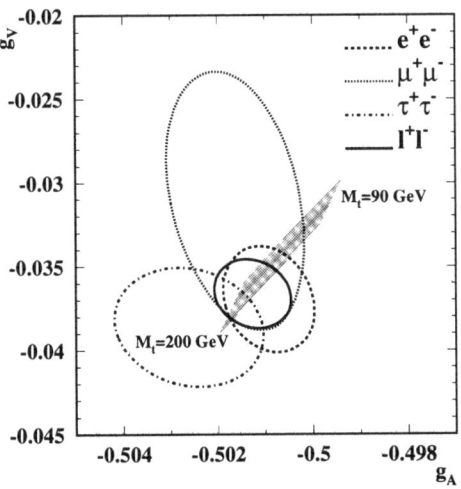

FIGURE 5. The 68% probability contours in the g_V-g_A plane (5). The solid contour results from a fit assuming lepton universality. The shaded band represents the Standard Model prediction.

The effective electroweak mixing parameter $\sin^2 \theta_{\text{eff}}^{\text{lept}}$ is deduced by interpreting the measured cross sections and asymmetries for leptonic and hadronic decays of the Z^0 within the framework of the Standard Model. The preliminary result is (5)

$$\sin^2 \theta_{\text{eff}}^{\text{lept}} = 0.2321 \pm 0.0004. \tag{14}$$

The final LEP 1 uncertainty will be $\Delta \sin^2 \theta_{\text{eff}}^{\text{lept}} = 0.00022$ to 0.00027.

CHARGED-CURRENT INTERACTIONS

Lorentz Structure of the τ Charged Current

The general derivative-free lepton-number-conserving matrix element for $\tau^- \to \ell^- \nu_\tau \bar{\nu}_\ell$ may be written (8)

$$\mathcal{M} = \frac{4 G_F}{\sqrt{2}} \sum_{i=S,V,T} \sum_{a,b=L,R} g^i_{ab} \langle \ell_a | \Gamma^i | \nu_\ell \rangle \langle \nu_\tau | \Gamma_i | \tau_b \rangle, \tag{15}$$

where i denotes the interaction type (scalar, vector, or tensor), and a and b are the chiralities of ℓ and τ. This representation has ten complex parameters

TABLE 3. Fit results from ALEPH 1990–1992 data, assuming lepton universality (10).

	Measured	SM
ρ_ℓ	$0.751 \pm 0.039 \pm 0.022$	0.75
η_ℓ	$-0.04 \pm 0.15 \pm 0.11$	0
ξ_ℓ	$1.18 \pm 0.15 \pm 0.06$	1
$(\delta\xi)_\ell$	$0.88 \pm 0.11 \pm 0.07$	0.75
ξ_h	$-1.006 \pm 0.032 \pm 0.019$	-1
P_τ	$-0.132 \pm 0.015 \pm 0.011$	

g^i_{ab} with one arbitrary overall phase. In the Standard Model, the only non-zero coefficient is $g^V_{LL} = 1$.

The Michel parameters ρ, η, δ, and ξ describe the ℓ energy spectrum in the lab rest frame in $\tau \to \ell\nu\bar{\nu}$ (9). These parameters may be expressed as bilinear combinations of the g^i_{ab}. The energy spectrum is then

$$\frac{1}{\Gamma}\frac{d\Gamma}{dz} = F_\ell(z) - P_\tau G_\ell(z), \qquad (16)$$

where $z = E_\ell/E_\tau$, and F and G are known functions of z, ρ, η, δ, and ξ; P_τ is the τ polarization.

Similarly, for hadronic τ decays one may write

$$\frac{1}{\Gamma}\frac{d\Gamma}{dz} = F_h(z) - P_\tau G_h(z); \qquad (17)$$

this time z represents the polarization-sensitive variable for the decay mode in question. F and G also depend on the decay mode. Furthermore,

$$G(z) = \xi_h g(z), \qquad (18)$$

where ξ_h is two times the average ν_τ helicity.

In $e^+e^- \to \tau^+\tau^-$ (via a photon or a Z^0), the τ^+ and τ^- have opposite helicities. ALEPH (10) has performed a global fit to the correlated z spectra, using $\tau \to e$, μ, π, ρ, a_1, and unidentified decays. For $\tau^- \to i$ and $\tau^+ \to j$, the distribution of z_i and z_j is given by

$$\frac{d^2\Gamma}{dz_i\,dz_j} = F_i(z_i)F_j(z_j) + G_i(z_i)G_j(z_j) - P_\tau\left[F_i(z_i)G_j(z_j) + G_i(z_i)F_j(z_j)\right]. \qquad (19)$$

To indicate the level of precision achieved, I give in Table 3 the results of a fit in which lepton universality is assumed. The ARGUS measurements reported in (11) are of similar precision.

Universality of Lepton Charged-Current Couplings

We may test the equality of the W-ℓ^--ν_ℓ couplings by comparing the decay rates for the reactions $\mu^- \to e^- \nu \bar{\nu}$, $\tau^- \to e^- \nu \bar{\nu}$, and $\tau^- \to \mu^- \nu \bar{\nu}$. Such tests are sensitive to violation of lepton universality at tree level, but they could also reveal, for example, the mixing of the τ neutrino with a very heavy, fourth-generation neutrino, or the effects of a charged Higgs boson exchanged in the decay (12). The measured τ properties that enter this type of universality test are the mass, mean lifetime, and leptonic branching fractions. As of the TAU94 workshop held in Montreux, the combined LEP measurements contribute the following fractions of the total weight in each world average: M_τ, 0.5%; τ_τ, 88%; $B(\tau^- \to e^- \nu \bar{\nu})$, 81%; and $B(\tau^- \to \mu^- \nu \bar{\nu})$, 92%. I therefore include no discussion of the LEP M_τ measurement (13) in my talk; I now describe the lifetime and branching fraction measurements.

τ Lifetime

The traditional decay length (or vertex) method is used by the LEP experiments to measure the mean τ lifetime from three-prong decays. Because the decay products of 45-GeV τ's are tightly collimated, some fraction of the three-prong decays cannot be reliably reconstructed, the inefficiency varying considerably among the experiments. Nevertheless, the precision vertex detectors yield precise decay length measurements for the candidates that are selected, and the lifetime uncertainty due to detector resolution is considerably smaller than that associated with the natural width of the exponential proper lifetime distribution. The size of the colliding beams contributes an even smaller uncertainty: the luminous region is typically $125\,\mu$m $\times$ $5\,\mu$m in the plane perpendicular to the beam axis, compared with $\beta\gamma c\tau_\tau = 2.2$ mm. Figure 6 shows the decay length distribution from OPAL (14).

The situation is considerably less straightforward for one-prong τ decays. It is not possible to reconstruct an individual τ decay length from a single charged daughter track; some information from the τ decay in the opposite hemisphere is necessary. A number of methods have been devised for analyzing the one-prong decays. The different methods suffer to different degrees from (1) the uncertainty on the τ direction, (2) the uncertainty on the τ production point, and (3) dependence of the result on the assumed impact parameter resolution. The multiplicity of methods in use reflects the fact that no method is completely immune to all of these sources of uncertainty.

It is beyond the scope of this brief review to describe all of the available one-prong analysis methods. Instead I merely touch on three examples: the classical Impact Parameter method, the Impact Parameter Difference method,

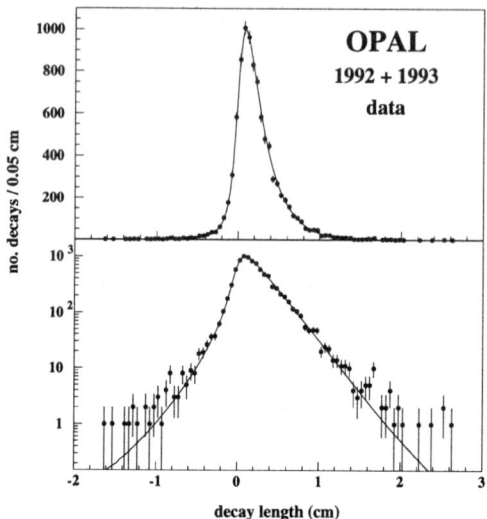

FIGURE 6. The distribution (on linear and logarithmic scales) of three-prong decay lengths which have decay length error less than 0.6 cm, together with a curve representing the result of a maximum likelihood fit.

and the Momentum-dependent Impact Parameter Sum method (15).

In the Impact Parameter method, the mean lifetime is extracted from the distribution of the lifetime-signed impact parameter of the daughter tracks. The lifetime sign is assigned under the assumption that the τ direction is given by the event thrust axis. The lifetime is then approximately proportional to the mean lifetime-signed impact parameter. (See (14), for example.)

In the Impact Parameter Difference method, the mean τ decay length is determined from the geometrical relation between the impact parameters and the acoplanarity of the daughter tracks in 1-1 topology events (16).

Finally, the Momentum-dependent Impact Parameter Sum method is applied to 1-1 topology events. The mean lifetime is extracted by means of a maximum likelihood fit to the sum of the impact parameters and the momenta of the daughter tracks. The fit is sensitive to the width of the impact parameter sum distribution, so the tracking resolution must be accurately parametrized (17).

The present world average τ lifetime is 291.6 ± 1.6 fs, corresponding to a relative uncertainty of 0.55%. The present LEP uncertainty is ±1.8 fs. The final LEP 1 uncertainty will reach ±0.9 to 1.2 fs (0.3 to 0.4%).

Leptonic Branching Fractions

The same basic strategy for measuring the τ leptonic branching fractions is employed by all four LEP experiments:

1. Select a sample of $\tau^+\tau^-$ candidate events.
   - Correct the number of events for non-$\tau\tau$ background.
   - Correct for the selection efficiency.

2. Select $\tau \to \ell\nu\bar{\nu}$ candidate decays in the $\tau^+\tau^-$ sample.
   - Correct the number of decays for background from other τ decays and from non-$\tau\tau$ events.
   - Correct for the lepton identification efficiency.
   - Correct for the selection efficiency (which depends slightly on the decay mode).

The measured leptonic branching fraction is then simply

$$B(\tau \to \ell\nu\bar{\nu}) = \frac{N(\tau \to \ell\nu\bar{\nu} \text{ decays})}{N(\tau)}. \qquad (20)$$

This determination is normalized to the observed number of $\tau^+\tau^-$ events, not to the theoretical cross section × integrated luminosity. Samples of $e^+e^- \to e^+e^-, \mu^+\mu^-$; $\gamma\gamma \to e^+e^-, \mu^+\mu^-$; and $\tau \to h + n\pi^0$ are used to check/measure the lepton identification and hadron misidentification probabilities as functions of momentum.

The status of the leptonic branching fraction measurements is summarized in Table 4.

Results of Charged-Current Universality Tests

The comparison of decay rates for $\mu^- \to e^-\nu\bar{\nu}$, $\tau^- \to e^-\nu\bar{\nu}$, and $\tau^- \to \mu^-\nu\bar{\nu}$, based on the world averages as of the TAU94 workshop, yields

$$\frac{g_\mu}{g_e} = 1.0007 \pm 0.0035, \qquad (21)$$

$$\frac{g_\tau}{g_\mu} = 0.9978 \pm 0.0037, \qquad (22)$$

$$\frac{g_\tau}{g_e} = 0.9986 \pm 0.0038. \qquad (23)$$

TABLE 4. τ leptonic branching fractions: results and predictions.

	$B(\tau \to e\nu\bar{\nu})$	$B(\tau \to \mu\nu\bar{\nu})$
Present world averages	0.17790 ± 0.00086 (0.48%)	0.17327 ± 0.00088 (0.51%)
Present LEP uncertainties	± 0.00096 (0.54%)	± 0.00091 (0.53%)
Final LEP 1 uncertainties	± 0.0005 to 0.0007 (0.3 to 0.4%)	± 0.0005 to 0.0006 (0.3 to 0.4%)

TABLE 5. Experimental uncertainties in lepton universality tests.

	Contributions to		
	$\Delta(g_\mu/g_e)$	$\Delta(g_\tau/g_\mu)$	$\Delta(g_\tau/g_e)$
τ lifetime	–	0.0027	0.0027
$B(\tau \to e\nu\bar{\nu})$	0.0024	0.0024	–
$B(\tau \to \mu\nu\bar{\nu})$	0.0025	–	0.0025
τ mass	–	0.0004	0.0004
Total	0.0035	0.0037	0.0038

No evidence for violation of lepton universality is observed. The breakdown of the experimental uncertainties in these tests is shown in Table 5. The precision of the universality tests is at present limited by the uncertainties on the τ lifetime and leptonic branching fractions.

Other interesting universality tests are based on the following trios of decay rates:

- $\tau^- \to \pi^-\nu$, $\pi^- \to \mu^-\bar{\nu}$, and $\pi^- \to e^-\bar{\nu}$;

- $\tau^- \to K^-\nu$, $K^- \to \mu^-\bar{\nu}$, and $K^- \to e^-\bar{\nu}$;

- $W^- \to e^-\bar{\nu}$, $W^- \to \mu^-\bar{\nu}$, and $W^- \to \tau^-\bar{\nu}$.

The results of these tests are also consistent with the hypothesis of lepton universality. In every case the theoretical uncertainties are much smaller than the experimental uncertainties.

At LEP 2 it may be possible to improve the precision of the universality tests involving $W \to \ell\nu$. We expect roughly 5000 produced $e^+e^- \to W^+W^-$ events, and hence 1100 $W \to \tau\nu$ decays, in each experiment during the LEP 2 program. The combined measurement of $B(W \to \tau\nu)$ might have a $\sim 2\%$ relative uncertainty, so universality tests with a sensitivity of 1 to 1.5% (on the ratios of g's) may be possible. (The present uncertainties on the W branching fractions are 4 to 9%, with large systematic components.)

Hadronic τ Decays

A global analysis of the τ branching fractions was performed by ALEPH (18). Candidate decays are classified in one of the following "quasi-exclusive" channels: e, μ, h, $h\pi^0$, $h2\pi^0$, $h \geq 3\pi^0$, $3h$, $3h\pi^0$, $3h2\pi^0$, $3h \geq 3\pi^0$, $5h$, and $5h\pi^0$. Clusters of hit towers in the electromagnetic calorimeter (ECAL) are subjected to a moments analysis, which is designed to detect the presence of overlapping photon showers resulting from high energy π^0's. For the quasi-exclusive analysis, π^0 candidates may consist of

o two neutral clusters in ECAL,

o a single cluster with substructure,

o a single cluster plus a conversion, or

o a single cluster.

The breakdown of π^0 candidates into these categories, as a function of energy, is depicted in Fig. 7.

A "mixing matrix" $E_{i \to j}$, corresponding to the probability of assigning a τ decay in class i to class j, is constructed by means of a Monte Carlo simulation. The sum of the τ branching fractions is constrained to 100%. Decays with $K^\pm$, K^0_S, and K^0_L are handled consistently. The precise results given in (18) are based on the 1991–1993 data. The branching fraction for $\tau \to 3h \geq 3\pi^0 \nu$ is measured for the first time: $0.20 \pm 0.07\%$.

Other ALEPH analyses which support the quasi-exclusive measurements:

o The branching fraction for τ decay into undetected modes is found to be less than 0.16% at the 95% CL.

o An "exclusive" analysis is performed in which single ECAL clusters are considered to be photons, not π^0's. The decay class h, $3h \geq 0\pi^0 \geq 1\gamma$ is added. In addition to unpaired clusters from π^0 decays, a small number of unpaired photons from $\omega \to \gamma\pi^0$ and $\eta \to \gamma\gamma$ are expected. The results of the exclusive analysis are consistent with the more precise quasi-exclusive measurements.

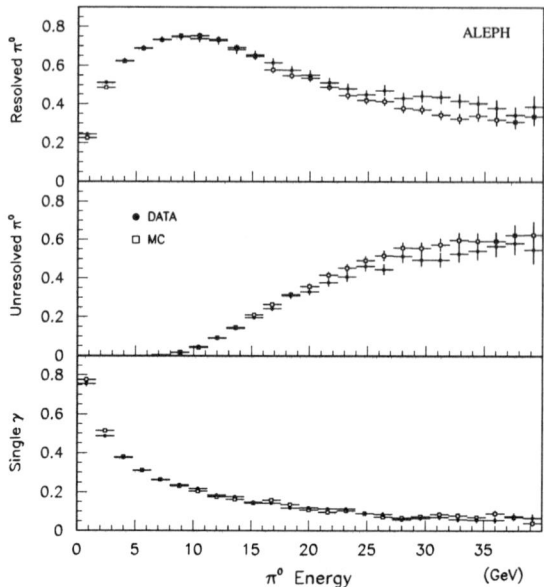

FIGURE 7. Fractions of π^0 candidates *vs.* energy: resolved (two-cluster) π^0's, unresolved (single-cluster) π^0's, and single photons.

The hadronic branching fraction measurements made at LEP and elsewhere are reviewed in (19). Recent investigations from LEP include a measurement of exclusive one-prong decays by L3 (20), a measurement of exclusive three-prong decays by OPAL (21), and a study of the resonant substructure in $\tau \to 3\pi\pi^0\nu$ by ALEPH (22).

The total inclusive branching fraction for modes with kaons is roughly 4%, so measurements of these channels are necessary for completing a consistent picture of the τ decays. All four LEP experiments have made contributions in this area (23).

The status of the topological branching fraction measurements is reviewed in (24). Some recent measurements are listed in Table 6.

TAU NEUTRINO MASS

Finally, I mention the recent attempts to measure the mass of the τ neutrino. The well-known technique of fitting to the invariant mass spectrum of high-multiplicity hadronic final states was applied by ARGUS (30) and CLEO-

TABLE 6. Recent τ topological branching fraction measurements.

B_3 (%)	
ALEPH[a] (25)	$14.79\,^{+0.13}_{-0.15}\,\pm 0.16$
DELPHI[a] (26)	$14.60 \pm 0.22 \pm 0.30$
L3 (27)	$14.4 \pm 0.6 \pm 0.3$
OPAL (21)	$14.96 \pm 0.09 \pm 0.22$
B_5 (%)	
ALEPH[a] (25)	$0.11\,^{+0.03}_{-0.02}\,\pm 0.02$
DELPHI[a] (26)	$0.33 \pm 0.11 \pm 0.10$
OPAL (28)	$0.26 \pm 0.06 \pm 0.05$
CLEO (29)	$0.097 \pm 0.005 \pm 0.011$

[a] Preliminary.

II (31) to set 95% CL upper limits limits on the ν_τ mass of $31\,\mathrm{MeV}/c^2$ and $32.6\,\mathrm{MeV}/c^2$, respectively. A new method, featuring a fit to the distribution of energy vs. invariant mass for the hadronic system, has led to limits of $74\,\mathrm{MeV}/c^2$ from five $\tau \to 5\pi\nu$ decays observed by OPAL (32), and $24\,\mathrm{MeV}/c^2$ from 25 $\tau \to 5\pi\pi^0\nu$ decays observed by ALEPH (33). The two-dimensional fit yields better sensitivity than the invariant mass method because the presence or absence of decays in which the hadronic system carries nearly the beam energy is indicative of the τ neutrino mass, even if the hadronic invariant mass is well below the endpoint.

CONCLUSIONS

The LEP experiments have obtained many precise results in the field of τ physics. Large samples of Z^0 decays have yielded impressive measurements of the neutral-current couplings of the different lepton flavors. Tests of universality of the charged-current couplings have been pushed to a new level of sensitivity. A variety of other properties of the τ and its neutrino have been measured. Yet in most analyses, a relatively small fraction of the ultimate LEP 1 sample has been studied. We may expect the statistical uncertainties on measurements of τ properties from LEP 1 to decrease by roughly a factor of 2 with respect to the results reported until now.

REFERENCES

1. D. Buskulic et al. (ALEPH Collaboration), Z. Phys. C **62**, 539 (1994).
2. P. Abreu et al. (DELPHI Collaboration), Nucl. Phys. B **417**, 3 (1994); **426**, 244(E) (1994).

3. M. Acciarri et al. (L3 Collaboration), Z. Phys. C **62**, 551 (1994).
4. P. Acton et al. (OPAL Collaboration), Z. Phys. C **58**, 219 (1993); R. Akers et al. (OPAL Collaboration), Z. Phys. C **61**, 19 (1994).
5. The LEP Collaborations and the LEP Electroweak Working Group, preprint CERN-PPE/94-187 (1994).
6. P. Abreu et al. (DELPHI Collaboration), Nucl. Phys. B **418**, 403 (1994).
7. D. Buskulic et al. (ALEPH Collaboration), preprint CERN-PPE/95-023 (1995), submitted to Z. Phys. C.
8. W. Fetscher, H.-J. Gerber, and K.F. Johnson, Phys. Lett. B **173**, 102 (1986).
9. L. Michel, Proc. Phys. Soc. **A63**, 514 (1950); C. Bouchiat and L. Michel, Phys. Rev. **106**, 170 (1957); T. Kinoshita and A. Sirlin, Phys. Rev. **108**, 844 (1957).
10. D. Buskulic et al. (ALEPH Collaboration), Phys. Lett. B **346**, 379 (1995).
11. M. Schmidtler (ARGUS Collaboration), Nucl. Phys. B (Proc. Suppl.) **40**, 265 (1995); I. Korolko (ARGUS Collaboration), Nucl. Phys. B (Proc. Suppl.) **40**, 275 (1995).
12. W. Marciano, Nucl. Phys. B (Proc. Suppl.) **40**, 3 (1995).
13. J. Timmermans, "Measurements of the τ Mass, Lifetime, and Leptonic Branching Ratio," in *Proceedings of the XXVII International Conference on High Energy Physics*, Glasgow, Scotland, 1994, p. 1077.
14. R. Akers et al. (OPAL Collaboration), Phys. Lett. B **338**, 497 (1994).
15. Other methods are introduced in D. Buskulic et al. (ALEPH Collaboration), Phys. Lett. B **297**, 432 (1992); P. Abreu et al. (DELPHI Collaboration), Phys. Lett. B **302**, 356 (1993); I. Ferrante, Nucl. Phys. B (Proc. Suppl.) **40**, 299 (1995).
16. D. Decamp et al. (ALEPH Collaboration), Phys. Lett. B **279**, 411 (1992).
17. H.Y. Kim, Ph.D. thesis, University of Washington (1994).
18. M. Girone, Nucl. Phys. B (Proc. Suppl.) **40**, 153 (1995).
19. B. Heltsley, Nucl. Phys. B (Proc. Suppl.) **40**, 413 (1995).
20. M. Acciarri et al. (L3 Collaboration), Phys. Lett. B **345**, 93 (1995).
21. R. Akers et al. (OPAL Collaboration), preprint CERN-PPE/95-070 (1995), submitted to Z. Phys. C.
22. P. Bourdon, Nucl. Phys. B (Proc. Suppl.) **40**, 203 (1995).
23. D. Buskulic et al. (ALEPH Collaboration), Phys. Lett. B **332**, 209 (1994); **332**, 219 (1994); P. Abreu et al. (DELPHI Collaboration), Phys. Lett. B **334**, 435 (1994); M. Acciarri et al. (L3 Collaboration), preprint CERN-PPE/95-042 (1995), submitted to Phys. Lett. B; H. Evans (OPAL Collaboration), Nucl. Phys. B (Proc. Suppl.) **40**, 361 (1995).
24. J.G. Smith, Nucl. Phys. B (Proc. Suppl.) **40**, 145 (1995).
25. ALEPH Collaboration, contribution to the XXVII International Conference on High Energy Physics, Glasgow, Scotland, ref. gls0569.
26. J.R. Patterson, "Weak and Rare Decays," in *Proceedings of the XXVII International Conference on High Energy Physics*, Glasgow, Scotland, 1994, p. 149.
27. B. Adeva et al. (L3 Collaboration), Phys. Lett. B **265**, 451 (1991).
28. P. Acton et al. (OPAL Collaboration), Phys. Lett. B **288**, 373 (1992).
29. D. Gibaut et al. (CLEO Collaboration), Phys. Rev. Lett. **73**, 934 (1994).
30. H. Albrecht et al. (ARGUS Collaboration), Phys. Rev. Lett. **291**, 221 (1992).
31. D. Cinabro et al. (CLEO Collaboration), Phys. Rev. Lett. **70**, 3700 (1993).
32. R. Akers et al. (OPAL Collaboration), Z. Phys. C **65**, 183 (1995).
33. D. Buskulic et al. (ALEPH Collaboration), Phys. Lett. B **349**, 585 (1995).

Prospects for High Precision Measurements at the τcF Including Current Achievements of BES

Tao Huang

Institute of High Energy Physics
Academia Sinica
P.O.Box 918
Beijing 100039, China

Abstract

Recent progress and the future prospects in the energy regions of the τ pair production and the charm meson thresholds is reviewed: Search for gluonia and hybrids, τ and ν_τ physics, Charmonium physics, Charmed meson and baryon physics, $D - \bar{D}$ mixing, and CP violation. Since data taking started in 1990, the Beijing Spectrometer (BES) at the Beijing Electron-Positron Collider (BEPC) has run at different energy settings covering the J/ψ and ψ' resonances and the τ pair and D_s production thresholds, leading to significant progress in these areas of particle physics. The merits of a τ-charm factory are investigated. It is concluded that a τ-charm factory is the ideal tool to explore the τ and ν_τ leptons and the physics of the charm quark providing a deeper understanding of the Standard Model of particle physics and exploring the frontier of its possible extensions.

I. Introduction

The success of the Standard Model (SM) in describing all available experimental data in undisputed. However, the SM leaves many questions unresolved and further experimental information is needed to develop a more complete picture of the laws of nature. Clearly, "New Physics" must exist! On one side, high-energy machines such as the LHC are required to study the mechanisms responsible for the spontaneous symmetry breaking of $SU(2) \times U(1)$; on the other side, higher precision measurements at low energies, e.g. at the ϕ, B, or τ-charm factory (τcF), complement the tests of the SM. In particular, the issues associated with flavor, such as particle masses, number of families, CP violation and the gluonic degree of freedom, should be studied at these factories.

The ν_τ, the τ lepton and the c-quark are important for understanding the fundamental aspects of the SM and beyond. Present machines have made significant progress on understanding the dynamics of these particles. For example, the Beijing Electron-Positron Collider BEPC[1,2,3] has contributed significantly to the progress of particle physics in the τ and charm energy regions. Since data taking started in 1990, the Beijing Spectrometer (BES) has run at different energy settings covering the J/ψ and ψ' resonances and the τ pair and D_s production thresholds.

The τcF [4] is designed for the precision study of the τ lepton, charm particles, charmonium and light flavor hadrons[5]. It will cover the energy range between 3 and 6 GeV, thus covering the J/ψ and ψ' resonances, and the τ pair and charm meson and baryon thresholds. Contamination from the decay of B mesons will be absent.

The concept of the proposed τcF involves an e^+e^- collider operating in the centre-of-mass energy range between 3 and 6 GeV and an integrated detector. The design peak luminosity of the collider, $10^{33} \text{cm}^{-2}\text{s}^{-1}$, represents an improvement by two orders of magnitude over BEPC, see Fig.1. However, the large statistics will not be the main advantage of the τcF compared to other low-energy factories. The threshold region provides a unique advantage, since backgrounds are both small and experimentally measurable. Near threshold, the $\tau^+\tau^-$ and $c\bar{c}$ are produced in simple particle-antiparticle final states. By observing the decay of one particle, its partner is cleanly tagged. Furthermore, the threshold region offers several kinematic advantages that result from the low particle velocities, such as the monochromatic spectra for two-body decays. Finally, the existence of two precisely known resonances, the J/ψ and

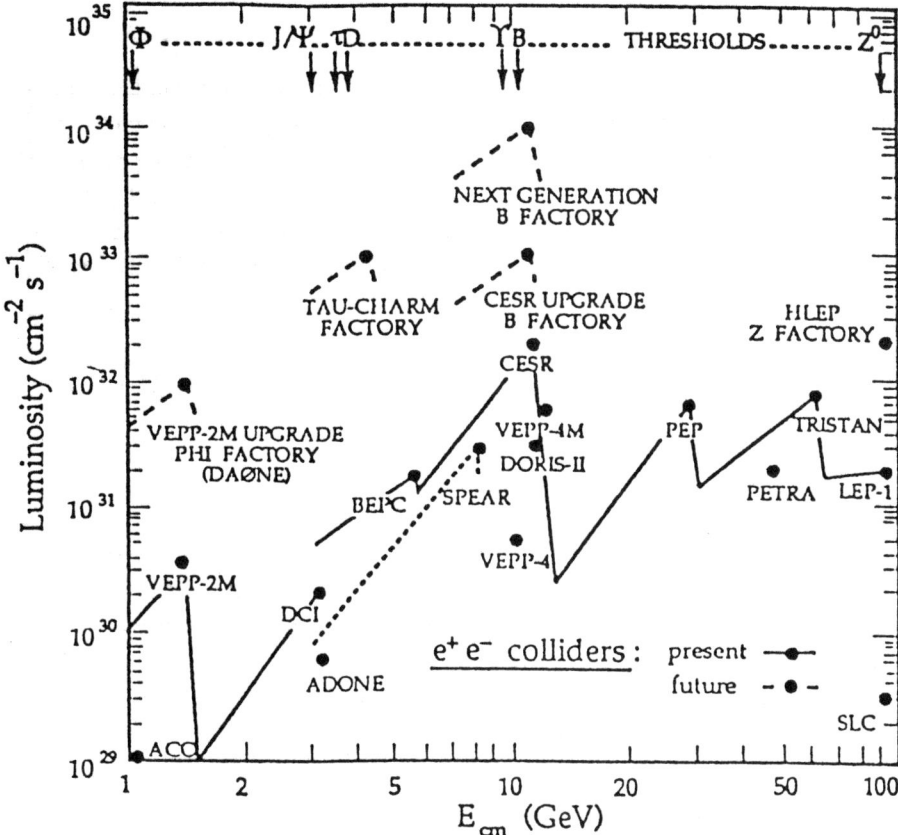

Figure 1: The luminosity of present and future e^+e^- colliders in the centre-of-mass energy range $1 \leq E_{cm} \leq 100$ GeV. The solid line represents the present envelope of the maximum luminosity. The long-dashed lines indicate the design luminosities of future factories. The short-dashed line shows the previous (SPEAR) luminosity in the $\tau^+\tau^-$ threshold region.

ψ', will provide a very high-rate ($\sim$ 1kHz) signal to calibrate and monitor the detector performance. Thus, experimenting at the τcF will benefit from very high statistics, low and measurable backgrounds, and reduced systematic errors. The coincidence of all these features creates an ideal facility for precision studies of the τ lepton and charm quark.

II. Charmonium and Search for New Hadrons

The existence of glueballs and hybrids is a direct consequence of QCD. In the 15 years since the introduction of QCD, several hadronic states have been found which are candidates for glueball or hybrid states, but have not been unambiguously identified as such: $\iota(1440)$, $\theta(1720)$, $\xi(2230)$, $E(1420)$, $f_0(975)$, $G(1590)$, g_T, g_T', g_T'', etc.

Recently, the BES collaboration announced new results concerning the $\xi(2230)$ state [6]. They confirmed the existence of the $\xi(2230)$ state in the decays $J/\psi \to \gamma K\bar{K}$ and also discovered two new, non-strange decay modes $\xi \to p\bar{p}$ and $\pi^+\pi^-$. Table I summarizes the preliminary results on the mass, width and decay branching ratios of the $\xi(2230)$ as measured by BES.

Table I: Results on the $\xi(2230)$

J/ψ decay	Mass [MeV]	Width [MeV]	$Br(J/\psi \to \gamma\xi) \times Br(\xi \to f.s.)$
$\gamma\pi^+\pi^-$	$2235^+_-4^+_-6$	$19^{+13}_{-11}{}^+_-12$	$(5.6^{+1.8}_{-1.6}{}^+_-1.4) \times 10^{-5}$
$\gamma K^+ K^-$	$2230^{+6}_{-7}{}^+_-12$	$20^{+20}_{-15}{}^+_-12$	$(3.3^{+1.6}_{-1.3}{}^+_-1.1) \times 10^{-5}$
$\gamma K_s K_s$	$2232^{+8}_{-7} \pm 15$	$20^{+25}_{-16} \pm 10$	$(2.7^{+1.1}_{-0.9} \pm 1.0) \times 10^{-5}$
$\gamma p\bar{p}$	$2235^+_-4^+_-5$	$15^{+12}_{-9}{}^+_-9$	$(1.5^{+0.6}_{-0.5}{}^+_-0.5) \times 10^{-5}$

Previous theoretical interpretations of the $\xi(2230)$ include identification as a high spin $s\bar{s}$ state[7], a multiquark state (such as a four quark state[8,9], a $\Lambda\bar{\Lambda}$ bound state[10], a neutral color singlet bound state[11], etc.), a hybrid state[8,12], and a glueball[13]. Since the Mark III collaboration only detected the strange decay modes of the $\xi(2230)$[14], the $s\bar{s}$ interpretation seemed plausible and the glueball interpretation was deemed unlikely. The recent detection of non-strange decay modes provides extremely important, additional clues to the exact nature of the $\xi(2230)$.

Compared with other mesons, the $\xi(2230)$ has many distinctive properties:

(1) flavor-symmetric decay: After removal of the phase space factor, the probability for $\xi \to \pi^+\pi^-$ is of the same order as the probability for $\xi \to K^+K^-$.

(2) Copious production in radiative J/ψ decays[15]: From the upper limit [16,17] of 1×10^{-4} for the decays $\xi \to p\bar{p}$ and $K\bar{K}$, where $K\bar{K}$ includes all kaon pairs, a lower bound of 3×10^{-3} for the $BR(J/\psi \to \gamma\xi)$ can be estimated.

(3) Narrow width: Both results from Mark III and BES show that the width of the $\xi(2230)$ is only about 20 MeV[6,14].

Assuming $\Gamma_\xi = 20$ MeV, one can easily estimate from (2) that the $BR(\xi \to K^+K^-)$ and $BR(\xi \to \pi^+\pi^-)$ are smaller than 2%, resulting in partial widths $\Gamma_{\pi^+\pi^-}$ and $\Gamma_{K^+K^-}$ smaller than 400 keV[15].

Combining information from both BES and PS185 results in striking features for the $\xi(2230)$: flavor symmetric couplings to $\pi\pi$ and $K\bar{K}$, a large production rate in radiative J/ψ decays, and very narrow partial decay widths to $\pi\pi$ and $K\bar{K}$. The $q\bar{q}$ model, multiquark model and hybrid model can not easily reproduce these observations. On the other hand, these properties are naturally explained by the identification of the $\xi(2230)$ as a bound glueball state[18].

It is interesting to note that a comprehensive lattice gauge calculation of SU(3) glueballs by the UKQCD collaboration suggests that the mass of the 2^{++} glueball be 2270 ± 100 MeV[17], emphasizing the need for a spin parity analysis of the observed $\xi(2230)$ state.

Since, according to the above discussion, the observed decay modes into $\pi\pi$, $K\bar{K}$, and $p\bar{p}$ are expected to be only a small portion of the decay modes of the $\xi(2230)$, searches for other decay modes are very important. A systematic test of the flavor-symmetric nature of the decays will be crucial for the glueball interpretation of the $\xi(2230)$. The $p\bar{p}$, $\pi\pi$ and $K\bar{K}$ modes are the easiest decay modes to be identified with high efficiency and low backgrounds by the BES detector. Other decay modes, such as $\eta\eta$, $\eta\eta'$, $\eta'\eta'$, $\rho\rho$, K^*K^*, $\omega\omega$, $\phi\phi$, $\pi\pi\pi\pi$, $\pi\pi KK$, etc., may suffer from too low detecting efficiencies, or too large backgrounds, or both.

Indeed, so far only 20-40 events for each decay channel of the $\xi(2230)$ have been detected by BES. Much more data is needed to determine its spin-parity quantum numbers and to identify other decay channels; a clear task for a future τcF.

The glueball and hybrid states are important in testing QCD. The τcF would be able to greatly increase the world's J/ψ sample and, therefore, might make a major impact in this area of physics. In one year, the τcF could produce $\sim 10^{10} J/\psi$ events, i.e. three orders of magnitude more than the presently accumulated statistic by BES. Thus, the τcF would allow for a systematic search for gluonia and hybrids in the decays of the J/ψ, ψ' and η_c.

The decay $J/\psi \to \gamma X$ might involve pure two-gluon intermediate states with masses $m_{gg} \leq 3.1 GeV$, like the following possible candidates: $\xi(2230)$, $\iota(1440)$ and $\theta(1700)$. Furthermore, the nature of the state X can be tested experimentally by comparing the different production mechanisms: $J/\psi \to \gamma X$, ωX and ϕX. At the ψ' peak, the large number of produced χ_c, via the process $\psi' \to \gamma \chi_c$, can be used to constrain the partial wave analysis of hadronic decays, by fixing the spin J of the initial χ_c, e.g. the decay $\chi_{c0} \to f_0(975) + M$ which enhances the scalar hadrons, M[19], and $\chi_{c1} \to \pi + H$ which is sensitive to the exotic hybrid sector $H(J^{PC} = 1^{-+})$[19].

Besides the J/ψ, ψ' and ψ'' resonances, there are four identified ψ states above 4 GeV decaying into D's, D^*'s and D^{**}'s. The relative fractions of D, D^* and D^{**} will provide important information and tests of the Heavy Quark Effective Theory, and can reveal the internal structure of these higher ψ states[19,20].

In lattice QCD and related models, hybrid charmonium, H_c, is predicted to exist within 300 MeV of the charm threshold[19]. These hybrids include vector states which can be formed directly in e^+e^- annihilation. Moreover, theoretical considerations suggest that charm is the optimal flavor to unambiguously identify vector hybrid states, since they are light enough to be produced in e^+e^- annihilation with a leptonic width of 0.1 keV, and heavy enough so that the conventional quarkonium states are well understood and extra states are readily identified. If such hybrid states exist it is likely that their discovery will have to wait for the on-coming of a τcF.

Detection of J/ψ's at any operating energy of the τcF may reveal the existence of further exotic states. An example is $J/\psi + \pi$, which might reveal isovector charmonium, such as $D\bar{D}$ or $D^*\bar{D}^*$ molecules. Another example is $J/\psi + \eta$, which might be a signature for the existence of ψ^* and complement the direct searches via $e^+e^- \to \psi^*$ leading to the clarification of the dynamical structure of the ψ^*[19].

Moreover, the $c\bar{c}$ wave function could be studied at the τcF via the decay $J/\psi \to 3\gamma$ and $\eta_c \to 2\gamma$, which constitute direct tests of the charmonium models. The huge production rate of η_c and χ at the τcF could be used to study their hadronic decays. Finally, the τcF would be ideal to study the properties of the barely-known $^1P_1$ and η'_c states.

III. τ and ν_τ Physics

No fundamental principle in the SM requires exactly massless neutrinos. The possibility of non-zero neutrino masses is a very fundamental question in particle physics. Such masses would allow for mixing between the lepton families and might provide answers to two of the major puzzles of cosmological physics: the deficiency of solar neutrinos and the composition of the dark matter of the universe. In fact, any ν_τ heavier than about 10 eV might decay,

$$\nu_\tau \to \nu_e + X$$
$$\nu_\tau \to \nu_\mu \, \nu_\mu \bar{\nu}_\mu$$
$$\nu_\tau \to \nu_\mu \, \nu_e \bar{\nu}_e,$$

thus, violating lepton flavor conservation. The ν_τ lifetime is given by

$$\tau_{\nu_\tau}^{-1} = \frac{2 G_{\nu_\tau}^2 m_{\nu_\tau}^5}{192 \pi^2}(1+y)$$

where the parameter y is model dependent. In order to be consistent with cosmological constraints on the mass density in the universe, the mass and lifetime are related by [21]

$$\tau_{\nu_\tau} \leq \left(\frac{100 \ keV}{m_{\nu_\tau}}\right)^2 (5.4 \times 10^{10}) \ s \ .$$

The present most stringent experimental upper limits on m_{ν_τ} are 29 MeV (BEPC) [22] and 23.8 MeV (ALEPH) [23].

For a conventional, charged lepton ℓ, the branching ratio for the decay into $e\nu\bar{\nu}$, B_ℓ^e, the lifetime, τ_ℓ, the mass, m_ℓ, and the weak coupling constant, $G_{\ell \to e\nu\bar{\nu}}$, are related up to small radiative and electroweak corrections, by

$$\frac{B_\ell^e}{\tau_\ell} = \frac{G_{\ell \to e\nu\bar{\nu}}^2}{192\pi^3} m_\ell^5 \ . \tag{1}$$

Equation (1) implies the following relationship among the above parameters for the τ and μ leptons:

$$\left[\frac{G_{\tau \to e\nu\bar{\nu}}}{G_{\mu \to e\nu\bar{\nu}}}\right]^2 = \left[\frac{m_\mu}{m_\tau}\right]^5 \frac{B_\tau^e}{\tau_\tau} \frac{\tau_\mu}{B_\mu^e} \ . \tag{2}$$

The averages for the above quantities quoted in the 1992 Review of Particle Properties (PDG) yield $(G_{\tau \to e\nu\bar{\nu}}/G_{\mu \to e\nu\bar{\nu}})^2 = 0.941 \pm 0.025$, implying a 2.4 standard deviation departure from lepton universality. Note that in this test of lepton universality the τ mass enters to the fifth power.

A greatly improved measurement of the τ mass can be provided by the measurement of the $\tau^+\tau^-$ production cross section in the region most sensitive to the τ mass, i.e. close to the production threshold. Such a measurement was recently made by BES at BEPC. The $\tau^+\tau^-$ pairs were identified by their $e\mu$ topology which provides the best combination of high detection efficiency and low background contamination. The value for the mass [24],

$$m_\tau = 1776.9^{+0.4}_{-0.5} \pm 0.2 \ MeV, \tag{3}$$

which is independent of the ν_τ mass, was obtained from a maximum likelihood fit to the energy dependence of the cross section. The above result is 7.2 MeV below the 1992 PDG average, $m_\tau = 1784.1^{+2.7}_{-3.6} MeV$, and has significantly smaller errors.

In a second stage, the following combination of decays were considered: $e\mu$, ee, $e\pi$, $e\rho$, $\mu\pi$, $\mu\mu, \pi\pi, eK$, $\mu K, \pi K$ [25]. The 64 observed events yield a substantially more precise measurement of the τ mass

$$m_\tau = 1776.96^{+0.18+0.20}_{-0.19-0.16} \ MeV. \tag{4}$$

Combining this measurement with recent values of the tau lifetime and the tau electronic branching fraction results in a weak coupling constant $G_{\tau \to e\nu\bar{\nu}}$ which is consistent with lepton universality, see Fig.2.

The present lack of understanding of the origin of CP violation emphasizes the need to search for CP violation in the lepton sector. For example, a measured non-zero electric dipole moment of the τ, d_τ [26], would unambiguously signal TCP violation. The current limit, $d_\tau < 5 \times 10^{-17} ecm$ (PDG), is orders of magnitude larger than the prediction of the SM [27].

Although in the SM CP violation can be introduced via a phase in the CKM matrix, other sources of CP violation may exist [28]. Recently, several proposals to search for CP violation in the lepton sector were made, considering CP violating effects in τ decays beyond the SM. C. Nelson et al. [29] define a stage-two spin-correlation function to detect CP violation in the decay $\tau \to (2\pi)\nu_\tau$. Y. S. Tsai [30] suggests to detect CP violation in $\tau \to (2\pi)\nu_\tau$ by using the longitudinally polarized electron and positron beams at the τcF. U. Kilan et al. [31] propose to study $\tilde{T}$-odd triple momentum correlations in $\tau \to K\pi\pi$ and $KK\pi$. T. D. Lee [32] suggests to measure the following

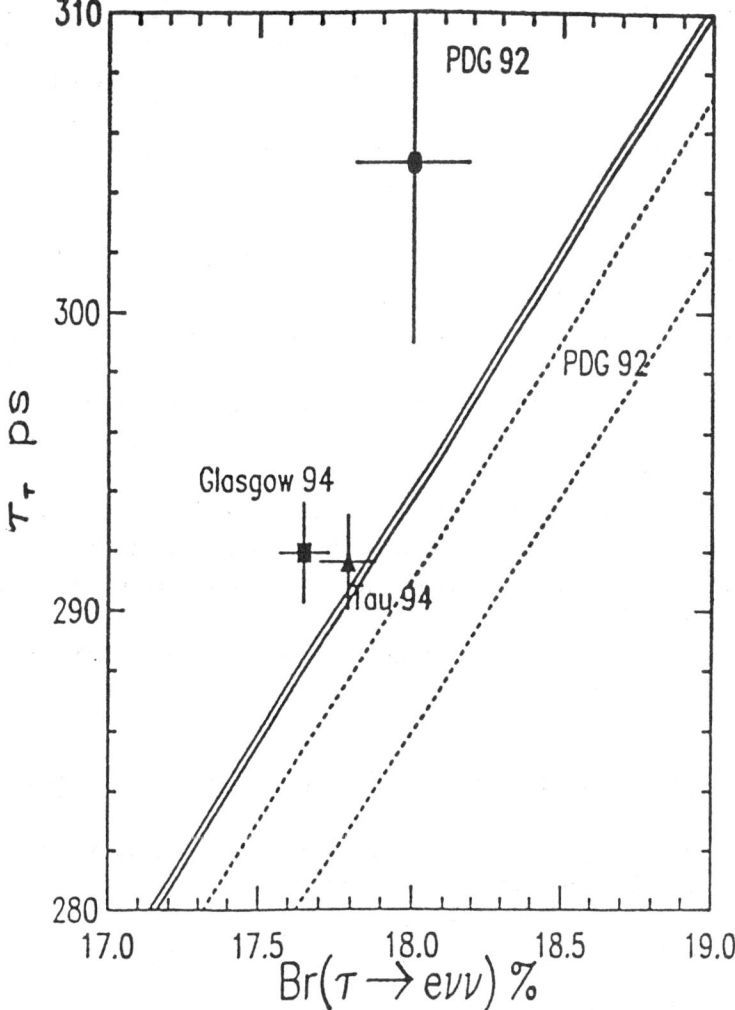

Figure 2: Recent values of the τ lifetime and its electronic branching fraction. The diagonal bands represent the region allowed by the τ mass measurement under the assumption of τ - μ universality: before (dashed) and after (solid) the new τ mass measurement from BES.

quantities $A \equiv < \hat{P}_e \cdot (\hat{P}_{e'} \times \hat{P}_\mu) >_{average}$ and $A' \equiv < \hat{\sigma}_\tau \cdot (\hat{P}_e \times \hat{P}_\tau >_{average}$ in $\tau^\mp \to e^\mp + 2\nu$ and $\tau^\pm \to \mu^\pm + 2\nu$. S. Y. Choi et al. [33] study CP violation in the decay $\tau \to (3\pi)\nu_\tau$ with the 3π forming either the 1^+ resonance, a_1, or the 0^- resonance $\pi(1300)$. They estimate the size of CP violation in the multi-Higgs doublet model and the left-right symmetric model and conclude that at least 10^7 τ pair events are needed to be able to measure a significant effect.

A systematic exploration of all τ decay channels in order to look for signs of discrepancies with the theoretical expectations is of great importance. The V-A structure of the charged current can be tested by studying the distribution of the final state charged lepton in the leptonic decay modes of the $\tau, \tau^- \to \nu_\tau e^- \bar{\nu}_e (\ell = e, \mu)$. The Lorentz structure of the charged weak current in terms of the so-called Michel parameters ρ, δ, η, ξ', and ξ [34] will be determined to high accuracy at the τcF.

The τ is the only known lepton heavy enough to decay into hadrons; its semileptonic decays are an ideal place to study strong interaction effects in very clean experimental conditions allowing for stringent tests of QCD at small momentum transfers.

Finally, at the τcF rare and forbidden τ decays will be searched for with unmatched sensitivity.

IV. Comprehensive Study of Charm Physics

Pure leptonic decays of the D_s and D mesons can be rigorously calculated within the SM. The predicted branching ratios are

$$BR(D_s^+ \to \ell^+ \nu_e) = \tau_D \frac{G_F^2}{8\pi} f_D^2 m_D \mid V_{cs} \mid^2 m_\ell^2 (1 - \frac{m_\ell^2}{m_D^2})^2 \;, \tag{5}$$

where f_D is the weak decay constant. This constant also enters the prediction for the rate of nonleptonic decays and second-order weak processes - including $D\bar{D}$ mixing and CP violation. The standard model predicts the following ratios of branching fractions

$$\frac{BR(D_s^+ \to \mu^+ \nu_\mu)}{BR(D_s^+ \to \tau^+ \nu_\tau)} = \frac{m_\mu^2 [1 - \frac{m_\mu^2}{m_{D_s}^2}]^2}{m_\tau^2 [1 - \frac{m_\tau^2}{m_{D_s}^2}]^2} = 0.102 \pm 0.001 \;, \tag{6}$$

and

$$\frac{BR(D \to \ell\nu_\ell)}{BR(D_s \to \ell\nu_\ell)} \simeq \frac{\tau_D}{\tau_{D_s}} \cdot \frac{f_D^2}{f_{D_s}^2} \cdot \frac{m_D}{m_{D_s}} \frac{\mid V_{cd} \mid^2}{\mid V_{cs} \mid^2} \simeq \frac{1}{10} \;, \tag{7}$$

where the decays into $e\nu_e$ are strongly suppressed within the SM.

QCD calculation give

$$\frac{f_{D_s}}{f_D} = 1.13 \pm 0.03 \qquad (8)$$

which is consistent with lattice gauge calculations [36].

Precise measurements of f_D, f_{D_s} also provide important tests of non-perturbative QCD calculations, thereby increasing confidence in extrapolations to the B system (since f_B is experimentally inaccessible) and allowing for a better theoretical prediction for $D^0\bar{D}^0$ mixing.

BES has recently collected a data sample at a center-of-mass energy of 4.03 GeV - slightly above the threshold for $D_s\bar{D}_s$ production - corresponding to about 25 pb^{-1}. A search for the following event topology was made where one D_s decays hadronically, $D_s \to$ 3 charged particles, and the other leptonically, $D_s \to \mu$ or e plus a missing ν.

Two candidates were found having a μ track and one candidate was found with an electron track. The masses of the tagging D_s's are reconstructed very closely to the nominal mass supporting the evidence for having identified leptonic decays of the D_s. Under the assumption of $e - \mu - \tau$ universality, we find [37]

$$BR(D_s \to \mu\nu_\mu) = (1.5^{+1.3+0.3}_{-0.6-0.2})\% \qquad (9)$$

$$BR(D_s \to \tau\nu_\tau) = (15^{+13+3}_{-6-2})\% \qquad (10)$$

and

$$f_{D_s} = (430^{+150+40}_{-130-40}) MeV \quad . \qquad (11)$$

This measurement is model-independent and is not affected by the branching fraction of $D_s^+ \to \phi\pi^+$. Precise measurements of pure leptonic D_s decays require large single-tagged event samples and are, therefore, uniquely accessible to the τcF.

Semileptonic decays of heavy mesons play a crucial role in the determination of the CKM mixing matrix. The CKM matrix elements V_{cs} and V_{cd} are poorly measured at present ($\pm 10-20\%$). A direct experimental determination with much better precision will be done at the τcF. Also a precise study of lepton spectra in inclusive and exclusive semileptonic charm decays is needed to improve our understanding of QCD. Such measurements will provide an important calibration for the methods used in extracting the CKM parameters V_{cb} and V_{ub} from semileptonic B meson decays. In fact, charm decays are a unique laboratory to probe QCD quantitatively at the boundary between perturbative and nonperturbative calculations.

A comprehensive analysis is required of not only the Cabibbo-allowed decay modes, but also of the singly and doubly Cabibbo-suppressed decay modes of the D and D_s mesons, in order to improve our understanding of the hadron dynamics in weak decays of heavy mesons.

Meson mixing has only been observed so far in the $K^0 - \bar{K}^0$ and $B^0 - \bar{B}^0$ systems where the second-order transition is associated with the down type quarks (s,b). The study of mixing in the $D^0 - \bar{D}^0$ system, which contains an up type heavy quark (c), is an important experiment to study the flavor structure of weak interactions. The rate for $D^0 - \bar{D}^0$ oscillations is expected to be quite small in the SM [38]

$$r_D \equiv \frac{BR(D^0 \to \bar{D}^0 \to \bar{f})}{BR(D^0 \to f)} \sim 10^{-5} - 10^{-4} \qquad (12)$$

and may be larger for non-standard flavor-changing neutral currents (FCNC).

In the SM, direct CP violation will give rise to asymmetries [39] of the order of $0.5 \times BR \approx 10^{-3}$ in certain Cabibbo-suppressed D decay channels. Indirect CP violation from $D^0 - \bar{D}^0$ mixing is expected to be very small. The τcF is ideally suited to provide an important first window on CP violation in the up-quark sector [40], thus opening up the exciting possibility of exploring the mechanism of CP violation in a system that is complementary to the B or K systems and, moreover, highly sensitive to physics beyond the SM.

Summary and Outlook

Since data taking started in 1990, BES has collected data at various energy settings, the J/ψ, ψ' resonances and the τ and D_s pair production thresholds. Several significant measurements have been made. In particular, the precise measurement of the τ lepton mass was performed in 1992 and helped resolve the long standing discrepancy in one-prong τ decay. In order to improve the performance a BES upgrade project was started in 1994 and will be finished by the beginning of 1996. At that time, the luminosity delivered by BEPC will have increased by a factor 2-3.

Many important issues in particle physics can only be resolved by a dedicated, high precision instrument providing large amounts of very high quality data. In our view, the planed τcF will address fundamental aspects of the SM in a way which is complementary to other facilities. The τcF combines the three ingredients required for making an accurate and exhaustive investigation of the τ lepton, ν_τ and c quark: high statistics, low and controlled backgrounds, and reduced systematic errors. In certain areas of particle physics, such as the search for glueballs, the measurement of the ν_τ mass and of CP violation in the lepton sector, the τcF will provide a unique opportunity to make a significant impact on our knowledge. In short, the τcF will be the ideal facility to explore the τ and ν_τ leptons and the charm quark and will be the natural continuation of the already very successful physics program pursued by BEPC.

Acknowledgments. I would like to thank K. T. Chao, S. Jin, J. Li, W. Lu, N. D. Qi, Q. X. Shen, Z. J. Tao, J. M. Wu, H. Yu, C. C. Zhang, D. H. Zhang and Z. P. Zheng for useful discussions.

References

[1] M. H. Ye Z. P. Zheng, Int. Journal of Modern Physics A2(1987)1707.

[2] M. H. Ye and Z. P. Zheng, Proceedings of the 1989 International Symposium on Lepton and Photon Interactions at High Energies (Stanford University, Stanford, 1989), p.122.

[3] T. Huang, Proceedings of the Charm Physics Workshop, Beijing, 1987, p.503.

[4] J. Kirkby, A τ-Charm Factory at CERN, CERN-EP/87-210(1987).

[5] A. Pich, Proceedings of the Marbella Workshop on the τ-Charm Factory, Marbella, June 1993.

[6] S. Jin, talk given at the International Workshop on Hadron Physics at Electron-Positron Colliders, Beijing. October 14-17, 1994;
J. Li talk given at the 27th International Conference on High Energy Physics, Glasgow, July 21-27, 1994.

[7] S. Godfrey, R. Kokoski, and N. Isgur, Phys. Lett. B141(1984)439.

[8] K. T. Chao, Phys. Rev. Lett. 60(1988)2579;
K. T. Chao, Commun. Theor. Phys. 3(1984)757.

[9] S. Pakvasa, M. Suzuki and S. F. Tuan, Phys. Lett. B145(1984)135.

[10] S. Ono, Phys. Rev. D35(1987)944.

[11] M. P. Shatz, Phys. Lett. B138(1984)209.

[12] M. S. Chanowitz and S. R. Sharpe, Phys. Lett. B132(1983)413;
M. Le Yaouanc et al., Z. Phys. C28(1985)309.

[13] B. F. L. Ward, Phys. Rev. D31(1985)2849.

[14] R. M. Baltrusaitis et al., Phys. Rev. Lett. 56(1986)107.

[15] K. T. Chao, talk given at the CCAST Workshop on the Tau-Charm Factory, Beijing, November 1994, PUTP-94-26 (hep-ph/9502408).

[16] P. D. Barnes et al., Phys. Lett. B309(1993)469.

[17] Particle Data Group, L. Montanet et al., Phys. Rev. D50 3-I(1994)1173.

[18] T. Huang, S. Jin, D. H. Zhang and K. T. Chao, BIHEP-TH-95-11.

[19] F. E. Close, Proceedings of the Marbella Workshop on the τ-Charm Factory, Marbella, June 1993.

[20] T. Barnes, Proceedings of the Marbella Workshop on the Tau-Charm Factory, Marbella, Spain, June 1993.

[21] E. Kolb and M. Turner, Phys. Rev. Lett. 67(1991)5; S. Bludman, Phys. Rev. D45(1992)4720.

[22] BEPC, preliminary result.

[23] ALEPH Collaboration, proceedings of Neutrino 94(1994).

[24] J. Z. Bai et al.,(BES Collaboration), Phys. Rev. Lett. 69(1992)3021.

[25] BES Collaboration, High Energy Physics and Nuclear Physics, (to be published).

[26] J. A. Grifols and A. Mendez, Phys. Lett. B255(1991)611 [Erratum: B259(1991)512]; F. del Aguila and M. Sher, Phys. Lett. B252(1990)116.

[27] S. M. Barr and A. Zee, Phys. Rev. Lett. 65(1990)21.

[28] T. D. Lee, Phys. Rev. D8(1973)1226; Phys. Rep. (1977), 9(1974)143.

[29] C. A. Nelson et al., Phys. Rev. D50(1994)4544.

[30] Y. S. Tsai, Phys. Rev. D51(1995)3172.

[31] U. Kilan et al., Z. Physik C62(1994)413.

[32] T. D. Lee, Proceedings of B-factory Workshop, Nagoya(1994).

[33] S. Y. Choi et al., KEK-TH-419(1994).

[34] A. Pich, Tau Physics, in Heavy Flavors, eds. A. J. Buras and M. Lindner, Advanced Series on Directions in High Energy Physics. Vol.10(World Scientific, Singapore, 1992), p.375; Nucl. Phys. B(Proc. Suppl.) 31(1993)213.

[35] T. Huang and C. W. Luo, Phys. Rev. D51(1994)5775.

[36] Bernard et al., Phys. Rev. D49(1994)2536; Alton et al., Nucl. Phys. B(Proc. Suppl.) 34(1994)456.

[37] J. Z. Bai et al., (BES Collaboration), Phys. Rev. Lett. 74(1995)4599; C. C. Zhang, (BES Collaboration), Proceedings of the 27-th International Conference on High Energy Physics, Glasgow, July, 1994.

[38] I. I. Bigi, Proceedings of τ-Charm Factory Workshop, SLAC(1989), ed. L. V. Beers, SLAC-Report-343(1989).

[39] A. Pugliese and P. Santorelli, Proceedings of the Marbella Workshop on the τ-Charm Factory, Marbella, Spain, June 1993.

[40] J. R. Fry and T. Ruf, Proceedings of the Marbella Workshop on the τ-Charm Factory, Marbella, Spain, June 1993.

Test of T and CP Violation in Leptonic Decay of $\tau^{\pm}$*

YUNG SU TSAI

Stanford Linear Accelerator Center
Stanford University, Stanford, California 94309

ABSTRACT

The $\tau^{\pm}$, highly polarized in the direction of the incident beam, can be obtained from the $e^{\pm}$ collider with the polarized incident e^- beam. This polarization vector $\vec{w}_i$ can be used to construct the T odd rotationally invariant product $(\vec{w}_i \times \vec{p}_\mu) \cdot \vec{w}_\mu$, where $\vec{p}_\mu$ and $\vec{w}_\mu$ are the momentum and polarization of the muon in the decay $\tau^- \to \mu^- + \bar{\nu}_\mu + \nu_\tau$. T is violated by the existence of such a term. CP can be tested by comparing it with a similar term in τ^+ decay. One can test whether T (and CP) violation is due to the charged Higgs boson exchange by doing a similar experiment for the $\mu^{\pm}$ decay.

In the Standard Model of Kobayashi and Maskawa [1] CP violation occurs as a result of a complex phase in the unitary matrix relating gauge eigenstates and mass eigenstates. The leptonic sector does not have CP violation if all neutrinos are massless. Both of these assumptions could be wrong. It is quite possible [2] that CP violation is due to exchange of some new particle such as a heavier W boson or a charged Higgs boson. If CP violation is milliweak or stronger in τ decay, one should be able to observe it in the proposed τ-charm factory where it is expected to have $1 \sim 3 \times 10^8$ highly polarized τ pairs per year [2].

The CP violation in τ production can be ignored because we are dealing with electromagnetic production. The radiative correction due to CP violation in the weak interaction is of order 10^{-5} if it is weak, but 10^{-8} if it is semiweak [3]. In contrast to the production, the decay of τ is weak, thus CP violation is of order 1 if it is weak and 10^{-3} if it is milliweak. Up to now, the only

---

*Work supported by the Department of Energy, contract DE-AC03-76SF00515.

CP violation is from K_L which is 2×10^{-3}. In a recent paper [2] we dealt with the CP violation in the semileptonic decay of τ with two or more final hadrons. For a single hadron in the semileptonic decay or a leptonic decay, the only rotationally invariant quantity we can form is $\vec{w}_i \cdot \vec{q}$ where $\vec{q}$ is the momentum of the final visible particle. But this term is T even so we cannot have CP violating effects from this term without violating TCP [4]. It is very desirable to measure CP violation in pure leptonic decay because in the semileptonic decay it is impossible to assign CP violation to the leptonic or hadronic vertex [2]. For the leptonic decay we have to measure the polarization of the muon and construct a rotationally invariant but T violating product

$$(\vec{w}_i \times \vec{p}_\mu) \cdot \vec{w}_\mu , \qquad (1)$$

where

$$\vec{w}_i = \frac{w_1 + w_2}{1 + w_1 w_2} \hat{e}_z , \qquad (2)$$

with w_1 and w_2 being the longitudinal polarization of the incident electron and positron respectively; $\vec{p}_\mu$ is the laboratory momentum, and $\vec{w}_\mu$ the polarization of the muon. The muon polarization is measured by the asymmetry in electron distribution coming from the term $\vec{w}_\mu \cdot \vec{q}_{e^-}$ where $\vec{q}_{e^-}$ is the electron momentum. Thus the correlation in Eq. (1) induces the correlation

$$c(\vec{w}_i \times \vec{p}_\mu) \cdot \vec{q}_{e^-} , \qquad (3)$$

which is also odd under T.

Equation (3) means that if one finds an asymmetry in the perpendicular component of $\vec{q}_{e^-}$ with respect to the plane formed by $\vec{w}_i$ and $\vec{p}_\mu$, one discovers the existence of T violating effect. Under CP we have $w_1 \leftrightarrow w_2$, $w_i \to w_i$, $\vec{p}_{\mu^-} \leftrightarrow -\vec{p}_{\mu^+}$, $\vec{q}_{e^-} \leftrightarrow -\vec{q}_{e^+}$. Thus if CP is conserved, we have for τ^+ decay

$$c'(\vec{w}_i \times \vec{p}_{\mu^+}) \cdot \vec{q}_{e^+} \qquad (4)$$

with $c' = c$. But since T is violated by both Eqs. (3) and (4), we better have $c' = -c$ in order to preserve TCP invariance. The discussion given above is completely model independent. Later we shall give a model which will illustrate all the above observations.

In order to measure the angular asymmetry in the decay electrons, we have to slow the muon down to almost at rest. For τ energy E_τ equal to 2.087 GeV, where the cross section is maximum [2], the maximum and minimum muon momenta are respectively:

$$p_3^{\max} = \frac{E_\tau}{2}\left[(1+\beta) - \frac{m^2}{M^2}(1-\beta)\right] = 1.589 \text{ GeV} \qquad (5)$$

$$p_3^{\min} = -\frac{E_\tau}{2}\left[(1-\beta) - \frac{m^2}{M^2}(1+\beta)\right] = -0.4904 \text{ GeV} \qquad (6)$$

where

$$m = 0.105658 \text{ GeV}$$
$$\beta = \sqrt{1.5 - \sqrt{1.5}}$$
$$M = 1.777 \text{ GeV}$$

p_3^{min} is negative means that it is going in the opposite direction to the τ momentum. Approximately 10% of muons are going backward. The asymmetry caused by the detector can be checked by reversing the polarization of the incident beam (or beams).

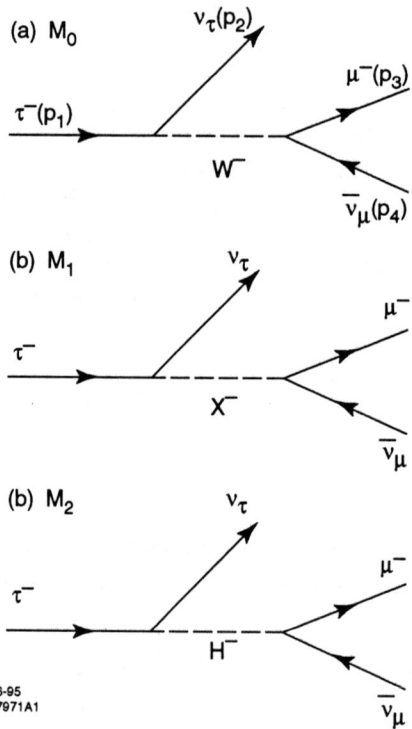

Figure 1: (a) M_0: Feynman diagram for $\tau^- \to \mu^- + \bar{\nu}_\mu + \nu_\tau$ in the Standard Model that conserve T and CP. (b) M_1: A possible T violating spin 1 exchange diagram that is shown not to contribute to the T violating effect. (c) M_2: A T violating spin 0 exchange diagram with a complex coupling constant.

In this paper we use the same model of T and CP violation as the previous paper [2]. It is shown there that if we limit the weak interaction to be transmitted by exchange of spin 1 and spin 0 particles, then we have only two possible choices of matrix elements denoted by M_1 and M_2 (see Fig. 1) that can interfere with the Standard Model matrix denoted by M_0.

$$M_0 = A\,\bar{u}(p_2)\gamma_\mu(1-\gamma_5)u(p_1)\bar{u}(p_3)\gamma_\mu(1-\gamma_5)v(p_4)\,, \tag{7}$$
$$M_1 = B\,\bar{u}(p_2)\gamma_\mu(1-\gamma_5)u(p_1)\bar{u}(p_3)\gamma_\mu(1-\gamma_5)v(p_4)\,, \tag{8}$$
$$M_2 = C\,\bar{u}(p_2)(1+\gamma_5)u(p_1)\bar{u}(p_3)(1-\gamma_5)v(p_4)\,, \tag{9}$$

where p_1, p_2, p_3 and p_4 are momenta of τ^-, ν_τ, μ^- and $\bar{\nu}_\mu$ respectively. We have assumed that m_{ν_μ}/m_μ and m_{ν_τ}/m_τ to be either zero or too small to be experimentally observable, so that possible terms such as $(1+\gamma_5)u(p_4)$ and $(1+\gamma_5)u(p_2)$ are ignored in M_1 and M_2. A is chosen to be real while B and C are allowed to be complex. Since there is no final state interaction the imaginary parts of B and C cause T violating effects. If TCP is conserved then B and C for the τ^+ decay must be the complex conjugate of B and C:

$$\overline{B} = B^* \quad \text{and} \quad \overline{C} = C^*\,. \tag{10}$$

Since the Standard Model is good to 10^{-3} to 10^{-2}, we can assume $M_1^+ M_1$ and $M_2^+ M_2$ to be at most 10^{-2} compared to $M_0^+ M_0$ and thus we shall ignore them. We shall also ignore M_1 completely because its interference with M_0 does not depend upon the imaginary part of B that causes the T violation:

$$M_0^+ M_1 + M_1^+ M_0 = (B + B^*) M_0^+ M_0 / A\,. \tag{11}$$

Writing $M_0 = A\mathcal{M}_0$ and $M_2 = C\mathcal{M}_2$, we have

$$M_0^+ M_2 + M_2^+ M_0 = A\,\mathrm{Re}\,C(\mathcal{M}_0^+\mathcal{M}_2 + \mathcal{M}_2^+\mathcal{M}_0) + iA\,\mathrm{Im}\,C(\mathcal{M}_0^+\mathcal{M}_2 - \mathcal{M}_2^+\mathcal{M}_0)\,. \tag{12}$$

Only the imaginary part of C contributes to T violation. The real part of B and C should be of order 10^{-2} or less compared with A, thus they will be ignored. We have therefore

$$(M_0 + M_1 + M_2)^+(M_0 + M_1 + M_2) \approx A^2 \mathcal{M}_0^+ \mathcal{M}_0 + iA\,\mathrm{Im}\,C(\mathcal{M}_0^+\mathcal{M}_2 - \mathcal{M}_2^+\mathcal{M}_0)\,.$$

Let $\vec{w}_3$ be the polarization vector of the muon in the rest frame of the muon. We are interested in the y component of $\vec{w}_3$ defined in Fig. 2(a) whose existence signifies the violation of T because $(\vec{w} \times \vec{p}_3) \cdot \vec{w}_3$ is odd under T. After averaging over the τ production angle the polarization vector of τ^-, $\vec{w}$, is replaced by the initial beam polarization $\vec{w}_i$. We note that since the y direction is perpendicular to $\vec{p}_3$ it is invariant under the Lorentz boost along

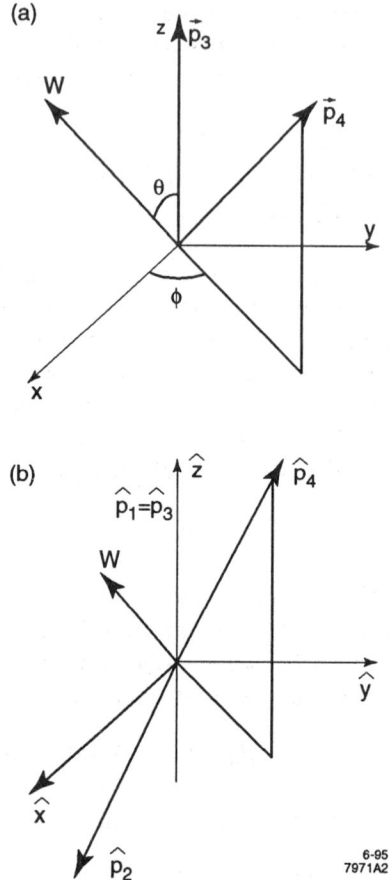

Figure 2: (a) The rest frame of τ^-, the coordinate system used in Eq. (18). (b) The rest frame of $u = p_2 + p_4 = p_1 - p_3$, the coordinate system used in integrating out the two unobserved neutrinos in Eqs. (13) and (14). This frame is obtained from the above diagram by boosting against the direction of $\vec{p}_3$.

$\vec{p}_3$. The y component of the muon polarization can be calculated using the formula

$$w_{3y} = \frac{i\,\mathrm{Im}\,C \int \left[(\mathcal{M}_0^+ \mathcal{M}_2 - \mathcal{M}_2^+ \mathcal{M}_0)_{\hat{s}=\hat{e}_y} - (\mathcal{M}_0^+ \mathcal{M}_2 - \mathcal{M}_2^+ \mathcal{M}_0)_{\hat{s}=-\hat{e}_y}\right] \times \frac{d^3 p_2}{2E_2} \frac{d^3 p_4}{2E_4} \delta^4(p_1 - p_2 - p_3 - p_4)}{A \int \sum_{\text{Spin of }\mu} (\mathcal{M}_0^+ \mathcal{M}_0) \frac{d^3 p_2}{2E_2} \frac{d^3 p_4}{2E_4} \delta^4(p_1 - p_2 - p_3 - p_4)}. \tag{13}$$

The second term inside the square bracket of the numerator is the negative of the first, thus the bracket is equal to twice the first term. The phase space integration with respect to the two undetected neutrinos is carried out in the rest frame of $u = p_2 + p_4$. We denote quantities in this frame by $\hat{\ }$.

$$\int \frac{d^3 p_2}{2E_2} \frac{d^3 p_4}{2E_4} \delta^4(p_1 - p_2 - p_3 - p_4) = \int d\hat{\Omega}_4 \frac{\hat{E}_4}{2} \delta(u^2 - 2u\hat{E}_4)\, d\hat{E}_4$$
$$= \frac{1}{8} \int d\hat{\Omega}_4, \tag{14}$$

$$(\mathcal{M}_0^+ \mathcal{M}_2 - \mathcal{M}_2^+ \mathcal{M}_0)$$
$$= \frac{\mathrm{Tr}}{4}(1+\gamma_5 \slashed{w})(\slashed{p}_1 + M)(1+\gamma_5)\gamma_\mu \slashed{p}_2$$
$$\times \frac{\mathrm{Tr}}{4} \slashed{p}_4(1+\gamma_5)\gamma_\mu(1+\gamma_5 \slashed{s})(\slashed{p}_3 + m)$$
$$- \frac{\mathrm{Tr}}{4}(1+\gamma_5 \slashed{w})(\slashed{p}_1 + M)(1-\gamma_5) \slashed{p}_2 \gamma_\mu$$
$$\times \frac{\mathrm{Tr}}{4} \slashed{p}_4(1+\gamma_5)(1+\gamma_5 \slashed{s})(\slashed{p}_3 + m)\gamma_\mu$$
$$= 4i\bigg[-(s \cdot p_4) EPS(w, p_1, p_3, p_4) + (p_3 \cdot p_4) EPS(s, w, p_1, p_4)$$
$$+ \frac{1}{2}(p_1 \cdot p_4) EPS(s, w, p_1, p_3)$$
$$- \frac{1}{2}\{(p_1 \cdot p_3) - m^2\} EPS(s, w, p_1, p_4) \bigg]. \tag{15}$$

In the above we have dropped all those terms that are odd in $\hat{p}_{4x}$, $\hat{p}_{4y}$, and $\hat{p}_{4z}$ because they yield zero after integration with respect to $d\hat{\Omega}_4$. The Levi-Civita EPS's are evaluated in the rest frame of τ and then $p_{4\mu}$ is Lorentz transformed to $\hat{p}_{4\mu}$ before the angular integration shown in Eq. (14). In the rest frame of $u = p_2 + p_4$ we have $\hat{p}_1 = \hat{p}_3$, thus the Lorenz boost is in the z direction [5]: $E_4 = \gamma \hat{E}_4 - \beta\gamma \hat{p}_{4z}$, $p_{4x} = \hat{p}_{4x}$, $p_{4y} = \hat{p}_{4y}$ and $p_{4z} = -\gamma\beta \hat{E}_4 + \gamma \hat{p}_{4z}$ with $\gamma = (M - E_3)/u$, $\beta = p_3/u$, $u = \sqrt{M^2 + m^2 - 2ME_3}$. The result of the angular integration is

$$\int (\mathcal{M}_0^+ \mathcal{M}_2 - \mathcal{M}_2^+ \mathcal{M}_0)_{\vec{s}=\hat{e}_y} d\hat{\Omega}_4 = \frac{4\pi i}{3} \left[3M^2 - 4E_3 M + m^2\right] EPS(\hat{e}_y, w, p_1, p_3), \tag{16}$$

with

$$EPS(\hat{e}_y, w, p_1, p_3) = M(\vec{w} \times \vec{p}_3)_y.$$

The denominator in Eq. (13) can be obtained similarly:

$$\int \sum_{\text{spin of}\,\mu} (\mathcal{M}_0^+ \mathcal{M}_0) d\hat{\Omega}_4 = \frac{32\pi M^2 E_3}{3}$$

$$\times \left[3M - 4E_3 - \frac{2m^2}{E_3} + \frac{3m^2}{M} + (\vec{w} \cdot \vec{p}_3)\left(\frac{M}{E_3} - 4 + \frac{3m^2}{E_3 M}\right)\right]. \tag{17}$$

Equation (17) agrees with the result of my previous paper [6] written several years before the discovery of the τ.

Putting everything together we have finally:

$$w_{3y} = \frac{-(\vec{w} \times \vec{p}_3)_y}{8E_3} \frac{\left[3M - 4E_3 + \frac{m^2}{M}\right] \text{Im}(C/A)}{3M - 4E_3 - \frac{2m^2}{E_3} + \frac{3m^2}{M} + (\vec{w} \cdot \vec{p}_3)\left(\frac{M}{E_3} - 4 + \frac{3m^2}{ME_3}\right)}. \tag{18}$$

For τ^+ decay we use the substitution $\vec{p}_3 \to -\vec{p}'_3$, $\vec{w} \to \vec{w}'$, $E_3 \to E'_3$, $C \to \overline{C}$. $\overline{C}$ is equal to C^* if TCP is conserved [2]. Under CP we have $w \to w'$, $\vec{p}_3 = -\vec{p}'_3$, thus it is opposite to the TCP conserved case. Thus CP must be violated in order to conserve TCP.

The polarization of the muon is analyzed by the decay electron momentum $\vec{q}$. Thus the measurement of the existence of the T violating term $(\vec{w} \times \vec{p}_3) \cdot \vec{w}_3$ can be done by measuring the existence of the T violating correlation $(\vec{w} \times \vec{p}_3) \cdot \vec{q}$, where $\vec{p}_3$ and $\vec{q}$ are momenta of muon and decaying electron in the rest frame of τ. Exactly at threshold such a correlation can be calculated using Eq. (18), but as energy is increased one must integrate over the τ production angle. The result must be proportional to the only T noninvariant correlation in the center-of-mass system $(\vec{w}_i \times \vec{p}_\mu) \cdot \vec{q}_e$, where $\vec{w}_i$ is the initial beam polarization defined in Eq. (2); $\vec{p}_\mu$ and $\vec{q}_e$ are center-of-mass momenta of the muon and electron respectively.

We have shown above that only the spin 0 exchange can produce T violating leptonic decay of the τ. By measuring a similar effect in $\mu^\pm \to e^\pm + \nu_e + \nu_\mu$ one should be able to decipher if the exchanged particle is the Higgs boson discussed by T. D. Lee [7] and S. Weinberg [8]. The test of T, CP, and charged Higgs boson exchange in the leptonic decay of τ proposed in this paper as well as the test of CP, TCP and CVC in the semileptonic decay of τ proposed in my previous paper [2] are mostly for the proposed tau-charm factory. However they can also be carried out in the B factories being constructed at SLAC, KEK and Cornell provided they add a capability to longitudinally polarize their initial electron (and preferably also positron) beam. It is regrettable that none of the B factories mentioned above have any plans to polarize their incident beam (or beams). At the B factory energy the cross section is about 1/6 that of the tau-charm factory for producing τ pairs and the polarization of produced τ is about 23% less favorable due to the reduced s wave dominance in the production [see Eq. (4.11) of Ref. 2]. However the luminosity of the machine is supposed to be roughly proportional to the energy that is a factor of three in favor of the B factory. Thus the tests proposed in this paper and Ref. [2] are still do-able with the B factories if they polarize their incident beams.

Acknowledgments

The author wishes to thank Professor W.K.H. Panofsky for reviving my interest in the Tau-Charm Factory. I also would like to thank Bill Dunwoodie, Charles Prescott and Francesco Villa for consultation on measurement of transverse polarization of the muon and electron.

References

[1] M. Kobayashi and T. Maskawa, *Prog. Theor. Phys.* **49**, 652, (1973).

[2] Y. S. Tsai, *Phys. Rev.* **D51**, 3172 (1995).

[3] The effect due to the electric dipole moment of τ can be regarded as the weak or milliweak radiative corrections to the electromagnetic vertex of τ and thus the correction must be of order $(m_\tau/m_w)^2 \alpha = 3 \times 10^{-6}$ if it is weak and another factor of 10^{-3} if it is milliweak. The interference of one γ exchange and one Higgs boson exchange is $(m_e/E)(E^2/m_H^2)\, m_\tau m_e/m_w^2 \approx (m_e^2 m_\tau^2)/(m_H^2 m_w^2) = 1.23 \times 10^{-10} \text{GeV}^2/m_H^2$.

[4] T. D. Lee made this valuable remark.

[5] For readers interested in the technical aspects of the calculation, the choice of the rest frame of $u = (p_1 - p_3) = (p_2 + p_4)$ to do the angular integration, the choice of the direction of $\vec{p}_3$ as the z axis to facilities Lorenz transformation and the choice of the rest frame of τ to express the final result are very important in expediting the whole calculation.

[6] Y. S. Tsai, *Phys. Rev.* **4D**, 2821 (1971). See Eq. (2.1′).

[7] T. D. Lee, *Phys. Rev.* **D8**, 1226 (1973); *Phys. Rep.* **9C**, 143 (1974).

[8] S. Weinberg, *Phys. Rev. Lett.* **37**, 657 (1976).

Addendum to the Test of CP Violation in Tau Decay [*]

YUNG SU TSAI

Stanford Linear Accelerator Center
Stanford University, Stanford, California 94309

ABSTRACT

We discuss the test of CP and CPT violation in τ decay without using the polarized electron beam by comparing partial fractions of τ^- and τ^+ decay into channels with strong final state interactions. For example, $\Gamma(\tau^- \to \pi^- + \pi^0 + \nu) \neq \Gamma(\tau^+ \to \pi^+ + \pi^0 + \nu)$ signifies violation of CP. The optimum energy to investigatge CP violation in τ decay is discussed. We conclude that this energy is a few MeV below $\psi(2s)$ in order to avoid the charm contribution and over abundance of hadrons at the $\psi(2s)$ peak.

1 Introduction

Understanding CP violation in the elementary particle system is a fascinating subject in itself. It is also a key to understanding the preponderance of matter over antimatter in our universe. Up to now the only evidence of CP violation on the elementary particle level is the decay of the K_L system and this is too meager to construct a credible standard theory for CP violation for all particles. In this paper we discuss measurement of CP violation in τ decay. This is an interesting subject because τ is the heaviest lepton and thus if a charged Higgs boson is responsible for CP violation we would most likely see the effect here among all the leptons. Also the Kobayashi-Maskawa theory [1] says that CP violation should not occur in the leptonic sector because the gauge eigenstate and mass eigenstate are identical in the lepton sector due to zero neutrino masses in the Standard Model. These basic assumptions of Kobayashi-Maskawa must be tested. CP violation in τ has been investigated

---

[*]Work supported by the Department of Energy, contract DE–AC03–76SF00515.

© 1996 American Institute of Physics

previously mainly in the production of τ pair coming from the possible existence of the electric dipole moment [2] of τ. However since the electric dipole moment of τ is induced by weak or semiweak corrections to the electromagnetic vertex of τ its effect is expected to be less than $(m_\tau/m_w)^2\alpha = 3 \times 10^{-6}$ and thus impossible to detect even with 10^8 τ pairs available in the Tau-Charm Factory. Similarly the interference between CP violating neutral Higgs boson exchange and the one photon exchange diagrams is also completely negligible [3, 4]. Thus CP can be assumed to be conserved in the production the τ pair; we need to consider only CP violation in the decay of τ.

Since the decay of τ is a weak interaction, if CP violation in τ is weak, then its effect should be of order 1 whereas if it is milliweak its effect should be of order 10^{-3} and detectable with 10^8 τ pairs available at the Tau-Charm Factory.

In my previous papers [3, 4] I have discussed how to use the polarized electron beam to investigate CP violation in τ decay by constructing rotationally invariant quantities such as $\vec{w}_i \cdot \vec{a}$, $(\vec{w}_i \times \vec{a}) \cdot \vec{b}$, $(\vec{w}_i \times \vec{\mu}) \cdot \vec{w}_\mu$ where $\vec{a}$ and $\vec{b}$ are momenta of hadrons in the semileptonic decay of τ; $\vec{\mu}$ and $\vec{w}_\mu$ are momentum and the polarization of muon in the decay $\tau^- \to \mu^- + \nu_\tau + \bar{\nu}_\mu$ or its charge conjugate; $\vec{w}_i$ is the initial beam polarization

$$\vec{w}_i = \frac{w_1 + w_2}{1 + w_1 w_2} \hat{e}_z , \qquad (1)$$

where $\hat{e}_z$ is the direction of the incident electron and w_1 and w_2 are polarization of the electron and positron in the z direction.

In section 2 I point out that $\Gamma(\tau^- \to \nu_\tau + a + b) \neq \Gamma(\tau^+ \to \bar{\nu}_\tau + \bar{a} + \bar{b})$ also signifies CP violation. We give the physical reason for it. We also compare the merits of this kind of measurement with those using polarized beams. In section 3 we discuss the optimum energy to do τ physics at the Tau-Charm Factory.

2 CP Violation in τ Decay using Branching Fractions

CPT conservation says that the total widths of τ^- and τ^+ must be equal. Also partial widths into those channels without final state interactions, such as $\tau^- \to \mu^- + \bar{\nu}_\mu + \nu_\tau$, $\tau^- \to e^- + \bar{\nu}_e + \nu_\tau$, $\tau^- \to \pi^- + \nu_\tau$, and $\tau^- \to k^- + \nu_\tau$, must be the same as the corresponding channels for τ^+ decay [5]. However for decay channels that contain final state interactions, such as $\tau^- \to \pi^- + \pi^0 + \nu_\tau$, $\pi^- + \pi^- + \pi^+ + \nu_\tau$, $\pi^- + \pi^0 + \pi^0 + \nu_\tau$, $\pi^- + k^0 + \nu_\tau$, and $\pi^0 + k^- + \nu_\tau$, the CP violation can show up as the inequality in partial widths for charge conjugate decay modes. For example, $\Gamma(\tau^- \to \pi^- + \pi^0 + \nu) \neq \Gamma(\tau^+ \to \pi^+ + \pi^0 + \bar{\nu})$ signifies violation of CP, but $\Gamma(\tau^- \to \mu^- + 2\nu) \neq \Gamma(\tau^+ \to \mu^+ + 2\nu)$ or $\Gamma(\tau^- \to$ all) $\neq \Gamma(\tau^+ \to$ all) will indicate that CPT is violated. The polarization vector $\vec{w}_i$ defined in Eq. (1) can be used to construct many rotationally invariant products to investigate T, CP, CVC, and charged Higgs boson exchange in leptonic [4] and semileptonic [3] decays of τ. The polarization dependent quantities will yield information on structure of CP violations whereas the polarization independent quantities such as the difference in partial widths between $\tau^- \to \nu_\tau + \pi^- + \pi^0$ and $\tau^+ \to \bar{\nu}_\tau + \pi^+ + \pi^0$ will merely indicate the existence and magnitude of the CP violation. As pointed out in Ref. [3] this difference in partial widths is due to the combined effects of CP violation and the inelastic final state interaction such as 2π going into 4π and vice versa. In the absence of CP violation the probabilities of 2π going into 4π and vice versa in the τ^- decay are equal to those in the τ^+ decay. However in the presence of CP violation the amplitudes for the decay is proportional to $\exp(i\delta_w + i\delta_s)$ for τ^- and $\exp(-i\delta_w + i\delta_s)$ for τ^+ and thus they become different.

3 Optimum Energy to do τ Physics

The energy of the machine should be set below charm threshold; i.e. E_{cm} = 1869.3 MeV for each beam. Near the threshold of τ pair production, τ pair events can uniquely be identified by e-hadron, e-μ, μ-hadron events. Above the charm threshold charm events produce the unwelcome leptonic background [6]. The best energy to run is either at $\psi(2s, 3,685$ MeV$)$ or slightly below it. The total cross section for $e^+e^- \to \psi(3.685) \to \tau^+\tau^-$ can be written as [7]

$$\sigma_r(w) = 12\pi \frac{\Gamma(\psi \to 2e)\Gamma(\psi \to 2\tau)}{(w^2 - M_R^2)^2 + \Gamma_t^2 M_R^2} , \qquad (2)$$

where $w = 2E$, $M_R = 3.685$ GeV, $\Gamma_t = 243$ keV, $\Gamma(\psi \to 2e) = 2.14$ keV, and $\Gamma(\psi \to 2\tau) = \Gamma(\psi \to 2e)(\beta(3 - \beta^2)/2)$, with $\beta = \sqrt{1 - ((2M_\tau)^2/w^2)} \approx 0.26426$, and $\beta(2 - \beta^2)/2 \approx 0.38717$. At the peak of the resonance we have

$$\sigma_r(M_R) = \frac{12\pi}{M_R^2} B^2(\psi \to 2e) \, 0.38717 = 32.40 \times 10^{-33} \text{cm}^2 . \qquad (3)$$

The peak cross section of $e^+e^- \to \tau^+\tau^-$ at continuum which occurs at $2E = 4.174$ GeV is [3, 4]

$$\sigma_c(4.174) = \frac{\pi\alpha^2}{6} \times 1.036 \frac{1}{M_\tau^2} = 3.562 \times 10^{-33} \text{cm}^2 . \qquad (4)$$

Thus

$$\frac{\sigma_r(3.685)}{\sigma_c(4.174)} = 9.096 . \qquad (5)$$

This number must be reduced because the machine width is much wider than the resonance width and the radiative corrections further broaden the effective machine width. This problem was first solved [7] by the author in 1974 immediately after the discovery of J/ψ. The most comprehensive account was given in Ref. [8] which we follow here. Qualitatively if the machine width is Δ and the resonance width is Γ_t, then only the fraction Γ_t/Δ of the beam is effective in producing the resonance peak if $\Delta \gg \Gamma_t$. The effect of radiative corrections can be estimated by the change in the height of the Gaussian peak of the machine energy by the radiative corrections because only the peak height matters when the resonance is narrower than the beam width. The result is [8]

$$\sigma_{\exp}(3.685) = \sigma_r(3.685) \left[\sqrt{\frac{\pi}{8}} \frac{\Gamma_t}{\Delta}\right] \left[\left(\frac{\sqrt{8}\Delta}{3.685}\right)^T \Gamma\left(\frac{T}{2}+1\right)\right] + \sigma_c(3.685) \quad (6)$$

where $\sigma_c(3.685) = 2.476 \times 10^{-33}$ cm is the continuum cross section, Δ is the Gaussian beam width defined by

$$G(w, w') = \frac{1}{\sqrt{2\pi}\Delta} \exp\left[-\frac{(w-w')^2}{2\Delta^2}\right], \quad (7)$$

and is related to the full width at a half maximum (FWHM) by

$$\Delta = \frac{(FWHM)}{2.3848}. \quad (8)$$

T is called the equivalent radiator thickness defined by

$$T = \frac{2\alpha}{\pi}\left[\ell n \frac{M_R^2}{m_e^2} - 1\right] = 0.14229. \quad (9)$$

Γ is the Gamma function and its value is

$$\Gamma\left(1 + \frac{T}{2}\right) = 0.96365. \quad (10)$$

The first square bracket shows that only a fraction of the incoming beam, Γ_t/Δ, is effective in producing the resonance. The factor $\sqrt{\pi/8}$ comes form the fact that Γ_t is the FWHM of the Breit-Wigner formula whereas FWHM of the Gaussian beam profile is given by Eq. (8). The second square bracket represents the reduction of the Gaussian peak height due to the photon emission whose effective cutoff is $\Delta E = \sqrt{8}\,\Delta$. The Gamma function, Eq. (10), comes from the folding of the Gaussian function with the photon straggling function [8]. At the Beijing Electron-Positron Collider $\Delta = 1.4$ MeV and thus from Eq. (6) we have

$$\sigma_{\exp}(3.685, \Delta = 1.4\,\text{MeV}) = 0.0411\,\sigma_r(3.685) + \sigma_c(3.685). \quad (11)$$

there is a scheme [9] to make Δ as small as 0.14 MeV using a monochromatizer; we have then

$$\sigma_{\rm exp}(3.685, \Delta = 0.14\,{\rm MeV}) = 0.286\,\sigma_r(3.685) + \sigma_c(3.685) \ . \qquad (12)$$

Since the branching fraction to τ pair is 0.34% in $\sigma_r(3.685)$ there are several hundred π's for each τ pair produced by $\sigma_r(3.685)$.

The BES Collaboration [6] has successfully carried out τ experiments using ψ' under the conditions shown in Eq. (11), where the first term is about 0.48 of the last term. For their experiment the hadron background did not cause any problem for four reasons: (1) most of the hadron backgrounds are multiprong events whereas τ events are mostly two-prong events. This fact can be used to eliminate the background. (2) They did not use the monochromatizer. (3) Particle ID has about 10^{-3} efficiency. (4) Accuracy of 10^{-2} is good enough for them, whereas CP experiment needs 10^{-3} accuracy.

An alternative to use Eq. (11) or (12) is to avoid ψ' all together and run the machine at a slightly lower energy, say at 3.680 GeV. From the consideration of background this is probably the ideal energy to run the Tau-Charm Factory. At $W = 3.680$ GeV the component of polarization of $\tau^\pm$ in the beam direction averaged over the production angle is slightly improved:

$$\overline{w}_z = \int_{-1}^{1} w_z \frac{d\sigma}{d\cos\theta} d\cos\theta \Big/ \sigma = \frac{w_1 + w_2}{1 + w_1 w_2} \frac{1 + 2a}{2 + a^2} \equiv \frac{w_1 + w_2}{1 + w_1 w_2} F(a) \ ,$$

where $a = 2M_\tau/W$. At $w = 4.174$ GeV we have $F = 0.992$, but at $w = 3.680$ GeV we have $F = 0.9996$. The cross section is reduced from $\sigma_c(4.174) = 3.562 \times 10^{-33}$cm^2 to $\sigma_c(3.680) = 2.44 \times 10^{-33}$cm^2. This energy is preferred in order to avoid both the charm background and overabundance of hadrons in the $\psi(2s)$ peak.

Acknowledgments

The author wishes to thank Bill Dunwoodie, W.K.H. Panofsky, and Martin Perl for discussions on the optimum energy for doing τ physics at the Tau-Charm Factory. I would also like to thank Karl Brown for explaining to me how the beam monochromatizer works.

References

[1] M. Kobayashi and T. Maskawa, *Prog. Theor. Phys.* **49**, 652 (1973).

[2] F. Hoogeveen and L. Stodolsky, *Phys. Lett.* **B212**, 505 (1988); S. Goozovat and C. A. Nelson *ibid.* **267**, 128 (1991); G. Couture, *ibid.* **B272** 404 (1991); W. Bernreuther and O. Nachtmann, *Phys. Rev. Lett.* **63**, 2787 (1989); W. Bernreuther, G. W. Botz, O. Nachtmann, and P. Overmann, *Z. Phys.* **C52**, 567 (1991); W. Bernreuther, O. Nachtmann, and P. Overmann, *Phys. Rev.* **D48**, 78 (1993); B. Aranthanarayan and S. D. Rindani, *Phys. Rev. Lett.* **73**, 1215 (1994).

[3] Y. S. Tsai, *Phys. Rev.* **D51**, 3172 (1995).

[4] Y. S. Tsai, *Test of T and CP Violation in Leptonic Decay of $\tau^\pm$,"* these proceedings.

[5] T. D. Lee, *Particle Physics and Introduction to Field Theory,* published by Harwood Academic Publishers, New York (1981), pg. 333, Problem 14.1.

[6] BES Collaborations (J. Z. Bai *et al.*), SLAC–PUB–95–6930, to be published.

[7] Y. S. Tsai, in *Notes from the SLAC Theory Workshop on the ψ,* SLAC–PUB–1515 (1974) unpublished.

[8] Y. S. Tsai, SLAC–PUB–3129 (1983) unpublished. See Eq. (5.13).

[9] Shuhong Wang, contribution to this Workshop.

Decay Rates, Structure Functions and New Physics Effects in Hadronic Tau Decays

Markus Finkemeier[1] and Erwin Mirkes[2]*

[1] Lyman Laboratory of Physics, Harvard University, Cambridge, MA 02148, USA
[2] Department of Physics, University of Wisconsin, Madison, WI 56706, USA

Abstract. Hadronic decays rates of the τ lepton into multi meson final states are presented. The structure of the hadronic matrix elements for various decay modes is discussed. The formalism of structure functions allows for a detailed test of these matrix elements. Various correlations are discussed which are sensitive to possible CP violation and new physics effects in the decay modes.

INTRODUCTION

The τ lepton is heavy enough to decay into a variety of hadronic final states. In particular, final states with kaons provide a powerful probe of the strange sector of the weak charged current. The Tau-Charm Factory operating at an e^+e^- cms energy of around 4 GeV and a luminosity of $L = 10^{33} \text{cm}^{-2}\text{s}^{-1}$ with good π/K separation [1] would allow for high precision measurements of the hadronic matrix elements in all decay modes. Rare decay modes could be searched for at the level of about 10^{-7} in branching fraction. Of particular interest would also be the search for possible CP violation in the hadronic matrix elements.

In the present paper, we specify the general structure of the matrix elements for τ decays into various multi meson final states. We study angular correlations in the exclusive decay modes and show that the formalism of structure functions allows for a detailed model independent test of the hadronic matrix elements. Furthermore, the structure functions allow for a systematic analysis of possible CP violation effects in the matrix elements, which would have to come from new non-Standard Model contributions.

It is shown that CP violation effects are in principle observable in a Tau-Charm Factory (without polarized beams) for three meson decay modes with

---

*talk presented by E. Mirkes

a nonvanishing vector *and* an axial vector current. CP violation effects originating from a charged Higgs could be detected only for decay modes with a nonvanishing vector current.

An observation of CP violation in two meson decays requires either polarized beams [2] or kinematical information from the second tau decay [3].

MATRIX ELEMENTS AND DECAY RATES

The matrix element $\mathcal{M}$ for the hadronic τ decay into n mesons $h_1, \ldots h_n$

$$\tau(l,s) \to \nu(l',s') + h_1(q_1, m_1) + \ldots h_n(q_n, m_n) , \tag{1}$$

can be expressed in terms of a leptonic (M_μ) and a hadronic current (J^μ) as

$$\mathcal{M} = \frac{G}{\sqrt{2}} \left(\begin{smallmatrix} \cos\theta_c \\ \sin\theta_c \end{smallmatrix} \right) M_\mu J^\mu . \tag{2}$$

In Eq. (2), G denotes the Fermi-coupling constant and θ_c is the Cabibbo angle. The leptonic current is given by

$$M_\mu = \bar{u}(l',s')\gamma_\mu(g_V - g_A\gamma_5)u(l,s) , \tag{3}$$

with $g_V = g_A = 1$ in the Standard Model. The hadronic current J^μ can in general be expressed in terms of a vector and an axial vector current

$$J^\mu(q_1,\ldots,q_n) = \langle h_1(q_1)\ldots h_n(q_n)|V^\mu(0) - A^\mu(0)|0\rangle . \tag{4}$$

The simplest decay mode into a pion or a kaon proceeds only through the axial vector current whereas all decays into an even number of pions are expected to proceed through the vector current. In fact, the decay rates for $\tau \to 2n\pi, KK$ can be related through the conserved vector current (CVC) hypothesis to $e^+e^- \to$ hadrons in the isovector state [4]. On the other hand, three body decay modes involving kaons allow for axial and vector current contributions at the same time. In the following, we specify the hadronic matrix elements for hadronic decays into multi meson final states as expected from the Standard Model.

One Meson Decays

The decay rate for the simplest decay mode with one pion or kaon is well predicted by the the pion or kaon kaon decay constants f_π and f_K defined by the matrix element of the axial vector currents

$$\langle \pi(q)|A^\mu(0)|0\rangle = i\sqrt{2}f_\pi q^\mu \tag{5}$$
$$\langle K(q)|A^\mu(0)|0\rangle = i\sqrt{2}f_K q^\mu . \tag{6}$$

Both decay constants can be determined using the precisely measured pion (kaon) decay widths $\Gamma(\pi(K) \to \mu\nu_\mu)$. Radiative corrections $\delta R_{\tau/\pi} = (0.16 \pm 0.14)\%$ and $\delta R_{\tau/K} = (0.90 \pm 0.22)\%$ to the ratios $\Gamma(\tau \to \pi\nu)/\Gamma(\pi \to \mu\nu)$ and $\Gamma(\tau \to K\nu)/\Gamma(K \to \mu\nu)$ have been calculated recently [5]. Using the recent world average $\tau_\tau = (291.6 \pm 1.6)\,\text{fs}$ for the tau lifetime [6] one obtains the following theoretical predictions for the branching ratios

$$\mathcal{B}(\pi\nu_\tau) = (10.95 \pm 0.06)\% \tag{7}$$
$$\mathcal{B}(K\nu_\tau) = (0.723 \pm 0.006)\% \tag{8}$$

These predictions agree within one standard deviation with the world averages as quoted in [7].

Two Meson Decays

The hadronic matrix element for the decay $\tau \to h_1 h_2 \nu_\tau$ can be written as ($Q^\mu = (q_1 + q_2)^\mu$)

$$\langle h_1(q_1) h_2(q_2) | V^\mu(0) | 0 \rangle = [(q_1 - q_2)_\nu T^{\mu\nu} F^{h_1 h_2} + Q^\mu F_4^{h_1 h_2}] \tag{9}$$

$T^{\mu\nu}$ is the transverse projector, defined by

$$T_{\mu\nu} = g_{\mu\nu} - \frac{Q_\mu Q_\nu}{Q^2}. \tag{10}$$

The form factor F_4 describes the two mesons h_1 and h_2 in an s wave. As mentioned before, the form factor $F^{\pi\pi}$ in $\tau^- \to \rho^-\nu \to \pi^-\pi^0\nu$ can be obtained (using the CVC theorem) from the iso-vector part of the electromagnetic current for $e^+e^- \to \pi^+\pi^-$ and the scalar form factor F_4 is expected to vanish. One has

$$F^{\pi^-\pi^0} = \sqrt{2} T_\rho^{(1)}, \tag{11}$$

where where $T_\rho^{(1)}$ is a normalized vector resonance form factor (two particle Breit-Wigner propagator) for the ρ resonance including the contribution from the radial excitations ρ' and ρ''. In general, the normalization of the form factors $F^{h_1 h_2}$ is fixed by chiral symmetry constraints, which determines the matrix elements in the limit of soft meson momenta. The strong interaction effects beyond the low energy limit are taken into account by vector resonance factors with the requirement $T_X^{(1)}(Q^2 = 0) = 1$ ($X = \rho, K^*$). The hadronic matrix elements for the Cabibbo suppressed decay modes $K^-\pi^0\nu_\tau$, $\overline{K^0}\pi^-\nu_\tau$ are dominated by the K^* resonance $T_{K^*}^{(1)}(Q^2)$ [8], whereas the one for the Cabibbo allowed mode $K^0 K^-$ is dominated by the high energy tail of the ρ. One has [10].

$$F^{\overline{K^0}\pi^-} = \frac{1}{\sqrt{2}} T_{K^*}^{(1)}(Q^2), \qquad (12)$$

$$F^{K^-\pi^0} = T_{K^*}^{(1)}(Q^2), \qquad (13)$$

$$F^{K^0 K^-} = T_{\rho}^{(1)}(Q^2). \qquad (14)$$

In the $\tau \to K\pi$ decay mode, F_4 gets a contribution from the off-shellness $(m_{K^*}^2 - Q^2)$ of the K^*. However, this scalar contribution is strongly suppressed compared to the contribution of F. As we will see in the last section, the form factor F_4 allows also for a possible contribution from a charged Higgs exchange and is therefore of special interest.

We use the following form for the two particle Breit-Wigner propagators with an energy dependent width $\Gamma_X(s)$ throughout this paper:

$$\mathrm{BW}_X[s] \equiv \frac{M_X^2}{[M_X^2 - s - i\sqrt{s}\Gamma_X(s)]}, \qquad (15)$$

where X stands for the various resonances of the two meson channels. The following parametrization is used for the ρ resonance:

$$T_{\rho}^{(1)}(s) = \frac{1}{1+\beta_\rho}\left[\mathrm{BW}_\rho(s) + \beta_\rho \mathrm{BW}_{\rho'}(s)\right], \qquad (16)$$

where $\beta_\rho = -0.145$, $m_\rho = 0.773\,\mathrm{GeV}$, $\Gamma_\rho = 0.145\,m_{\rho'} = 1.370\,\mathrm{GeV}$, $\Gamma_{\rho'} = 0.510\,\mathrm{GeV}$. These are the values which have been determined from $e^+e^- \to \pi^+\pi^-$ in [9]. The parameterization for $T_{K^*}^{(1)}(Q^2)$ allows for a contribution of the first excitation $K^{*\prime}(1410)$ in analogy to Eq. (16):

$$T_{K^*}^{(1)}(s) = \frac{1}{1+\beta_{K^*}}\left[\mathrm{BW}_{K^*}(s) + \beta_{K^*} \mathrm{BW}_{K^{*\prime}}(s)\right], \qquad (17)$$

where $\beta_{K^*} = -0.135$, $m_{K^*} = 0.892\,\mathrm{GeV}$, $\Gamma_{K^*} = 0.050\,\mathrm{GeV}$, $m_{K^{*\prime}} = 1.412\,\mathrm{GeV}$, $\Gamma_{K^{*\prime}} = 0.227\,\mathrm{GeV}$. The parameter β_{K^*} was fixed in [10] by comparing the theoretical results to the recent experimental branching ratio for $\mathcal{B}(K^*\nu_\tau) = 1.36 \pm 0.08$ [7]. The value $\beta_{K^*} = -0.135$ is remarkably close to the strength of the ρ' contribution to the ρ Breit-Wigner, supporting the use of approximate $SU(3)$ flavour symmetry.

The branching ratios based on these parametrizations are $\mathcal{B}(\pi^-\pi^-\nu_\tau) = 23.5\%$, $\mathcal{B}(\overline{K^0}\pi^-\nu_\tau) = 0.45\%$, $\mathcal{B}(K^-\pi^0\nu_\tau) = 0.9\%$, For the decay into two kaons we obtain $\mathcal{B}(K^0 K^-\nu_\tau) = 0.11\%$, in good agreement with the recent world average $\mathcal{B}(K^0 K^-\nu_\tau) = 0.13 \pm 0.04\%$ [7].

Three Meson Decays

The hadronic matrix elements for three meson final states have a much richer structure. The decay modes involving kaons allow for axial and vector current contributions at the same time [11,12].

$$J^\mu(q_1, q_2, q_3) = \langle h_1(q_1)h_2(q_2)h_3(q_3)|V^\mu(0) - A^\mu(0)|0\rangle \,. \tag{18}$$

The most general ansatz for the matrix element of the quark current J^μ in Eq. (18) is characterized by four form factors F_i [13], which are in general functions of Q^2, $s_1 = (q_2 + q_3)^2$, $s_2 = (q_1 + q_3)^2$ and $s_3 = (q_1 + q_2)^2$

$$J^\mu(q_1, q_2, q_3) = V_1^\mu F_1 + V_2^\mu F_2 + iV_3^\mu F_3 + V_4^\mu F_4 \,, \tag{19}$$

with

$$\begin{aligned}
V_1^\mu &= (q_1 - q_3)_\nu T^{\mu\nu} \,, \\
V_2^\mu &= (q_2 - q_3)_\nu T^{\mu\nu} \,, \\
V_3^\mu &= \epsilon^{\mu\alpha\beta\gamma} q_{1\alpha} q_{2\beta} q_{3\gamma} \,, \\
V_4^\mu &= q_1^\mu + q_2^\mu + q_3^\mu = Q^\mu \,.
\end{aligned} \tag{20}$$

$T^{\mu\nu}$ denotes again the transverse projector as defined in Eq. (10). The form factors F_1 and $F_2(F_3)$ originate from the axial vector hadronic current (vector current) and correspond to a hadronic system in a spin one state, whereas F_4 is due to the spin zero part of the axial current matrix element. In the limit of vanishing quark masses, the weak axial-vector current is conserved and this implies that the scalar form factor F_4 vanishes. The massive pseudoscalars give a contribution to F_4, however, the effect is very small [14] and we will neglect this contribution in the subsequent discussion of this section. Note however that the form factor F_4 in the $\tau \to (3\pi)\nu_\tau$ decay mode could receive a sizable contribution due to the $J^P = 0^-$ resonance of the π' [15,13]. Furthermore, the form factor F_4 allows also for a possible contribution from a charged Higgs exchange. We will consider this in more detail in the last section.

The form factors F_1 and F_2 can be predicted by chiral lagrangians, supplemented by informations about resonance parameters. Parametrizations for the 3π final states based on this model can be found in [9,13,16]. In this case the vector form factor is absent due to the G parity of the pions. On the other hand, the decay mode $\tau^- \to \eta\pi^-\pi^0\nu_\tau$ has a vanishing contribution from the axial vector current [17,18,12]. The vector form factor is related to the Wess-Zumino anomaly [19,17] whereas the axial-vector form factors are again predicted by chiral Lagrangians as mentioned before. A general parameterization of the form factors for various three meson decays modes with pions and kaons was proposed in [12]. The parameterization has been extensively reanalyzed in [10] which lead to sizable differences in the predictions of the decay rates compared to [12]. Furthermore, a parameterization for the final states with two neutral kaons $\tau^- \to K_S\pi^- K_S\nu_\tau$, $\tau^- \to K_L\pi^- K_L\nu_\tau$, and $\tau^- \to K_S\pi^- K_L\nu_\tau$ was derived in [10]. The results for the form factors F_i in Eq. (19) for the decay modes $\tau \to abc\nu_\tau$ are summarized by

$$F_1^{(abc)}(Q^2, s_2, s_3) = \frac{2\sqrt{2}A^{(abc)}}{3f_\pi} G_1^{(abc)}(Q^2, s_2, s_3) \,, \tag{21}$$

TABLE I. Parameterization of the form factors F_1 and F_2 in Eqs. (21,22) for the matrix elements of the weak axial-vector current for the various channels.

channel (abc)	$A^{(abc)}$	$G_1^{(abc)}(Q^2, s_2, s_3)$	$G_2^{(abc)}(Q^2, s_1, s_3)$
$\pi^-\pi^-\pi^+$	$\cos\theta_c$	$BW_{A_1}(Q^2)T_\rho^{(1)}(s_2)$	$BW_{A_1}(Q^2)T_\rho^{(1)}(s_1)$
$\pi^0\pi^0\pi^0$	$\cos\theta_c$	$BW_{A_1}(Q^2)T_\rho^{(1)}(s_2)$	$BW_{A_1}(Q^2)T_\rho^{(1)}(s_1)$
$K^-\pi^-K^+$	$\dfrac{-\cos\theta_c}{2}$	$BW_{A_1}(Q^2)T_\rho^{(1)}(s_2)$	$BW_{A_1}(Q^2)T_{K^*}^{(1)}(s_1)$
$K^0\pi^-\overline{K^0}$	$\dfrac{-\cos\theta_c}{2}$	$BW_{A_1}(Q^2)T_\rho^{(1)}(s_2)$	$BW_{A_1}(Q^2)T_{K^*}^{(1)}(s_1)$
$K_S\pi^-K_S$	$\dfrac{-\cos\theta_c}{4}$	$BW_{A_1}(Q^2)T_{K^*}^{(1)}(s_3)$	$-BW_{A_1}(Q^2)\times [T_{K^*}^{(1)}(s_1) + T_{K^*}^{(1)}(s_3)]$
$K_S\pi^-K_L$	$\dfrac{-\cos\theta_c}{4}$	$BW_{A_1}(Q^2)\times [2T_\rho^{(1)}(s_2) + T_{K^*}^{(1)}(s_3)]$	$BW_{A_1}(Q^2)\times [T_{K^*}^{(1)}(s_1) - T_{K^*}^{(1)}(s_3)]$
$K^-\pi^0K^0$	$\dfrac{3\cos\theta_c}{2\sqrt{2}}$	$BW_{A_1}(Q^2)\times \left[\dfrac{2}{3}T_\rho^{(1)}(s_2) + \dfrac{1}{3}T_{K^*}^{(1)}(s_3)\right]$	$\dfrac{1}{3}BW_{A_1}(Q^2)\times [T_{K^*}^{(1)}(s_1) - T_{K^*}^{(1)}(s_3)]$
$\pi^0\pi^0K^-$	$\dfrac{\sin\theta_c}{4}$	$T_{K_1}^{(a)}(Q^2)T_{K^*}^{(1)}(s_2)$	$T_{K_1}^{(a)}(Q^2)T_{K^*}^{(1)}(s_1)$
$K^-\pi^-\pi^+$	$\dfrac{-\sin\theta_c}{2}$	$T_{K_1}^{(a)}(Q^2)T_{K^*}^{(1)}(s_2)$	$T_{K_1}^{(b)}(Q^2)T_\rho^{(1)}(s_1)$
$\pi^-\overline{K^0}\pi^0$	$\dfrac{3\sin\theta_c}{2\sqrt{2}}$	$\dfrac{2}{3}T_{K_1}^{(b)}(Q^2)T_\rho^{(1)}(s_2)$ $+\dfrac{1}{3}T_{K_1}^{(a)}(Q^2)T_{K^*}^{(1)}(s_3)$	$\dfrac{1}{3}T_{K_1}^{(a)}(Q^2)\times [T_{K^*}^{(1)}(s_1) - T_{K^*}^{(1)}(s_3)]$

$$F_2^{(abc)}(Q^2, s_1, s_3) = \frac{2\sqrt{2}A^{(abc)}}{3f_\pi}G_2^{(abc)}(Q^2, s_1, s_3), \qquad (22)$$

$$F_3^{(abc)}(Q^2, s_1, s_2, s_3) = \frac{A^{(abc)}}{2\sqrt{2}\pi^2 f_\pi^3}G_3^{(abc)}(Q^2, s_1, s_2, s_3). \qquad (23)$$

The Breit-Wigner functions $G_{1,2}$ (G_3) and the normalizations $A^{(abc)}$ are listed in Tab. I (II) for the various decay modes. Note that by convenient ordering of the mesons, the two body resonances in F_1 (F_2) occur only in the variables s_2, s_3 (s_1, s_3).

Let us briefly discuss the three particle resonances in Tab. I and II (for details see [10]). We use the A_1 resonance in the non-strange case with energy

TABLE II. Parameterization of the form factor F_3 in Eq. (23) for the matrix elements of the weak vector current for the various channels.

channel (abc)	$A^{(abc)}$	$G_3^{(abc)}(Q^2, s_1, s_2, s_3)$
$K^-\pi^-K^+$	$-\cos\theta_c$	$T_\rho^{(2)}(Q^2)(\sqrt{2}-1)\left[\sqrt{2}T_\omega(s_2) + T_{K^*}^{(1)}(s_1)\right]$
$K^0\pi^-\overline{K^0}$	$\cos\theta_c$	$T_\rho^{(2)}(Q^2)(\sqrt{2}-1)\left[\sqrt{2}T_\omega(s_2) + T_{K^*}^{(1)}(s_1)\right]$
$K_S\pi^-K_S$	$\dfrac{-\cos\theta_c}{2}$	$T_\rho^{(2)}(Q^2)(\sqrt{2}-1)\left[T_{K^*}^{(1)}(s_1) - T_{K^*}^{(1)}(s_3)\right]$
$K_S\pi^-K_L$	$\dfrac{\cos\theta_c}{2}$	$T_\rho^{(2)}(Q^2)(\sqrt{2}-1)\left[2\sqrt{2}T_\omega(s_2) + T_{K^*}^{(1)}(s_1) + T_{K^*}^{(1)}(s_3)\right]$
$K^-\pi^0K^0$	$\dfrac{-\cos\theta_c}{\sqrt{2}}$	$T_\rho^{(2)}(Q^2)(\sqrt{2}-1)\left[T_{K^*}^{(1)}(s_3) - T_{K^*}^{(1)}(s_1)\right]$
$\eta\pi^-\pi^0$	$\dfrac{\cos\theta_c}{\sqrt{3}}$	$T_\rho^{(2)}(Q^2)T_\rho^{(1)}(s_1)$
$\pi^0\pi^0K^-$	$\sin\theta_c$	$\dfrac{1}{4}T_{K^*}^{(2)}(Q^2)\left[T_{K^*}^{(1)}(s_1) - T_{K^*}^{(1)}(s_2)\right]$
$K^-\pi^-\pi^+$	$\sin\theta_c$	$\dfrac{1}{2}T_{K^*}^{(2)}(Q^2)\left[T_\rho^{(1)}(s_1) + T_{K^*}^{(1)}(s_2)\right]$
$\pi^-\overline{K^0}\pi^0$	$\sqrt{2}\sin\theta_c$	$\dfrac{1}{4}T_{K^*}^{(2)}(Q^2)\left[2T_\rho^{(1)}(s_2) + T_{K^*}^{(1)}(s_1) + T_{K^*}^{(1)}(s_3)\right]$

dependent width $\text{BW}_{A_1}(s) = \dfrac{m_{A_1}^2}{m_{A_1}^2 - s - im_{A_1}\Gamma_{A_1}g(s)/g(m_{A_1})}$, with $m_{A_1} = 1.251$ GeV, $\Gamma_{A_1} = 0.475$ GeV. The function $g(s)$ has been calculated in [9]. The three particle resonances with strangeness are

$$T_{K_1}^{(a)}(s) = \frac{1}{1+\xi}\left[\text{BW}_{K_1(1400)}(s) + \xi\text{BW}_{K_1(1270)}(s)\right],$$

$$T_{K_1}^{(b)}(s) = \text{BW}_{K_1(1270)}(s). \qquad (24)$$

with $\xi = 0.33$ [10]. The three body vector resonances $T_\rho^{(2)}$ and $T_{K^*}^{(2)}$ include the higher radial excitations ρ' and ρ'' and $K^{*\prime}$ and $K^{*\prime\prime}$

$$T_\rho^{(2)} = \frac{1}{1+\lambda+\mu}\left[\text{BW}_\rho(s) + \lambda\text{BW}_{\rho'}(s) + \mu\text{BW}_{\rho''}(s)\right],$$

$$T_{K^*}^{(2)} = \frac{1}{1+\lambda+\mu}\left[\text{BW}_{K^*}(s) + \lambda\text{BW}_{K^{*\prime}}(s) + \mu\text{BW}_{K^{*\prime\prime}}(s)\right], \qquad (25)$$

with $\lambda = -0.25, \mu = -0.038$. The ω resonance $T_\omega(s) = \dfrac{1}{1+\epsilon}[\text{BW}_\omega(s) + \epsilon\text{BW}_\Phi(s)]$ in the vector form factor F_3 in Tab. II allows for a contribution of the ϕ with a relative strength $\epsilon = 0.05$ [10].

TABLE III. Predictions for the normalized decay widths $\Gamma(abc)/\Gamma_e$ and the branching ratios $\mathcal{B}(abc)$ for the various channels. The contribution from the vector current is listed in column 3 and available experimental data are listed in column 5. The later are taken from [7,20,21].

channel (abc)	$\left(\dfrac{\Gamma(abc)}{\Gamma_e}\right)^{(pred.)}$	$\left(\dfrac{\Gamma(abc)}{\Gamma_e}\right)^{(pred.)}_V$	$\mathcal{B}(abc)^{(pred.)}$	$\mathcal{B}(abc)^{(expt.)}$
$\pi^-\pi^-\pi^+$	0.48	0.	8.6%	$(8.64\pm 0.24)\%$
$\pi^0\pi^0\pi^-$	0.48	0.	8.6%	$(9.09\pm 0.14)\%$
$K^-\pi^-K^+$	0.011	0.0045	0.20%	$(0.20\pm 0.07)\%$
$K^0\pi^-\overline{K^0}$	0.011	0.0045	0.20%	
$K_S\pi^-K_S$	0.0027	0.0008	0.048%	$(0.021\pm 0.006)\%$
$K_S\pi^-K_L$	0.0058	0.0029	0.10%	
$K^-\pi^0 K^0$	0.0090	0.0032	0.16%	$(0.12\pm 0.04)\%$
$\eta\pi^-\pi^0$	0.0108	0.0108	0.19%	$(0.170\pm 0.028)\%$
$\pi^0\pi^0 K^-$	0.0080	0.0007	0.14%	$(0.09\pm 0.03)\%$
$K^-\pi^-\pi^+$	0.043	0.0043	0.77%	$(0.40\pm 0.09)\%$
$\pi^-\overline{K^0}\pi^0$	0.054	0.0058	0.96%	$(0.41\pm 0.07)\%$

Numerical results for the hadronic decay widths $\Gamma(abc)$ normalized to the leptonic width Γ_e and for the branching ratios in Tab. III based on this model for the form factors. The predictions for the branching ratios use $\Gamma_e/\Gamma_{tot} = 17.8\%$, as calculated from the experimental values for the tau mass $m_\tau = 1.7771$ GeV and lifetime $\tau_\tau = 291.6$ fs [6].

Four Pion Decays

In order to predict the two tau decays into four pions, $\tau^- \to \nu_\tau \pi^-\pi^-\pi^+\pi^0$ and $\tau^- \to \nu_\tau \pi^0\pi^0\pi^0\pi^-$, there are two possible approaches.

The first approach is based on the fact that these tau decays are again related through CVC to corresponding e^+e^- annihilation channels, namely to $e^+e^- \to 2\pi^+2\pi^-$ and $e^+e^- \to \pi^+\pi^-2\pi^0$. And so by using the measured e^+e^- cross sections as input, the tau decays can be predicted [4,8,28], and the results are in good agreement with the τ data [7,23–25]. This approach, however, allows only to predict the integrated decay rates and the four pion invariant mass distributions. In order to predict the various two and three pion differential distributions, or in order to understand angular distributions, a dynamical model is need.

Such a dynamical model has be constructed in [26] which uses the other possible approach. One follows along the lines which have been used above to obtain the hadronic current in the three meson modes. Again one starts from the structure of the hadronic current in the chiral limit and then implements low lying resonances in the various channels (ρ, ρ', ρ'', A_1 and ω mesons). There are a few free parameters, which are fixed using the experimental $e^+e^- \to 2\pi^+2\pi^-$ cross sections and the measured decay rate of the $\tau \to \omega\pi\nu_\tau$ submode. After parameter fixing, predictions for $e^+e^- \to \pi^+\pi^-2\pi^0$ and for the four pion decay modes of the τ are obtained, including detailed two, three and four pion differential mass distributions. The various predictions agree well with the available experimental data.

The $\omega\pi$ contribution to the 4π final state is expected to proceed via a vector current. However, a violation of G-parity would allow the $\omega\pi$ system to be in an axial vector state, which could be revealed by an analysis of the angular distribution in the $\omega\pi$ mode as introduced in [27]

ANGULAR DISTRIBUTIONS AND STRUCTURE FUNCTIONS IN TWO AND THREE MESON DECAY MODES

In this section, we study angular distribution of the hadronic system of two and three meson final states which are accessible in a future τ-charm factory. We will assume that the direction of the τ in the hadronic rest frame is known and that no spin informations of the decaying τ can be used in the analysis. Of particular interest in the three meson case are the distributions of the normal to the Dalitz plane and the distributions around this normal. It is shown that the most general distribution in the three meson case can be characterized by 16 structure functions most of which can be determined under the conditions mentioned above. The study of angular correlations of the hadronic system allows for much more detailed studies of the hadronic charged current than it is possible by rate measurements alone. Special emphasis is put on T-odd triple momentum correlations, which allow for the observation of CP-violating contributions beyond the Standard Model.

Two Body Decays

Of particular interest in the two body decays is the distribution of the direction of h_1 ($\hat{q}_1 = \vec{q}_1/|\vec{q}_1|$) and the direction of the τ (denoted by $\vec{n}_\tau$) viewed from the hadronic rest frame $\cos\beta = \vec{n}_\tau \cdot \hat{q}_1$. After integration over the unobserved neutrino direction, the differential decay rate for a two meson final state is given by

$$d\Gamma(\tau \to 2h) = \left\{ \bar{L}_B W_B + \bar{L}_{SA} W_{SA} + \bar{L}_{SF} W_{SF} + \bar{L}_{SG} W_{SG} \right\} \times \quad (26)$$

$$\frac{G^2}{4m_\tau}(g_V^2 + g_A^2)\binom{\cos^2\theta_c}{\sin^2\theta_c}\frac{1}{(4\pi)^3}\frac{(m_\tau^2 - Q^2)^2}{m_\tau^2}|\vec{q}_1|\frac{dQ^2}{\sqrt{Q^2}}\frac{d\cos\beta}{2}$$

with $\vec{q}_1^* = \frac{1}{2\sqrt{Q^2}} \left([Q^2 - m_1^2 - m_2^2]^2 - 4m_1^2 m_2^2 \right)^{1/2}$. The hadronic structure functions W_X can be expressed in terms of the form factors F and F_4 as defined in Eqs. (9) as follows:

$$W_B = 4(\vec{q}_1)^2 |F|^2 \tag{27}$$
$$W_{SA} = Q^2 |F_4|^2 \tag{28}$$
$$W_{SF} = 4\sqrt{Q^2}|\vec{q}_1| \operatorname{Re}[F F_4^*] \tag{29}$$
$$W_{SG} = -4\sqrt{Q^2}|\vec{q}_1| \operatorname{Im}[F F_4^*] \tag{30}$$

The leptonic coefficients are

$$\begin{aligned} \bar{L}_B &= K_1 \sin^2 \beta + K_2 \\ \bar{L}_{SA} &= K_2 \\ \bar{L}_{SF} &= -K_2 \cos \beta \\ \bar{L}_{SG} &= 0 \end{aligned} \tag{31}$$

with

$$K_1 = 1 - (m_\tau^2/Q^2); \quad K_2 = (m_\tau^2/Q^2); \tag{32}$$

Note that the coefficient $\bar{L}_{SG}$ vanishes, if only the β dependence of the decay is analyzed. In the case of a polarized τ (as it is the situation at LEP) one can use the direction of the τ spin-vector $\vec{s}$ in the lab to define a further angle α by $\cos \alpha = \frac{(\vec{n}_\tau \times \vec{s}) \cdot (\vec{n}_\tau \times \hat{q}_1)}{|\vec{n}_\tau \times \vec{s}| |\vec{n}_\tau \times \hat{q}_1|}$ (see also Fig. 5 in [13]). Taking into account the distribution with respect to this angle would allow to measure also the structure function W_{SG}. Note that the structure function W_{SG} is proportional to the imaginary part of the form factors $(F F_4^*)$ and requires nontrivial phases of the amplitudes resulting from final state interactions. These strong interaction phases are essential for the observation of possible CP violation effects in the hadronic decay amplitudes. However, in our case the angle α is not observable and has to be averaged out. Hence, the T-odd correlation $\bar{L}_{SG} W_{SG}$ vanishes and no test of CP violation is possible. However, a nonvanishing contribution to the distributions $\bar{L}_{SA} W_{SA}$ or $\bar{L}_{SF} W_{SF}$ would be a clear signal of a scalar contribution (parametrized by F_4) to the two meson decay modes.

Three Body Decays

Like in the two body case, the three meson decay modes are most easily analyzed in the hadronic rest frame $\vec{q_1} + \vec{q_2} + \vec{q_3} = 0$. The orientation of the hadronic system is in general characterized by three Euler angles (α, β and γ) as introduced in [13,16]. Performing the analysis of $\tau \to \nu_\tau + 3$ mesons in the hadronic rest frame has the advantage that the product of the hadronic and the leptonic tensors reduce to a sum [13] $L^{\mu\nu}H_{\mu\nu} = \sum_X \bar{L}_X W_X$. In this system the hadronic tensor $H_{\mu\nu}$ is decomposed into 16 hadronic structure functions W_X corresponding to 16 density matrix elements for a hadronic system in a spin one [contributions proportional to $V_1^\mu F_1, V_2^\mu F_2, V_3^\mu F_3$ in Eq.(20)] and spin zero state [$V_4^\mu F_4$] (nine of them originate from a pure spin one and the remaining are pure spin zero or interference terms). The 16 structure functions contain the dynamics of the three meson decay and depend only on the hadronic invariants Q^2 and the Dalitz plot variables s_i. The leptonic factors $\bar{L}_X$ factorize the dependence on the Euler angles and also depend on the chirality parameter $\gamma_{VA} = \dfrac{2 g_V g_A}{g_V^2 + g_A^2}$. In our case, one can measure two Euler angles β and γ defined by $\cos\beta = \vec{n}_\tau \cdot \vec{n}_\perp$, $\cos\gamma = -\dfrac{\vec{n}_\tau \cdot \hat{q}_3}{|\vec{n}_\tau \times \vec{n}_\perp|}$, $\sin\gamma = \dfrac{(\vec{n}_\tau \times \vec{n}_\perp) \cdot \hat{q}_3}{|\vec{n}_\tau \times \vec{n}_\perp|}$. The vector $\vec{n}_\tau$ denotes the τ direction in the hadronic rest frame. The (x, y) plane is aligned with the hadron momenta, i.e. $\vec{n}_\perp = (\vec{q}_1 \times \vec{q}_2)/|\vec{q}_1 \times \vec{q}_2|$ (the normal to the hadronic plane) pointing along Oz. The Ox axis is defined by the direction of $\hat{q}_3 = \vec{q}_3/|\vec{q}_3|$. In the three pion case $\pi^-\pi^-\pi^+$ we choose $\vec{q}_3 = \vec{q}_{\pi^+}$ and $|\vec{q}_2| > |\vec{q}_1|$.

The differential decay rate with respect to these two angles is then given by

$$d\Gamma(\tau \to 3h) = \frac{G'^2}{2m_\tau}\binom{\cos^2\theta_c}{\sin^2\theta_c}\left\{\sum_X \bar{L}_X W_X\right\} \times \qquad (33)$$

$$\frac{1}{(2\pi)^5}\frac{1}{64}\frac{(m_\tau^2 - Q^2)^2}{m_\tau^2}\frac{dQ^2}{Q^2}ds_1\,ds_2\,\frac{d\gamma}{2\pi}\frac{d\cos\beta}{2}.$$

The leptonic coefficients $\bar{L}_X$ will be discussed below. The dependence of the structure functions on the form factors F_i reads [13]:

$$W_A = (x_1^2 + x_3^2)|F_1|^2 + (x_2^2 + x_3^2)|F_2|^2 + 2(x_1 x_2 - x_3^2)\,\mathrm{Re}\,(F_1 F_2^*)$$
$$W_B = x_4^2|F_3|^2$$
$$W_C = (x_1^2 - x_3^2)|F_1|^2 + (x_2^2 - x_3^2)|F_2|^2 + 2(x_1 x_2 + x_3^2)\,\mathrm{Re}\,(F_1 F_2^*)$$
$$W_D = 2\left[x_1 x_3 |F_1|^2 - x_2 x_3 |F_2|^2 + x_3(x_2 - x_1)\,\mathrm{Re}\,(F_1 F_2^*)\right]$$
$$W_E = -2x_3(x_1 + x_2)\,\mathrm{Im}\,(F_1 F_2^*)$$

$$W_F = 2x_4 [x_1 \operatorname{Im}(F_1 F_3^*) + x_2 \operatorname{Im}(F_2 F_3^*)]$$
$$W_G = -2x_4 [x_1 \operatorname{Re}(F_1 F_3^*) + x_2 \operatorname{Re}(F_2 F_3^*)]]$$
$$W_H = 2x_3 x_4 [\operatorname{Im}(F_1 F_3^*) - \operatorname{Im}(F_2 F_3^*)] \qquad (34)$$
$$W_I = -2x_3 x_4 [\operatorname{Re}(F_1 F_3^*) - \operatorname{Re}(F_2 F_3^*)]$$
$$W_{SA} = Q^2 |F_4|^2$$
$$W_{SB} = 2\sqrt{Q^2} [x_1 \operatorname{Re}(F_1 F_4^*) + x_2 \operatorname{Re}(F_2 F_4^*)]$$
$$W_{SC} = -2\sqrt{Q^2} [x_1 \operatorname{Im}(F_1 F_4^*) + x_2 \operatorname{Im}(F_2 F_4^*)]$$
$$W_{SD} = 2\sqrt{Q^2} x_3 [\operatorname{Re}(F_1 F_4^*) - \operatorname{Re}(F_2 F_4^*)]$$
$$W_{SE} = -2\sqrt{Q^2} x_3 [\operatorname{Im}(F_1 F_4^*) - \operatorname{Im}(F_2 F_4^*)]$$
$$W_{SF} = -2\sqrt{Q^2} x_4 \operatorname{Im}(F_3 F_4^*)$$
$$W_{SG} = -2\sqrt{Q^2} x_4 \operatorname{Re}(F_3 F_4^*)$$

The variables x_i are defined by $x_1 = V_1^x = q_1^x - q_3^x$, $x_2 = V_2^x = q_2^x - q_3^x$, $x_3 = V_1^y = q_1^y = -q_2^y$, $x_4 = V_3^z = \sqrt{Q^2 x_3 q_3^x}$, where q_i^x (q_i^y) denotes the x (y) component of the momentum of meson i in the hadronic rest frame. They can easily be expressed in terms of s_1, s_2 and s_3 [13,16].

Note that the first 9 structure functions originate from the hadronic system in a spin one state (W_A, W_C, W_D, W_E from the axial vector current, W_B from the vector current and W_F, W_G, W_H, W_I from the interference of the axial vector and vector current). W_{SA} originates only from a hadronic system in a spin zero state and the remaining six structure functions are interference terms between the spin one and spin zero states.

An inspection of Eq. (34) shows also that the structure functions $W_E, W_F, W_H, W_{SC}, W_{SE}, W_{SF}$ require nontrivial phases of the amplitudes resulting from final state interactions. Only the T-odd correlations $\bar{L}_X W_X$, $X \in \{E, F, H, SC, SE, SF\}$ allow in principle for a measurement of CP violating effects in the hadronic matrix elements (see next section).

The leptonic coefficients $\bar{L}_X$ depend on the two angles β, γ and on γ_{VA}:

$$\begin{aligned}
\bar{L}_A &= 1/2\ K_1(1+\cos^2\beta) + K_2\,; & \bar{L}_{SA} &= K_2\,; \\
\bar{L}_B &= K_1\sin^2\beta + K_2\,; & \bar{L}_{SB} &= K_2\sin\beta\cos\gamma\,; \\
\bar{L}_C &= -1/2\ K_1\sin^2\beta\cos 2\gamma\,; & \bar{L}_{SC} &= 0\,; \\
\bar{L}_D &= 1/2\ K_1\sin^2\beta\sin 2\gamma\,; & \bar{L}_{SD} &= -K_2\sin\beta\sin\gamma\,; \\
\bar{L}_E &= \gamma_{VA}\cos\beta\,; & \bar{L}_{SE} &= -0\,; \qquad (35)\\
\bar{L}_F &= 1/2\ K_1\sin 2\beta\cos\gamma\,; & \bar{L}_{SF} &= -K_2\cos\beta\,; \\
\bar{L}_G &= -\gamma_{VA}\sin\beta\sin\gamma\,; & \bar{L}_{SG} &= 0\,; \\
\bar{L}_H &= -1/2\ K_1\sin 2\beta\sin\gamma\,; & & \\
\bar{L}_I &= -\gamma_{VA}\sin\beta\cos\gamma\,; & &
\end{aligned}$$

The coefficients K_i are defined in Eq. (32). Note that the coefficients $\bar{L}_{SC}, \bar{L}_{SE}, \bar{L}_{SG}$ vanish if only the two Euler angles β and γ are considered. It has been shown in [13] that in the case of a polarized τ (as it is the situation at LEP) one can use the direction of the τ spin-vector in the lab to define a further Euler angle α. If this additional angle is considered, all 16 coefficients $\bar{L}_X$ in Eqs. (35) are nonvanishing enabling the measurement of all 16 structure functions W_X.

The coefficients $\bar{L}_{SC}$ $\bar{L}_{SE}$ are of particular importance for the detection of possible CP violation originating from a charged Higgs exchange (see below).

Numerical results for the nonvanishing structure functions in the 3π decay mode are discussed in [13,16]. Furthermore, it has been shown in [13] that the technique of the structure functions allows for a model independent test of possible spin zero components (parametrized by F_4) in the hadronic current by analyzing the structure functions W_{SB} and W_{SD}. Note that the $\cos\beta$ distribution allows already for a model independent separation of the axial-vector and the vector current contribution in the decay modes with different mesons, i.e. the structure functions W_A and W_B in Eq. (34) can be disentangled due to the different β dependence of $\bar{L}_A$ and $\bar{L}_B$. Numerical results of the structure functions for several three meson decay modes with different mesons based on the model in [12] are discussed in [29]. A more detailed analysis (including the full Q^2 and s_i dependence of the structure functions) based on the parameterization in [10] is in preparation [30].

CP VIOLATION EFFECTS

Currently CP violation has been experimentally observed only in the K meson system. The effect can be explained by a nontrivial complex phase in the Kobayashi-Maskawa flavour mixing matrix. However, the fundamental origin of this CP violation is still unknown. CP-odd correlations of the τ and τ^+ decay products, which originate from an electric dipole moment in the τ pair production, have been discussed in [31]. In this paper, we investigate the effects of possible non-Kobayashi-Maskawa-type of CP violation, i.e. CP violation effects beyond the Standard Model. Such effects could originate for example from multi Higgs boson models [32], scalar leptoquark model [33] or left-right symmetric models [34].

Any possible observation of these CP violation effects needs not only a CP-violating complex phase (parametrized as η and χ below) in the hadronic matrix elements but also the interference with a CP conserving phase resulting from final state interactions. Therefore, only the correlations involving structure functions proportional to the imaginary part of the form factors F_i allow in principle for an observation of CP violation effects by taking the difference of $d\Gamma[\tau^-] - d\Gamma[\tau^+]$ of the corresponding T-odd correlations (see below).

In the two meson decay modes, the only structure function which is sensitive to CP-violation effects is W_{SG} in Eq. (30) [proportional to $\text{Im}[FF_4^*]$]. Unfortunately, this structure function is not observable if only distributions of the angle β are considered, i.e. the coefficient $\bar{L}_{SG}$ vanishes. However, W_{SG} could in principle be measured by taking into account additional distributions with respect to the τ spin vector (assuming polarized incident beams).

CP violation effects in the $\tau \to 2\pi\nu$ decay mode from the scalar sector (e.g. the multi Higgs boson models) have recently been discussed in terms of "stage-two spin correlation functions" in [3] and in the case of polarized electron-positron beams at τ charm factories in [2]. In [3], the decay products of the second tau decay are used to define a T-odd correlation whereas the τ polarization (assuming a polarized incident electron beam) is used in [2] to define a T-odd triple correlation. In fact, the correlations in [2] are equivalent to the product $\bar{L}_{SG}W_{SG}$ as discussed before in the two meson case, if the angle α is defined with respect to the τ spin as described after Eq. (32).

In the three meson case, the structure functions $W_E, W_F, W_H, W_{SC}, W_{SE}, W_{SF}$ in Eq. (34) require nontrivial phases of the amplitudes resulting from final state interactions. Only the T-odd correlations $\bar{L}_X W_X$, $X \in \{E, F, H, SC, SE, SF\}$ allow therefore in principle for a measurement of CP violating effects in the hadronic matrix elements. As can be seen from Eq. (35) the coefficients $\bar{L}_{SC}, \bar{L}_{SE}, \bar{L}_{SG}$ vanish if only the two Euler angles β and γ are considered. However, the structure functions W_E, W_F, W_H, W_{SF} can be measured through the β and γ dependence encoded in the coefficients $\bar{L}_E, \bar{L}_F, \bar{L}_H, \bar{L}_{SF}$.

Let us therefore parametrize possible CP violation effects in the hadronic decay amplitudes by replacing Eqs. (19) by

$$J^\mu(q_1, q_2, q_3) = [(V_1^\mu F_1 + V_2^\mu F_2)(1 + \chi_A) + V_4^\mu F_4 (1 + \chi_A + \eta) + iV_3^\mu F_3 (1 + \chi_V)] \tag{36}$$

where V_i^μ are given in Eq. (20).

The term proportional to η parametrizes the effect of a possible charged Higgs boson [32], whereas the complex numbers χ_A and χ_V parametrize any new physics that would arise from vector or scalar boson exchange motivated by left-right symmetric models [34]. The Standard Model prediction is obtained from Eq. (36) by setting χ and η to zero. Let us now assume that the complex numbers χ_A, χ_V and η transform like

$$\chi_A \xrightarrow{CP} \chi_A^*; \quad \chi_V \xrightarrow{CP} \chi_V^*; \quad \eta \xrightarrow{CP} \eta^*. \tag{37}$$

The hadronic structure functions $\tilde{W}_X$, which include the new physics effects parametrized by the numbers η and χ are easily obtained from Eq. (34) using the transformation

$$F_1 \to \tilde{F}_1 = F_1(1 + \chi_A), \tag{38}$$
$$F_2 \to \tilde{F}_2 = F_2(1 + \chi_A), \tag{39}$$
$$F_3 \to \tilde{F}_3 = F_3(1 + \chi_V), \tag{40}$$
$$F_4 \to \tilde{F}_4 = F_4(1 + \chi_A + \eta). \tag{41}$$

The hadronic structure functions are affected by the sign change in the weak phases under CP transformation as described in Eq. (37). Note that the strong (complex) phases due to final state interactions [given by Breit-Wigner propagators for the two body resonances] are not changed, because the strong interaction is invariant under charge conjugation. Besides of the sign change in the weak phases, the structure functions $\tilde{W}_F, \tilde{W}_G, \tilde{W}_H, \tilde{W}_I, \tilde{W}_{SF}, \tilde{W}_{SG}$, which originate from the interference of the axial vector and vector current, change sign. Furthermore, the amplitude for the CP conjugated process τ^+ can be obtained from the results for τ^- by reversing all momenta and spins of the particles. Thus, $\cos\beta \to -\cos\beta$ and $\gamma_{VA} = -\gamma_{VA}$. CP invariance therefore relates the differential decay rates for τ^+ and τ^- as:

$$d\Gamma[\tau^-](\cos\beta, \gamma_{VA}, \tilde{W}_X) \stackrel{CP}{=} d\Gamma[\tau^+](-\cos\beta, -\gamma_{VA}, a_X\tilde{W}_X) \tag{42}$$

with $a_X = -1$ for $X \in \{\tilde{W}_F, \tilde{W}_G, \tilde{W}_H, \tilde{W}_I, \tilde{W}_{SF}, \tilde{W}_{SG}\}$ and $a_X = 1$ else.

If CP is not violated, the difference $d\Gamma[\tau^-] - d\Gamma[\tau^+]$ should vanish. From the T-odd correlations $\bar{L}_X \tilde{W}_X$, $X \in \{E, F, H, SC, SE, SF\}$, one can construct CP-violating quantities by taking the difference of these correlations for τ^- and τ^+.

$$\Delta_X = \frac{1}{2}\left(\bar{L}_X(\cos\beta, \gamma_{VA})\tilde{W}_X[\tau^-] - \bar{L}_X(-\cos\beta, -\gamma_{VA})a_X\tilde{W}_X[\tau^+]\right)$$
$$= \bar{L}_X(\cos\beta, \gamma_{VA})\left(\tilde{W}_X[\tau^-] - \tilde{W}_X[\tau^+]\right) \equiv \bar{L}_X \Delta\tilde{W}_X, \tag{43}$$

where

$$\Delta\tilde{W}_X = \tilde{W}_X[\tau^-] - \tilde{W}_X[\tau^+] \tag{44}$$

The nonvanishing CP-violating differences can be calculated from Eqs. (34,38-41) and expressed in terms of the form factors F_i and the complex numbers χ_A, χ_V and η as follows:

$$\Delta \tilde{W}_F = 2x_4 \left[x_1 \operatorname{Re}\left(F_1 F_3^*\right) + x_2 \operatorname{Re}\left(F_2 F_3^*\right) \right] \operatorname{Im}\left(\chi_A - \chi_V + \chi_A \chi_V^*\right) , \qquad (45)$$

$$\Delta \tilde{W}_H = 2x_3 x_4 \left[\operatorname{Re}\left(F_1 F_3^*\right) - \operatorname{Re}\left(F_2 F_3^*\right) \right] \operatorname{Im}\left(\chi_A - \chi_V + \chi_A \chi_V^*\right) , \qquad (46)$$

$$\Delta \tilde{W}_{SF} = -2\sqrt{Q^2} x_4 \operatorname{Re}\left(F_3 F_4^*\right) \operatorname{Im}\left(\chi_V - \chi_A - \eta + \chi_V (\chi_A^* + \eta^*)\right) . \qquad (47)$$

An observed nonzero values for these differences would signal a true CP-violation. Note that all CP-violating differences are proportional to the imaginary part η and χ. Note also that $\Delta \tilde{W}_E$ vanishs, because the form factors F_1 and F_2 multiply the same complex weak phase. Eqs. (45,46) show that CP violation effects parametrized by χ_A and χ_V are in principle observable in a Tau-Charm Factory for three meson decay modes with a nonvanishing vector (proportional to F_3) *and* and axial vector current (proportional to F_1, F_2). CP violation effects from a charged Higgs could be detected through $\Delta \tilde{W}_{SF}$ only for decay modes with a nonvanishing vector current. Therefore, CP-violation tests in the three pion decay mode are not possible, if only the decay distribution with respect to the angles β and γ are taken into account.

As mentioned before, it has been shown in [13] that in the case of a polarized τ one can use the direction of the τ spin-vector in the lab to define a further Euler angle α. This additional angular dependence allows in principle for the measurement of the two additional CP-violating differences

$$\Delta \tilde{W}_{SC} = 2\sqrt{Q^2} \left[x_1 \operatorname{Re}\left(F_1 F_4^*\right) + x_2 \operatorname{Re}\left(F_2 F_4^*\right) \right] \operatorname{Im}\left(-\eta + \chi_A \eta^*\right) ,$$

$$\Delta \tilde{W}_{SE} = 2\sqrt{Q^2} x_3 \left[\operatorname{Re}\left(F_1 F_4^*\right) - \operatorname{Re}\left(F_2 F_4^*\right) \right] \operatorname{Im}\left(-\eta + \chi_A \eta^*\right) .$$

and hence for CP violation tests originating from a charged Higgs in the three pion decay mode.

The authors in [35] studied the effects of T-odd triple correlations (as derived in [13]) in the decay modes $\tau \to K\pi\nu$ and $\tau \to KK\pi\nu$ using the model for the hadronic form factors as suggested in [12]. They found that CP violation effects in some extensions of the Standard Model could be as big as 0.1%. CP violating effects in the $\tau \to 3\pi\nu$ decay mode have also been discussed in [36].

ACKNOWLEDGEMENTS

We would like to thank J.H. Kuehn for collaboration on part of the work presented here.

The work of E. M. was supported in part by the U. S. Department of Energy under Grant No. DE-FG02-95ER40896. Further support was provided by the University of Wisconsin Research Committee, with funds granted by the Wisconsin Alumni Research Foundation. The work of M.F. has been supported in part by the National Science Foundation under Grant PHY-9218167.

---

[1] J. Kirkby, these proceedings.
[2] Y. S. Tsai, Phys. Rev. D **51**, 3172 (1995); see also these proceedings.
[3] C.A. Nelson et al., Phys. Rev. D **50**, 4544 (1994).
[4] Y.S. Tsai, Phys. Rev. D **4**, 2821 (1971);
    H.B. Thacker and J.J. Sakurai, Phys. Lett. B **36**, 103 (1971).
[5] R. Decker and M. Finkemeier, Phys. Lett. B **334**, 199 (1994);
    R. Decker and M. Finkemeier, Nucl. Phys. B **438**, 17 (1995).
[6] M. Davier, in: Proceedings of the Third Workshop on Tau Lepton Physics (Montreux, September 1994), Nucl. Phys. B (Proc. Suppl.) 40 (1995).
[7] B. Heltsley, in: Proceedings of the Third Workshop on Tau Lepton Physics (Montreux, September 1994), Nucl. Phys. B (Proc. Suppl.) 40 (1995).
[8] F. J. Gilman and S. H. Rhie, Phys. Rev. D **31**, 1066 (1984).
[9] J. H. Kühn and A. Santamaria, Z. Phys. C **48**, 445 (1990).
[10] M. Finkemeier and E. Mirkes, MAD/PH/882, LNF-95/015(P) (1995), to appear in Z. Phys. C.
[11] E. Braaten, R. J. Oakes and S. Tse, Int. J. Mod. Phys. A **14**, 2737 (1990).
[12] R. Decker, E. Mirkes, R. Sauer and Z. Was, Z. Phys. C **58** 445 (1993).
[13] J. H. Kühn and E. Mirkes, Z. Phys. C **56**, 661 (1992).
[14] R. Decker, M. Finkemeier and E. Mirkes, Phys. Rev. D **50** , 3197 (1994).
[15] N. Isgur, C. Morningstar and C. Reader, Phys. Rev. D **39** , 1357 (1989).
[16] J. H. Kühn and E. Mirkes, Phys. Lett. B **286**, 281 (1992).
[17] G. Kramer, W.F. Palmer and S. Pinsky, Phys. Rev. D **30**, 89 (1984);
    G. Kramer, W.F. Palmer, Z. Phys. C **25**, 195 (1984); ibid. C **39**, 423 (1988).
[18] A. Pich, Phys. Lett. B **196**, 561 (1987);
    F.J. Gilman, Phys. Rev. D **35** 3541 (1987).
[19] J. Wess and B. Zumino, Phys. Lett B **37**, 95 (1971);
    E. Witten, Nucl. Phys. B **223**, 422 (1983); ibid. **223**, 433 (1983).
[20] J.G. Smith, in: Proceedings of the Third Workshop on Tau Lepton Physics (Montreux, September 1994), Nucl. Phys. B (Proc. Suppl.) 40 (1995).
[21] Review of Particle Properties, L. Montanet et al., Phys. Rev. D **50** 1173 (1994).
[22] J.J. Gomez-Cadenas, M.C. Gonzales-Garcia and A. Pich, Phys. Rev. D **42**, 3093 (1990).
[23] ARGUS Collab., H. Albrecht et al., Phys. Lett. B 185 (1987) 223; B 260 (1991) 259.
[24] CLEO–Collab., M. Procario et al., Phys. Rev. Lett. 70 (1993) 207; Douglas F.

Cowen, CALT-68-1934.
- [25] P. Bourdon, in: Proceedings of the Third Workshop on Tau Lepton Physics (Montreux, September 1994), Nucl. Phys. B (Proc. Suppl.) 40 (1995) 203.
- [26] R. Decker, P. Heiliger, H.H. Jonsson, M. Finkemeier, TTP-94-13 (1994), to appear in Z. Phys. C.
- [27] R. Decker and E. Mirkes, Z. Phys. C **57**, 495 (1993).
- [28] S.I. Eidelman and V.N. Ivanchenko, Phys. Lett. B **257**, 437 (1991);
  ibid. in: Proceedings of the Third Workshop on Tau Lepton Physics (Montreux, September 1994), Nucl. Phys. B (Proc. Suppl.) 40 (1995) 203.
- [29] R. Decker and E. Mirkes, Phys. Rev. D **47** 4012 (1993).
- [30] M. Finkemeier and E. Mirkes, in preparation.
- [31] W. Bernreuther and O. Nachtmann, Phys. Rev. Lett. **63**, 2787 (1989); erratum: **64**, 1072 (1990);
  W. Bernreuther, G.W. Boltz, O. Nachtmann and P. Overmann, Z. Phys. C **52**, 567 (1991);
  C.A. Nelson, Phys. Lett. B **267**, 128 (1991);
  S. Goozovat and C.A. Nelson, Phys. Rev. D **44**, 2818 (1991).
- [32] Y. Grossman, Nucl. Phys. B **426**, 355 (1994); J. F. Donoghue and E. Golowich, Phys. Rev. D **37**, 2542 (1988);
  J. F. Donoghue, J. Hagelin and B.R. Holstein, Phys. Rev. D **25**, 195 (1982);
  J. F. Donoghue B.R. Holstein, Phys. Rev. D **32**, 1125 (1985);
  C. H. Albright, J. Smith and S,-H.H. Tye, Phys. Rev. D **21**, 711 (1980);
  G.C. Branco, Phys. Rev. Lett. **44**, 504 (1980).
- [33] A.J. Davis and X.-G. He, Phys. Rev. D **43**, 225 (1991);
  W. Buchmüller, R. Rückl and D. Wyler, Phys. Lett. B **191**, 44 (1987).
- [34] J.C. Pati and A. Salam, Phys. Rev. D **10**, 275 (1974);
  R.N. Mohapatra and J.C. Pati, Phys. Rev. D **11**, 566 (1975).
- [35] U. Kilian, J .G. Körner, K .Schilcher and Y. L. Wu, Z. Phys. C **62**, 413 (1994).
- [36] S. Y. Choi, K. Hagiwara and M. Tanabashi, KEK-TH-419 (1994).

ACCELERATOR DESIGNS

Tau/charm Factory Collider Design at BEPC

Y.Z. Wu, N. Huang, L.H. Jin, D. Wang

Institute of High Energy Physics, Chinese Academy of Sciences
P.O. Box 918-9, Beijing 100039
People's Republic of China

Abstract

Main goals and parameters of the design of the Beijing tau-charm factory (τcF) are presented. Three schemes, the standard, the monochromator and the crossing angle scheme, are implemented into one versatile lattice. In addition, a beam polarization scheme is discussed. Some systems of the factory are briefly described.

I. Introduction

Several international workshops on the τcF since 1989 concluded that the τcF is a unique facility to contribute significantly to several basic research areas of particle physics. Construction of a τcF in China bears many advantages due to the existing site of the BEPC complex [1]. Experience from the existing e^+e^- storage ring provides a firm foundation on which to base a new τcF design. The feasibility study of the Beijing τcF has been officially approved by the government of China. It will be completed within one and a half years [2]. The τcF storage rings are planed to be located in the east section of the site along the BEPC linac, BEL. Fig.1 shows a possible layout of the Beijing τcF.

II. Design Goals

Following is a list of the goals of our design effort:

1. Rings capable of operating over the energy range from 1.5 to 3.0 GeV.
2. Maximum peak luminosity $\geq 10^{33} cm^{-2} s^{-1}$ at an energy of 2.0 GeV.
3. Construction and commissioning of the machine in three phases:
   - Phase 1: conventional scheme,
   - Phase 2: crossing angle scheme & monochromator scheme,
   - Phase 3: longitudinal beam polarization scheme.

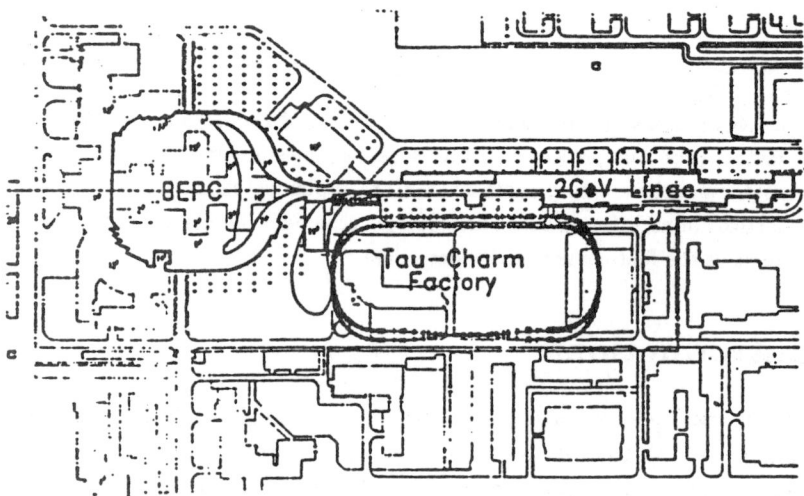

Figure 1. General layout of the τcF on the BEPC site

III. General Description

Assuming an optimum coupling, the luminosity can be expressed as,

$$\mathcal{L} = \frac{I\gamma\xi_y}{2er_e\beta_y^*}\left(1+\frac{\sigma_y^*}{\sigma_x^*}\right)$$

For flat beams, the expression reduces to

$$\mathcal{L} \propto \left(I\xi_y/\beta_y^*\right) E$$

The key ingredients to achieve high luminosity are common to other factory designs: high current, multibunch operation, and separate rings. A comparison of the performance between the τcF and BEPC is shown in Table 1.

	BEPC	τcF
Number of bunches K_b	1	32
Bunch spacing S_b (m)	240.4	11.48
Particles/bunch N	2.16×10^{11}	1.32×10^{11}
Bunch current I_b (mA)	40	17.16
Beam Current I (mA)	40	550
β-function β_y^* (cm)	5	1
Bunch length σ_z (cm)	4.5	< 1.0
B-B parameters ξ	0.04	0.04
Luminosity $\mathcal{L}$ (cm^{-2}s^{-1})	1.5×10^{31}	1.0×10^{33}

Table 1. Comparison between τcF and BEPC operating at 2.0 GeV

Compared with BEPC, the luminosity in the τ is increased by

a. A factor of 13.8 from the increased beam current,

b. A factor of 5 from the micro-β insertion.

These gain factors are based on known collider design principles and result in an overall improvement of the luminosity by a almost two orders of magnitude compared to BEPC. This factor is sufficient to achieve our goal of $\mathcal{L} = 10^{33} cm^{-2} s^{-1}$.

Recent machine studies show that the beam-beam parameter ξ has exceeded 0.045 in BEPC. Since BEPC operates in the same energy range as the τcF, we believe that a τcF design with ξ equal to 0.04 is quite conservative. Below 2.0 GeV the design luminosity is expected to decrease proportionally to E^2.

IV. The Lattice Design of the Storage Ring

The collider consists of two rings, one above the other with one single IP. Each ring is 61.4 m wide and 143.4 m long with a circumference of 367.5 m. There are 32 bunches in each ring for both the standard and the monochromator operation. A schematic layout of the storage rings is shown in Fig.2 and the layout of half a ring in Fig.3 [3].

The crossing angle scheme has been implemented into the conventional design by simply adding a pair of horizontal deflection dipoles. The parameters related to the crossing angle scheme are the following:

- Crossing angle θ: ± 3.0 mrad, which satisfies $\theta \sigma_l \leq \sigma_x^*/10$
- Bunch spacing: 3.83 m
- Number of bunches: 96
- Beam current: 1.65 A
- Luminosity: $(2 \sim 3) \times 10^{33} cm^{-2} s^{-1}$

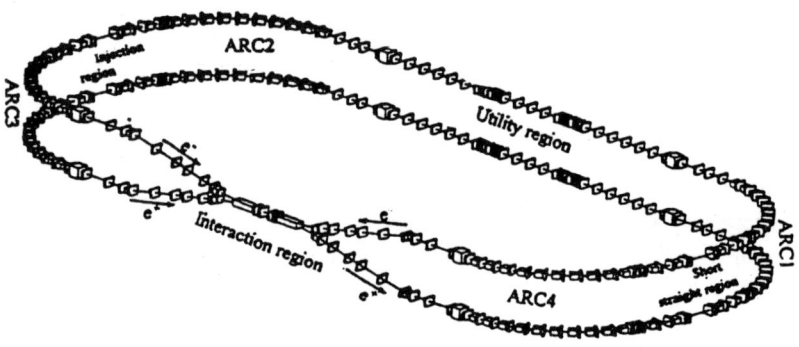

Figure 2. Schematic of the storage ring of the Beijing τcF.

The pair of horizontal deflection dipoles are placed symmetrically at the point where the phase advance from the IP is π, as shown in Fig.4. The orbit perturbation outside the pair of dipoles vanishes. The vertical beam separation is kept the same as in the case of the conventional scheme.

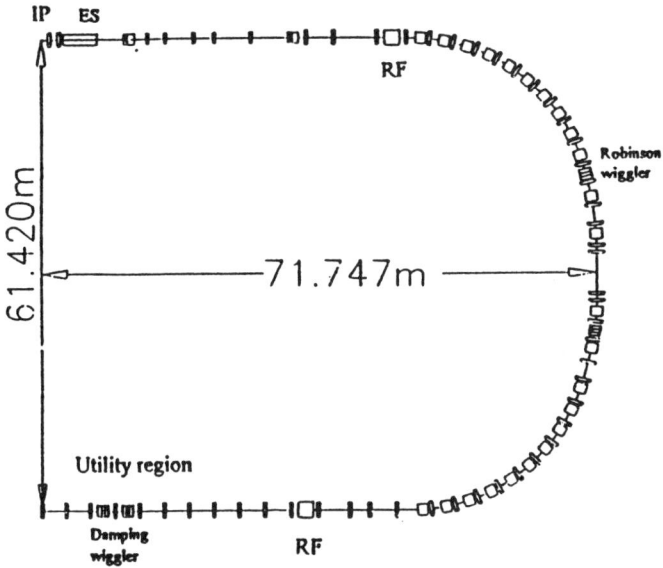

Figure 3. Layout of half the ring.

To separate the two beams horizontally by over $8 \sim 10\ \sigma_x$ at the parasitic crossing point, we take $\beta_x^*=0.8$ m, since β_x tends toward larger values at the IP. General parameters for the standard, monochromator and crossing angle schemes are listed in Table 2.

Interaction Region & Separation Scheme

As shown in Fig.4, a micro-β insertion, an e^+e^- orbit separation section and a β-function matching section are placed at each side of the IP. A superconducting quadrupole doublet is used to achieve a low β-function at the IP. After the collision at the IP the two beams are vertically separated by an electrostatic separator ES, two vertical septum magnets BV1 and BV2 and a vertical bend magnet BV3.

Scheme	Standard	Monochr.	Crossing Angle
Nominal energy E(GeV)	2.0	1.5	2.0
Ring circumference C(m)	367.5	367.5	367.5
Crossing angle at IP θ (mrad)	0.0	0.0	3.0
β-function at IP β_x^*/β_y^* (m)	0.2/0.01	0.01/0.15	0.8/0.01
Dispersion at IP D_y^* (m)	0.00	0.35	0.00
Momentum compaction α_p	0.022	0.008	0.022
Natural emittance ϵ_x (nm rad)	251	20	251
Emittance with wiggler		10 (J_x=2)	
Vertical emittance ϵ_{yc}	12	2	4.8
Energy spread σ_E	5.4×10^{-4}	8×10^{-4}	5.4×10^{-4}
Energy loss per turn U_0 (keV)	142.6	45.0	427.8
Damping time $\tau_x/\tau_y/\tau_e$ (ms)	34/34/17	41/80/80	34/34/17
RF frequency (MHz)	500	500	500
RF voltage (MV)	9.00	9.00	9.00
Numbers of bunches K_B	32	32	32×3
Bunch spacing S_B	11.48	11.48	11.48/3
Total current per beam (A)	0.55	0.215	1.65
Particles per bunch	1.32×10^{11}	5.14×10^{10}	1.32×10^{11}
Natural bunch length (cm)	1.0	0.78	1.0
Impedance $\mid Z/n \mid_{\parallel}$ (Ω)	0.32	>0.32	0.32
Beam-Beam effect ξ_x/ξ_y	.04/.04	.031/.015	.04/.04
Beam life time τ (hours)	4.8	1.5	
Transverse tune $Q_x/$	11.192/	13.18/	11.15/
Q_y	10.192	9.24	10.18
Synchrotron tune Q_s	0.098	0.068	0.098
Natural chromaticity $Q_x'/$	−26.6/	−35.9/	−20.0/
Q_y'	−32.0	−44.5	−35.09
Luminosity $\mathcal{L}$ (cm^{-2}s^{-1})	1×10^{33}	2.2×10^{32}	3×10^{33}
CM energy spread σ_W (MeV)	1.53	0.105	1.53

Table 2. Main Parameters of the τcF in Beijing.

Six quadrupoles are used to complete focusing and vertical dispersion matching between BV2 and BV3. In the monochromator scheme Q7 will be switched off and Q1, Q2, Q8 will change polarity. The two beams are separated by $5\sigma_x$ in the standard scheme and $22\sigma_y$ in the monochromator scheme at the parasitic crossing point which is 5.74 m ($S_B/2$) away form the IP.

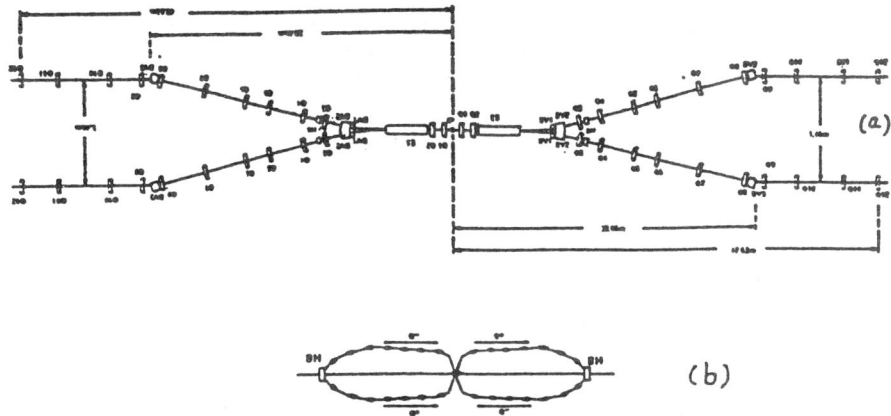

Figure 4. (a) IR and separation in the vertical plane, (b) the small collision crossing angle in the horizontal plane.

Fig.5 shows the lattice function of the IR and the separation region for different schemes. The following topics need to be studied in the future:

- Studies of beam-beam effects with different conditions at the IP.
- Parasitic beam-beam effects
- Background control
- Solenoid compensation

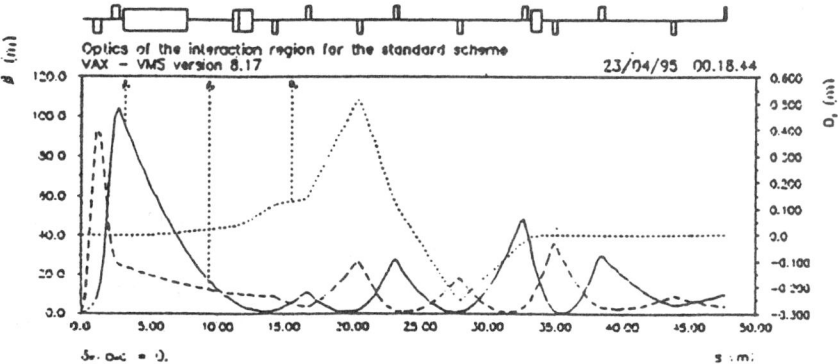

Figure 5a. Lattice function in IR and separation region for the standard scheme.

145

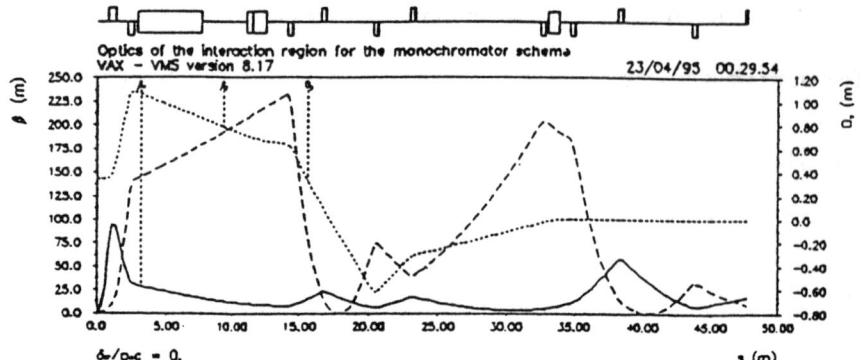

Figure 5b. Lattice function in IR and separation region for the monochromator scheme.

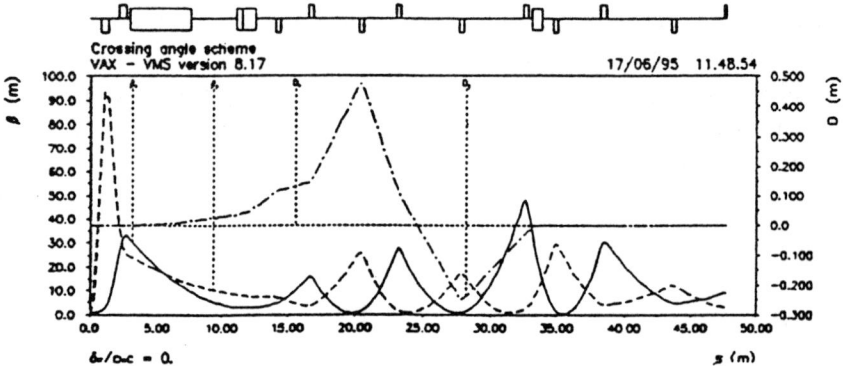

Figure 5c. Lattice function in IR and separation region for a crossing angle scheme

The Arc

Each arc consists of seven non-standard FODO cells, where independent power supplies of quadrupoles will be used to adjust the lattice functions. Experience with BEPC shows that the non-standard cells make the lattice very flexible. Each bending magnet deflects the beam by 7.5°. One bending magnet will be left out in the sixth cell and be replaced by a Robinson wiggler. The following strategies to control the emittance will be applied:

- Variation of the phase advance per cell,
- Adjustment of the maximum dispersion D_x,

- Use of Robinson wigglers and damping wigglers.

The emittance can be varied from 550 nm at 2.0 GeV to 10 nm at 1.55 GeV.

Utility Section

One long utility section will be located opposite to the IR and the separation region. RF cavities, damping wigglers as well as some diagnostic instruments will be installed there. The working point can be adjusted in this region.

V. Alternative Lattice Designs

We also study another versatile lattice which will allow for a very large range in emittances. The major feature of this design is the wide range of natural horizontal emittances which can vary from several nm.rad to several hundred nm.rad without the use of wigglers. The arc of each ring consists of 36 dipoles (32 cells and 4 dispersion suppressor cells). Fig.6 shows the cell structure in the arc of the τcF storage ring. It consists of 1 dipole and 4 quadrupoles. The lattice function inside the dipole is optimized to get low emittance to meet the requirement of the monochromator scheme. Fig.7 shows the tuned lattice function for the standard high luminosity scheme with high emittance. In this case, two quadrupoles are simply switched off.

VI. Longitudinal Polarized Beam Scheme

Longitudinally polarized e^+ and e^- are required in the energy range between 1.55 GeV (J/ψ resonance) and 2.0 GeV. The study of the polarization scheme shows that the beams should be transversely polarized before they are filled into the main ring. A compact polarization ring with high magnetic field is needed (for the e^+ at least). But since the optimum beam energy for this small ring is constrained to a certain value, it is difficult to get transversely polarized beams at 1.55 GeV and 2.0 GeV using the same storage ring. Recent physics considerations show that collisions of polarized e^- with unpolarized e^+ are sufficient to study CP violation in the lepton sector. Therefore, in an initial stage, we plan to run with polarized electrons only.

We started to study the spin rotation scheme in τcF main ring. The two major candidates for our design are a symmetric solenoid and a mini-rotator scheme. The symmetric solenoid scheme consists of two 90° solenoids and two 90° horizontal bends. A detailed design is available. Some modifications in the arc lattice are needed but the overall performance will not be affected. The main difficulty of the mini-rotator scheme lies in the large bending angle in both the horizontal and the vertical plane. This produces a large dispersion and the large orbit excursions will complicate the design of the dispersion suppression and of the ring layout.

VII. RF system

For minimizing both the impedance contributions from the RF cavities to the machine and the RF power consumption, we have chosen to use 400~ 500 MHz superconducting RF cavities. The maximum RF voltage, 9 MV, is dominated by the bunch length requirements in the τcF storage ring. The total energy loss due to SR and HOMs at an energy of 2.0 GeV is about 120 kW for the standard scheme. We have also investigated existing RF cavities in other laboratories. After calculation of the growth time of the multibunch coupled instability to different types of cavities, we recommend to use the CESR-B cavities for the Beijing τcF.

VIII. Vacuum system

The τcF storage ring requires an operating pressure better than 10^{-9} Torr. To be able to handle the high thermal loads and to minimize the photodesorbed gas load, the vacuum system for the τcF is based on extruded aluminum chambers with conventional Distributed Ion Pumps (DIP) and with discrete copper masks.

The possibility of augmenting the DIPs with non-evaporable getter (NEG) pumps is being investigated. The welding and stretch-forming to fabricate the chambers are under study. The chamber aperture is kept constant at 70 × 70 mm^2 along the whole length. Great attention will be paid to obtain a smooth chamber surface.

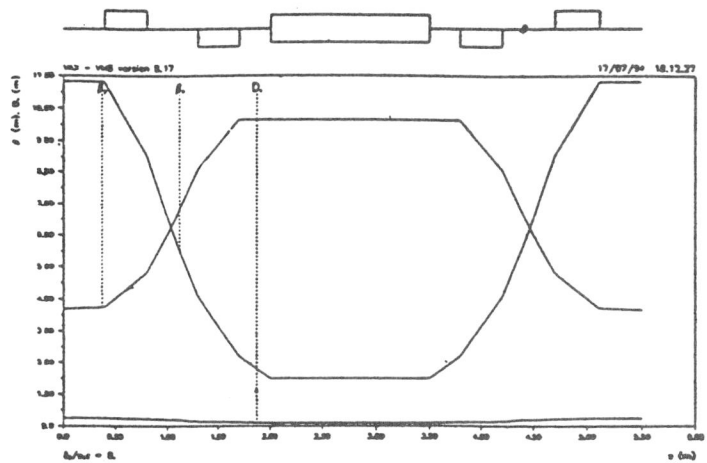

Figure 6. Cell structure for low emittance (4nm.rad at 1.55 GeV)

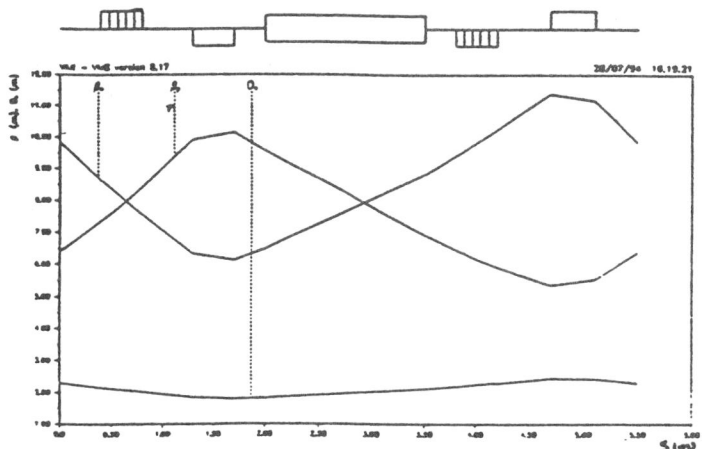

Figure 7. Cell structure for high emittance (320nm.rad at 2.0 GeV)

IX. Instability Control

Single bunch and multibunch effects will ultimately limit the performance of the rcF. Close attention should be paid to these issues in the design.

Multibunch Instabilities

The coupled bunch instabilities are potentially a serious problem in the τcF. To overcome the instabilities, several solutions will be considered:

- Designing of a cavity with small HOM form impedance, R_s/Q and R_t/Q,

- Adjusting of the HOM frequencies to the area with low instability growth rate,

- Using HOM dampers to reduce Q_r to less than 100,

- Even if the HOMs in RF cavities are heavily damped, there are some unstable modes which have a rise-time faster than the natural damping time. Therefore, a powerful fast feedback system is still necessary.

Bunch Lengthening

In order to match the low β function at the IP, the bunch length must be shorter than 1 cm. The Keil-Schnell-Boussard criterion for the absence of turbulent bunch-lengthening indicates that the longitudinal broad band impedance must be less than 0.32 Ω. Careful designs of a smooth chamber, kickers, separator, and bellows with low impedances are required. Experimental studies in BEPC on the bunch length vs longitudinal broad band impedance are under way. Removing two redundant kickers and shielding the bellows with sliding contact would reduce the ring impedance, and thus decrease the bunch lengthening effect.

X. Injector Scheme

The BEPC linac system will be upgraded to a full energy injector which will provide the ability to inject the necessary positron current inabout 5 minutes [4]. The following modifications to the existing injector system are planed:

- Replacing the existing (30MW) with a more powerful klystron (60MW) resulting in a higher positron energy above 2.0 GeV.

- Replacing the existing (4A) by a new electron gun (8A).

- Moving the positron target to the next station. The incident electron energy will be increased to 500 MeV.

- Increasing the repetition frequency from 12.5Hz to 50Hz.
- Use of the subharmonic buncher to increase the positron yield.

This improved linac will reduce the "full-fill" injection time to about 5 minutes. Another option using the 8 backward wave traveling accelerating sections to increase the incident electron energy on the positron target is also under investigation [5].

XI. Conclusions

The conventional design of the τcF is based on existing technology and on demonstrated successful collider design principles, although there are some challenging technological problems to be resolved. It is possible to achieve the luminosity goal of $10^{33} cm^{-2} s^{-1}$. Some new schemes, such as monochromatic collision and polarization beam collision, will be carefully studied and implemented in successive phases, thus limiting the risks. Some challenging aspects of the design of the τcF need to be studied in detail by a vigorous R&D program.

Acknowledgments.

We would like to thank J.Wang, Y.X Luo, Y.Z Lin, and the colleagues from the theory group for useful discussions. We especially thank Maury Tigner for very helpful discussions.

References

[1] Memorandum on the tau-charm Factory in the era of B-factories and CESR, SLAC Stanford, Aug. 15-16, 1994.

[2] S.H.Wang, " BEPC Status and Plans" in Proceedings of 1995 Particle Accelerator Conference and International Conference on High-Energy Accelerators.

[3] L.Jin et al., "A Preliminary Lattice Design of a Tau-Charm Factory Storage Ring in Beijing" in Proceedings of 1995 Particle Accelerator Conference and International Conference on High-Energy Accelerators.

[4] J.Wang, Y.X.Luo, private communication.

[5] Y.Z.Lin, private communication.

JINR Tau/charm Factory: Status and Perspectives

E. Perelstein

*Joint Institute for Nuclear Research
141980 Dubna, Russia*

Abstract. A review of the Tau/charm Factory in JINR (Dubna) is presented. The structure, the working regime, the parameters, and the open issues of the injector system are reviewed. A versatile magnet lattice is used in the Tau/charm collider which can realize a conventional flat beam scheme and a monochromatization or crossing angle scheme. The list of parameters of the Tau/charm Factory is given. Technical solutions for magnet elements and their power supplies used in the booster and in the collider, for RF power supplies of the collider, and for the vacuum system of the periodic cells are presented.

1. Plans for the JINR Accelerator Complex

The project for a new electron-positron storage complex is currently being investigated at JINR. It will provide significant contributions to the Institute's traditional research activities: elementary particle physics, nuclear physics, condensed matter physics, and applied investigations.

The project discussed involves: a high resolution neutron source (IREN), a Tau/charm Factory (TcF), and an 8-10 GeV positron (electron) storage ring (NK-10) for synchrotron radiation studies. The layout of the proposed JINR storage ring complex is shown in Fig.1.

The construction of the complex is expected to involve three stages. In a the first stage, the high resolution neutron source IREN will be built. The decision for the construction of IREN was taken at the 76th session of the JINR Scientific Council in June, 1994. The second stage will be the construction of the Tau/charm Factory. A successful discussion of the TcF proposal took place at the 75th session of the JINR Scientific Council in 1994. For the successful realization of the TcF project a wide international collaboration including a broad spectrum of foreign institutions is expected to be formed.

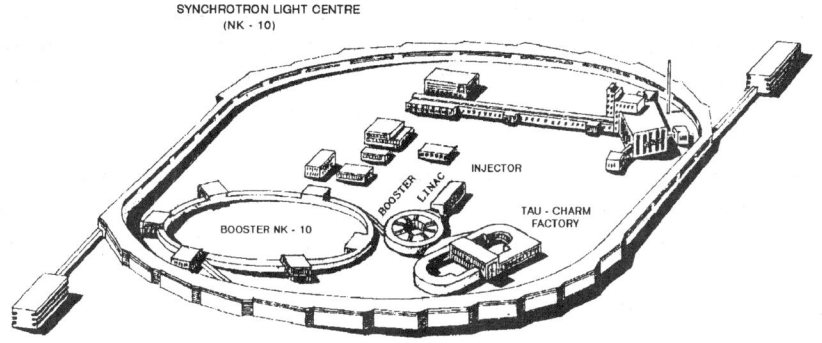

Figure 1. The layout of the JINR storage ring complex.

We believe that Dubna has many advantages for the location of a TcF or a similar facility. It would create great opportunities to advance High Energy Physics in Eastern European Countries where JINR is a well-known scientific centre, to develop new, high technologies, to raise the education level of the younger generations in the region, and to create an international collaboration in this field of physics. On the basis of the existing Training Centre at JINR it would be possible to establish an International University at Dubna. Students of particle physics would then have the opportunity to gain experience at a modern facility such as the TcF.

Presently the JINR staff counts about 500 High Energy physicists. The development of an international collaboration around the JINR TcF will be based on: a) open access to the TcF for a large number of visiting scientists and students, b) progress with JINR's computer communications, c) the geographical location of Dubna (near Moscow and near the international airport Sheremetyevo), d) the available infrastructure of JINR as an international scientific centre.

2. Tau/charm Factory Physics Programme

The broad spectrum of physics topics and the requirements on the machine and detector of a TcF have been discussed in many workshops [1-5]. The scientific research programme for the TcF includes tau-lepton and tau-neutrino physics, charmonium spectroscopy, CP violation searches, and charmed meson and baryon physics.

The core of the experimental programme for TcF should be: the study of the properties of the second-generation quarks and the third-generation leptons through investigations of tau-lepton physics, charmed meson and baryon physics, and charmonium physics. This energy region is currently being explored by the BEPC facility in Beijing (China). However, the experimental accuracy is largely dominated by the lack of statistics. Therefore, the major reason for building a TcF is to reach higher luminosities and a higher level of precision. When operating at the design luminosity, the TcF will produce about 10^7 τ and D meson pairs per year leading to precise measurements with a statistical accuracy $\leq$ 1%.

The TcF allows to study the properties of the tau-lepton and charmed particles near their production threshold and, thus, to obtain high-quality data with uniquely low background contaminations. The TcF is an ideal instrument to investigate the most interesting features of the Standard Model and of the hadrons as composite systems [6].

The physics goals of a TcF impose difficult constraints on the machine [3]. They are summarized as follows:

1. The peak luminosity around the tau pair production threshold, $E_{beam} \simeq 2$ GeV, must be at least 10^{33} cm^{-2}s^{-1},

2. The TcF must provide a high average luminosity, of the order of half the peak luminosity,

3. High luminosity must be provided in a wide energy range from $E_{beam} \simeq$1.5 GeV (J/ψ physics) up to $\simeq$2.85 GeV (charmed baryon physics),

4. At specific energies (i.e. at the J/ψ resonance and the tau pair production threshold) a centre-of-mass energy resolution of 100 keV or less is desirable. This requires beam monochromatization,

5. The availability of polarized beams together with monochromatization is of special interest.

3. Present Status of the JINR TcF Project

Recently, two different TcF projects with a centre-of-mass energy range of 3 to 5.7 GeV were investigated. The first one aimed at the CERN-ISR site, taking advantage of the existing, powerful injector, and the second one was directed towards Dubna. Apart from the site, these projects have many common features. This is due to the well defined constraints and the strong collaboration between the designers of the two projects.

We foresee three possible phases for the TcF project [7]. The first phase of operation of the JINR TcF will be conventional. The corresponding TcF main ring is presented in Fig. 2. The next phase will implement either a monochromatization [8] or a crossing angle scheme [7, 9] to provide a luminosity of $3 \div 5 \cdot 10^{33}$ cm^{-2} s^{-1}.

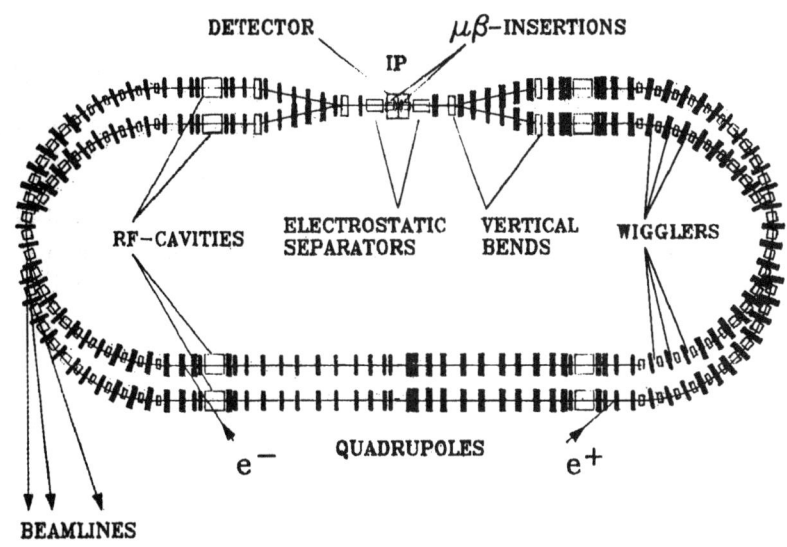

Figure 2. The scheme of the main ring of the Tau/charm Factory.

The injection complex (see Fig. 3) consists of a preinjector and a fast booster synchrotron, where electrons and positrons are accelerated up to the main ring energy [10]. The preinjector energy, 500 MeV, will be suitable for initial acceleration of the particles for the planned Synchrotron Radiation (SR) Source NK-10.

Assuming that the average luminosity reaches the level of 80 % of the peak luminosity, the number of positrons in the TcF must be $\geq 4.8 \cdot 10^{12}$ to achieve the required luminosity in the conventional scheme. Taking in account that the transfer efficiency from the injection complex through the booster into the TcF is about 10%, and the filling time is ideally chosen to be 15 min, a production rate of $5.4 \cdot 10^{10} e^+/s$ by the injection complex is required. The positron production efficiency limited by the positron energy spread and the emittance acceptable by the booster is estimated at 0.3 %. Therefore, the electron flux impinging on the conversion target must be about $2 \cdot 10^{13} e^-/s$. Taking into account a bunching efficiency of the order of 50% the electron flux provided by the gun must be $\geq 3.7 \cdot 10^{13} e^-/s$.

The booster synchrotron will be used for the acceleration of 500 MeV electrons and positrons, injected from the preinjector, up to the full energy of the TcF. Its circumference is of 189 m and allows to inject 15 bunches per single turn into the main ring. With a repetition rate of 25 Hz and a 2 turn injection the booster will provide a 0.6 A positron current which will be stored in the TcF to maintain a peak luminosity of 10^{33} cm^{-2}s^{-1}.

The magnetic structure of the booster consists of 6 superperiods, each containing 6 FODO-type cells.

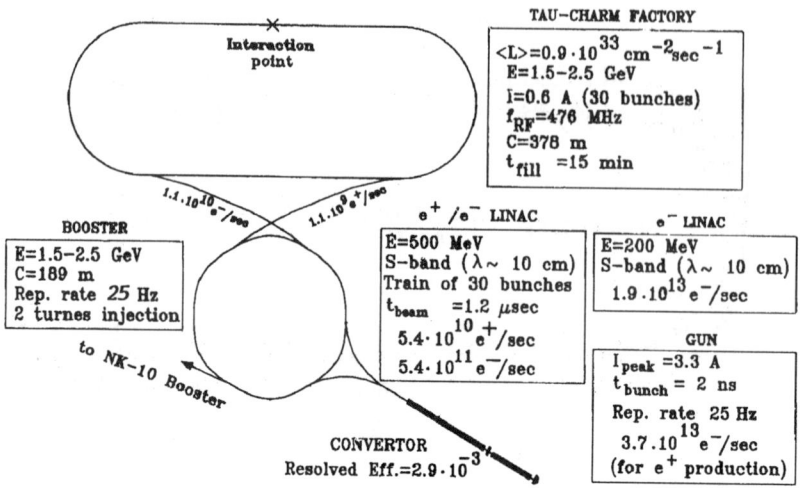

Figure 3. The layout of the Tau/charm Factory.

The hexagonal shape of the booster is determined by the location of the injection channels in the configuration chosen for the complex. Two long straight sections contain the injection devices, three additional ones are used for extraction into the injection channels of the TcF and the NK-10 booster. The sixth section houses the RF station.

Each superperiod consists of 2 standard FODO cells and 4 cells with the suppressed dispersion. To avoid a time varying sextupole component, created by the rising magnetic field, we intend to use ceramic vacuum chambers in the dipole magnets.

The conventional scheme is considered the basic scheme for the JINR TcF. The versatile design of the collider provides the possibility to implement monochromatization for the experiments requiring a small energy resolution [8].

The horizontal crossing angle option with minimum modifications to the conventional scheme storage ring is discussed in [9]. Only the interaction and separation regions have to be modified, while the arcs and the long straight section opposite to the interaction point are kept untouched. The layout of the interaction region and its horizontal and vertical views are shown in Figs. 4 and 5, respectively. Among the various horizontal crossing angle solutions, the one which uses the first micro-beta insertion quadrupole Q1 in common for the electron and positron beams while separating the other magnetic elements, gives a maximum luminosity with a minimum total current [9]. Calculations show that it is possible and advantageous to use an additional compact permanent quadrupole QPM located just in front of the quadrupole Q1 to improve the vertical focussing and to amplify the horizontal beam separation. Like the quadrupole Q1, the quadrupole QPM is common to both beams. The insertion optics has been chosen similar to the one presented in the Cornell B-Factory proposal. The long drift space after quadrupole Q1 provides enough horizontal beam deviation to have two vacuum chambers starting at the horizontally focussing quadrupole Q2.

The value of the crossing angle, $\phi = \pm 12$ mrad, has been chosen as a compromise between a fast beam separation necessary to separate the vacuum chambers in quadrupole Q2 and a reasonable orbit excursion in quadrupole Q1. After Q1, the deflecting angles become ± 37 mrad and the horizontal distance between the beam axes at the entrance of Q2 is 118 mm. The list of TcF parameters for the three options is given in Table 1.

The main features of the design prepared by JINR (Dubna), SRIEA (St. Petersburg), and RIPR (St. Petersburg) on the basis of a versatile lattice were discussed in [8].

The aluminum vacuum chamber of the TcF is designed such that the SR passes the next straight section and is absorbed at the end of the bending magnet. After chemical cleaning and heating of the vacuum chamber the outgassing rate of aluminum is much less than stimulated desorption. Using the combined pumps one gets the pressure to about $2 \cdot 10^{-7}$ Pa at the absorber location, which corresponds to a beam lifetime of 30 h.

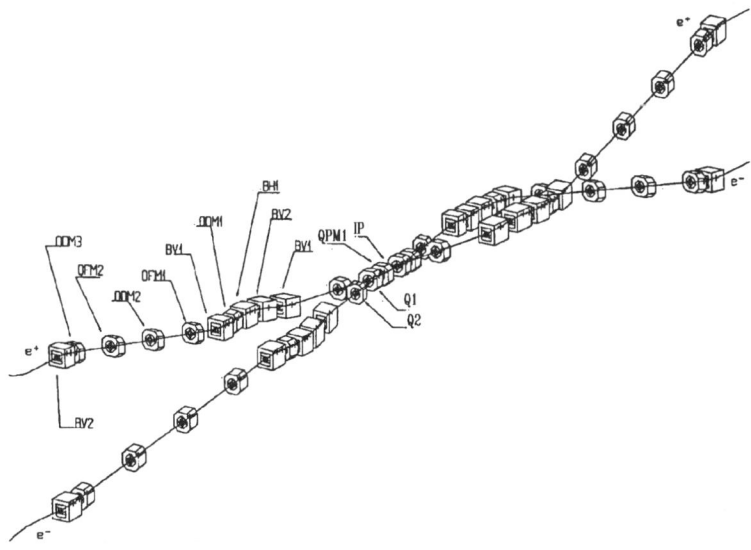

Figure 4. The interaction region layout in the crossing angle scheme.

TABLE 1. List of Parameters of the Tau/charm Factory

		Standard scheme	Monochr. scheme	Cros. angle scheme
Beam energy, GeV	E	2.0	2.0	2.0
Luminosity, cm^{-2}s^{-1}	L	$1.0 \cdot 10^{33}$	$0.9 \cdot 10^{33}$	$3.5 \cdot 10^{33}$
C.M. energy resolution, MeV	σ_w	1.9	0.14	1.7
Circumference, m	C	377.8	377.8	377.8
Natural emittance, nm	ε_0	426	17.0	299
Damping partition numbers	$J_x/J_y/J_\varepsilon$	0.6/1/2.4	2/1/1	0.6/1/2.4
Bending radius in arc, m	ρ	10.5	10.5	10.5
Damping times, msec	$\tau_x/\tau_y/\tau_s$	37/22/9	18/35/34	41/25/11
Momentum compaction	α	$1.58 \cdot 10^{-2}$	$8.02 \cdot 10^{-3}$	$1.59 \cdot 10^{-2}$
Energy spread	σ_E/E	$6.66 \cdot 10^{-4}$	$7.32 \cdot 10^{-4}$	$5.89 \cdot 10^{-4}$
Total current, A	I	0.566	0.479	2.0
Number of particles per bunch	N_b	$1.49 \cdot 10^{11}$	$1.26 \cdot 10^{11}$	$1.05 \cdot 10^{11}$
Number of bunches	k_b	30	30	150
RF voltage, MV	V	8	5	7
RF frequency, MHz	f_{RF}	476	476	476
Harmonic number	q	600	600	600
Energy loss per turn, kV	U_0	226	143	199
Bunch length, mm	σ_s	8.15	8.06	7.72
Bunch spacing, m	S_b	12.6	12.6	2.52
Required long. impedance, Ohm	$|Z_n/n|$	0.25	0.18	0.27
Beta functions at I.P., m	β_x^*/β_y^*	0.20/0.01	0.01/0.15	0.50/0.01
Vertical dispersion at I.P., m	D_y^*	0.	0.36	0.
Beam-beam parameters	ξ_x/ξ_y	0.04/0.04	0.04/0.03	0.04/0.04

An additional pump is used for the pumping of the remaining part of the vacuum volume and provides a pressure at the level of $2 \cdot 10^{-8}$ Pa.

The resistive part of the vacuum chamber contributes 70 mΩ to the broadband longitudinal impedance. The contribution from the interaction region, slots, absorbers, and RF-cavities are about 25 mΩ.

We plan to use 500 MHz superconducting RF cavities. At a beam energy of 2 GeV the total energy loss due to SR and Higher Order Mode losses is of the order of 300 kW. The maximum RF voltage is 16 MV for one ring. The RF power supply scheme for the TcF is based on the principle of separate supplies for each cavity.

The main remaining issues are the choice of an adequate final stage amplifier and the feeder line design. Klystrons developed at "SVETLANA" (St.Petersburg) satisfy the TcF requirements and have the following parameters: output power - 80 kW, frequency - 500 MHz, efficiency - 0.58, amplification - 45 dB, collector voltage - 16 kV, collector current - 8.6 A. The RF power supplies consist of 4 independent feeder lines with a total output power of 320 kW.

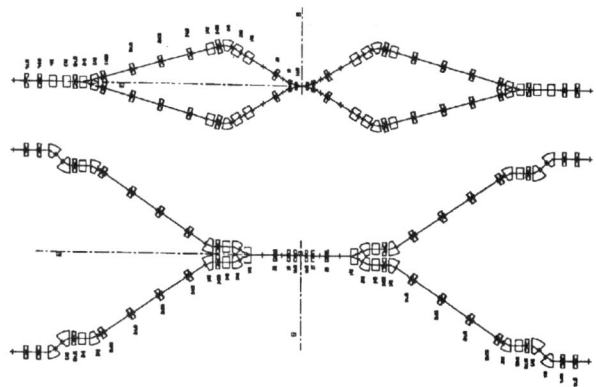

Figure 5. Horizontal and vertical view of the interaction region in the crossing angle scheme.

REFERENCES

1. Beers L. V. (ed.), *Proceedings of the Tau-Charm Factory Workshop*, SLAC-Report-343, 1989.
2. Kirkby J. and Quesada J. M. (ed.), *Proceedings of the Meeting on the Tau-Charm Factory Detector and Machine*, University of Seville, Andalucia, Spain, 29 April-2 May 1991.
3. Kirkby J. and Kirkby R. (ed.), *Proceedings of the Third Workshop on the Tau-Charm Factory*, 1-6 June 1993, Marbella, Spain, Editions Frontieres, 1994.
4. *Proceedings of the Workshop on JINR c-tau Factory*, 29-31 May 1991, JINR, Dubna, D1, 9, 13-92-98.
5. *Proceedings of the II Workshop on JINR c-tau Factory*, 27-29 April 1993, JINR, Dubna, D1, 9, 13-93-459.
6. Kuraev E., "Physical program of C-tau Factories", in *Proceedings of the 2nd Workshop on JINR Tau-Charm Factory*. D1,9,13-93-459, Dubna, 1993.
7. Beloshitsky P., Le Duff J., Mouton B., Perelstein E., "Modern view on Tau-Charm Factory Design Principles", Preprint LAL RT/94-05, Orsay, 1994.
8. Perelstein E. A. et al.,"Further study of JINR Tau-Charm Factory", in *Proceedings of the IEEE Conference on Particle Accelerators*, Washington, 1993, p. 2042.
9. Beloshitsky P. F.and Perelstein E. A., "Horizontal Crab-Crossing Scheme for Tau-Charm Factory", in *Proceedings of the EPAC 94*, Singapore, New Jersey, Hong Kong, World Scientific, 1994, p.470.
10. Perelstein E. A. e.a., "JINR Tau-Charm Factory Study", in *Proceedings of the XV International Conference on High Energy Accelerators*, Hamburg,1992)

Argonne Tau-charm Factory Collider Design Study

L. C. Teng, E. A. Crosbie, J. Norem, J. Repond

Argonne National Laboratory
9700 S. Cass Avenue
Argonne, Illinois 60439

Abstract. The design approach and design principles for a Tau-charm Factory at Argonne were studied. These studies led to a set of preliminary parameters and tentative component features as presented in this paper.

INTRODUCTION

Since 1993 members of Argonne's High Energy Physics (HEP) division have been considering the possibility of constructing a Tau-charm Factory (TcF) on site. Delegates from HEP participated in both the 1989 and the 1994 workshops on the TcF at SLAC. An evaluation of the physics case (1) for the facility was performed by an *ad hoc* committee formed in October 1993 within the HEP Division. The committee recommended a serious consideration of a TcF for the future plans of the division.

In addition, a small number of interested accelerator physicists from Argonne's HEP and Advanced Photon Source (APS) divisions have devoted part-time effort to the study of the design of a TcF collider. First proposed in 1987 (2), the TcF was the subject of four international workshops (3). Given the large amount of detailed studies already performed by other groups (4) and the rather demanding specification on the performance, the design options for the TcF collider are tightly confined. No great departures from the "standard" design seem worth consideration. A re-examination remains to be done with a possible redirection of the style and the principle adopted for the design, and potentially different engineering choices. For the latter we have leaned heavily on our up-to-date experiences derived from the APS project.

We adopted the following starting principles in our design study:

1. The phased approach—This is based both on economic and operational considerations. The complexity and precision requirements of the machine are such that operational experiences at one stage will greatly enhance the probability of success in the construction and operation of the succeeding, more sophisticated phases.

2. Simplicity of the design—The design should be as simple as possible while satisfying the following demanding specifications: a) the total center-of-mass energy is adjustable from 3 to 6 GeV, b) the luminosity is greater than 10^{33}/cm²/s over the entire energy range, c) the energy spread at 3.1 GeV is within the width of the J/ψ, and d) collisions of longitudinally polarized (pure helicity state) beams are possible.

First we shall describe briefly our concept of the different phases of construction and operation. Then we shall describe the design of the more difficult and unconventional component systems of the collider.

PHASE I: ZERO-DISPERSION INTERACTION

As the initial phase we shall consider a single-ring collider which satisfies the energy and luminosity requirements a) and b) but not necessarily the others. The adjusted gross beam parameters are

$$\text{Emittance} = \varepsilon \cong 560 \text{ nm-rad},$$
$$\text{Energy spread} = \sigma_E/E < 4 \times 10^{-4}.$$

To equalize the horizontal (x) and the vertical (y) beam-beam limits, we set at the Interaction Point (IP):

$\beta_x^* = 0.06$ m, $\qquad \beta_y^* = 0.01$ m
and $\quad \varepsilon_x = 480$ nm-rad, $\qquad \varepsilon_y = 80$ nm-rad $\quad$ (coupling = 1/6).

This results in beam widths

$\sigma_x^* = 0.17$ mm, $\qquad\qquad \sigma_y^* = 0.028$ mm.

For head-on collisions of beam bunches with 1.5×10^{11} particles/bunch the beam-beam limits at a beam energy of E = 1.55 GeV are then $\xi_x = \xi_y \cong 0.04$. With a bunch spacing of 10 m (current I = 0.72 A) dictated by the minimum distance needed to separate the beams, we get the luminosity 1.14×10^{33}/cm²/s, satisfying the specification b).

The ring is composed of two Arcs and two long Straight Sections. The Arcs are composed of FODO cells with horizontal dispersion suppressors at both ends. Away from the single IP situated at the midpoint of the Interaction Straight the beams are separated by electrostatic (ES) separators using the "pretzel" geometry. The maximum beam width outside of the Interaction Straight is $\sigma \cong 3$ mm. Thus, a ring aperture diameter of $\cong 80$ mm is adequate. More details of this arrangement will be discussed later.

PHASE II: HIGH-ENERGY RESOLUTION

The conventional scheme to reduce the center-of-mass energy spread is to impart to the colliding beams large equal and opposite vertical dispersions at the IP. This necessitates storing the beams in separate rings (see Fig. 1). The beams are brought together vertically by septum magnets and ES separators (combiners) to collide head-on. The orbit functions at the IP are adjusted to

$$\beta_x^* = 0.01 \text{ m}, \quad \beta_y^* = 0.076 \text{ m},$$
$$D_x^* = 0, \quad D_y^* = \pm 0.4 \text{ m}.$$

With beam emittances $\varepsilon_x = 68.32$ nm-rad, $\varepsilon_y = 1.63$ nm-rad and a beam intensity of 1.6×10^{11} particles/bunch (0.76 A at 10-m bunch spacing), the luminosity reaches 1.16×10^{33}/cm²/s at the beam-beam limit $\xi_x = \xi_y = 0.04$. The phase I ring can be retained to be used for one beam, and an identical second ring is built for the other beam. The emittance change between the two phases can be accomplished simply by changing the partition number J_x using Robinson wigglers (see below).

With finite dispersion at the IP the beam-beam force excites coupled synchrobetatron oscillations. Indeed, numerical simulations show broadenings of the betatron resonances due to synchrotron satellites. The effect on the beam-beam limit, however, is not clear (5). Even the zero-dispersion, hence no synchrobetatron coupling, beam-beam limit of $\xi \cong 0.04$ is derived only from operational experience and not from theory or simulation. We will pursue the study of this effect in the future, but for the present study we will keep the vertical beam-beam limit at $\xi_y = 0.04$. If compelling arguments are made to change the beam-beam limit we will readjust the ring parameters accordingly.

We designed the Arcs to yield an emittance at $J_x = 1$ of $\varepsilon \cong 121$ nm-rad. For phase I, four Robinson wigglers with positive gradients are placed in the four straight FODO cells at the ends of the two Arcs forming the dispersion suppressors. They reduce J_x to 0.216, thereby raising the emittance to 560 nm-rad. For phase II, each rectangular Robinson wiggler magnet is rotated 180° about a central vertical axis to produce a negative gradient which leads to $J_x = 1.73$ and a corresponding emittance of 70 nm-rad. We have made rough designs of the Robinson wiggler magnet to ensure that it can be manufactured easily and operated reliably.

After phase II is implemented, there will be no need to go back to the zero-dispersion operation of phase I. The excellent energy spread and the vertical spreading of the event vertices are presumably advantageous or acceptable for all experiments at all energies. Indeed, one may consider skipping phase I altogether. In this case, one can reduce the ring aperture somewhat, since only a single beam will ever be stored in one ring. The use of Robinson wigglers and more detailed parameters for phases I and II are given in Table 1.

TABLE 1: Beam and Performance Parameters for Phases I and II

	Phase I (Zero Dispersion)	Phase II (High Resolution)
Number of rings	1	2
Dispersion at IP	$D_x = D_x' = 0$	$D_x = D_x' = D_y' = 0$
	$D_y = D_y' = 0$	$D_y = \pm 0.4$ m
Without Robinson Wiggler		
Total emittance	$\varepsilon = 121$ nm-rad	$\varepsilon = 121$ nm-rad
Energy spread	$\sigma_E/E = 0.4 \times 10^{-3}$	$\sigma_E/E = 0.4 \times 10^{-3}$
With Robinson Wiggler		
Wiggler bend radius	$\rho = 10$ m	$\rho = 10$ m
Wiggler gradient	$B'/B = 4.30$ m^{-1}	$B'/B = -4.30$ m^{-1}
Total emittance	$\varepsilon = 560$ nm-rad	$\varepsilon = 70$ nm-rad
x and y emittances		
y/x Coupling	C = 16.7%	C = 1.6%
Horizontal ε	$\varepsilon_x = 480$ nm-rad	$\varepsilon_x = 68.32$ nm-rad
Vertical ε	$\varepsilon_y = 80$ nm-rad	$\varepsilon_y = 1.63$ nm-rad
Energy spread in beam	$\sigma_E/E = 0.33 \times 10^{-3}$	$\sigma_E/E = 0.49 \times 10^{-3}$
Amplitude function at IP	$\beta_x = 0.06$ m	$\beta_x = 0.01$ m
	$\beta_y = 0.01$ m	$\beta_y = 0.076$ m
Center-of-mass energy spread at 2E = 3.1 GeV (for J/ψ)	$2\sigma_{E0} = 1.03$ MeV	$2\sigma_{E0} = 0.086$ MeV
No. of particles per bunch	$N = 1.5 \times 10^{11}$	$N = 1.6 \times 10^{11}$
Beam-beam parameter	$\xi_x \cong \xi_y \cong 0.04$	$\xi_x \cong \xi_y \cong 0.04$
Bunch spacing	10 m	10 m
Beam current	0.72 A	0.76 A
Luminosity (head-on collision)	1.14×10^{33}/cm^2/s	1.16×10^{33}/cm^2/s

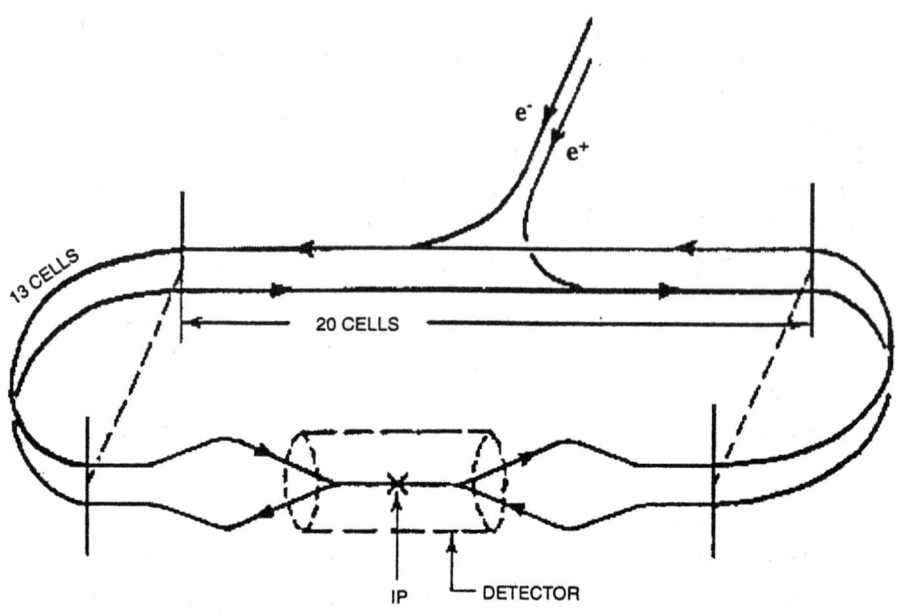

FIGURE 1. Overall geometry of the two collider rings for Phase II. The extra vertical bends in the Interaction Straight are introduced to provide flexibility.

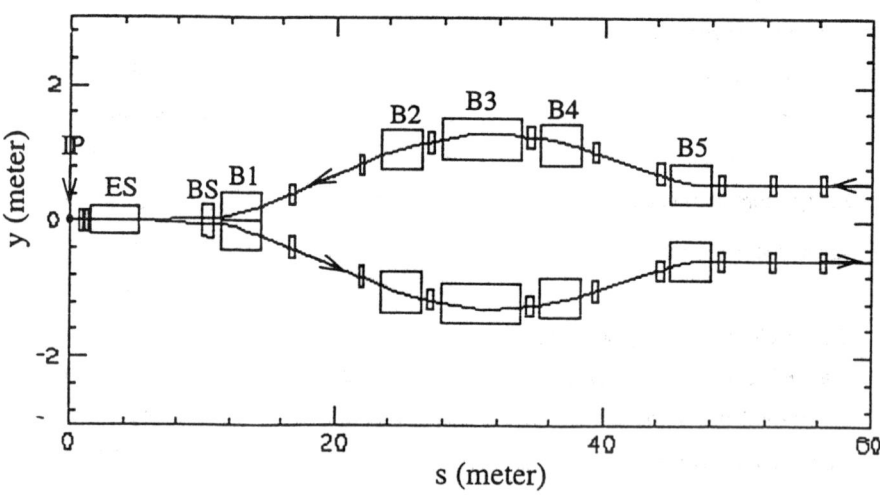

FIGURE 2. The vertical geometry of the orbits of the colliding beams in Phase II. The orbit-separating elements are: ES = electrostatic separator; BS = septum dipole; B1, B2, B3, B4, B5 = separate dipoles on separated beams (shown as singles).

Straight-section Matching

The requirements of orbit-function matching in the straight sections are the same for phases I and II. In the Utility Straight the matching depends on the detailed geometrical arrangement of the injection lines and elements, the rf cavities, the beam scrapers and aborts, etc. Since detailed geometrical arrangements cannot be finalized at this time, and since the matching is relatively straightforward, no effort has been spent on designing this section.

In the Interaction Straight the matching consists mainly of producing the orbit separation of the two beams and matching the orbit functions at the end of the Arc to those desired at the IP. We have performed many matching exercises with various assumed matching conditions just to assure that the assumed geometry and transport elements provide adequate flexibility to take care of all requirements. To form the orbit separation, moving away from the IP we first have a strong superconducting quadrupole doublet used commonly by both beams to produce the mini-β values at the IP. For phase I, the beams are then separated into the "pretzel" geometry by two sets of ES separators, one vertical and one horizontal, spaced by ~ 90° betatron phase advance. For a bunch spacing of 10 m, the next bunch crossing point is 5 m away from the IP. At that point the beams are already separated by more than 16 mm so that the electromagnetic interaction between them is tolerable. The mini-β quadrupole doublet and the "pretzel" forming ES separators can all be accommodated in a straight section space of ~ 25 m. However, since the β-function matching requires a longer distance, a length of about 60 m is allocated on either side of the IP. Neglecting the rather small dispersion introduced by the ES separators, both the horizontal and the vertical dispersions remain essentially zero throughout the straight section.

For phase II with two vertically separated rings, after the first set of vertical ES separators a 5-m drift space increases the beam separation to a value sufficiently large to clear the septum of a center-septum magnet with equal and opposite fields on opposite sides. After the septum magnet, the beams go into separate transport lines having the same optics and opposite vertical geometry and dispersions. In addition to the minimum dog-leg, extra vertical bends are added to provide more flexibility in the matching. Many studies and sample matchings were performed. A typical example of a matched geometry and orbit functions are shown in Figs. 2 and 3. In Fig. 3 the values $\beta_x = 10.0$ m, $\beta_y = 4.0$ m, $D_x = D_y = 0$ at the end of the Arc are matched to the specified values $\beta_x^* = 0.01$ m, $\beta_y^* = 0.076$ m, $D_x^* = 0$, $D_y^* = \pm 0.4$ m at the IP.

In general for colliders with mini-β IPs, the maximum β values inside the common quadrupole doublet are so large that the large chromaticities generated by these quadrupoles must be corrected by sextupoles placed close by, in order that the

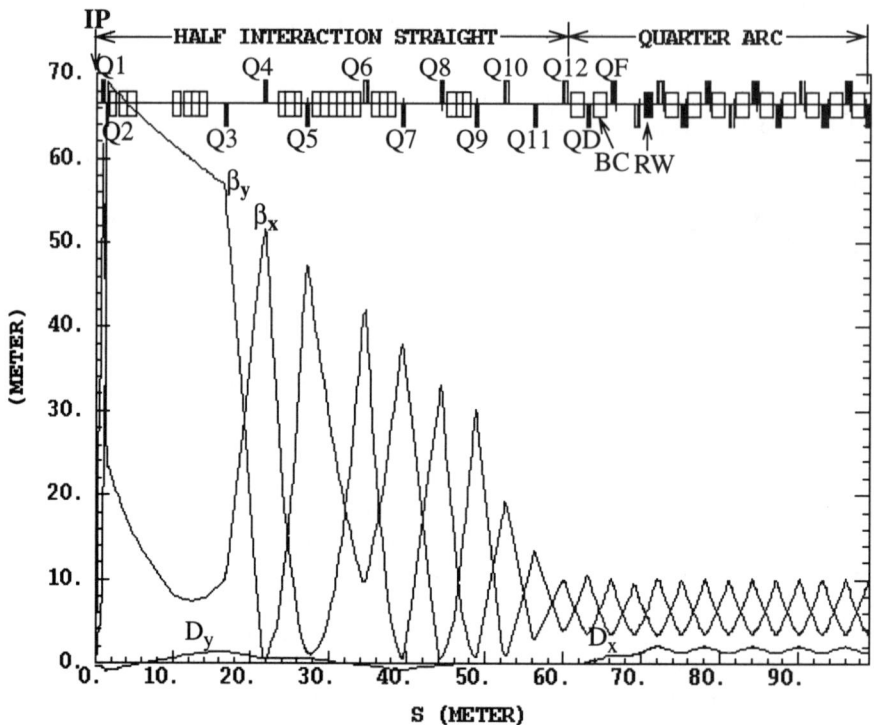

FIGURE 3. Orbit functions over a quarter of the ring. The Interaction Straight matching quadrupoles Q1 to Q12 are as shown in Fig. 2. The Arc elements are: cell dipoles BC; cell quadrupoles QF, QD; and Robinson wigglers RW. Not shown are the sextupoles.

dynamic aperture limitations imposed by the sextupoles are not overly restrictive. When matching the Interaction Straight we strive to keep the maximum β values in the mini-β quadrupoles as low as possible taking advantage of the additional flexibility introduced. This allows the chromaticity corrections to be provided by sextupoles distributed only in the Arcs next to the cell quadrupoles. The dynamic apertures remain ample over the entire energy spread in the beam, as shown in Fig. 4. Not having to correct the chromaticities immediately next to the mini-β quadrupoles greatly simplifies the Interaction Straight matching and the matching transport elements.

We also looked into approximate first-order designs for all the transport elements—superconducting mini-β quadrupoles, ES separators, septum vertical dipoles, etc.—to make sure that these transport elements can be fabricated with available material and technology and operated with the required performance and reliability.

Beyond phases I and II we visualize two further phases, although we have not been able so far to spend any effort on their design.

PHASE III: SUPER-HIGH LUMINOSITY

To further increase the luminosity by, say, a factor of 3, one can reduce the bunch spacing by a factor of 3 to some 3 m, and make the beams cross at a finite angle so that they are naturally separated at the next crossing point 1.5 m away from the IP. In this case, to reap the full luminosity gain of a factor of 3, the beam bunches must be canted to form the "crab-crossing" geometry. This geometry has been extensively studied (6) and can presumably be adopted rather straightforwardly. However, more detailed efforts are needed to investigate the possibility of incorporating the large vertical dispersions in the "crab-crossing" geometry to reduce the center-of-mass energy spread. If the reduced energy spread arrangement is incompatible with this geometry, the capability of going back to phase II for charmonium production should be retained.

PHASE IV: POLARIZED BEAMS

To study symmetry-violating reactions (7), collisions of longitudinally polarized particles are desired. The possibility of providing polarized beam collisions has been studied in some detail by A. Zholents (8). In this scheme the beams are first transversely polarized by the Sokolov-Ternov mechanism in a storage ring with very strong bending fields, hence very small radius. The typical polarization time is some tens of minutes. Spin rotators are added to the TcF ring

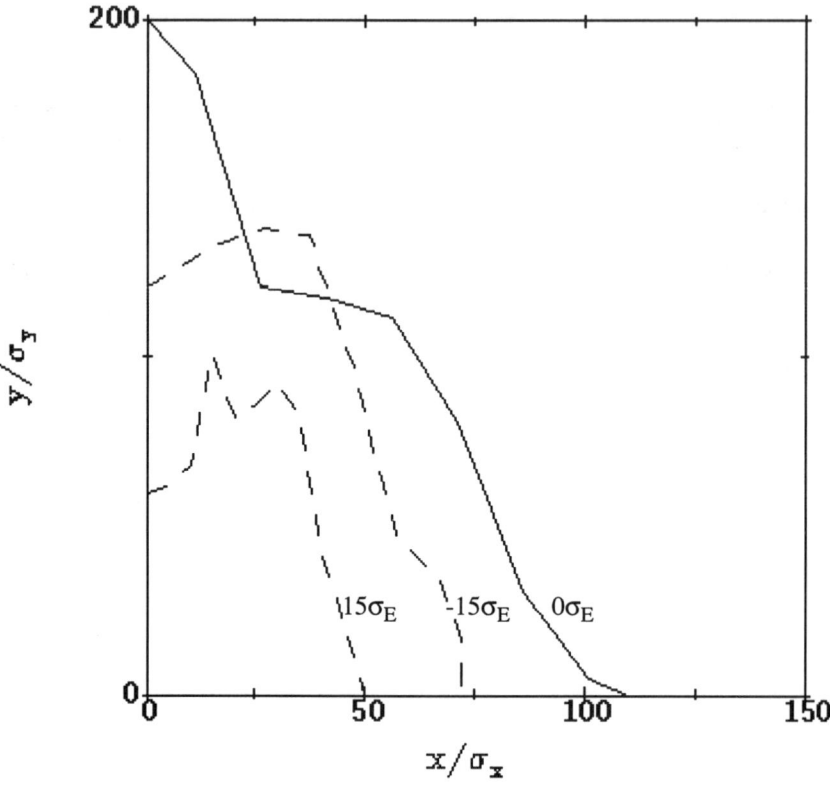

FIGURE 4. Dynamic apertures at various energy deviations for the lattice shown in Fig. 3.

so that the eigen-spin direction at the IP becomes longitudinal while the eigen-spin direction at the injection point remains transverse. The injected transversely-polarized beam will then collide in pure helicity states. We intend to pursue the design of this phase in detail in the near future. To provide the additions and modifications necessary for implementing this phase will be difficult and costly. In any case, even after phase IV is implemented one should retain the capability of colliding unpolarized beams.

Beyond the overall beam design and the approximate design of the ring transport elements, we also investigated the design of other component systems of the ring. For all these we draw heavily on our up-to-date experience with the corresponding components in the Advanced Photon Source.

COMPONENT SYSTEMS

Vacuum System

Extruded aluminum chambers with button-type beam position monitors (BPM) mounted directly on the chambers have performed very well in the past. Aluminum is easy to extrude and has good thermal conductivity; thereby it is easy to cool and has a low gas desorption coefficient ($\eta_\gamma = 2 \times 10^{-7}$ mol./photon after 150 Ampere-hour of synchrotron radiation cleaning) and a low bake-out temperature ($< 200°C$). The aluminum-to-stainless steel transition is by now a well-known and reliable technology. Distributed pumping is best provided by non-evaporative getter (NeG) pumping strips. The Zr/V/Fe NeG strip has a pumping speed of 300 liter/s for a 100 cm^2 surface area. These strips are mounted over the circumference of the entire ring on the inner-radius side of the vacuum chamber and separated from the main beam chamber by a wall with pump-out slots. With flush-mounted BPM buttons and with all chamber discontinuities, protrusions, and irregularities carefully shielded, one should be able to keep the coupling impedance low enough to avoid single bunch instabilities in the beam at the required high currents. Lumped ion or getter pumps are added at locations with high outgassing rates, such as the rf cavities.

RF System

The highest frequency should be used to give the shortest beam bunch. To realize the maximum luminosity the bunches should not be much longer than the smallest β at the IP, namely, ~ 1 cm. The highest frequency cw klystron available operates at 500 MHz. At this frequency, to achieve a 1-cm short bunch length, one needs a rather high rf voltage, some 4 MV. Operating close to the transition energy would be useful. However, this requires a lattice with very high transition energy which can be adjusted over the operating beam energy range of 1.5 GeV to 3.0 GeV. This is obviously difficult to achieve. Furthermore, at the very high bunch current discussed, a great deal of bunch lengthening can be expected. Clearly some further thinking and new ideas are needed.

Single cavities are likely to have the fewest higher-order modes (HOMs). These HOMs must be deQed so as not to excite the bunch-to-bunch instabilities in the beam. In case the deQing of the HOMs alone is not adequate to stabilize all coupled bunch modes, the beam can be stabilized by the use of properly designed feedback systems. Recent developments suggest that for cw operation at this frequency and voltage, superconducting cavities would be more advantageous.

Magnet Power Supplies

All horizontal Arc dipoles are connected in series and powered by a 12-phase rectifier power supply. The vertical dipoles in the Interaction Straight are individually powered and controlled. Furthermore, it is convenient to have all the ring quadrupoles, sextupoles, and orbit-corrector dipoles individually powered and individually adjustable. The "chopper" type of d.c. supplies as used on the APS and other synchrotron radiation rings is convenient, reliable, and available at reasonable cost.

CONCLUSION

The rather demanding performance requirements for the TcF almost uniquely defines the configuration of the machine. The Argonne design study incorporated only minor departures and shifts in emphasis from the "standard" design. The study offered a review and critique of the design rationale and features. It is understood that substantially more effort is needed to reach the level of a realistic Conceptual Design Report.

REFERENCES

1. Berger E. et al., "A Tau-charm Factory at Argonne," ANL-HEP-TR-94-12, (1994).

2. Kirkby, J., "A τ-Charm Factory at CERN," CERN-EP/87-210 (1987).
   Jowett, J. M., "Initial Design of a τ-Charm Factory at CERN," CERN LEP-TH/87-56 (1987).

3. The four workshops were held at SLAC (May 1989); Seville, Spain (May 1991); Marbella, Spain (June 1993); and SLAC (August 1994).

4. Jowett, J. M., "Frontiers of Particle Beams: Factories with e⁺ e⁻ Rings," edited by Dienes, M., Month, M., Strasser, B., Turner, S. (Springer Verlag, 1994); and the many references given in this paper.

5. Gerasimov, A. L., Zholents, A. A., "Beam-beam Effects in Storage Rings with a Monochromator Scheme," Proc. of XIII Int'l. Conf. on High Energy Accel., Vol. 1, p. 82 (1987).
Gerasimov, A. L., Shatilov, D. N., and Zholents, A. A., "Beam-Beam Effects with Large Dispersion Function at the Interaction Point," Nucl. Inst. and Methods, A305, pp. 25-29 (1991).

6. Palmer, R., "Energy Scaling, Crab Crossing and the Pair Problem," SLAC-PUB-4707 (1988). Beloshitsky, P. F. and Perelstein, E. A., "Horizontal Crab-Crossing Scheme for Tau-charm Factory, " JINR Reprint E9-94-13, Dubna, 1994.

7 Tsai, Y.-S., paper in these Proceedings.

8. Zholents, A., "Polarized J/ψ Mesons at a Tau-Charm Factory with a Monochromator Scheme," CERN SL/92-27 (AP), (1992).

B Factory Collider Designs and Future Plans[†]

Michael S. Zisman

Accelerator & Fusion Research Division
Lawrence Berkeley National Laboratory
Berkeley, CA 94720

Abstract. Typical parameters of *B* factory colliders are presented, along with their justification. Design challenges that arise from these parameter choices are indicated. These challenges appear in both the physics design of the collider and its technological implementation. An overview of the three active *B* factory projects (PEP-II, KEK-B, and the CESR upgrade) is briefly given, and technical approaches adopted by the projects to deal with the design challenges are outlined. Project status and plans for the various *B* factory projects are also indicated. Because the problems faced by the designers of *B* factories are closely related to those that will be faced in the design of a Tau-Charm Factory (τcF), the solutions adopted by the *B* factory designers can in many cases be carried over to the τcF essentially unchanged.

I. INTRODUCTION

Starting from a concept proposed by Pier Oddone in 1987 (1), there developed a strong interest in the design of a high-luminosity e⁺e⁻ collider to serve as a *B* factory (2). This led to the development of several proposals (3–9) to build such a collider at a number of high-energy physics laboratories worldwide. The primary motivation for building such a facility is to study the origin of the phenomenon referred to as "*CP* violation" in the $B\bar{B}$ system.

Compared with existing colliders, a *B* factory has several novel and interesting design requirements. Firstly, the key to successful studies of *CP* violation is to have a moving center of mass for the $B\bar{B}$ system. This implies the use of an "asymmetric" collider in which the two beam energies are different. Clearly, this requires a two-ring collider. (As is true for a τcF, a *B* factory has other features— small bunch spacing and high beam currents—that lead to a two-ring design.) Secondly, a *B* factory must produce a high luminosity, 2–3 × 10^{33} cm⁻² s⁻¹, to study *CP* violation. In practice, of course, the appropriate figure of merit for a *B* factory is not the peak luminosity but the *integrated* luminosity, as this is what is

---
[†]Supported by the U.S. Department of Energy under Contract No. DE-AC03-76SF00098.

needed to provide the required large sample of $B\bar{B}$ pairs. The design integrated luminosity for a B factory is

$$\int \mathcal{L} \, dt = 3 \times 10^{40} \, \text{cm}^{-2} = 30 \, \text{fb}^{-1} \tag{1}$$

based on a canonical Snowmass "year" of 10^7 seconds. Thus, we see that both of the novel requirements of a B factory are very similar to those of the τcF.

II. TYPICAL PARAMETERS

In general, we define the luminosity of a collider as

$$\mathcal{L} = \frac{N_+ N_- f_c}{4\pi \sigma_x^* \sigma_y^*} \tag{2}$$

where N_+ (N_-) is the number of positrons (electrons) per bunch, f_c is the collision frequency, and σ_x^* (σ_y^*) is the horizontal (vertical) beam size at the interaction point (IP). However, for purposes of accelerator design, we write the luminosity in terms of machine parameters as

$$\mathcal{L} = 2.17 \times 10^{34} \, \xi \, (1+r) \left(\frac{IE}{\beta_y^*}\right)_{+,-} \quad [\text{cm}^{-2} \, \text{s}^{-1}] \tag{3}$$

where ξ is the beam-beam tune shift parameter, r is the beam aspect ratio (σ_y^*/σ_x^*), I is the total beam current (A), E is the beam energy (GeV), β_y^* is the vertical beta function at the IP (cm), and the subscripts (+,−) indicate evaluating the parameters for the positron or electron beam, respectively. The reason for writing the luminosity expression as in Eq. (3) is to indicate explicitly the empirical fact that the luminosity of a collider is limited by the beam-beam tune shift parameter.

For a B factory, many of the parameters in Eq. (3) are not freely adjustable but are constrained by various things. In particular, the beam energy is constrained by the need to operate at a fixed center-of-mass energy to populate the Υ(4S) resonance, the beam aspect ratio is (loosely) constrained by background considerations, and the beam-beam parameter is limited by beam blowup to a maximum value (typically $\xi \approx 0.03$–0.05). Thus, it is clear from Eq. (3) that the key to achieving high luminosity is to use a high beam current and a small beam size at the IP.

Though it is easy to identify what is needed to reach high luminosity, it is less easy to achieve it for many practical reasons. For one thing, to take full advantage of the low β_y^* requires that the bunch length be comparable to or less than the beta value. Producing short bunches requires a high RF voltage (which, in turn, means substantial impedance-producing RF hardware); a practical limit corresponds to a beta function of about 1 cm. Another issue is that the high beam current gives rise to a large photodesorption gas load. We then require substantial pumping speed to maintain adequately low pressure in the beam pipe to have a long beam lifetime and to limit the detector background arising from scattered particles near the IP. Finally, the combination of high beam current and short bunches produces large amounts of higher-order-mode (HOM) power that can cause major thermal problems.

Given that we need high beam current, there remains the question of how to distribute it into individual bunches. It turns out that single-bunch instabilities preclude the possibility of a few-bunch high-current operating mode. The only way to mitigate these instabilities is to reduce the broadband impedance of the ring ($Z_\parallel/n$ for longitudinal instabilities; $Z_\perp$ for transverse instabilities). This is more easily said than done if big gains from today's colliders are needed, as modern machines have already reduced the broadband impedance to quite low levels. Thus, consideration of the above issues argues for a collider design where the beam is stored in many low-intensity bunches, keeping the instability "battleground" strictly in the multibunch arena. With this approach, the single-bunch parameters of the collider (including those related to the beam-beam interaction) can be kept in the normal operating range familiar from today's colliders. Our freedom to increase the number of bunches is limited by the need to avoid parasitic collisions as the bunches enter and leave the IP, as illustrated schematically in Fig. 1. As a rule of thumb, providing a clearance between beams in excess of 7σ at the parasitic crossing (PC) points is adequate to avoid beam lifetime problems.

For the B factory projects presently under way, the minimum bunch separation contemplated is 0.6 m. For typical beam sizes, a separation of 1–2 mm is needed at the PCs. Clearly, even a small crossing angle at the IP is a great help in providing this separation (compared with the alternatives of separation dipoles or

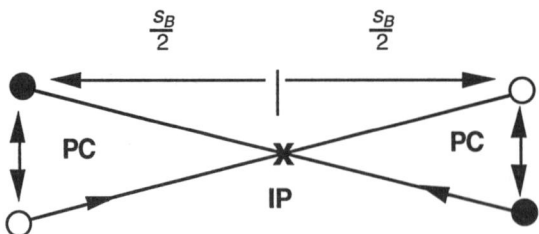

FIGURE 1. Schematic of parasitic crossing (PC) points at half of the bunch separation distance.

electrostatic separators), as an angle of about 10 mrad suffices. Whether such a crossing angle can be used without the technique (10) of "crab crossing" the beams is presently uncertain.

Based on the above discussion, the typical B factory parameter range for a collider with a design luminosity of 2–3 $\times 10^{33}$ cm^{-2} s^{-1} is summarized in Table 1.

III. DESIGN CHALLENGES

As should be obvious from inspection of the parameters in Table 1, the design of a high-luminosity collider to serve as a B factory gives rise to both physics and technology challenges. Here we will briefly cover some of the major issues. Specific approaches chosen by various groups to handle these design challenges will be indicated in Section V.

Physics Challenges

Some of the physics challenges stem directly from the need for a two-ring collider. For example, there is no "automatic" mechanism to keep the two beams in collision when they reside in separate rings. Drifts in power supplies, magnet vibrations, etc., do not operate on both beams simultaneously and it is likely that an active feedback system will be needed to maximize the luminosity. Indeed, the whole question of optimizing the luminosity for a two ring collider (especially an asymmetric one) is nontrivial. There is a much larger parameter space available in the two-ring case (separate tunes, bunch lengths, beam emittances, beta functions, ...). While these offer the potential of improving the luminosity beyond that of a single-ring collider, the large parameter space makes it difficult to know *ab initio* what to design the rings for. To deal with this difficulty, the typical approach has been to constrain most parameters to nominal values consistent with what a single-ring would permit, while (hopefully) providing sufficient adjustment range to optimize performance in the operating collider.

TABLE 1. Typical B Factory Beam Parameter Range

Parameter	Range
Bunch current [mA]	1–10
No. of bunches	100–5000
Total current [A]	1–3
Horizontal emittance [nm-rad]	≈100
Bunch length, rms [mm]	10

Lattice Design

The first design challenge for the lattice is to provide the required low values of β_y^* (on the order of 1 cm) while maintaining an acceptable dynamic aperture. This typically requires sophisticated sextupole schemes to correct high-order chromaticity in order to attain stable particle motion for off-momentum particles. While this challenge is common to the single-ring case, the constraint of having to focus two beams (of different energies) to a common spot greatly increases the difficulty of achieving the desired result. The optics in the interaction region (IR) must also accommodate the separation of the two beams rapidly enough to avoid parasitic crossings, as required to maintain acceptable beam lifetime. Moreover, the region near the IP must be kept free of machine components beyond a specified "detector-stay-clear" region (defined for a B factory as a roughly 300-mrad cone around the beam axis, centered at the IP).

In addition to the optics issues, the IR must also provide room for the masks that shield the detector elements from synchrotron radiation and lost-particle backgrounds. Minimizing the latter source of backgrounds also requires quite low pressure near the IP, which means that substantial vacuum pumping must be accommodated in an already crowded region. An indication of the spatial constraints that must be overcome is shown in Fig. 2.

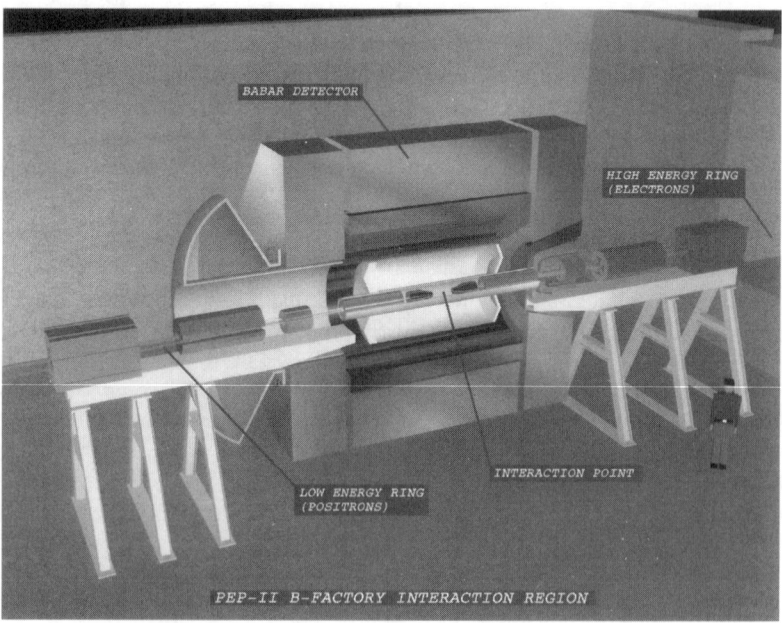

FIGURE 2. Cut-away view of BABAR detector of PEP-II.

Vacuum-Related Issues

There are vacuum-related issues in a two-ring collider that are often unimportant in a single-ring machine. These arise because the two beams circulate in separate pipes, so that each produces a large beam potential well.

In the electron ring, ion trapping or "dust" trapping may occur. This phenomenon is seen in many single-beam storage rings, such as synchrotron light sources. Usually it is seen only sporadically, which makes it difficult to deal with. Typical "cures" for ion trapping involve leaving a few-percent gap in the bunch train and/or adding clearing electrodes; maintaining a good vacuum is also beneficial. For the high beam current of a B factory, however, it is not easy to provide effective clearing electrodes to overcome the beam potential well. Dust trapping involves the capture of micron-sized charged particles by the beam. In many rings (11), there is circumstantial evidence that such particles are generated by sputtering from ion pumps or by particulate matter from non-evaporable getter (NEG) strips. Careful design of the pumping screen that separates the beam duct from the distributed ion pump (DIP) section should reduce the migration of dust into the beam duct.

In the positron ring, possible phenomena include the so-called "pressure-bump" instability (12), beam-induced multipactoring (13), and a coupled-bunch instability driven by the interaction between the positron beam and the cloud of low-energy electrons that fill the beam duct (14).

The pressure bump instability, first observed many years ago at the CERN ISR, is due to ion-induced desorption of weakly bound gas molecules on the vacuum chamber walls. The ions, which are created by beam (and synchrotron radiation) collisions with the residual gas in the vacuum chamber, are accelerated to the walls by the beam potential, releasing more gas molecules. If the rate is high enough, an avalanche can occur, leading to beam loss. The threshold current depends on the local pumping speed and the cleanliness of the beam pipe wall (as reflected by the ion desorption coefficient, η, analogous to the photodesorption coefficient). The best way to reduce the likelihood of a pressure bump problem is to remove contamination of the walls by means of glow-discharge cleaning.

Multipactoring is a resonant phenomenon caused by electrons being accelerated into the chamber walls by the beam potential. If the beam current is suitable and if the secondary electron emission coefficient is above unity, the electron current crossing the chamber between bunch passages can grow to the point where the local pressure increases, leading to beam loss. In general, this is only a problem close to threshold. Furthermore, many common vacuum chamber materials (stainless steel and copper) have low secondary emission coefficients. One common material—aluminum—has a high secondary emission coefficient, however, so care must be taken in this case. Using a coating in the beam duct, such as TiN, is expected to eliminate the problem.

The last effect mentioned above, interaction with the copious cloud of low energy photoelectrons in the beam duct, is presently under study at several laboratories. The motivation is the observation (14) at the KEK Photon Factory of fast-growing coupled-bunch instabilities seen with positron beams but not with electron beams under the same conditions. The bunch-to-bunch coupling in this case is through the electron cloud, which moves rapidly in response to transverse beam motion. As with multipactoring, the best recourse is to minimize the number of electrons in the positron beam chamber. An antechamber design, in which most of the photons land outside the beam duct, will be helpful; reducing the secondary electron yield by means of a suitable coating should also be beneficial.

Coupled-Bunch Instabilities

Coupled-bunch instabilities are expected to be very severe for a B factory (or a τcF). These are driven by wakefields in high-Q resonant objects, of which the higher-order modes (HOMs) of the RF cavities are the prime example. Because the resultant growth rates scale with the *total* beam current, the problem is hard to avoid in high-current rings. Moreover, as noted earlier, the need for short bunches forces the use of many RF cavities to provide the required voltage.

Technology Challenges

For all B factory designs, the main technological challenges lie in the areas of vacuum system, RF system, and feedback system design and implementation. Though it might easily be overlooked, it is worth noting here that injection requirements are nontrivial and must be dealt with carefully.

Vacuum System

The vacuum system in a high-luminosity collider faces several difficult problems. The chamber walls (or photon stops if an antechamber design is chosen) must withstand the high thermal flux from the intense synchrotron radiation fans. Furthermore, the system must be designed to provide low pressure (below 10 nTorr) in the face of copious synchrotron radiation induced gas desorption. The IR is an especially difficult area in this context, because there is an acute need for low pressure to avoid detector backgrounds but there is little, if any, room for pumps.

A more subtle issue of vacuum system design is that of impedance. It is imperative to maintain an RF-smooth chamber profile, with minimal gaps or sudden changes in cross section, in order to avoid problems with HOM heating by the beam. The scaling for HOM power is

$$P_{\text{HOM}} \propto \hat{I} I_{\text{avg}} \qquad (4)$$

where $\hat{I}$ is the peak current and I_{avg} the average current in the ring. Thus, the combination of short bunches (high peak current) and large beam current can lead to difficulties. In present B factory designs, a peak current in excess of 100 A is often specified. Many storage rings have had problems, especially with bellows RF shields, at much lower currents.

RF System

The RF system of a high-current storage ring is one of the most challenging technical systems to build. It must provide the high voltage needed to maintain a short bunch length and replenish the beam power lost to synchrotron radiation. These demands can require transmitting up to 500 kW of power through the cavity input window with high reliability. In order to minimize the HOM impedance seen by the beam (and thus reduce the growth of coupled-bunch instabilities), it is beneficial to keep the number of cavities as low as possible. Because we need a high voltage to produce short bunches, this results in the cavity voltage being high and thus, for a room temperature (RT) cavity, thermal losses into the cavity walls are substantial—up to 100–150 kW.

Even after minimizing the number of cavities, the HOM impedance is much too high to maintain beam stability, with or without feedback systems. The HOM impedance of each cavity must be damped by roughly a factor of 100 to result in practical parameters for a multibunch feedback system. For both RT and superconducting (SC) cavities this damping is a difficult problem, though, as discussed below, reasonable solutions have been found.

Feedback System

As noted earlier, the high beam currents in a B factory or τcF give rise to fast-growing coupled-bunch instabilities. Longitudinal instabilities are typically driven by HOM impedances from the RF cavities and have growth times of about 1 ms for typical B factory parameters. Although transverse HOMs in the cavities can likewise drive coupled-bunch growth, the more difficult source of impedance is due to the "resistive wall" (associated with image currents in the vacuum chamber walls). Even for a high-conductivity material such as copper or aluminum, the resistive wall instability growth time can be on the order of 1 ms. The impedance decreases rapidly as the beam duct dimensions increase, but practical considerations such as magnet bore size limit the realizable gains.

In addition to the need to handle high growth rates, the short bunch spacing typical of these rings means that fast processing is required. In the frequency domain, the required bandwidth to damp all unstable coupled-bunch modes is given by

$$\Delta = \frac{c}{2s_B} \tag{5}$$

where c is the velocity of light and s_B is the bunch spacing. For example, with a bunch spacing of 1 m, the bandwidth must be 150 MHz. Because of the large number of potentially unstable modes, B factory designers have generally chosen to operate in the time domain, that is, to provide a bunch-by-bunch feedback system capable of detecting and correcting the offset of each individual bunch (separated by only a few nanoseconds).

For a room-temperature RF system, the RF cavities must be considerably detuned in frequency to compensate for beam loading in the high-current rings. In this situation, it is inevitable that the cavity fundamental mode (the accelerating mode) drives coupled-bunch instabilities. To deal with this strong driving term, the klystrons and cavities themselves are made to serve as part of the feedback system to give extra power for controlling certain modes. This application benefits from klystrons with large bandwidth and low group delay compared with typical tubes; klystrons with these properties are being developed commercially.

Injection System

Despite the fact that the injection process for a storage ring is well understood, there are significant issues that must be addressed in the context of a B factory or τcF. For one thing, it is necessary to provide a uniform fill for hundreds—or even thousands—of bunches. This requires the ability to monitor the charge in each individual bunch at the level of a few percent and then provide a signal to tell the injection system whether to add more charge to any particular bunch.

The other major issue for injection is to minimize the detector backgrounds associated with the injection process. Though this sounds easy, many facilities have had the experience that half of all the radiation dose to the sensitive detector components comes during the (relatively brief) injection periods. For a B factory, it is envisioned that the rings will operate in "top-up" mode, that is, the beam current will be replenished without dumping the already stored beam. This approach clearly increases the average luminosity of the collider for a given peak luminosity. The logical limit of running in top-up mode is to inject small amounts of charge in each ring essentially continuously ("trickle charging"). If this mode of operation can be shown to be practical from a detector standpoint, it would be a real benefit to the facility. Designing a detector with this in mind seems a worthwhile, albeit difficult, goal.

IV. PROJECT OVERVIEWS

In this section we describe the features of a typical B factory project, the PEP-II project undertaken by a SLAC, LBNL, and LLNL collaboration. Approaches taken by other design groups are indicated to illustrate the range of solutions contemplated. In general, B factory design parameters are chosen to keep the challenges primarily in the engineering rather than the accelerator physics arena. In effect, one acts on the tacit assumption that thermal effects can be calculated more reliably than the beam-beam effect. Experience to date confirms this view.

The design philosophy of the PEP-II project is to employ single-bunch parameters (bunch current, emittance, bunch length, and beam-beam tune shift) that are essentially standard, that is, parameters used routinely in today's successful colliders. In addition, the PEP-II design is based on a value for the vertical beta function at the IP, β_y^*, of a few centimeters. The high design luminosity (see Table 2) comes from using a greatly increased number of bunches compared with present practice.

The PEP-II collider comprises two storage rings—a high-energy ring (HER) containing 9 GeV electrons (15) and a low-energy ring containing 3.1 GeV positrons (16). As indicated in Fig. 3, the LER will be mounted above the HER in the arcs. At the collision point, the LER beam is brought down into the plane of the HER. The injection system for PEP-II is based on the present Stanford Linear Collider (SLC) injector. Beams are extracted from the two-mile linac at the energies required for injection and transported to the collider rings in bypass lines that run alongside the linac. The demonstrated capabilities of the SLC linac exceed the requirements for PEP-II. The SLC produces pulses of about 3×10^{10}

TABLE 2. Main PEP-II Collider Parameters

	LER	HER
Energy, E [GeV]	3.1	9
Circumference, C [m]	2199.32	2199.32
$\varepsilon_y/\varepsilon_x$ [nm·rad]	2.6/64	1.9/48
β_y^*/β_x^* [cm]	1.5/50	2.0/66
$\xi_{0x,0y}$	0.03	0.03
f_{RF} [MHz]	476	476
V_{RF} [MV]	5.1	18.5
Bunch length, σ_ℓ [mm]	10	10
Number of bunches, k_B	1658†	1658†
Bunch separation, s_B	1.26	1.26
Damping time, τ_E/τ_x [ms]	26.2/52.5	18.4/37.2
Total current, I [A]	2.14	0.99
U_0 [MeV/turn]	0.87	3.58
Luminosity, $\mathcal{L}$ [cm^{-2}s^{-1}]	\multicolumn{2}{c}{3×10^{33}}

†includes gap of ≈5% for ion clearing

FIGURE 3. PEP-II prototype arc cell installed in the tunnel. The LER magnets, shown mounted on a raft above the HER dipole, are only mockups.

electrons and positrons per pulse at 120 Hz, whereas the PEP-II injection scheme calls for only $0.2-1 \times 10^{10}$ per pulse. The PEP-II design calls for top-up injection in about 3 minutes, with a fill time from zero current of 6 minutes.

The KEK-B project (17) comprises an HER operating with 8 GeV electrons and an LER with 3.5 GeV positrons. Because the project is sited in the large TRISTAN tunnel, the rings are side-by-side, with collisions in the horizontal plane. This layout is somewhat more convenient than that of PEP-II because it avoids the optics required to bring the two rings into the same collision plane. (The use of a vertical crossing configuration for PEP-II was avoided because of well-known difficulties with vertical crossing, associated with synchrobetatron resonances excited by the beam-beam interaction, that would preclude a future upgrade to a non-zero crossing angle configuration.) As indicated in Table 3, KEK-B is based on a higher estimate of the maximum beam-beam tune shift parameter than is PEP-II and it uses a 1 cm vertical beta function at the IP and very short bunches (4 mm). With a bunch separation of 0.6 m (every RF bucket filled), the ultimate luminosity goal for the KEK-B collider is $1 \times 10^{34}\,\text{cm}^{-2}\,\text{s}^{-1}$.

The CESR upgrade (18) involves increasing the luminosity of the existing CESR single-ring collider—already the highest luminosity collider in the world. This machine will continue to operate as a symmetric collider, which limits its ability to study *CP* violation. The upgrade consists of improving the CESR IR design, modifying the storage ring optics to provide a "pretzel" separation scheme, replacing the RF system with one based on SCRF cavities, and improving the vacuum system to handle the higher intensity beams.

TABLE 3. Main KEK-B Collider Parameters

	LER	HER
Energy, E [GeV]	3.5	8
Circumference, C [m]	3016	3016
$\varepsilon_y/\varepsilon_x$ [nm·rad]	0.4/18	0.4/18
β_y^*/β_x^* [cm]	1.0/33	1.0/33
$\xi_{0x,0y}$	0.04/0.05	0.04/0.05
f_{RF} [MHz]	509	509
V_{RF} [MV]	5–10	10–20
Bunch length, σ_ℓ [mm]	4	4
Number of bunches, k_B	5000†	5000†
Bunch separation, s_B	0.59	0.59
Damping time, τ_E/τ_x [ms]	23/46	23/46
Total current, I [A]	2.6	1.1
U_0 [MeV/turn]	1.5	3.5
Luminosity, $\mathcal{L}$ [cm^{-2}s^{-1}]	1 x 10^{34}	

†includes gap of ≈2% for ion clearing

V. DESIGN APPROACHES

In this section we discuss some of the design approaches used by the various projects to handle the challenges outlined in Section III. The topics covered, IR design, vacuum system design, RF system design, and feedback system design serve to illustrate the key features of a B factory design and the range of possible solutions that exist. It should be clear that the approaches taken will be equally valid for the τcF case.

IR Design

PEP-II makes use of a head-on collision geometry, as is used in nearly all operating colliders. As illustrated in Fig. 4, the B1 separation dipoles on either side of the IP have opposite polarities, which we refer to as an "S-bend" configuration. (Note that the S-bend geometry lends itself to an eventual future upgrade to a non-zero crossing angle.) To maintain maximum solid angle for the detector, the B1 magnets are tapered such that they remain within a 300-mrad cone. The beam separation at the nearest PCs provided by this separation scheme is more than $11\sigma_x$, which is adequate to ensure good beam-beam lifetime. The Q1 magnet is the only common quadrupole. It is offset with respect to the low-energy beam (LEB) axis and thus contributes to the rapid separation of the two beams. The Q2 magnet is a septum quadrupole that focuses only the LEB.

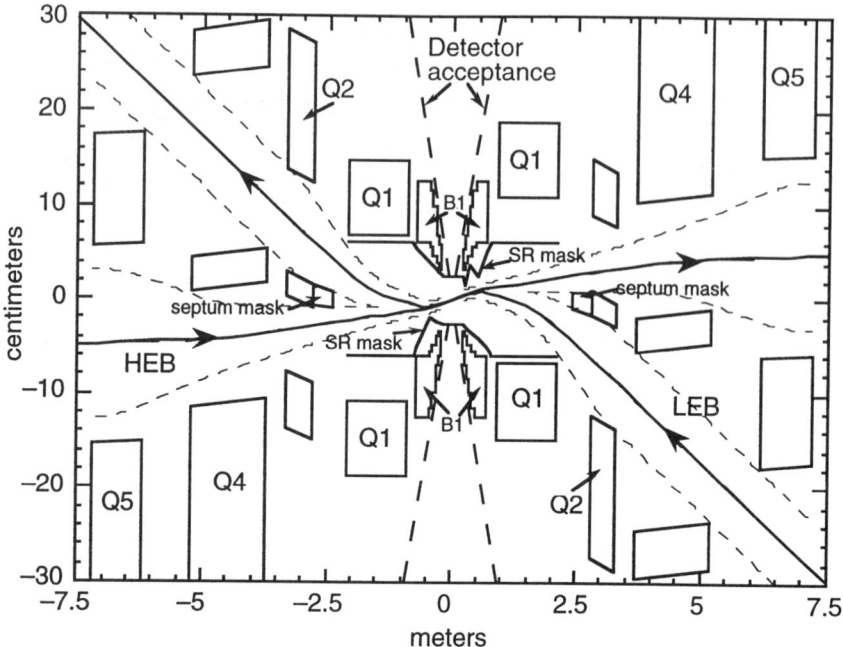

FIGURE 4. Anamorphic plan view of the PEP-II IR. The B1 dipoles and Q1 quadrupoles are common to both beams. Because they are immersed in the field of the detector solenoid (see Fig. 2), these magnets are based on permanent-magnet technology.

The KEK-B project has adopted a crossing angle of ±11 mrad as its initial configuration, in order to avoid the need for separation dipoles very close to the IP (as used in the PEP-II case). Such a large crossing angle suffices to easily separate the beams at the nearest PC, even in the case where every RF bucket is filled (in which case the first PC is only 30 cm from the IP).

To aid in beam separation, the upgraded CESR ring will operate with a small crossing angle, ±2.1 mrad. Experiments at CESR have already demonstrated (18) that this configuration does not degrade the beam-beam performance of the ring for symmetric collisions. In the final IR configuration, a superconducting quadrupole doublet will provide a β_y^* in the range of 7–10 mm.

Vacuum System Design

The two PEP-II rings use different technology for their vacuum systems. The HER chamber is extruded copper and the LER chamber is extruded aluminum. The 5.5 m long HER dipole chamber has a simple cross section, shown in Fig. 5, to accommodate the beam duct, a duct for the distributed ion pumps (DIPs), a

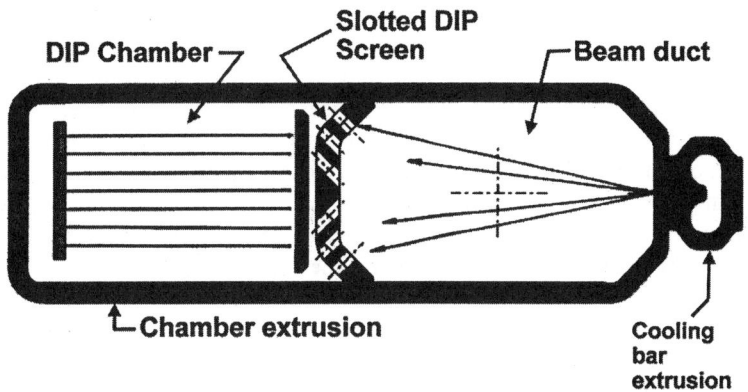

FIGURE 5. Cross section of extruded copper arc dipole chamber of the PEP-II HER.

screen to separate the two chambers, and an external cooling channel. Because of limitations with copper extrusion technology, the roughly rectangular vacuum chamber is made from one extrusion and the cooling bar from a separate extrusion. The screen is a separate piece that slides into place in the dipole extrusion. Note that the screen slots shown in Fig. 5 are angled to eliminate line-of-sight connection between DIP chamber and beam duct to avoid sputtered material from the pump entering the beam duct. The screen design also includes holes at the bottom of the slots to avoid TE modes. The cooling bar and flanges are joined by electron beam welding rather than brazing to maintain the mechanical strength of the chamber. Though copper is more difficult to work with than aluminum, it has several advantages for the HER. Copper has excellent thermal and electrical conductivity properties, good photodesorption properties (it cleans up rapidly and typically provides a lower photodesorption coefficient than does aluminum), and its higher atomic number provides improved shielding for the synchrotron radiation. The HER quadrupole extrusion (also copper) has a smaller cross section than that shown in Fig. 5; it has the same profile as the octagonal beam chamber portion of the dipole chamber to minimize beam impedance.

The PEP-II LER uses a different approach, dictated in part by the very different layout of the magnets. In the LER, the dipoles are short (0.45 m) and all of the magnets for each FODO half-cell are clustered together on a common support raft. The regions between rafts are spanned by vacuum chambers ("pumping chambers") that, in effect, are simply mini-straight sections. Because the synchrotron radiation from a dipole does not strike the chamber wall until well downstream, and because the dipole packing factor is very low (≈0.05), there is no benefit to DIPs. Instead, the synchrotron radiation fans are intercepted at discrete locations on copper photon stops located in an extruded aluminum antechamber (Fig. 6 upper) that communicates with the beam duct via a 15 mm tall continuous slot. As shown in Fig. 7, each photon stop sits directly above a titanium sublimation pump (TSP) that provides very high pumping speed in the region near the main source of gas. In the short region where the magnets are located, a smaller aluminum extrusion is employed (see Fig. 6 lower). Here too, a slot is present so that the photon fan does not hit the chamber wall before it reaches the photon stop. With this arrangement, we expect an average pressure in the LER arcs of about 1 nTorr at the maximum beam current of 3 A. Because a low-energy beam is quite sensitive to gas scattering processes, the low pressure in the LER is very beneficial to ensuring a long beam lifetime.

The KEK-B project employs an approach similar to that of the PEP-II HER, using copper extrusions. For their LER, they use a round copper pipe (94 mm ID) to minimize the growth rate of the strong resistive-wall instability. The CESR ring utilizes aluminum extrusions. They will maintain the present chamber but provide additional cooling for flanges and bellows to handle the higher heat loads due to the increased beam current.

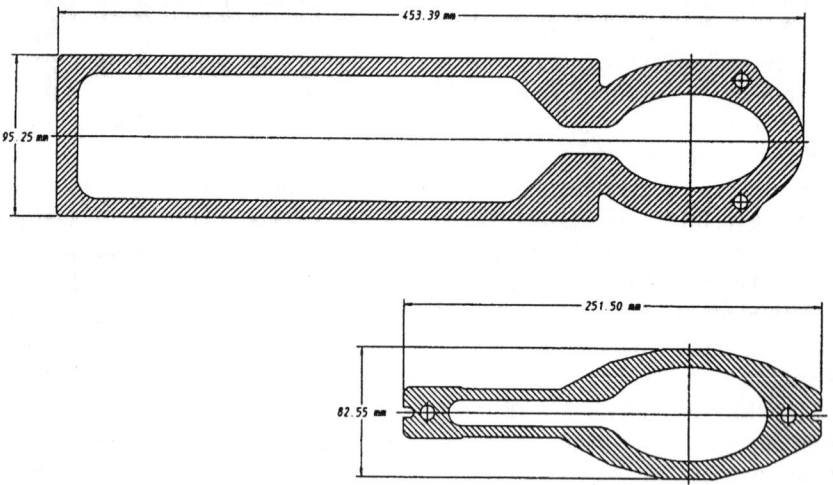

FIGURE 6. Cross section of aluminum extrusion of PEP-II LER pumping chamber (upper); cross section of aluminum extrusion of PEP-II LER magnet chamber (lower).

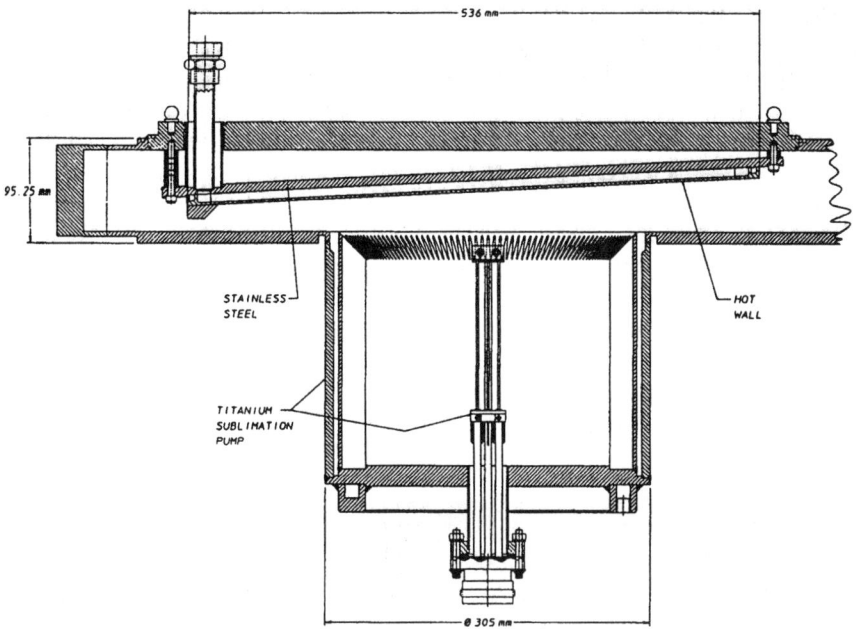

FIGURE 7. Side view of PEP-II LER photon stop and TSP assembly. These items are mounted in the antechamber section shown in the upper portion of Fig. 6.

RF System Design

As noted earlier in this paper, the RF system of a B factory or τcF is one of the most complex and expensive systems to design. The choice of technology is itself a complicated issue, depending not only on the parameters of the storage rings but also on the experience and expertise of the group that will implement the design and the skills of the groups that will provide the complementary feedback systems (to be discussed below). It is no surprise, then, that the choices of the various design groups show the most diversity in this area.

One of the key decisions to make is whether to adopt RT or SC technology, and each has its strong proponents. To make a proper choice for a τcF, it is important to understand the reasons of each B factory group for making its particular choice and then assess the relevance of those reasons for the τcF case. In simple terms, the SCRF approach is most beneficial if the limiting parameter is the required voltage (to produce short bunches, for example) and is less necessary (and probably less cost-effective) if the limit is due to the beam power that must be provided. For PEP-II, RTRF is favored because the voltage requirements are modest. For the CESR upgrade, SCRF will be used to provide short bunches in a ring having relatively little space for RF equipment. In the KEK-B case, both options are being considered; the present plan is to begin with RTRF.

The PEP-II RTRF cavity (19) is shown in Fig. 8. The required HOM damping uses three rectangular waveguides whose dimensions are such that the HOM fields propagate to loads while the fundamental frequency is below cut-off. The use of three waveguides guarantees that transverse modes cannot shift such as to avoid being damped. The cavity is designed to dissipate 150 kW of power into the walls. Nominal operating parameters call for a maximum wall dissipation of about 100 kW. This has already been demonstrated in the high-power test cavity shown in Fig. 8. Demonstration of the efficacy of the damping technique was accomplished with a low-power test cavity some time ago. A drawback of the RTRF approach is that there must be considerable detuning of the cavity on account of the beam loading, and thus that cavity fundamental drives coupled-bunch motion. A special cavity feedback system is part of the PEP-II design to compensate for this effect.

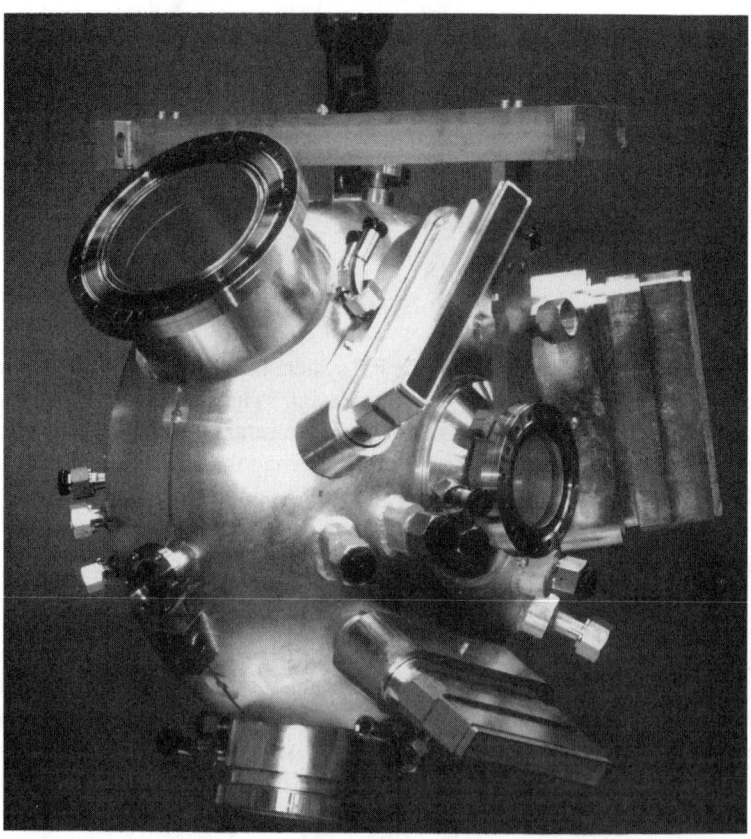

FIGURE 8. High-power test cavity for PEP-II.

At KEK-B, a novel approach has been developed (20) to mitigate the beam loading problem. They resonantly couple an energy storage cavity to the accelerating cavity, as illustrated in Fig. 9; the resultant system has enough stored energy to limit the detuning to a fraction of the rotation frequency, so the cavity fundamental does not drive coupled-bunch modes. The accelerating cavity is also a novel design based on the so-called "choke mode" approach. As a parallel effort, the KEK-B RF group is looking at a SCRF cavity patterned after the CESR design concept (shown below). In this approach the HOM damping is done by using the beam pipes as HOM waveguides leading to coaxial loads. To ensure the damping of transverse modes, the two beam pipes are different diameter, as shown in Fig. 10. (In Fig. 10, the beam pipes are termed large and small, LBP and SBP, though large and very large seem more accurate designations.)

The Cornell approach (21) focuses on SCRF technology. Their cavity design is shown in Fig. 11. In this design, one of the beam pipes is fluted to propagate some modes that would otherwise be trapped. CESR plans to upgrade to four single-cell SCRF cavities to handle beam currents up to 500 mA per beam. Initial beam tests of the system (21) were very encouraging.

Feedback Systems

As explained earlier, broadband feedback systems are required for the operating parameter regime typical of a B factory or τcF. Because of the

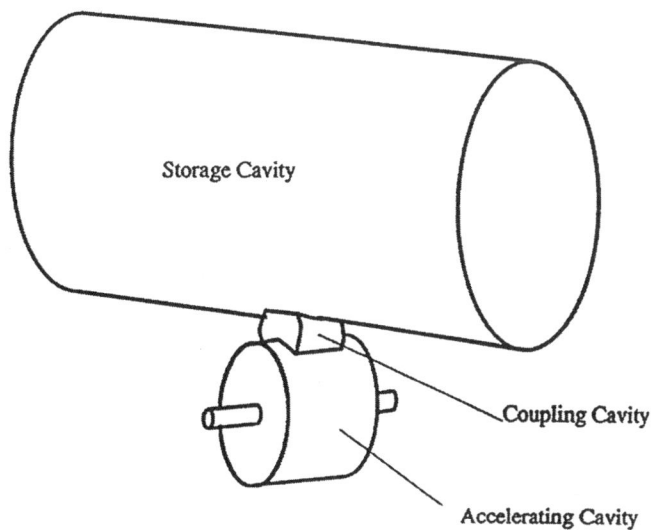

FIGURE 9. KEK-B "ARES" cavity.

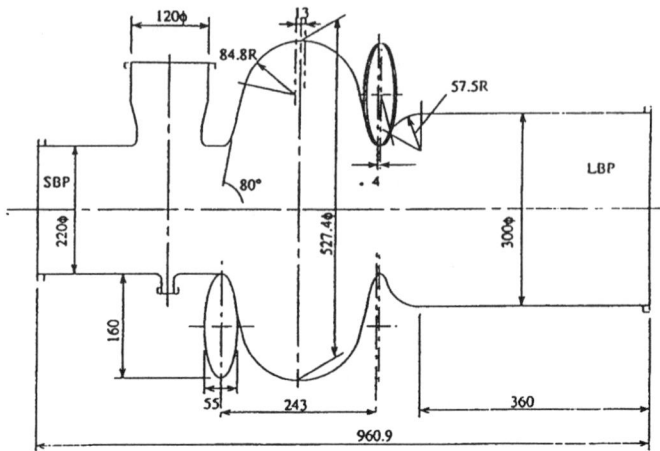

FIGURE 10. KEK-B SCRF cavity.

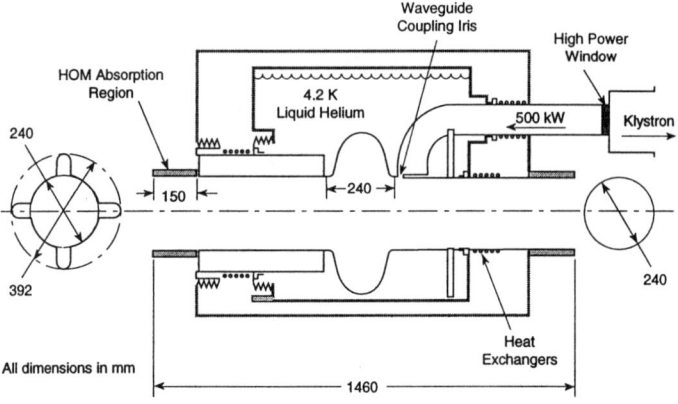

FIGURE 11. CESR SCRF cavity.

difficulty of knowing exactly what the frequencies of unstable modes will be, the standard approach is to use a bunch-by-bunch feedback system operating in the time domain. In this approach, the offset of each bunch is individually measured and corrected. One advantage of this technique is that the systems combat bunch motion from any source, not just standard coupled-bunch motion induced by cavity. In particular, they help with damping injection transients.

For the longitudinal case, one can either measure a phase offset or, in a dispersive region, an energy offset. The former is typically easier to implement. For the transverse case, one measures position offsets in a non-dispersive region (to avoid coupling with synchrotron motion). To avoid wasting broadband power, the system must be able to eliminate the DC offset signal (due to any static closed-orbit distortion).

The PEP-II longitudinal system (22) uses a very flexible architecture based on digital signal processing (DSP) technology. The principle of operation has now been demonstrated successfully in beam tests at SPEAR and at the Advanced Light Source (ALS). A full longitudinal system has recently been commissioned at the ALS that can damp few-millisecond growth times in a beam with a 2 ns bunch spacing (demonstrating a bandwidth capability that equals or exceeds what is required for a *B* factory or τcF). Figure 12 shows the beam size in a dispersive region of the ALS with (Fig. 12c) and without (Fig. 12b) longitudinal feedback; transverse feedback is on in both cases. The damping of longitudinal motion results in a smaller spot size because the energy excursions of the beam are reduced.

Transverse feedback systems (23) in PEP-II (note that two systems are needed, damping horizontal and vertical motion, respectively) are simpler to implement than the longitudinal system. Though the transverse systems share many components with the longitudinal system, such as the digital delay, they do not need the DSPs. ALS results for the vertical feedback system are included in Fig. 12. With the horizontal and longitudinal feedback systems on, the beam size is shown with (Fig. 12 c) and without (Fig. 12 a) the vertical feedback operating.

The KEK-B feedback system (24) uses a similar approach to that of PEP-II, but a different implementation has been chosen. They create a two-tap finite-impulse-response filter using CMOS logic and fast memory that can run at the full KEK-B bunch repetition rate of 2 ns.

CESR does not appear to require longitudinal feedback to ensure beam stability. A transverse feedback system is now operational (25).

VI. STATUS AND PLANS

Here we briefly summarize the status of the various *B* factory projects. The information here is collected primarily from papers (15–18) given at the 1995 Particle Accelerator Conference in Dallas. Considerably more detailed technical information on all of the projects may be found in those proceedings.

For the PEP-II project, the fabrication activities are well along. Essentially all of the HER magnets (originally used in PEP-I) are now refurbished and are being reinstalled in the tunnel. Copper extrusions for the vacuum chambers are in production and an electron-beam welder has been delivered for chamber production. The main LER magnets (dipoles and quadrupoles) are being built as part of a collaboration with the Institute of High Energy Physics (IHEP) in Beijing. The prototype quadrupole has been delivered to LBL and is now undergoing magnetic measurement. Orders for the vacuum chamber extrusions have been placed. Installation of the injection bypass lines is nearly completed

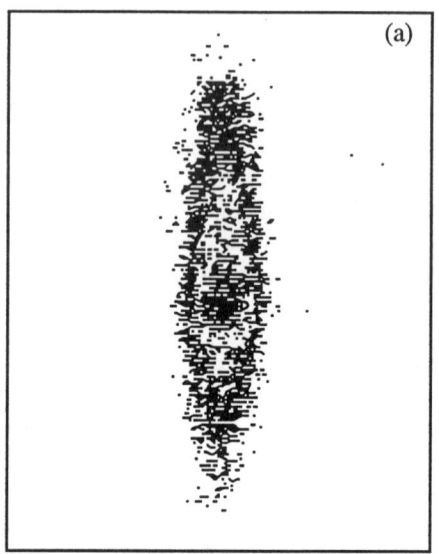

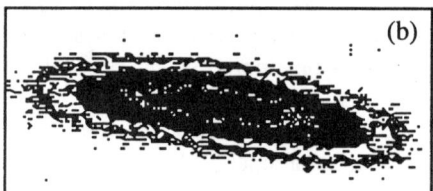 

FIGURE 12. ALS beam size images under various feedback system conditions: (a) horizontal and longitudinal feedback on, vertical feedback off; (b) horizontal and vertical feedback on, longitudinal feedback off; (c) horizontal, vertical, and longitudinal feedback on.

and tests will commence on the electron channel soon.

The KEK-B project does not gain access to the TRISTAN tunnel until the end of 1995 when the TRISTAN run terminates, so they cannot start installation until January 1997. They anticipate procuring LER magnets and vacuum components in 1995–1996 with the HER magnets and vacuum components following about one year later, that is, in 1996–1997.

The CESR upgrade project is being done in phases. The Phase I project was to add pretzel optics; this is done and commissioned. The Phase II upgrade, now under way, has a luminosity goal of 6×10^{32} cm^{-2} s^{-1}. To achieve this, a new IR optics is being installed. This will comprise a longer permanent-magnet QD along with an electromagnet for QF. The electrostatic beam separators will be replaced with new, lower impedance models. In addition, the RF cavity coupling will be adjusted to optimize cavity performance for the higher beam currents being developed, and new diagnostic devices, such as common beam position monitors (BPMs) at the IP, will be developed. In Phase III, scheduled to begin in late 1997, the goal is to deliver a luminosity of 1×10^{33} cm^{-2} s^{-1}. This final phase will entail a further upgrade of the IR optics (pushing the permanent magnet quadrupoles closer to the IP and adding a superconducting quadrupole doublet to obtain lower β_y^* values. They will also upgrade to four single-cell SCRF cavities, which will support shorter bunch length and higher beam currents.

VII. SUMMARY

After some five years of workshops and R&D activity, two asymmetric B factory projects, PEP-II and KEK-B, were funded in fiscal year 1994. The CESR upgrade program to produce a very high luminosity single-ring collider is also funded. All three projects are now well under way. As should be clear from the discussions in this paper, the issues studied for a B factory, and the design solutions proposed, should be generally applicable to the design of a τcF. Indeed, the quest for very high luminosity and high reliability is a common thread of *all* the various "factories" (B, τc, and Φ).

Despite the fact that solutions to all of the anticipated problems appear to be at hand, it is clear that making a large jump in luminosity performance for a collider has never been an easy undertaking. The best way to ensure our success in this exciting endeavor is to treat the challenges we face with proper respect and not become complacent.

That having been said, I'm sure I speak for my colleagues at SLAC, LBNL, LLNL, and KEK in stating that we are greatly looking forward to the start of asymmetric B factory commissioning activities at SLAC and KEK in 1998.

ACKNOWLEDGMENTS

The designs and concepts reported here are the work of many accelerator physicists and engineers at SLAC, LBNL, LLNL, KEK, Cornell, and elsewhere. The very existence of the projects described herein is due in large measure to their dedicated efforts. I wish to thank all of these colleagues for their enthusiasm, especially during the bleak years when funding looked improbable. Particularly noteworthy in this regard are Pier Oddone and Swapan Chattopadhyay, both of LBNL, the first for having the vision to see the possibilities of an asymmetric collider, and the second for having the audacity to believe it could be done. What started in the minds of a few imaginative people is now well on the way to fruition. I hope the Tau-Charm community takes heart from this fact and continues the fight to see a machine for them built somewhere in the world.

REFERENCES

1. Oddone, P., in *Proceedings of UCLA Workshop on Linear Collider BB Factory Conceptual Design,* 1987, p. 243.
2. Chattopadhyay, S., "Physics and Design Issues of Asymmetric Storage Ring Colliders as B-Factories," *Part. Accel.* **31**, 121 (1990).
3. Wacker, K., et al., *Proposal for an Electron-Positron Collider for Heavy Flavour Particle Physics and Synchrotron Radiation*, PSI PR-88-09, July, 1988.
4. Nakada, T. ed., *Feasibility Study for a B Meson Factory in the CERN ISR Tunnel,* CERN 90-02, PSI PR-90-08, March, 1990.
5. Berkelman, K., et al., *CESR-B Conceptual Design for a B Factory Based on CESR* (Updated June, 1993), Cornell Report.
6. Dubrovin, A. N. and Zholents, A. A., in *Proceedings of the 1991 IEEE Particle Accelerator Conference*, San Francisco, May 6-9, 1991, p. 2835.
7. Zisman, M. S. ed., *PEP-II An Asymmetric B Factory,* Conceptual Design Report, LBL PUB-5379, SLAC-418, CALT-68-1869, UCRL-ID-114055, UC-IIRPA-93-01, June 1993.
8. Kurokawa, S., Satoh, K. and Kikutani, E., eds., *Accelerator Design of the KEK B Factory*, KEK Report 90-24, March, 1991.
9. Albrecht, H., et al., *HELENA: A Beauty Factory in Hamburg*, DESY 92-041, March, 1992.
10. Oide, K., Yokoya, K., SLAC-PUB-4832, 1989.
11. Kelly, D. R. C., Bialowons, W., Brinkmann, R., Ehrlichmann, H., Kouptsidis, J., "The Electron Beam Lifetime Problem in HERA," presented at the Particle Accelerator Conference, Dallas, May 1-5, 1995; Rogers, J., "Photoelectron Trapping Mechanism for Horizontal Coupled Bunch Mode Growth in CESR," CBN95-2, March, 1995.
12. Calder, R. S., "Ion Induced Gas Desorption Problems in the ISR," *Vacuum* **24**, 437 (1974).
13. Gröbner, O., "Bunch Induced Multipactoring," in *Proceedings of International Conference on High Energy Accelerators*, Serpukhov, 11–17 July, 1977, p. 277.
14. Izawa, M., Sato, Y., Toyomasu, T., *Phys. Rev. Lett.* **74**, 5044 (1995).
15. Wienands, U., Reuter, E., Seeman, J. T., Davies-White, W., Fisher, A., Fox, J. Genova, L., Gracia, J., Perkins, C., Pietryka, M., Schwarz, H., Taylor, T., Jackson, T., Belser, C., Shimer, D., "Status of the High Energy Ring of the PEP-II B-Factory," presented at the Particle Accelerator Conference, Dallas, May 1–5, 1995.
16. Zisman, M. S., Yourd, R. B., Hsieh, H., "Design of the PEP-II Low Energy Ring," presented at the Particle Accelerator Conference, Dallas, May 1–5, 1995.
17. Kurokawa, S. "KEKB Status and Plans", presented at the Particle Accelerator Conference, Dallas, May 1–5,1995.
18. Rubin, D. L., "CESR Status and Plans," presented at the Particle Accelerator Conference, Dallas, May 1–5, 1995.
19. Rimmer, R., Allen, M., Saba, J., Schwarz, H., Belser, F. C., Berger, D., Franks, R. M., "Development of a High-Power RF Cavity for the PEP-II B Factory," presented at the Particle Accelerator Conference, Dallas, May 1–5, 1995.
20. Yamazaki, Y., Kageyama, T., *Part. Accel.* **44**, 107 (1994); Akai, K., Yamazaki, Y., *Part. Accel.* **46**, 197 (1994).
21. Moffat, D., Barnes, P., Kirchgessner, J., Padamsee, H., Sears, J., "Preparation and Testing of a Superconducting Cavity for CESR-B," in *Proceedings of the 1993 Particle Accelerator Conference*, May 1993, p. 763; Padamsee, H. *et al.,* "Beam Test of a Superconducting Cavity

for the CESR Luminosity Upgrade," presented at the Particle Accelerator Conference, Dallas, May 1–5, 1995.
22. Fox, J., Eisen, N., Hindi, H., Linscott, I., Oxoby, G., Sapozhnikov, L., Serio, M., "Feedback Control of Coupled-Bunch Instabilities," in *Proceedings of the 1993 Particle Accelerator Conference*, May 1993, p. 2076; Fox, J. *et al.*, "Operation and Performance of the PEP-II Prototype Longitudinal Damping System at the ALS," presented at the Particle Accelerator Conference, Dallas, May 1–5, 1995.
23. Barry, W., *et al.* "Commissioning and Operation of the ALS Transverse Coupled-Bunch Feedback System," presented at the Particle Accelerator Conference, Dallas, May 1–5, 1995.
24. Kikutani, E., Kasuga, T., Minagawa, Y., Obina, T., Tobiyama, M., "Recent Progress in the Development of the Bunch Feedback Systems for KEKB," presented at the Particle Accelerator Conference, Dallas, May 1–5, 1995.
25. Rogers, J., *et al.* "Operation of a Fast Digital Transverse Feedback System in CESR," presented at the Particle Accelerator Conference, Dallas, May 1–5, 1995.

Round Table Discussion of Collider Designs

S. Kurokawa[1], E.Perelstein[2], L.Teng[3],
M.Tigner[4](recorder), Y.Z. Wu[5], M.Zisman[6]

*1. KEK, Tsukuba, Japan, 2. JINR, Dubna, Russia, 3. ANL, Argonne, IL,
4. Cornell, Ithaca, NY, 5. IHEP, Beijing, P.R.China, 6. LBL, Berkeley, CA.*

Abstract

A number of technical aspects of τcF colliders are discussed including comparisons of the current concepts with each other and with B Factories, luminosity limits, operating energy limits, needed R/D and next steps in pushing forward with the designs.

I. The Questions Discussed

1. Comparison of the three designs presented with each other and with B Factories.

2. Luminosity($\mathcal{L}$) limits in the high $\mathcal{L}$ mode.

3. Luminosity limits in the Monochrometer mode.

4. Luminosity limits in the Polarization mode.

5. Limits to operating energy.

6. Next steps in developing the designs.

7. Important R/D tasks remaining.

8. Need for a real design workshop involving both detector and accelerator physicists.

II. Responses

1. At the present concept level the τcF designs share the same basic ideas, and are thus quite similar. In comparison with the B Factories, they appear to be both easier and more difficult. Perhaps the principal contrast is in the variety of operating modes needed for the τcF in comparison with the rather narrow range of operating modes now conceived for the B Factories. The need for operating with high emittance for high luminosity, very low emittance for monochromatization, and with polarized beams as well, is a unique challenge for the τcF. There are, however, technical solutions for each of the operating modes foreseen. On the other side of the ledger, the lower energy of the τcF will make the IP focus and separation somewhat easier.

2. Assuming that the per bunch luminosity will be limited, as now, by the beam-beam tune spread, ξ, the limit on luminosity will be given by the maximum achievable total beam current. One might ask, for example, whether it will be possible to push a τcF using today's accelerator technology to $\mathcal{L} = 10^{34} \text{cm}^{-2}\text{s}^{-1}$. This would entail ten-fold increase in the beam current to, say, 5 A with something like 1 m between bunches. (This is to be compared with the 3 A design current for the 3 GeV Low Energy Rings of the asymmetric B-factories.) At some current, probably in the low Ampere range, present impedance control technology will no longer be adequate to prevent the excitation of coupled bunch internal modes. These are very difficult to suppress with today's feedback technology. In addition, one turn ion effects will limit the current at some value as well. Neither of these effects is fundamental but avoiding them will require new technologies not available now. Note for example that the beam pipe, masks and beam position monitor technologies now in use enforce a minimum impedance. New designs and methods of construction could be developed to lower this minimum in the future.

In conclusion, it does not seem unreasonable to expect that with operating experience and first round upgrades one might reach luminosities of $3 \times 10^{33} \text{cm}^{-2}\text{s}^{-1}$ in a τcF built with current technology.

3. The limits in the Monochromer Mode are less certain than in the High $\mathcal{L}$ mode since no experience in this configuration has yet been accumulated to reveal its peculiar challenges. By comparison with the High $\mathcal{L}$ Mode, there are some additional factors which could limit the luminosity more stringently. They are: i) limit to the achievable ξ due to the synchrobetatron resonances

driven by the high dispersion at the IP; ii) Touschek effect limitation to the single bunch current; iii) maintenance of the low emittance required. With regard to iii) we can say that such low emittance has been achieved at the ALS at 1.5 GeV, the relevant energy, thus giving us an optimistic outlook for that particular limit.

While one cannot be really quantitative without operating experience, past experience with other electron storage ring colliders would indicate that the lowering of luminosity in the Monochromer mode with respect to the High $\mathcal{L}$ Mode would not exceed a factor of 2 or 3 and perhaps be much less.

4. Again, lack of operating experience in Polarization Mode colliders precludes a precise statement of $\mathcal{L}$ limits. What is particularly not known is the influence of the beam-beam effect on the polarization lifetime. Because the beam-beam effect is so highly non-linear, however, a small decrease in ξ can make a large decrease in its deleterious effects. Thus one might estimate that any luminosity degradation due to beam-beam polarization killing would not exceed a factor of 2 or 3, even taking into account the slower injection if radiation polarization is used.

5. Before one may discuss limits to the operating energy range, the question must be put more precisely. On the low energy side, the luminosity will be limited by beam current. The radiation damping rate for instabilities goes like the 3rd power of the beam energy. If one operates at the stability limit at energy E_0, then the $\mathcal{L}$ will fall as $(E/E_0)^3$ unless special damping wigglers and/or feedback equipment is added. On the high side, the energy and luminosity are limited by the voltage and power of the RF system and the pumping speed and heat removal capacity of the vacuum system. The dynamic aperture of the lattice will also eventually bind the upper energy. Thus one cannot speak of absolute limits to the operating energy. The correct approach is to name the energy range and the $\mathcal{L}$ as a function of energy required and ask what design (and cost) is needed to achieve it.

6. Next steps in pursuing the designs must focus on achieving an integrated detector/IR design including analysis of beam generated backgrounds. This needs to be done for each of the planned operating modes separately before one could be ambitious enough to explore the possibility of a universal design. From that one may proceed to make conceptual designs for the remainder of the facility, again one for each of the proposed operating modes before at-

tempting to integrate them onto one grand design. Part of this exploration should includes a study of alternate methods of polarization.

7. While one may imagine many R/D topics, it is important to note that much relevant work is now taking place vis-a-vis the B-Factories under construction so that the list for the τcF can be shortened correspondingly. Perhaps of paramount importance now is the development and test of a detector background simulator flexible enough to be applied to test various designs. This simulator could be tested by using it to design background-decrease methods in BEPC/BES and then compare the calculated backgrounds with the actual results. R/D concerning various vacuum effects and calibration of long range beam-beam effects at 1.5 GeV are also a high priority.

8. It seems plain that a *working* Workshop would be of high value. Experience such as that gained at Snowmass 88 showed the great value of a joint Detector-Accelerator workshop. In that case detector and accelerator physicists studying B-Factories came together to work on design integration. Background issues were clarified and the physics requirements for the detector and accelerator were defined clearly.

In the current instance where we are discussing three quite different operating modes over a now somewhat vague operating energy range, the need for such an exercise seems obvious.

DETECTOR STUDIES

Monte Carlo Simulation of the Tau/charm Factory at IHEP

Y.Z.Huang, H.M.Liu, W.J.Xiong, J.L.Hu, B.S.Cheng, D.H.Zhang, S.Jin, S.M.Chen, X.L.Fan, A.M.Ma, and S.Z.Ye

Institute of High Energy Physics
Academia Sinica
B.O.Box 918
Beijing 100039, China

Abstract. We report new results of a Monte Carlo study to assess the merits of the Tau/charm factory. We investigate several key physics issues, such as the identification of glueballs and the determination of the upper limit of the τ neutrino mass. We conclude with an overview of our plans for the near future.

Introduction

Monte Carlo (MC) simulations are a powerful tool to study the merits of the Tau/charm factory (τcF) in addressing important issues in particle physics. The technique can predict the expected accuracy of the results and, thus, is essential in establishing the physics case for such a facility. Simulating the same physics processes with different detector parameters allows to optimize important detector parameters leading to an improved overall resolution of the detector and to improved designs of the detector subsystems.

The present study includes an estimate of the expected systematic errors. The statistical errors will be greatly reduced at the τcF, so that the systematic errors will be dominant. For instance, the statistical error on the mass of the τ lepton can easily be reduced to less than 0.1 MeV. In this case, the systematic errors coming from the spread of the beam energy, the uncertainty on the width of the spread, the acceptance of the detector, the uncertainty of the luminosity measurement, etc., will become dominating. Our fast MC simulation was designed such that the main detector parameters are adjustable. We have used this feature to compare different detector setups in order to minimize the systematic error.

The History of τcF MC Studies at IHEP

We have created a Fast MC Simulation Method which does not consider the detailed structure of each subdetector. After generation of the hard process and subsequent particle decays, the particle parameters are smeared according to the performances of a realistic state-of-the-art detector. The method and the first physics results obtained with it were presented during the Tau/charm factory workshop [1] at SLAC in 1994. These first results include:

1. Measurement of the τ mass: $\sigma_{m_\tau} < 0.1$ MeV

2. Measurement of the ν_τ mass: Upper limit of $m_{\nu_\tau} \approx 3$ MeV (95% C.L.)

3. Measurement of the Michel parameter ρ: $\sigma_\rho \approx 0.5\%$

4. Measurement of the decay constant of the D meson: $\sigma_{f_D} \approx 1 \sim 2\%$

5. Measurement of the decay constant of the D_s meson: $\sigma_{f_{D_s}} \approx 1 \sim 2\%$

Recent Results and Progress

1. Study of Glueball Candidates

Recent Lattice QCD calculations [2] predict the following masses for the 0^{++} and 2^{++} glueballs:

$$m(0^{++}) = 1550 \pm 50 \text{ MeV}$$
$$m(2^{++}) = 2270 \pm 100 \text{ MeV}$$

New results [3] obtained by the Beijing Spectrometer Collaboration (BES) concerning the decays $\xi(2230) \to \pi^+\pi^-$, $p\overline{p}$, and $K\overline{K}$ as observed in J/ψ radiative decays strongly support the glueball interpretation of the $\xi(2230)$ state.

1.1 Study of the decay $\xi(2230) \to \pi^0\pi^0$ at the τcF

Due to the poor energy resolution and the insufficient luminosity, neutral decays, such as $\xi(2230) \to \pi^0\pi^0$, can not be studied with BES. Detailed studies of the various decay modes of the $\xi(2230)$ can only be performed at the τcF with high rates and a state-of-the-art detector. The main contribution to the background to $\xi(2230) \to \pi^0\pi^0$ comes from the decay $J/\psi \to \omega\pi^0\pi^0 \to \gamma\pi^0\pi^0\pi^0$. In our study we used the following event selection criteria:

a) At least 5 identified γ's and no charged tracks,

b) At least 2 reconstructed π^0's with $|m_{\gamma\gamma} - m_{\pi^0}| < 10$ MeV,

c) No double counting for used γ's,

d) $U_{miss} < 0.1$ GeV, where $U_{miss} = E_{miss} - |P_{miss}|$,

e) $P_{t_\gamma}^2 < 0.001$ GeV^2, with $P_{t_\gamma}^2 = 4 \cdot |P_{miss}|^2 \cdot \sin^2\theta$, where θ is the angle between the γ and the missing momentum of the $2\pi^0$ system.

The last two cuts are powerful to eliminate multi-π^0 backgrounds such as $J/\psi \to \omega\,\pi^0\,\pi^0$ etc. The efficiency for identifying $\xi(2230) \to \pi^0\pi^0$ is about 33%. Using 10^8 J/ψ events a clear ξ signal can be seen above a smooth background, see Fig. 1. The high luminosity is essential to the study of ξ decays to neutral channels, as e.g. no significant ξ signal can be seen using only $10^7 J/\psi$ events.

1.2 Study of the decay $G(1590) \to \pi^+\pi^-$

We have identified the following main background channels to the search for the decay $G(1590) \to \pi^+\pi^-$:

$$\begin{aligned} J/\psi &\to \rho\pi \to \pi^+\pi^-\pi^0 \\ &\to \omega\pi^+\pi^- \to \gamma\pi^0\pi^+\pi^- \\ &\to \omega\pi^0\pi^0 \to 3\pi^0\pi^+\pi^- \\ &\to \pi^+\pi^-\pi^0 \\ &\to \omega\eta \to 2\gamma\pi^+\pi^- \\ &\to \gamma\pi^+\pi^-2\pi^0 \end{aligned}$$

The following event selection criteria have been applied:

a) 2 charged tracks with $\sum Q = 0$,

b) $U_{miss} < 0.05$ GeV,

c) $P_{t_\gamma}^2 < 0.0005$ GeV^2,

d) $|P_{miss} - P_\gamma| < 0.2$ GeV,

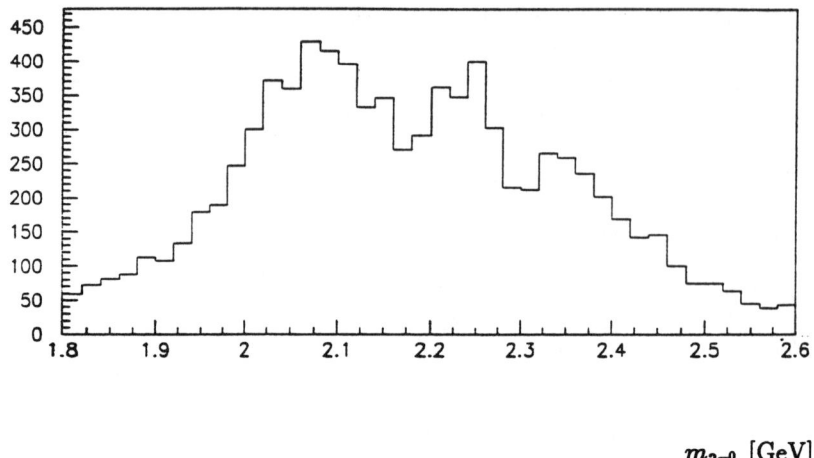

Figure 1: Invariant mass of 2 π^0's from selected J/ψ events

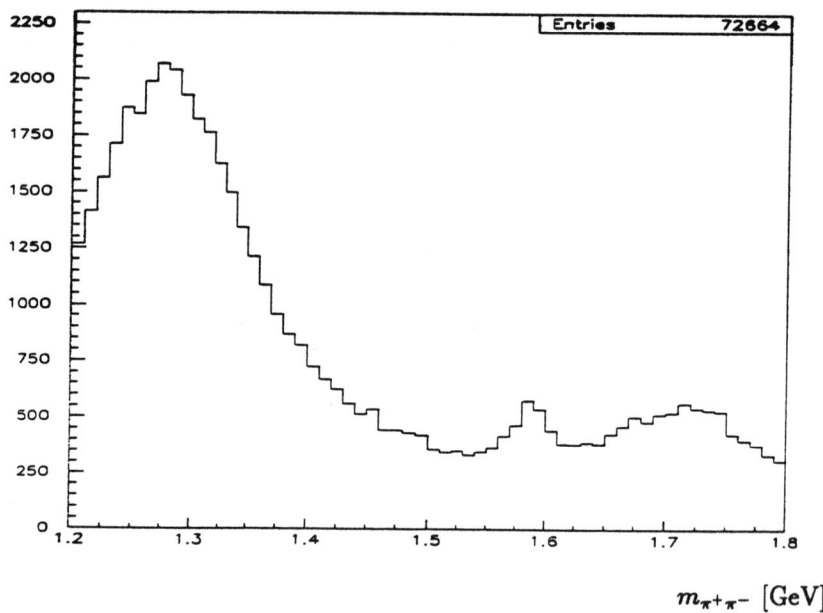

Figure 2: Invariant mass of $\pi^+\pi^-$ from selected J/ψ events

e) For events with π^0's, the γ's must satisfy an α_{π^0} cut: $\alpha_{\pi^0} \geq 8°$.

The α_{π^0} cut is very powerful to remove events with π^0's. For two gammas, γ_1 and γ_2, coming from a π^0, we can calculate the direction of γ_2 using the π^0 momentum, $p_{\pi^0} = \vec{P}_{miss}$, and the experimentally observed direction of γ_1. The angle between the calculated and experimentally observed direction of γ_2 is called $(\alpha_{\pi^0})_2$. Similarly, we can calculate $(\alpha_{\pi^0})_1$. The smaller of the two, called α_{π^0}, is required to be greater or equal 8°.

Assuming the width of the G(1590) is 20 MeV, we observe a clear signal based on an 10^8 event sample, see Fig. 2. Varying the mass of the G(1590) showed that it can be efficiently detected in the mass region between 1500 − 1600 GeV.

2. Upper limit on the ν_τ mass based on a 2-dimensional fit

Based on our previously used 1-dimensional fit [1], we developed a 2-dimensional fit method to extract an improved ν_τ mass limit. The main differences between the 2-dimensional and the 1-dimensional fit are the use of a different probability function and an extended range of sensitivity. The probability function is defined as

$$P_i(m_\nu) = \frac{1}{N(m_\nu)} \int_{x_0}^{x_1(m_\nu)} dx \int_{m_\tau}^{E_{beam}} dE_\tau g(E_\tau) \int_{y_0(E_\tau,m_\nu)}^{y_1(E_\tau,m_\nu)} dy \times$$

$$R(x - x_i, y - y_i)\epsilon(x,y)\frac{d^2\Gamma(x,y,m_\nu^2)}{dxdy},$$

where

- $x = \dfrac{m_{had}}{m_\tau}$,

- $y = \dfrac{E_{had}}{E_{beam}}$,

- $\dfrac{d^2\Gamma(x,y,m_\nu^2)}{dxdy}$ is the theoretical distribution for a given decay mode as a function of x and y,

- $\epsilon(x,y)$ is the selection efficiency as a function of x and y,

- $R(x - x_i, y - y_i)$ is the normalized resolution function determined from the experiment,

- $g(E_\tau)$ is the initial state radiation function,

- $(y_0, y_1) = \left(\dfrac{E^* \left(1 - \beta\sqrt{1 - (m_{had}/E^*)^2}\right)}{m_\tau}, \dfrac{E^* \left(1 + \beta\sqrt{1 - (m_{had}/E^*)^2}\right)}{m_\tau} \right)$

are the lower and upper limit of the energy range with the hadronic energy in the τ rest frame given as $E^* = \dfrac{m_\tau^2 + m_{had}^2 - m_\nu^2}{2m_\tau}$ and $\beta = \dfrac{p_\tau}{E_\tau}$.

In this study we investigated the decay channel $\tau \to \rho''(1750)\nu_\tau \to KK\pi\nu_\tau$. Both leptonic τ decays, $\tau \to e\nu_e\nu_\tau$, and $\mu\nu_\mu\nu_\tau$, where used to tag τ pair events. The global selection efficiency was determined to be $\epsilon_{lKK\pi}(\sqrt{s} \approx 4\text{ GeV}) \approx 15\%$ and the the fraction of events in the sensitive range of the fit $f_{sensitive} \approx 12\%$. The sensitive range, see Fig. 3, was selected somewhat larger than for the 1-dimensional fit:

- $m_{KK\pi} > 1.75$ GeV, or

- $m_{KK\pi} > 1.65$ GeV and $E_{KK\pi} > 0.99 y_1 \cdot E_{\text{beam}}$.

In the case of $\sqrt{s} = 4.03$ GeV, the number of events are determined as $N_{lKK\pi} \approx 2N_{\tau\tau} \cdot (B_e + B_\mu) \cdot B_{KK\pi} \cdot f_{sensitive} \cdot \epsilon_{lKK\pi} \approx 1000$ events/year.

Shown in Fig. 4 is the probability function versus m_{ν_τ}. From this distribution, we estimate the upper limit of m_{ν_τ} to be 0.875 MeV (95% C.L.), without including systematic errors.

There are two major sources of systematic errors, the beam energy spread and the error on the τ lepton mass. The spread is dependent on the operational mode of the collider [4]: $\sigma_E \approx 1$ MeV for standard and $\sigma_E \approx 0.1$ MeV for monochromatic operation.

Similarly the precision of the determination of the mass of the τ lepton depends on the mode of operation: $\sigma_{m_\tau} \approx 0.2$ (0.03) MeV for standard (monochromatic) operation. Including contributions from systematic uncertainties the best upper limits on the ν_τ mass are estimated to be

standard: $m_{\nu_\tau} < 1.0$ MeV

monochromatic: $m_{\nu_\tau} < 0.88$ MeV.

3. τcF Detector Simulation with GEANT

The simulation of τcF detector uses a computer program [5] based on GEANT [6]. The simulation includes all major detector elements, such as the inner tracking detectors, the crystal calorimeter, the coil, and the muon detection systems. The code runs both on the VAX and on HP workstations

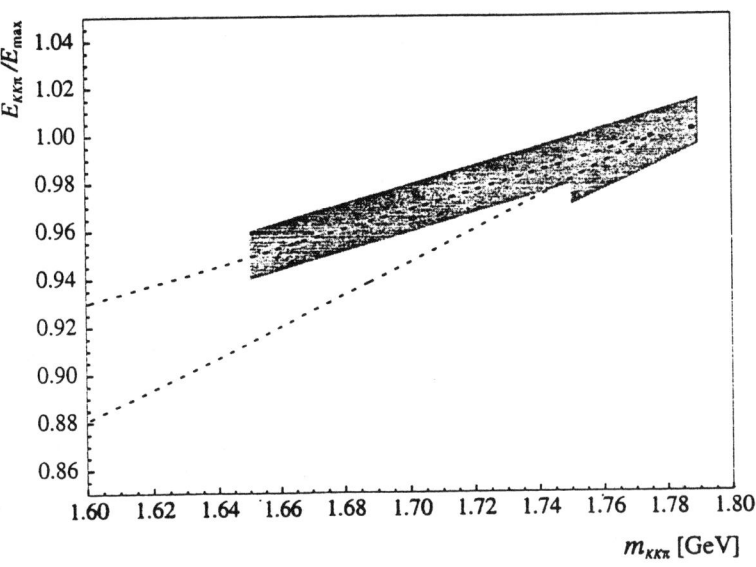

Figure 3: Region included in 2-dimensional fit (shaded) and the kinematically allowed region of τ decays (enclosed by dashed lines).

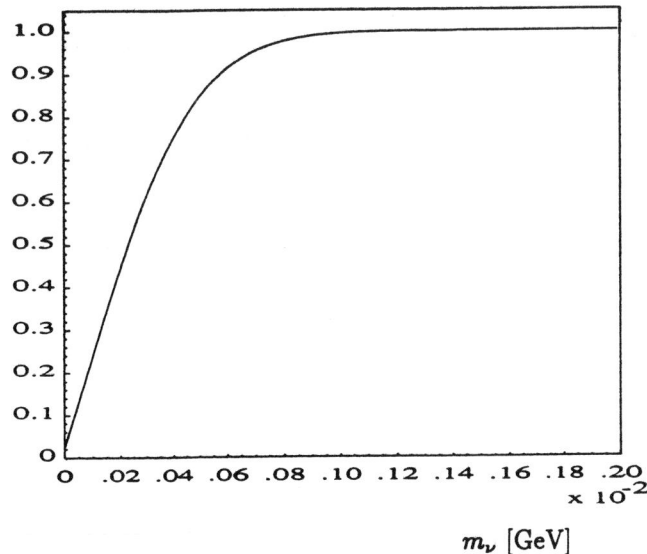

Figure 4: Probability as a function of m_ν.

and can be set-up to generate single, monochromatic particles into fixed directions. The particles are tracked through the detector materials simulating all their decay modes and interactions in the materials.

Plans for the Future

In the future, we plan to include other important physics processes into the study of the physics capability of the τcF, to improve our understanding of systematic errors, and to create a flexible detector simulation code which can be adapted to different design options.

The study of the $\xi(2230)$ will remain our main goal. We will study the probability for identifying new decay modes such as $\rho\rho$, K^*K^*, $\omega\omega$, $\pi\pi\pi\pi$, $\pi\pi KK$, etc. based on high statistics and good particle identification. We will study the sensitivity to determine its J^{PC} assignment. We will investigate other decay channels of the $G(1590)$ resonance, such as $\eta\eta'$, which are essential in providing more information about its nature. According to new results from BES, the $\iota(1440)$ consists of three resonances, two 1^{++} states on the lower and upper side and one 0^{-+} state in the centre. We will concentrate on the study of the 0^{-+} state in order to determine the sensitivity of the τcF.

Futhermore, we plan to study CP violation in τ decay and $D\overline{D}$ mixing. These studies will also include a complete estimate of the systematic errors.

Finally, before using the full detector simulation to study different physics processes, we plan to improve the detector simulation in the following way:

1. Develop an interface to be able to use different event generators,

2. Modify the geometry to fit our own detector design and physics interest,

3. Optimize the tracking cuts in order to improve the speed of the detector simulation.

Summary

We can draw the following conclusions from the MC studies done so far:

1. The τcF appears to be ideal for the study of glueballs.

2. The τcF will provide a stringent upper limit on the mass of the τ neutrino. In this case, running the collider at any center of mass energy between 3.55 and 4.25 GeV, except on the ψ' resonance at $\sqrt{s} = 3.68$ GeV, would be useful.

3. The τcF has the potential of studying CP violation in the lepton and the up-quark sector.

4. An improved detector simulation based on GEANT will allow us to make more accurate predictions of the physics reach of the τcF.

5. A systematic study of different physics processes will be essential to optimize the parameters of both detector and collider.

References

[1] Proc. of "The Tau-Charm Factory in the Era of B-Factories and CESR", Eds. L.V.Beers and M.L.Perl, SLAC-451 (unpublished).

[2] T.Barnes, these proceedings.

[3] T.Huang, these proceedings.

[4] Y.Z.Wu et al., E.Perelstein, and L.C.Teng et al., these proceedings.

[5] The original code was received from Dr. E.Gonzalez.

[6] GEANT 3.13: R.Brun et al., CERN DD/EE/84-1 (1987).

A Fast Time-of-Flight Detector also Used as a Tracker

Ting-Yang Chen

Department of Physics

Nanjing University

Nanjing, 210008, China

Mao He, Nai-Jian Zhang, and Xue-Yao Zhang

High Energy Physics Group

Shandong University

Jinan, 250100, China

Abstract

We present a possible scheme for a fast Time-of-Flight detector which could also serve as part of the charged particle tracking system. The detector consists of scintillating fibers read out by newly developed microchannel plate photomultipliers. The performance of the detector in the experimental environment of the τ/charm factory was investigated by a detailed Monte Carlo simulation. The time and space resolutions of the detector are expected to be about 92 ps and 140 μm, respectively.

I. Introduction

Particle identification will be one of the major challenges facing the detector collaboration of the τ/charm factory (τcF) [1]. With the development of fast photomultiplier tubes, Time-of-Flight detectors (TOF) have become a reliable approach to particle identification. For a given momentum p, measured by the tracking detectors, the flight times of different particle types differ depending on their masses. Since the difference between flight times decreases with increasing momentum, the momentum range of particle identification of TOF detectors depends strongly on the time resolution σ_t. For example, in order to reach a π/K separation at 1 GeV/c at the 3σ level, the TOF system is required to have at least a precision of 100 ps.

It has been pointed out that the overall time resolution of a TOF detector is dominated by the contributions from the scintillator, the phototube, and the readout electronics [2]. By replacing the bulk scintillator by scintillating fibers, the time spread from the scintillator can be significantly reduced [2]. in addition, coupling the fibers to microchannel plate photomultiplier tubes (MCP PMT) [3] decreases the contribution from the phototube, due to the fast pulse rise time [5]. Since the MCP PMTs are insensitive to magnetic fields, these tubes are suitable for the high magnetic field environment of the future τcF.

The scintillating fibers can also serve as an additional tracking detector [7,8] provided the signals are read out individually. The two essential requirements for tracking detectors would be fulfilled [6,7]: a fast time response and a good spacial resolution.

The experimental environment of the planned Beijing τcF has been simulated by detailed Monte Carlo calculations. The time (space) resolution is expected to be 92 ps (140 μm). Since the TOF will be located close to the EM calorimeter, the spatial information will serve as an additional handle for γ/e separation.

II. Simulation of the TOF/Tracker Detector

A charged particle which moves in the magnetic field is subject to the Lorentz force. The equation of motion can be written as

$$\frac{d\mathbf{P}}{dt} = e\mathbf{v} \times \mathbf{B}. \tag{1}$$

If we define the origin of the coordinate system to be at the interaction point, the particle is produced at

$$x|_{t=0} = y|_{t=0} = z|_{t=0} = 0 \tag{2}$$

with an initial momentum of

$$p_x|_{t=0} = p_x^0, \quad p_y|_{t=0} = p_y^0, \quad p_z|_{t=0} = p_z^0. \tag{3}$$

At the planned τcF, the magnetic field of the detector will be parallel to the axis of the solenoid, defined as the z direction

$$B_x = B_y = 0, \quad B_z = B \tag{4}$$

Integrating eq. (1) with respect to t over the interval $t = 0$ to $t = t$, taking into account the initial conditions (2) and (3) we have

$$p_x = p_x^0 + q\mathbf{B}y \tag{5}$$

$$p_y = p_y^0 - q\mathbf{B}x \tag{6}$$

$$p_z = p_z^0. \tag{7}$$

Since the presence of a magnetic field does not change the energy E of a particle, eqs. (5) to (7) can be rewritten as

$$E\frac{dx}{dt} = p_x^0 + q\mathbf{B}y \tag{8}$$

$$E\frac{dy}{dt} = p_y^0 - q\mathbf{B}x \tag{9}$$

$$E\frac{dz}{dt} = p_z^0. \tag{10}$$

Solving eqs. (8) to (10) for the spatial coordinates gives

$$x = -\frac{p_y^0}{qB}\cos(\omega t) + \frac{p_x^0}{qB}\sin(\omega t) + \frac{p_z^0}{qB} \tag{11}$$

$$y = \frac{p_y^0}{qB}\sin(\omega t) + \frac{p_x^0}{qB}\cos(\omega t) - \frac{p_z^0}{qB} \tag{12}$$

$$z = \frac{p_z^0}{E}t, \tag{13}$$

where $\omega = qB/E$ is the cylindrical frequency. Combining the above formulae, it is easy to see that

$$\left(x - \frac{p_y^0}{qB}\right)^2 + \left(y + \frac{p_x^0}{qB}\right)^2 = \frac{1}{q^2B^2}\left(p_x^{0^2} + p_y^{0^2}\right) \tag{14}$$

In a cylindrical coordinate system, we have

$$p_x^0 = p_r^0 \cos\phi_0 \tag{15}$$

$$p_y^0 = p_r^0 \sin\phi_0 \tag{16}$$

and eqs. (11) to (13) become

$$x = \frac{p_r^0}{qB}[(1 - \cos z)\sin\phi_0 + \sin z \cos\phi_0] \tag{17}$$

$$y = \frac{p_r^0}{qB}[\sin z \sin\phi_0 + (\cos z - 1)\sin\phi_0], \tag{18}$$

where

$$z = \omega t = \frac{qB}{E}t = \frac{qB}{p_z^0}z \tag{19}$$

Eqs. (17) and (18) are similar to the eqs. derived in ref. [4]. Using the above equations, eq. (14) can be rewritten as

$$r^2 = \frac{2p_r^{0^2}}{q^2B^2}(1 - \cos z) \tag{20}$$

215

or

$$r = \left| \frac{2p_r^0}{qB} \sin \frac{z}{2} \right| \qquad (21)$$

The layout of the detector and the TOF system are shown in Figs. 1 and 2, respectively.

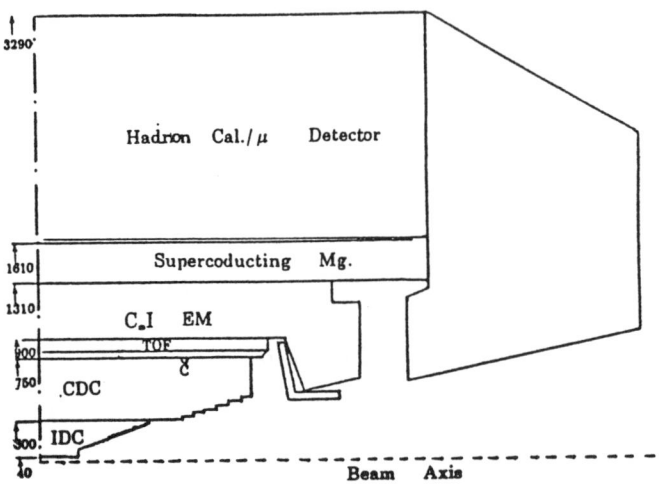

Figure 1. The schematic drawing of the planned Beijing τcF detector.

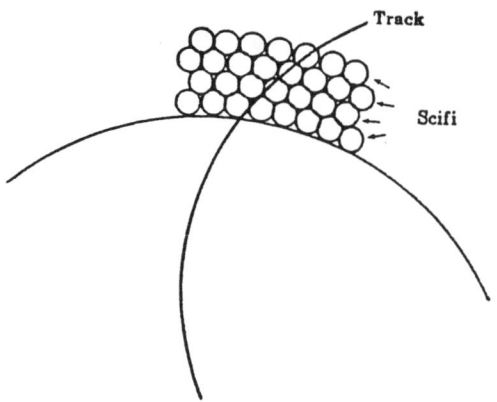

Figure 2. The layout of SCIFI Time-of-Flight/Tracking detector.

The trajectory of a charged track will cross one of the scintillating fibers centered at $(x_i, y_i), i = 1, 2, ..., n$ in each layer with radius R, as shown in Fig. 3.

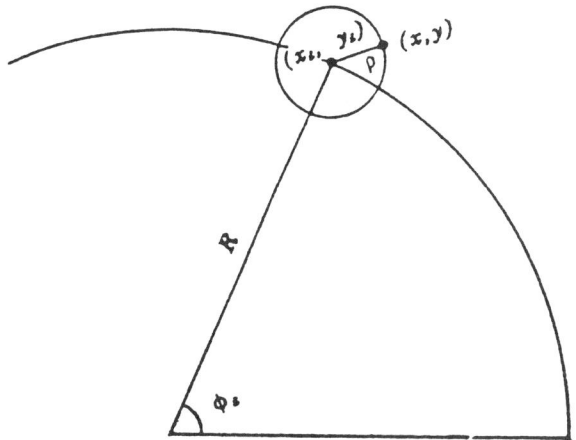

Figure 3. Position of the center of each SCIFI.

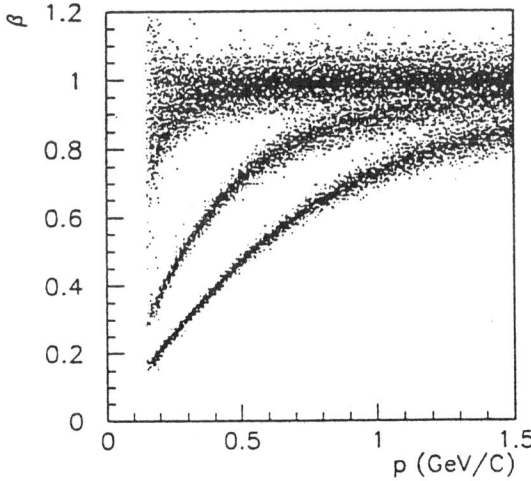

Figure 4. Particle identification by the TOF detector.

The equation for the boundary of the fiber is

$$(x - x_i)^2 + (y - y_i)^2 = \rho^2 \qquad (22)$$

or

$$r^2 - 2(xx_i + yy_i) = \rho^2 - R^2 \qquad (23)$$

where the x_i, and y_i satisfy the following equations (Fig. 3)

$$x_i^2 + y_i^2 = R^2 \qquad (24)$$

$$x_i = R\cos\phi_i \qquad (25)$$

$$y_i = R\sin\phi_i \qquad (26)$$

Solving eqs. (17), (18) and (22) simultaneously, we have

$$\left[\frac{2p_r^{0^2}}{q^2\mathbf{B}^2} + \frac{2Rp_r^0}{q\mathbf{B}}\sin(\phi_i - \phi_0)\right]\cos z + \frac{2Rp_r^0}{q\mathbf{B}}\cos(\phi_i - \phi_0)\sin z =$$
$$= \frac{2p_r^{0^2}}{q^2\mathbf{B}^2} + R^2 - \rho^2 + \frac{2Rp_r^0}{q\mathbf{B}}\sin(\phi_i - \phi_0) \qquad (27)$$

With the following substitutions

$$U = \frac{2p_r^{0^2}}{q^2\mathbf{B}^2} + \frac{2Rp_r^0}{q\mathbf{B}}\sin(\phi_i - \phi_0) \qquad (28)$$

$$V = \frac{2Rp_r^0}{q\mathbf{B}}\cos(\phi_i - \phi_0) \qquad (29)$$

$$W = \frac{2p_r^{0^2}}{q^2\mathbf{B}^2} + R^2 - \rho^2 + \frac{2Rp_r^0}{q\mathbf{B}}\sin(\phi_i - \phi_0). \qquad (30)$$

eq. (27) becomes

$$U\left(\cos z + \frac{V}{U}\sin z\right) = W \qquad (31)$$

The intersection at which the trajectory of a charged particle crosses the i-th fiber can be found by solving eq. (31)

$$z = \alpha \pm \arccos\left(\frac{W \cos \alpha}{U}\right) \qquad (32)$$

where

$$\alpha = \arccos\frac{V}{U}. \qquad (33)$$

The light yield is proportional to the length of the trajectory inside the fiber. The light will be attenuated exponentially along the fiber [2,9]

$$N = N_o \exp^{d/\lambda} \qquad (34)$$

In our simulation the attenuation length λ is taken from reference[2]. A signal crossing the fixed threshold of a leading edge discriminator will be subject to a time spread given by

$$\sigma_s \cong \frac{\sigma_{noise}}{dV/dt} \qquad (35)$$

where dV/dt is the slope of the leading edge of the signal. The MCP PMT typically have $\sigma_{noise} = 23$mV and $dV/dt \cong 450$ mV/ns. An additional contribution to the time dispersion is expected to come from the readout electronics. We assumed $\sigma_e = 0.075 \pm 0.009$ ns [2].

We have generated 10k events to simulate the performance of the TOF detector. The time (spatial) response is shown in Fig. 5 (6). Fits to Gaussian forms yield $\sigma_t = 92$ ns and $\sigma_\phi = 140$ μm.

III. Signal Readout

Fibers will be fused and read out at both ends. One end will be used for the TOF signal and will be read out by a single PMT. The other end will be read out individually, to provide the spatial resolution. Each fiber will be coupled to the anode of a PMT. The signal will be stored in a memory element [6]. The binary data from the memory will be clocked sequentially as 16(32)-bit words via parallel shift registers into an external memory.

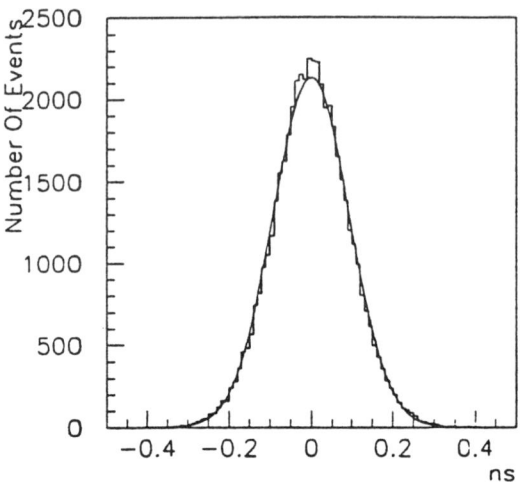

Figure 5. The time resolution of the TOF detector.

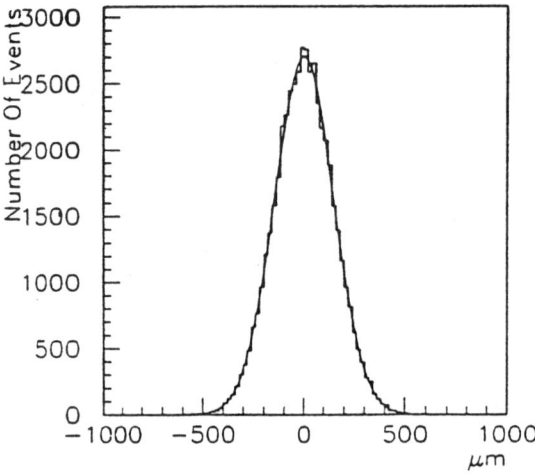

Figure 6. The spatial resolution of SCIFI.

Acknowledgments. The authors wish to express their thanks to the colleagues at High Energy Institute for their valuable discussions. The work is supported by BCPC National Lab of China.

References

[1] J. Kirkby, "Experimental Aspects of the Tau-Charm Factory", CERNPPE/94-37, 25 February 1994.

[2] M. Kuhlen et al, Nucl. Instr. and Meth. A301(1991)223.

[3] T.Y. Chen et al., "A New Position Sensitive Photomultiplier Tube for the SCIFI Readout Device", 1994 Beijing International Calorimeter Symposium.

[4] T. Alther et al. Nucl. Instr. and Meth. A332(1993)284.

[5] J.A. Hauger et al. Nucl. Instr. and Meth. A337(1994)362.

[6] F. Anselmo et al., Nucl. Instr. and Meth. A349(1994)398.

[7] D. Autiero et al., Nucl. Instr. and Meth. A336(1993)521.

[8] V. Agoritsas et al., Nucl. Instr. and Meth. A357(1995)78.

[9] X.H.Yang et al., Nucl. Instr. and Meth. A354(1995)270.

Program of the Beijing TcF Feasibility Study

Shu-Hong Wang

Institute of High Energy Physics, Academia Sinica
P.O.Box 918, Beijing 100039
E-mail: wangsh@bepc2.ihep.ac.cn

Abstract. We present the status of the Tau/charm factory design study in China. The report consists of a detailed description of the different phases of the study and the local organization established to carry it out. A list of workshops intended to regularly review the status of the design study is included.

Introduction

On August 15 - 16, 1994 an international workshop entitled "The Tau/charm Factory in the Era of B-factories and CESR" was held at SLAC, California. The meeting participants discussed the current status of tau lepton and charm meson and baryon, and charmonium physics, the prospects from future experiments at B-factories and fixed target experiments, and defined goals to be achieved within the next two decades. The conclusions reached at the meeting can be summarized as follows:

"A Tau/charm factory with a design luminosity of 10^{33}cm^{-2}s^{-1} and covering the energy range between 3 and 6 GeV is a unique facility for greatly expanding our knowledge in tau lepton, charm, and charmonium physics. The possibility to construct such a facility in China is recognized as a unique opportunity. The establishment and construction of a TcF in China in the near future is strongly encouraged."

In the following I shall briefly summarize our plans for the feasibility study and present the schedule for future workshops on the TcF. I shall conclude by describing the main features of the TcF effort in China.

Feasibility Study

The Chinese government has officially approved an extensive feasibility study for a TcF to be performed in the coming 1.5 to 2 years. The decision was taken in February, 1995 and the study was fully funded with 5 million RMB (about 0.6 million U.S.$). The goals of the feasibility study are defined as:

- Firm establishment of the necessity for a TcF
- Given the general framework of international collaboration, demonstration of the feasibility of its construction in China
- Preparation of a Letter of Intent
- Preparation of a proposal for R&D studies, the next stage toward full approval and construction of a TcF

A preliminary hearing concerning the proposal of R&D efforts for the TcF was held in the context of prioritizing the long term scientific projects to be carried out in China over the next five years. The proposal for R&D studies for the TcF received a favourable response from both the review committee and the Chinese authorities. The official approval of the R&D study will come after successful completion of the feasibility study, expected by the end of 1996. The R&D phase will last 2 - 3 years and will lead to the final decision on the construction of the TcF by the end of the century.

The physics considerations are the leading factors of the feasibility study. While concentrating on several important physics topics the study will address the following points:

- A quantitative estimate of the sensitivity of the TcF
- An estimate of the integrated luminosity required for a given measurement
- An evaluation of the expected sensitivity for physics beyond the Standard Model.

Furthermore, the requirements on the machine and detector performances will be identified leading to the definition of the main parameters of the machine and the detector:

- Establishment of the nominal luminosity and the operating energy range
- Determination of the priority of different machine operation modes, such as high luminosity, monochromatic or polarized modes, or combinations of the above
- Identification of technical difficulties in achieving the above operation modes and search for possible solutions
- Selection of detector technology to match the high precision goal of the facility
- Understanding of the technical potential available in China and identification of tasks requiring international collaboration. Identification of interested international collaborators and definition of the mode of collaboration.
- Estimation of the cost for construction in China paying special attention to the major cost driving items and retaining the balance between performance and cost.
- Identification of construction stages with well defined physics goals and costs.

The first construction stage will be defined by considering important physics goals to be achieved while remaining within a limited budget.

Schedule of Workshops

Following is a list of workshops on the TcF which were held in the past or are planned to be held in the near future:

- **SLAC**, August 15 - 16, 1994: The workshop firmly established the uniqueness of the TcF for greatly expanding our knowledge in tau lepton, charm meson and baryon, and charmonium physics.

- **Argonne**, June 21 - 23, 1995: The workshop assessed the progress made in studies related to the TcF. The physics goals, the preliminary designs of both machine and detectors were discussed.

- **Beijing**, September, 1995: This domestic workshop will assess the progress made by the collaborating institutions in China regarding the Beijing feasibility study. The workshop will concentrate on the physics goal and the parameters of both machine and detector.

- **Beijing**, January, 1996: This workshop will be organized jointly by IHEP and the Center of Chinese Advanced Science and Technology (CCAST), headed by Prof. T.D. Lee. The status of the TcF studies will be reviewed. A preliminary draft of the conceptual design including the results of the feasibility studies concerning the physics, the machine, and the detector will be discussed. Experts of the physics, the machine, and the detector from abroad will be invited to participate.

- **Beijing**, October 1996: This workshop will review the results from the three feasibility studies, the complete conceptual design for the TcF, and the proposal for R&D studies including technical goals, research topics, and estimation of costs.

Local Organization

Currently the following institutions and universities in China are involved in the feasibility study. Their main responsibilities are quoted in parenthesis:

- Institute of High Energy Physics, IHEP, Beijing
- Beijing University (Superconducting cavities)
- Qinhua University (Linac injector, impedance, polarization)
- Univ. of Sciences and Technology of China, Hefei (J/ψ physics, detector)
- Shandong University (Detector)
- Institute of Theoretical Physics, Beijing (Tau and charm physics)
- Nanjing University (TOF, detector)
- Shanghai Jiaotong University (GaAs system, detector)

A "Domestic Institution Collaboration Board" has been formed with one member from each participating institute to establish and regularly review the status of the collaboration. In a first stage the board consists only of Chinese institutions and universities. In a next stage collaborations with foreign institutions will be added. I would like to emphasize that all interested institutions are very welcome to join the collaboration.

A "Scientific Board" including several prestigious physicists has been formed to oversee the progress of the different topics of the feasibility study.

An "Executive Board" responsible for the management of the feasibility study has been formed. The board consists of the following members with their specific tasks mentioned in parenthesis:

Chairman	Zhi-Peng Zheng (Overview, detector)
Deputy	Shu-Hong Wang (Progress, machine)
	Tao Huang (Physics)
Advisor	Maury Tigner
	Jialin Xie

Finally, a "Working Group" formed by the Executive Board to help in the management of the the feasibility study is in charge of producing the Letter of Intent. The group consists of the following members: Weiguo Li (head), Xin Ju, Lihue Jin, and Zhijian Tao.

Conclusion

Recently an extensive feasibility study for a TcF in Beijing has been approved by the Chinese government. We outlined the major stages of the program of the Beijing TcF, such as the feasibility study, the R&D phase, and the construction phase and described the established organization to carry it out. A series of workshops are planned to regularly assess the progress of the feasibility study and to produce the Letter of Intent and the Conceptual Design Report.

CHARMONIUM PHYSICS
AND
HADRONIC SPECTROSCOPY

Charmonium Theory

Aida X. El-Khadra*

Physics Department, Ohio State University, 174 W 18th Ave, Columbus, OH 43210, U.S.A.

Abstract. Recent theoretical progress in calculations of the spectrum and decays of charmonium is reviewed. Traditionally, our understanding of charmonium was based on potential models. This is now being replaced by first principles.

INTRODUCTION

The charmonium system is at present among the theoretically best understood hadronic systems. The charm quark mass is large compared to the typical QCD scale, Λ_{QCD}. The $c\bar{c}$ bound states are therefore governed by non-relativistic dynamics. Historically, while the QCD potential was not known from first principles, relatively simple guesses for phenomenological potentials had proven quite successful in describing the experimentally measured bound state spectrum of charmonium [1]. Likewise, for annihilation decays of charmonium, an intuitive factorization *ansatz*, involving the wave function at the origin (calculable in potential models) worked quite well, for the most part.

This model-based theoretical understanding is now being replaced by first principles. For charmonium spectroscopy, the pogress comes from using lattice QCD. Charmonium decays can be treated rigorously whith the help of non-relativistic QCD (or NRQCD).

Lattice field theory offers a systematic first principles approach to solving QCD. It has been argued by Lepage [2] that charmonium is one of the easiest systems to study with lattice QCD, with the potential of complete control over all systematic errors. Finite-volume errors are much easier to control for quarkonia than for light hadrons. Lattice-spacing errors, on the other hand, can

---

*Address after Sept. 1, 1995: Physics Department, University of Illinois, 1110 W. Green St., Urbana, IL 61801

be larger for quarkonia and need to be considered. An alternative to reducing the lattice spacing in order to control this systematic error is improving the action (and operators). For quarkonia, the size of lattice-spacing errors in a numerical simulation can be *anticipated* by calculating expectation values of the corresponding operators using potential model wave functions. They are therefore ideal systems to test and establish improvement techniques. Most of the work of phenomenological relevance is done in what is generally referred to as the "quenched" (and sometimes as the "valence") approximation. In this approximation gluons are not allowed to split into quark - anti-quark pairs (sea quarks). In the case of charmonium, potential model phenomenology can be used to estimate this systematic error. Control over systematic errors in turn allows the extraction of Standard Model parameters from the quarkonia spectra.

Non-relativistic systems, like charmonium, are best described with an effective field theory, non-relativistic QCD (or NRQCD). It was shown in Ref. [3] that the application of NRQCD to the problem of annihilation decays of quarkonium leads to a general factorization formula. It reproduces the earlier (*ad hoc*) factorization *ansatz* for S-wave decays while putting it on a firm theoretical footing. In the case of P-wave decays, the previous *ansatz* is modified.

Lattice QCD and NRQCD are introduced in the following subsections. The remainder of the talk is organized in two parts. The first part reviews recent progress in calculations of the charmonium spectrum based on lattice QCD. The second part reviews progress in understanding charmonium decays based on NRQCD.

An Introduction to Lattice QCD

Lattice Field theory is formulated using the Feynman path integral in Euclidean space. The quantities that are actually calculated are expectation values of Greens functions ($\mathcal{G}$), which are products of gauge and fermion fields. The physical quantities of interest, hadron masses, matrix elements, etc., are then extracted from these Greens functions.

The discretization of space-time (with lattice spacing a) regulates the path integral at short distances or in the ultraviolet. A finite volume (of length L) is necessary for numerical techniques and also introduces an infrared cut-off or momentum-space discretization. The vacuum expectation of a Greens function, $\mathcal{G}$, is defined as:

$$\langle \mathcal{G} \rangle = \lim_{L \to \infty} \lim_{a \to 0} \langle \mathcal{G} \rangle_{L,a} \; , \quad \langle \mathcal{G} \rangle_{L,a} = Z_{L,a}^{-1} \int \mathcal{D}\psi \mathcal{D}\bar{\psi} \mathcal{D}U \, \mathcal{G} \, e^{-S_{lat}} \; . \quad (1)$$

$Z_{L,a}$ normalizes the expectation value. I have omitted spin and color indices for compactness. The gauge degrees of freedom are written as (path ordered) exponentials of the gauge field, A_μ:

$$U_\mu(x) = e^{i\int_x^{x+a} dx' A_\mu(x')} \simeq e^{iaA_\mu(x)} \;, \tag{2}$$

which makes it easy to maintain gauge invariance. The link fields, U, are $SU(3)$ matrices. The (Euclidean) QCD action,

$$S = S_g + S_f \;, \quad S_g = \frac{1}{4g^2}\int d^4x\, F_{\mu\nu}F^{\mu\nu} \;, \quad S_f = \int d^4x\, \bar\psi(x)(\slashed{D}+m)\psi(x) \;. \tag{3}$$

is discretized, such that Eq. (3) is recovered in the the continuum ($a \to 0$) limit:

$$S_{\text{lat}} = S + \mathcal{O}(a^n) \;, \quad n \geq 1 \;. \tag{4}$$

I will not go into the explicit formulations of S_{lat} here, but instead refer the reader to pedagogical introductions [4]. The most common form for the gauge action is Wilson's [5], written in terms of plaquettes – products of U fields around the smallest closed loop on a lattice. Wilson's gauge action has discretization errors of $\mathcal{O}(a^2)$.

For fermions the situation is more complicated. The discretization of

$$M \equiv \slashed{D} + m \;, \tag{5}$$

is a sparse, finite dimensional matrix. Two different approaches are in use. In Wilson's formulation [6] chiral symmetry is explicitly broken, but restored in the continuum limit. The pay-off is a solution of the so-called fermion doubling problem. Staggered fermions [7] keep a $U(1)$ chiral symmetry at the expense of dealing with 4 degenerate flavors of fermions.

Eq. (1) emphasizes that QCD is a limit of lattice QCD. However, in numerical calculations these limits cannot be taken explicitly, only by extrapolation. This is feasible, because theoretical guidance for both limits is available. The zero-lattice-spacing limit is guided by asymptotic freedom, since the lattice spacing is related to the gauge coupling by the renormalization group. Quantum field theories in large but finite volumes have also been analyzed theoretically [8].

In a numerical calculation the limits are taken by considering a series of lattices, as illustrated in Figure 1. While keeping the physical volume (or L) fixed, the lattice spacing is successively reduced; then, keeping the lattice spacing fixed the volume is increased. The calculation is in the continuum (infinite volume) limit once the hadron spectrum or matrix elements of interest become independent of the lattice spacing (volume).

In practice, however, limitations in computational resources do not permit the ideal lattice QCD calculation just described. In particular, the computational cost of reducing the lattice spacing naively scales like $(L/a)^4$. (The

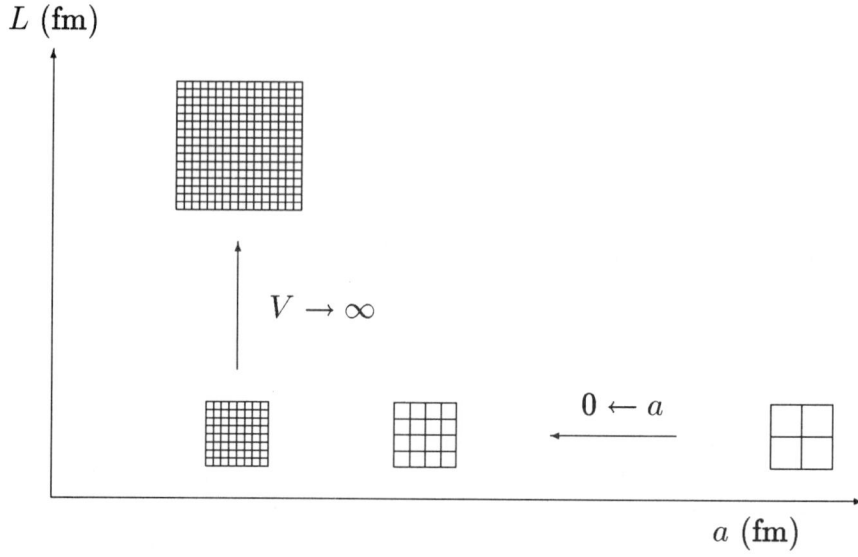

Figure 1: Illustration of the continuum and infinite-volume limits.

computational cost is really higher, because of numerical problems at smaller lattice spacings.) Eq. (4) illustrates an alternative. By improving the discretization errors in the lattice action (and operators), the continuum limit can be reached at coarser lattice spacings than before. Simulations with improved actions can come at only a slightly higher computational price. The ideas underlying improvement were developed some time ago [9, 10, 11], and have since been revitalized [12, 13, 14, 15].

If the quark mass is large compared to the typical QCD scale, Λ_{QCD}, effective theories (such as NRQCD) are most adequate in describing the physics [16]. In that case, the lattice spacing cannot be taken to zero. Lattice-spacing errors can, however, be systematically reduced by improvement [17].

The problem is now (more or less) set up. I again refer the reader to the literature [4] for more details on the organization of typical lattice QCD calculations.

An Introduction to NRQCD (Non-Relativistic QCD)

In non-relativistic systems, the velocity v of the heavy quark inside the bound state is a small parameter; in charmonium $v^2 \sim 0.3$. The important momentum scales with regard to the structure of the bound state are the heavy quark momentum (mv) and its kinetic energy (mv^2). Momenta of the order of the heavy quark mass (m), or above, are relatively unimportant in the bound state dynamics. If v is small enough, then all the different momentum scales are well separated, in particular:

$$\Lambda_{\text{QCD}} , \, mv^2 , \, mv \ll m \tag{6}$$

The most adequate description of such systems is in terms of an effective field theory, non-relativistic QCD (or NRQCD). The theory has a cut-off $\Lambda \sim m$. The effective lagrangian can be written as an expansion in powers of $1/m$. The NRQCD lagrangian is [16]

$$\mathcal{L}_{\text{NRQCD}} = \mathcal{L}_{\text{light}} + \mathcal{L}_{\text{heavy}} + \delta\mathcal{L} . \tag{7}$$

$\mathcal{L}_{\text{light}}$ is the fully relativistic lagrangian for the light quarks and gluons. $\mathcal{L}_{\text{heavy}}$ is the heavy quark (and anti-quark) lagrangian,

$$\mathcal{L}_{\text{heavy}} = \psi^\dagger \left(iD_0 + \frac{\boldsymbol{D}}{2m} \right) \psi + \chi^\dagger \left(iD_0 - \frac{\boldsymbol{D}}{2m} \right) \chi , \tag{8}$$

where ψ and χ are 2-component Pauli spinors describing quark and anti-quark degrees of freedom.

Relativistic effects of full QCD introduce corrections to Eq. (8) which appear in $\delta\mathcal{L}$ as local, non-renormalizable interactions, with coefficients that are calculable in perturbation theory [16, 17]. In principle, infinitely many terms must be considered to reproduce full QCD. In practice, however, only a finite number is needed, since every operator scales with a certain power of v, as shown in Ref. [17]. These power counting rules (e.g., $\psi \sim (mv)^{(3/2)}$, $\boldsymbol{D} \sim mv$, etc.) effectively order the terms in the NRQCD lagrangian by powers of v.

The annihilation of a $Q\bar{Q}$ pair occurs at momenta of order m. This short distance physics cannot be treated directly in NRQCD. However, the annihilation contribution to low-energy $Q\bar{Q} \to Q\bar{Q}$ scattering can be incorporated in NRQCD by adding local 4-fermion operators to $\delta\mathcal{L}$ [3],

$$\delta\mathcal{L}_{\text{4-fermion}} = \sum_i \frac{f_i}{m^{d_i - 4}} \mathcal{O}_i . \tag{9}$$

For example,

$$\mathcal{O}_1(^1S_0) = \psi^\dagger \chi \chi^\dagger \psi . \tag{10}$$

The f_i are again calculable in perturbation theory as expansions in $\alpha_s(m)$.

CHARMONIUM SPECTROSCOPY

Two different formulations for fermions have been used in lattice calculations of these spectra. Lepage and collaborators [17] have adapted the NRQCD formalism to the lattice regulator. Several groups have performed numerical calculations of quarkonia in this approach. In Ref. [18] the NRQCD action is used to calculate the charmonium spectrum, including terms of $\mathcal{O}(mv^4)$ and $\mathcal{O}(a^2)$. In addition, this group has calculated the $b\bar{b}$ spectrum in the quenched approximation ($n_f = 0$) [19] and also using gauge configurations that include 2 flavors of sea quarks [21, 20].

The Fermilab group [13] developed a generalization of previous approaches, which encompasses the non-relativistic limit for heavy quarks as well as Wilson's relativistic action for light quarks. Lattice-spacing artifacts are analyzed for quarks with arbitrary mass. Ref. [22] uses this approach to calculate the $c\bar{c}$ (and $b\bar{b}$) spectra in the quenched approximation. We considered the effect of reducing lattice-spacing errors from $\mathcal{O}(a)$ to $\mathcal{O}(a^2)$.

The two groups mentioned above use gauge configurations generated with the Wilson action, leaving $\mathcal{O}(a^2)$ lattice-spacing errors in the results. The lattice spacings, in this case, are in the range $a \simeq 0.05 - 0.2$ fm. Ref. [15] uses an improved gauge action (to $\mathcal{O}(a^4)$) together with a non-relativistic quark action improved to the same order (but without spin-dependent terms) on coarse ($a \simeq 0.4 - 0.24$ fm) lattices.

The first step in any lattice QCD calculation is the determination of the two free free parameters of the theory, the gauge coupling and quark mass, from experiment. This is discussed in the following subsection. The results for the charmonium spectrum from all groups are summarized in Figure 2.

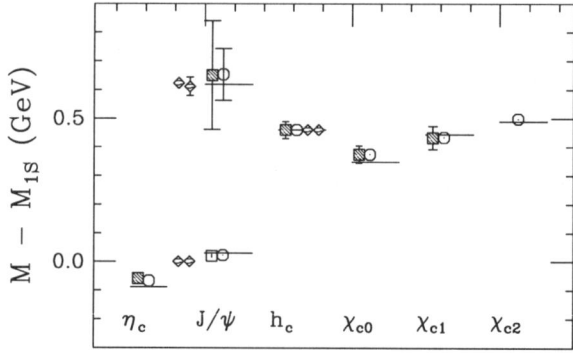

Figure 2: A comparison of lattice QCD results for the $c\bar{c}$ spectrum using the quenched approximation ($n_f = 0$). The error bars are statistical only. —: Experiment; □: FNAL [22]; ○: NRQCD [18]; ◇: ADHLM [15].

The agreement between the experimentally-observed spectrum and lattice QCD calculations is respectable. As indicated in the preceding paragraphs, the lattice artifacts are different for all groups. Figure 2 therefore emphasizes the level of control over systematic errors. I should also note, however, that the theoretical errors (from the Monte Carlo integration alone) are still much larger than the experimental ones. They also grow with every level of excitation considered.

The first quarkonium results with 2 flavors of degenerate sea quarks have appeared [21, 23, 24] with lattice-spacing and finite-volume errors similar to the quenched calculations, significantly reducing this systematic error. In Refs. [23, 24] the 1P-1S splitting in charmonium is calculated, while Ref. [21] considers the $b\bar{b}$ spectrum.

Several systematic effects associated with the inclusion of sea quarks must still be studied. They include the dependence on the quarkonium spectrum of the number of flavors of sea quarks and the sea-quark action (staggered vs. Wilson). The inclusion of sea quarks with realistic light-quark masses is very difficult. However, quarkonia are expected to depend only very mildly on the masses of the light quarks. This systematic error has not been included yet and should be checked numerically.

The first (and second) generation of lattice QCD calculations, as described here, is focused on the simplest physical quantities, like the low lying states or simple (decay) matrix elements, to establish the method. Once first principles calculations have been achieved, this technology can and will (given sufficient motivation) be used to look at more complicated problems, like higher excited states, mixing with glueballs, hybrids, etc.

Standard Model Parameters from Charmonium

The first step, the determination of the lattice gauge coupling and quark mass, follows from comparing appropiate quantities calculated on the lattice with the corresponding experimental measurements. The lattice parameters can then be converted to their counterparts in continuum QCD with perturbation theory.

Precise determinations of Standard Model parameters are an interesting by-product of lattice QCD calculations of charmonium (and bottomonium). However, the theoretical uncertainties must be reduced by an order of magnitude before they become comparable to the present experimental errors. After discussing the determination of the lattice spacing (which sets the scale in Figure 2), I will summarize the results for the strong coupling and the charm quark mass from charmonium.

Determination of the Lattice Spacing, a

The input gauge coupling sets the lattice spacing, a, which is determined in physical units by comparing a suitable quantity on the lattice with its ex-permimental value. For this purpose, one should identify quantities that are

Table 1: Spin-averaged splittings in the J/ψ and Υ systems in comparison.

	$c\bar{c}$ (MeV)	$b\bar{b}$ (MeV)
$m(1P - 1S)$	456.8	452
$m(2S - 1S)$	596	563
$m(2P - 1P)$	—	359.7

insensitive to lattice errors. In quarkonia, spin-averaged splittings are good candidates. The experimentally observed 1P-1S and 2S-1S splittings depend only mildly on the quark mass (for masses between m_b and m_c), as shown in table 1. Figure 3 shows the observed mass dependence of the 1P-1S splitting in a lattice QCD calculation. The comparison between results from different lattice actions illustrates that higher-order lattice-spacing errors for these splittings are small[21, 22]. In contrast, Figure 4 shows the hyperfine splitting as an example of a quantity that strongly depends on both the mass and the lattice action. It would therefore be a poor choice for a determination of the lattice spacing.

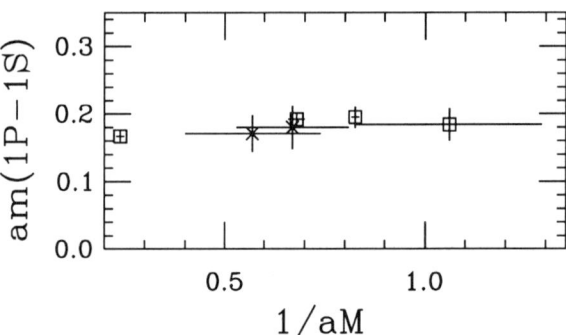

Figure 3: The 1P-1S splitting as a function of the 1S mass (statistical errors only) from Ref. [22]; □: $\mathcal{O}(a^2)$ errors; ×: $\mathcal{O}(a)$ errors.

The Strong Coupling, α_s

Within the framework of lattice QCD the conversion from the bare to a renormalized coupling can, in principle, be made non-perturbatively [25]. An alternative is to define a renormalized coupling through short distance lattice quantities [26]. The size of higher-order corrections associated with the above

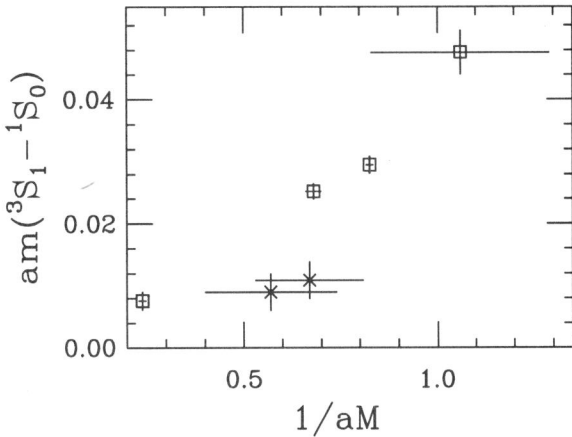

Figure 4: The hyperfine splitting as a function of the 1S mass (statistical errors only) from Ref. [22]; □: $\mathcal{O}(a^2)$ errors; ×: $\mathcal{O}(a)$ errors.

defined coupling constant can be tested by comparing perturbative predictions for short-distance lattice quantities with non-perturbative results [26].

At this point the relation to the $\overline{\rm MS}$ coupling is known to 1-loop, leading to a 5% uncertainty. It has recently been calculated to 2-loops [27] in the quenched approximation (no sea quarks, $n_f = 0$). The extension to $n_f \neq 0$ will significantly reduce the uncertainty due to the use of perturbation theory.

Sea Quark Effects. Calculations that properly include all sea-quark effects do not yet exist. If we want to make contact with the "real world", these effects have to be estimated phenomenologically or extrapolated away.

The phenomenological correction necessary to account for the sea-quark effects omitted in calculations of quarkonia that use the quenched approximation gives rise to the dominant systematic error in these calculations [28, 29]. Similar ideas were used to correct for sea-quark effects in early calculations of quarkonia spectra from the heavy-quark potential calculated in quenched lattice QCD [30].

By demanding that, say, the spin-averaged 1P-1S splitting calculated on the lattice reproduce the experimentally observed one (which sets the lattice spacing, a^{-1}, in physical units), the effective coupling of the quenched potential is in effect matched to the coupling of the effective 3 flavor potential at the typical momentum scale of the quarkonium states in question. The difference in the evolution of the zero flavor and 3,4 flavor couplings from the effective low-energy scale to the ultraviolet cut-off, where α_s is determined, is the perturbative estimate of the correction.

For comparison with other determinations of α_s, the $\overline{\text{MS}}$ coupling can be evolved to the Z mass scale. An average [31] of Refs. [22, 29] yields for α_s from calculations in the quenched approximation:

$$\alpha_{\overline{\text{MS}}}^{(5)}(m_Z) = 0.110 \pm 0.006 \quad . \tag{11}$$

The experimental error in this determination from the quarkonium mass splitting is much smaller than the theoretical uncertainty and does not contribute to the total.

The phenomenological correction described in the previous paragraphs has been tested from first principles in Refs. [21, 23, 24]. All groups calculate quarkonium splittings with 0 and 2 flavors of sea quarks. After extrapolating to the physical 3 flavor case and evolving the coupling to m_Z, Refs. [23, 24] find for the strong coupling from charmonium

$$\alpha_{\overline{\text{MS}}}^{(5)}(m_Z) = 0.111 \pm 0.005 \tag{12}$$

in good agreement with the previous result in Eq. (11). The total error is now dominated by the rather large statistical errors and the perturbative uncertainty.

At present, the result of Ref. [21] has the smallest statistical and systematic errors for the strong coupling (in this case from the $b\bar{b}$ spectrum):

$$\alpha_{\overline{\text{MS}}}^{(5)}(m_Z) = 0.115 \pm 0.002 \quad . \tag{13}$$

Phenomenological corrections are a necessary evil that enter most coupling constant determinations. In contrast, lattice QCD calculations with complete control over systematic errors will yield truly first-principles determinations of α_s from the experimentally observed hadron spectrum.

At present, determinations of α_s from the experimentally measured quarkonia spectra using lattice QCD are comparable in reliability and accuracy to other determinations based on perturbative QCD from high energy experiments. They are therefore part of the 1994 world average for α_s [31]. The phenomenological corrections for the most important sources of systematic errors in lattice QCD calculations of quarkonia are now being replaced by first principles, which will significantly increase the accuracy of α_s determinations from quarkonia. In particular, the systematic errors associated with the inclusion of sea quarks into the simulation have to be checked.

The Heavy Quark Masses

Because of confinement, the quark masses cannot be measured directly, but have to be inferred from experimental measurements of hadron masses, and depend on the calculational scheme employed. In lattice QCD quark masses are determined non-perturbatively, by tuning the input lattice quark mass (m_Q^{lat}) so that, for example, the experimentally observed J/ψ mass is reproduced by the calculation.

Phenomenologically useful quark masses are the perturbatively defined pole and $\overline{\text{MS}}$ masses, which the bare lattice mass can be related to by perturbation theory:

$$m_Q^{\text{pole}} = Z_m^{\text{pole}} m_Q^{\text{lat}} \quad , \qquad m_Q^{\overline{\text{MS}}}(m_Q) = Z_m^{\overline{\text{MS}}} m_Q^{\text{lat}} \quad . \qquad (14)$$

Of course, as always, all systematic errors arising from the lattice QCD calculation need to be under control for a phenomenologically interesting result; in particular, the systematic error introduced by the (partial) omission of sea quarks has to be removed. The short-distance corrections that introduced the dominant uncertainty to the α_s determination from quarkonia are absent for the pole mass determination, because this effective mass does not run for momenta below its mass. An analysis of the b-quark mass from the $b\bar{b}$ spectrum with and without sea quarks is consistent with this estimate [20].

Ref. [32] analyzes the the charm quark mass from the charmonium spectrum with the preliminary result, $m_c^{\text{pole}} = 1.5(2)$ GeV.

The $\overline{\text{MS}}$ mass for the charm quark has also been determined from a compilation of D meson calculations in the quenched approximation [33], with $m_c^{\overline{\text{MS}}}(2\,\text{GeV}) = 1.47(28)$ GeV. The error includes statistical errors from the original calculations and the perturbative error. However sea-quark effects cannot, in this case, be estimated phenomenologically, leaving this systematic error uncontrolled.

CHARMONIUM DECAYS

Historically [34, 1], charmonium annihilation decays were treated with a factorization *ansatz*, which divides the decay rate into a short distance, perturbative part, and a long distance non-perturbative part, usually parametrized as the wave function at the origin, $R(0)$. This *ansatz* worked quite well for S-wave decays, even after including radiative corrections at next-to-leading order [35]. For P-wave decays, the long distance piece was identified as the derivative of the wave function at the origin, $R'(0)$. However, the radiative corrections were found to have infra-red divergences [36] for $J = 1$ P-wave states already at leading order, and for $J = 0, 2$ P-wave states at next-to-leading order, preventing reliable theoretical predictions. This *ansatz* also did not allow for a systematic inclusion of higher Fock states, or relativistic corrections.

If the effective field theory framework of NRQCD is used to describe annihilation decays, it was shown [3] that a general factorization formula holds for the decay rates. The annihilation decay rate of a quarkonium state A can be

written as
$$\Gamma(A) = 2\,\text{Im}\langle A|\delta\mathcal{L}_{4-\text{fermion}}|A\rangle \; . \qquad (15)$$

Using Eq. (9) this gives

$$\Gamma(A) = \sum_i \frac{F_i}{m^{d_i-4}} \langle A|\mathcal{O}_i|A\rangle \; . \qquad (16)$$

The coefficients F_i are the imaginary parts of the f_i in Eq. (9). Because of the power counting rules of Ref. [17], the matrix elements scale with some power of the velocity,

$$\langle A|\mathcal{O}_i|A\rangle \sim v^{2n} \; , \qquad (17)$$

which usually increases as higher dimensional operators are considered in the sum of Eq. (16). Effectively, this approach expresses the quarkonium decay rates as expansions in the the short distance parameter $\alpha_s(m)$ and the long distance parameter v^2. The desired accuracy of the theoretical prediction thus serves as a truncation criterion for the (infinite) sum in Eq. (16).

The matrix elements in Eq. (16) are calculable from first principles using lattice NRQCD. These calculations are in progress by the ANL group [37]. In the meantime, the matrix elements can also be extracted from experimental measurements of charmonium decay rates.

It was shown in Ref. [3] that heavy-quark spin symmetry and the vacuum-saturation approximation reduces the number of independent matrix elements that have to be determined non-perturbatively or phenomenologically.

The matrix elements of the 4-fermion operators, which contribute to the charmonium decays into light hadrons can be simplified using the vacuum saturation approximation. This approximation is valid in NRQCD up to relative order v^4. For example,

$$\langle \eta_c|\psi^\dagger\chi\chi^\dagger\psi|\eta_c\rangle = |\langle 0|\chi^\dagger\psi|\eta_c\rangle|^2(1+\mathcal{O}(v^4)) \; . \qquad (18)$$

This relates the matrix elements appearing in electro-magnetic charmonium decays to those in strong decays (into light hadrons).

Heavy-quark spin symmetry relates matrix elements of states in the same radial and orbital levels but different spins. For example,

$$\epsilon^* \cdot \langle 0|\chi^\dagger\boldsymbol{\sigma}\psi|J/\psi\rangle = \langle 0|\chi^\dagger\psi|\eta_c\rangle(1+\mathcal{O}(v^2)) \; . \qquad (19)$$

The following two subsections discuss the application of this approach to S- and P-wave decays, using specific examples.

S-Wave Decays

I discuss the theoretical knowledge of S-wave decays using $\eta_c \to \gamma\gamma$ as an example. At leading order in v^2 the rate is

$$\Gamma(\eta_c \to \gamma\gamma) = \frac{F_{\gamma\gamma}(^1S_0)}{m^2} |\overline{R_S}|^2 , \qquad (20)$$

where by spin symmetry (see Eq. (19)) $\overline{R_S}$ can be taken as $\overline{R_{\eta_c}}$ or the spin-averaged combination of $\overline{R_{\eta_c}}$ and $\overline{R_\psi}$, with

$$\overline{R_{\eta_c}} = \sqrt{\frac{2\pi}{3}} \langle 0|\chi^\dagger \psi|\eta_c\rangle . \qquad (21)$$

The other η_c and J/ψ decays differ from Eq. (20) in the short distance coefficient but, using Eqs. (18) and (19), have the same long distance parameter, $\overline{R_S}$. Identifying $\overline{R_S}$ with the wave function at the origin, $R_S(0)$, leads back to the old factorization *ansatz*. The short distance coefficients, F_i, are known to next-to-leading order in α_s.

At next to leading order in v^2, the decay rate becomes

$$\Gamma(\eta_c \to \gamma\gamma) = \frac{F_{\gamma\gamma}(^1S_0)}{m^2} |\overline{R_{\eta_c}}|^2 + \frac{G_{\gamma\gamma}(^1S_0)}{m^4} \text{Re}(\overline{R_S}^* \overline{\nabla^2 R_S}) . \qquad (22)$$

Now the difference between $\overline{R_{\eta_c}}$ and $\overline{R_S}$, which is of order v^2 (see Eq. (19)) has to be taken into account for consistency.

$$\langle 0|\chi^\dagger(\boldsymbol{D}^2)\psi|\eta_c\rangle = \sqrt{\frac{3}{2\pi}} \overline{\nabla^2 R_S} \left(1 + \mathcal{O}(v^2)\right) . \qquad (23)$$

$\overline{\nabla^2 R_S}$ can be interpreted as a laplacian of the wave function at the origin. The short distance coefficients, G_i, are known to leading order in α_s.

The long distance parameters, $\overline{R_{\eta_c}}$, $\overline{R_\psi}$ and $\overline{\nabla^2 R_S}$ can be calculated in lattice NRQCD, estimated using potential models or extracted from a phenomenological analysis of experimental data. The relative errors (from the truncation of the perturbative and non-relativistic series) are $\alpha_s(m_c)^2$, $\alpha_s(m_c)v^2$ and v^4, adding up to $\lesssim 15$ %.

It is not inconceivable that at least for the electro-magnetic S-wave decays some (or all) of the higher order calculations will be performed, reducing the theoretical error accordingly.

P-Wave Decays

The structure of P-wave decay rates is more complicated than that of S-waves. I shall start with a simple example first, the decay $\chi_{cJ} \to \gamma\gamma$.

$$\Gamma(\chi_{cJ} \to \gamma\gamma) = \frac{F_{\gamma\gamma}(^3P_J)}{m^4} |\overline{R'_P}|^2 , \qquad (24)$$

where again, by spin symmetry $\overline{R'_P}$ can be taken from the matrix elements of any of the P-wave states (or their spin-averaged combination). For example,

$$\langle 0|\chi^\dagger(\frac{1}{2}\boldsymbol{D}\cdot\boldsymbol{\sigma})\psi|\chi_{c0}\rangle = \sqrt{\frac{27}{2\pi}}\,\overline{R'_{\chi_{c0}}}\,(1+\mathcal{O}(v^2))\;. \tag{25}$$

The obvious interpretation of $\overline{R'_P}$ as the derivative of the wave function at the origin, $R'_P(0)$, again connects this formalism to the old factorization *ansatz*. The coefficients, $F_{\gamma\gamma}$, are known to next-to-leading order in α_s. The relative error is v^2 or $\sim 25\,\%$ of the decay rate.

The situation is different for strong decays of P-wave states. Taking as an example $h_c \to LH$ (light hadrons), Eq. (16) gives at leading order in v^2,

$$\Gamma(h_c \to LH) = \frac{F_1(^1P_1)}{m^4}\langle\mathcal{O}_1\rangle + \frac{F_8(^1S_0)}{m^4}\langle\mathcal{O}_8\rangle\;, \tag{26}$$

where

$$\begin{aligned}\langle\mathcal{O}_1\rangle &= \langle h_c|\psi^\dagger(\frac{1}{2}\boldsymbol{D}\cdot\boldsymbol{\sigma})\chi\chi^\dagger(\frac{1}{2}\boldsymbol{D}\cdot\boldsymbol{\sigma})\psi|h_c\rangle \\ &= \frac{9}{2\pi}|\overline{R'_P}|^2\,(1+\mathcal{O}(v^2))\end{aligned} \tag{27}$$

and

$$\langle\mathcal{O}_8\rangle = \langle h_c|\psi^\dagger T^a\chi\chi^\dagger T^a\psi|h_c\rangle\;. \tag{28}$$

The two matrix elements $\langle\mathcal{O}_1\rangle$ and $\langle\mathcal{O}_8\rangle$ enter at the same order in v^2, because they are both suppressed by v^2 with respect to the leading order S-wave matrix elements. $\mathcal{O}_1$ is a dimension eight operator; the two powers of $\boldsymbol{D}$ give the v^2 suppression. $\mathcal{O}_8$ is of dimension six, but its matrix element picks the $|c\bar{c}g\rangle$ Fock state, where the cc pair is in a color-octet $^1S_0$ state. The dominant Fock state of charmonium is the color-singlet $|c\bar{c}\rangle$.

$$|A\rangle = \Psi_{c\bar{c}}|c\bar{c}\rangle + \Psi_{c\bar{c}g}|c\bar{c}g\rangle + \ldots\;, \tag{29}$$

with $\Psi_{c\bar{c}g} \sim \mathcal{O}(v)$. This gives the v^2 suppression.

This departure from the old factorization *ansatz* solves the problem of infrared divergences, and thus leads to a consistent treatment of this case similar to the other charmonium decays.

The short distance coefficients, F_1, are known to next-to-leading order, while the F_8's are only known to leading order in α_s. At present, the dominant relative error is thus α_s and v^2, adding up to $\sim 40\,\%$.

It should be possible to extend the (perturbative) calculations of P-wave decays in this frame work to the same order as what is presently available for S-wave decays, reducing the theoretical error to $\lesssim 15\,\%$ (assuming knowledge of the relevant matrix elements).

A comparison of theory and experiment for P-wave decays shows fair agreement, albeit with rather large errors from both sides [38].

CONCLUSIONS

Quarkonia were, upon their discovery, called the hydrogen atoms of particle physics. Their non-relativistic nature justified the use of potential models, which gave a nice, phenomenological understanding of these systems. This phenomenology is at present useful to control systematic errors in lattice QCD calculations of the charmonium spectrum. However, we are quickly moving towards truly first-principles calculations of quarkonia using lattice QCD, thereby testing QCD non-perturbatively. In this sense, quarkonia are still the hydrogen atoms of particle physics. Precise determinations of the Standard Model parameters, α_s, m_c (and m_b), are by-products of this work.

Still lacking for a first-principles result is the proper inclusion of sea quarks. The most difficult problem in this context is the inclusion of sea quarks with physical light quark masses. At present, this can only be achieved by extrapolation (from $m_q \simeq 0.3 - 0.5 m_s$ to $m_{u,d}$). If the light quark mass dependence of the quarkonia spectra is mild, as anticipated, the associated systematic error can be controlled. First-principles calculations of quarkonia could then be performed with currently available computational resources.

The present theoretical status of charmonium annihilation decays is rather promising. The frame-work developed in Ref. [3] leads to a systematic expansion in $\alpha_s(m_c)$ and v^2, with controllable uncertainties. Until first-principles calculations of the non-perturbative matrix elements become available, this formalism can still be tested phenomenologically, using experminental data. Theoretical predictions for the decay rates should be available with uncertainties of $\lesssim 10\,\%$, in most cases before the Tau/Charm factory turns on.

It is conceivable, that by the time a Tau/Charm factory turns on, the theory of charmonium will be solidly based upon first principles with accurate predictions for the spectrum and decays of the low-lying states. We will have moved to the next stage in theoretical (first-principles) calculations concerning, for example, the properties of hybrid states or mixing with glueballs. Experimental information gathered at the Tau/Charm factory and earlier experiments [39] will then give us *precision* tests of perturbative and non-perturbative QCD in the charmonium system.

ACKNOWLEDGEMENTS

I thank the organizers for an enjoyable conference, and G. Bodwin, E. Braaten and J. Shigemitsu for discussions while preparing this talk.

REFERENCES

1. W. Kwong, J. Rosner and C. Quigg, *Annu. Rev. Nucl. Part. Sci.* **37** (1987) 325.

2. P. Lepage, *Nucl. Phys.* **B** (Proc. Suppl.) **26** (1992) 45; B. Thacker and P. Lepage, *Phys. Rev.* **D43** (1991) 196; P. Lepage and B. Thacker, *Nucl. Phys.* **B** (Proc. Suppl.) **4** (1988) 199.

3. E. Braaten, G. Bodwin and P. Lepage, *Phys. Rev.* **D46** (1992) 1914; *Phys. Rev.* **D51** (1995) 1125; E. Braaten, NUHEP-TH-94-22, hep-ph/9409286.

4. For pedagogical introductions to Lattice Field Theory, see, for example: M. Creutz, *Quarks, Gluons and Lattices* (Cambridge University Press, New York 1985); A. Hasenfratz and P. Hasenfratz, *Annu. Rev. Nucl. Part. Sci.* **35** (1985) 559; A. Kronfeld, in *Perspectives in the Standard Model*, R. Ellis, C. Hill and J. Lykken (eds.) (World Scientific, Singapore 1992), p. 421; see also A. Kronfeld and P. Mackenzie, *Annu. Rev. Nucl. Part. Sci.* **43** (1993) 793; A. El-Khadra, in *Physics in Collision 14*, S. Keller and H. Wahl (eds.) (Editions Frontieres, Cedex - France 1995), p. 209; for introductory reviews of lattice QCD.

5. K. Wilson, *Phys. Rev.* **D10** (1974) 2445.

6. K. Wilson, in *New Phenomena in Subnuclear Physics*, A. Zichichi (ed.) (Plenum, New York 1977).

7. L. Susskind, *Phys. Rev.* **D16** (1977) 3031; T. Banks, J. Kogut and L. Susskind, *Phys. Rev.* **D13** (1976) 1043.

8. M. Lüscher, *Comm. Math. Phys.* **104** (1986) 177; *Comm. Math. Phys.* **105** (1986) 153.

9. See for example, T. Bell and K. Wilson, *Phys. Rev.* **B11** (1975) 3431; K. Symanzik, *Nucl. Phys.* **B226** (1983) 187; *ibid.* 205.

10. P. Weisz, *Nucl. Phys.* **B212** (1983) 1; M. Lüscher and P. Weisz, *Nucl. Phys.* **B212** (1984) 349; *Comm. Math. Phys.* **97** (1985) 59; (E) **98** (1985) 433.

11. B. Sheikholeslami and R. Wohlert, *Nucl. Phys.* **B259** (1985) 572.

12. C. Heatlie, *et al.*, *Nucl. Phys.* **B352** (1991) 266.

13. A. El-Khadra, A. Kronfeld and P. Mackenzie, Fermilab PUB-93/195-T.

14. P. Hasenfratz, *Nucl. Phys.* **B** (Proc. Suppl.) **34** (1994) 3; P. Hasenfratz and F. Niedermayer, *Nucl. Phys.* **B414** (1994) 785; U. Wiese, *Phys. Lett.* **B315** (1993) 417; W. Bietenholz and U. Wiese, *Nucl. Phys.* **B** (Proc. Suppl.) **34** (1994) 516.

15. P. Lepage, in *The Building Blocks of Creation*, S. Raby and T. Walker (eds) (World Scientic, Singapore 1994), hep-lat/9403018; M. Alford, *et al.*, *Nucl. Phys.* **B** (Proc. Suppl.) **42** (1995) 787; hep-lat/9507010.

16. E. Eichten and F. Feinberg, *Phys. Rev.* **D23** (1981) 2724; W. Caswell and P. Lepage, *Phys. Lett.* **B167** (1986) 437.

17. P. Lepage and B. Thacker, *Phys. Rev.* **D43** (1991) 196; P. Lepage, *et al.*, *Phys. Rev.* **D46** (1992) 4052.

18. C. Davies, *et al.*, hep-lat/9506026.

19. C. Davies, *et al.*, *Phys. Rev.* **D50** (1994) 6963.

20. C. Davies, *et al.*, *Phys. Rev. Lett.* **73** (1994) 2654.

21. C. Davies, *et al.*, *Phys. Lett.* **B345** (1995) 42.

22. A. El-Khadra, G. Hockney, A. Kronfeld, P. Mackenzie, T. Onogi and J. Simone, Fermilab PUB-94/091-T.

23. S. Aoki, *et al.*, *Phys. Rev. Lett.* **74** (1995) 22.

24. M. Wingate, *et al.*, hep-lat/9501034.

25. For a review of α_s from the heavy-quark potential, see K. Schilling and G. Bali, *Nucl. Phys.* **B** (Proc. Suppl.) **34** (1994) 147; M. Lüscher, R. Sommer, P. Weisz, and U. Wolff, *Nucl. Phys.* **B413** (1994) 481; G. de Divitiis, *et al.*, *Nucl. Phys.* **B433** (1995) 390; *Nucl. Phys.* **B437** (1995) 447; C. Bernard, C. Parrinello and A. Soni, *Phys. Rev.* **D49** (1994) 1585.

26. P. Lepage and P. Mackenzie, *Phys. Rev.* **D48** (1992) 2250.

27. M. Lüscher and P. Weisz, *Phys. Lett.* **B349** (1995) 165; hep-lat/9505011.

28. A. El-Khadra, G. Hockney, A. Kronfeld and P. Mackenzie, *Phys. Rev. Lett.* **69** (1992) 729; A. El-Khadra, *Nucl. Phys.* **B** (Proc. Suppl.) **34** (1994) 141

29. The NRQCD Collaboration, *Nucl. Phys.* **B** (Proc. Suppl.) **34** (1994) 417.

30. D. Barkai, K. Moriarty and C. Rebbi, *Phys. Rev.* **D30** (1984) 2201; M. Campostrini, *Phys. Lett.* **B147** (1984) 343.

31. For reviews on the status of α_s determinations, see, for example: B. Webber, ICHEP'94; I. Hinchliffe, DPF'94 and Phys. Rev. **D50** (1994) 1173, p.1297.

32. A. El-Khadra and B. Mertens, *Nucl. Phys.* **B** (Proc. Suppl.) **42** (1995) 406.

33. C. Allton, et al., *Nucl. Phys.* **B431** (1994) 667.

34. V. Novikov, et al., *Phys. Rep.* **C41** (1978) 1.

35. R. Barbieri, G. Curci, E. d'Emilio and E. Remiddi, *Nucl. Phys.* **B154** (1979) 535; K. Hagiwara, C. Kim, T. Yoshino, *Nucl. Phys.* **B177** (1981) 461; P. Mackenzie and P. Lepage, *Phys. Rev. Lett.* **47** (1981) 1244.

36. R. Barbieri, R. Gatto and R. Kögerler, *Phys. Lett.* **B60** (1976) 183; R. Barbieri, R. Gatto and E. Remiddi, *Phys. Lett.* **B61** (1976) 465; R. Barbieri, M. Caffo, R. Gatto and E. Remiddi, *Phys. Lett.* **B95** (1980) 93; *Nucl. Phys.* **B192** (1981) 61.

37. G. Bodwin, S. Kim and D. Sinclair, *Nucl. Phys.* **B** (Proc. Suppl.) **34** (1994) 434.

38. M. Mangano and A. Petrelli, *Phys. Lett.* **B352** (1995) 445.

39. C. Ginsburg (E760 collaboration), these proceedings.

Fermilab E760 and E835
Charmonium Formation in $\bar{p}p$ Annihilation

C.M. Ginsburg
Northwestern University, Evanston, IL 60208, USA
(for the E760/E835 Collaboration)[†]

Abstract. Fermilab experiment E760 was dedicated to the study of charmonium spectroscopy by the resonant formation of charmonium states in $\bar{p}p$ annihilations. Charmonium resonances were formed in the collision of a stochastically cooled antiproton beam in the Fermilab Antiproton Accumulator with an internal hydrogen gas jet target. Several charmonium measurements from E760 are described and compared to previous results. The upgrade experiment of E760, Fermilab E835, is described, and its physics goals outlined.

1. INTRODUCTION

In November 1974, two experimental groups simultaneously discovered an unusually narrow new particle with mass 3.1 GeV and width less than 1.3 MeV. This particle appeared in the e^+e^- invariant mass spectrum in $p + \text{Be} \rightarrow e^+e^- + X$ at Brookhaven [1] and was called J. The same particle appeared in $e^+e^- \rightarrow$ leptons and hadrons at SLAC [2] and was called ψ. Experimenters at several other e^+e^- colliders soon confirmed the discovery. Because the J/ψ production in e^+e^- annihilation must proceed through a virtual photon, its quantum numbers had to be those of the photon, i.e., $J^{PC} = 1^{--}$.

The SLAC experiment soon discovered a higher-mass state, ψ' [3]. Its direct production in e^+e^-, large mass (3.7 GeV), and narrow width (≤ 2.7 MeV), were consistent with it being the first radial excitation of the J/ψ. Theorists soon predicted states corresponding to a $\bar{c}c$ bound system. The $\chi(^3P)$ states were soon identified in radiative decays of the ψ'; and it became completely clear that all these states were indeed bound states of charmonium.

The role of QCD in the charmonium system is analogous to the role of QED in positronium. For small $\bar{q}q$ separation, r, the theoretical formalism maintains this analogy. In the theory, the interaction between quarks at small separations is assumed to be mediated by the exchange of a gluon. As in the exchange of a photon in QED, this gives rise to a potential which is proportional to $1/r$. Unlike electrons and positrons in positronium, however, free quarks and gluons do not exist, so the QCD potential must contain a confinement term as well. At large distances, multiple gluon exchange is expected to generate an

© 1996 American Institute of Physics

effective confining potential, which is commonly assumed to be proportional to r.

The charmonium spectrum is shown in Figure 1. It exhibits a clear positronium-like set of states. The states are often designated by the common spectroscopic notation $n^{2S+1}L_J$, where S is the spin, L the relative angular momentum, $J = L + S$ the total angular momentum and n the radial quantum number. Like all fermion-antifermion systems, the charmonium state has parity $P = (-1)^{L+1}$ and charge-conjugation $C = (-1)^{L+S}$.

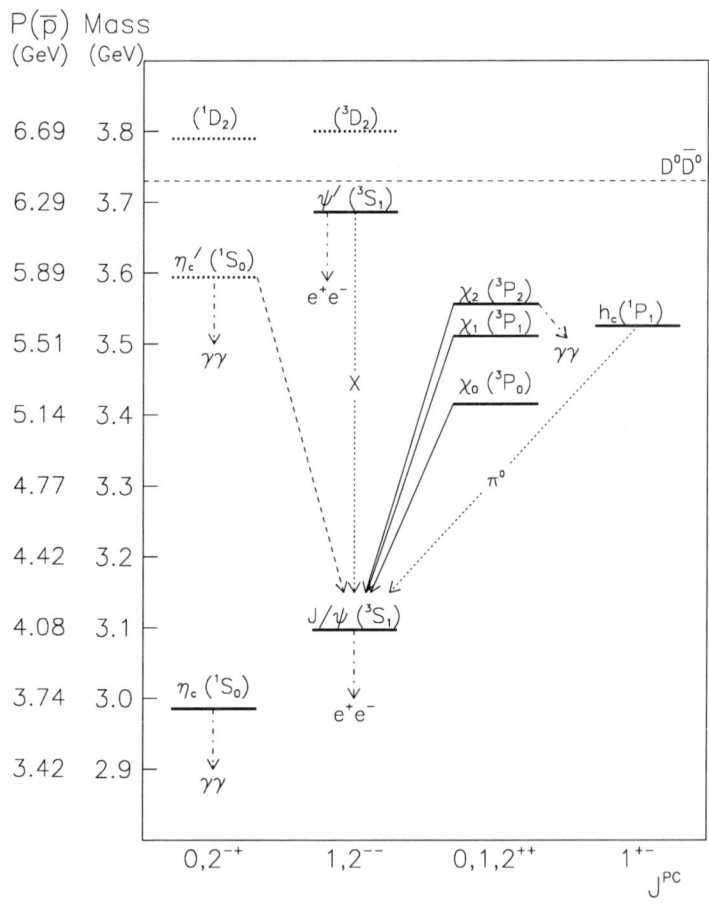

FIGURE 1: The Charmonium Spectrum.

2. CHARMONIUM IN e^+e^- ANNIHILATION

The pioneering SLAC experiment with the Mark I detector mentioned above, and subsequent experiments with improved detectors (Mark II, Mark III and Crystal Ball) have provided most of the experimental data on charmonium spectroscopy. Experiments at DESY, Orsay and Frascati have also made significant contributions to the global knowledge of the charmonium system.

These e^+e^- experiments have the common disadvantage that they can directly form only $\bar{c}c$ resonances with $J^{PC} = 1^{--}$ (see Figure 2). These vector states are produced with peak cross sections which are large (1 mb $\lesssim \sigma_{peak} \lesssim$ 100 mb) compared to the continuum e^+e^- cross section ($\sim$ 30 nb). For these resonances, the FWHM (Γ) mass resolution depends only on the energy resolution of the electron and positron beams, and is about $\Gamma \sim$ 3.5 MeV. However, states with $J^{PC} \neq 1^{--}$ must be formed through the decay of one of the 1^{--} states, usually through radiative decay. For states accessed through radiative decay, the mass resolution depends on the energy resolution of a photon detector, and is generally $10 \lesssim \Gamma \lesssim 25$ MeV. Finally, in the electron experiments, the radiative corrections of the line shapes are very large, $\sim$ 30% for J/ψ. Therefore the knowledge of the true widths of such states is seriously compromised.

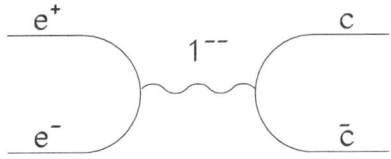

FIGURE 2: e^+e^- production of charmonium through a virtual photon.

3. CHARMONIUM IN $\bar{p}p$ ANNIHILATION

In contrast to e^+e^- annihilation, $\bar{p}p$ annihilation proceeds via two or three gluons (see Figure 3), so that states with any J^{PC} can be directly formed in $\bar{p}p$ annihilations. Thus the mass resolution for all states is determined by the beam energy resolution. The difficulty in detecting charmonium resonances in $\bar{p}p$ annihilation is in extracting the small charmonium signals (10 pb $\lesssim \sigma_{peak} \lesssim$ 360 nb) from the huge $\bar{p}p$ total cross section ($\sim$ 70 mb). It is possible, however, to distinguish charmonium from the large hadronic background by detecting only the electromagnetic decays of charmonium. The first experiment to study charmonium via electromagnetic final states in $\bar{p}p$ annihilations was experiment R704 at the CERN ISR.

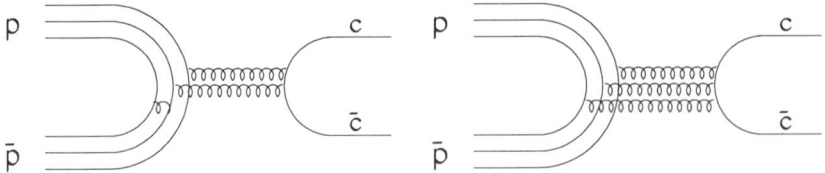

FIGURE 3: $\bar{p}p$ production of charmonium through two or three gluons.

In 1985, R704 pioneered the technique in which a stochastically-cooled antiproton beam collided with an internal hydrogen gas target, to form a variety of charmonium states. With a very limited run time, this experiment made several measurements which demonstrated the excellent capabilities of the technique. The mass resolution achieved was $\Gamma \sim 2.5$ MeV, and the instantaneous luminosity $\sim 10^{30}$ cm^{-2}s^{-1}. The R704 detector was a two-arm spectrometer, with an acceptance in the polar angle range $17° < \theta < 66°$, and a coverage of $45°$ in the azimuthal angle (ϕ). In 1990, Fermilab E760 refined and enhanced the R704 technique by providing a superior mass resolution, $\Gamma \sim 0.5$ MeV, and a higher instantaneous luminosity, $\sim 10^{31}$ cm^{-2}s^{-1}. A larger detector acceptance was achieved by building a detector with full azimuthal coverage, and a polar angle range $2° < \theta < 70°$.

4. FERMILAB E760

E760 was located within the Fermilab Antiproton Accumulator ring, where a hydrogen gas jet target transversely intersected the variable-energy antiproton beam (see Figure 4).

To extract the small charmonium signals from the huge $\bar{p}p$ total cross section, E760 implemented a distinctive electromagnetic trigger to select only electromagnetic decays. Measuring the more difficult hadronic final states was not a primary goal of E760; however, the feasibility of such measurements will be increased in the upgrade experiment E835.

4.1 The Antiproton Source

The sequence for obtaining a cooled circulating beam of antiprotons in the Accumulator is as follows. Protons are accelerated to a kinetic energy of 120 GeV in the Main Ring and are steered onto a copper target. In the collision, a variety of particles are produced, including antiprotons. The antiprotons are produced with a broad and fairly flat momentum distribution with maximum production at 8.9 GeV/c. The produced particles are focussed by a lithium lens. Negatively charged particles with momentum 8.9 GeV/c are then selected by a bending magnet into the Debuncher Ring. In the Debuncher, the beam is transversely pre-cooled and RF-rotated to fit into the smaller Accumulator beam pipe. By this point, all particles other than antiprotons have decayed out of the beam. In the Accumulator the antiproton beam is stochastically cooled both transversely and longitudinally, and decelerated to the energy of interest to E760. Once the beam is at the desired energy, the internal hydrogen gas target is turned on and data taking started.

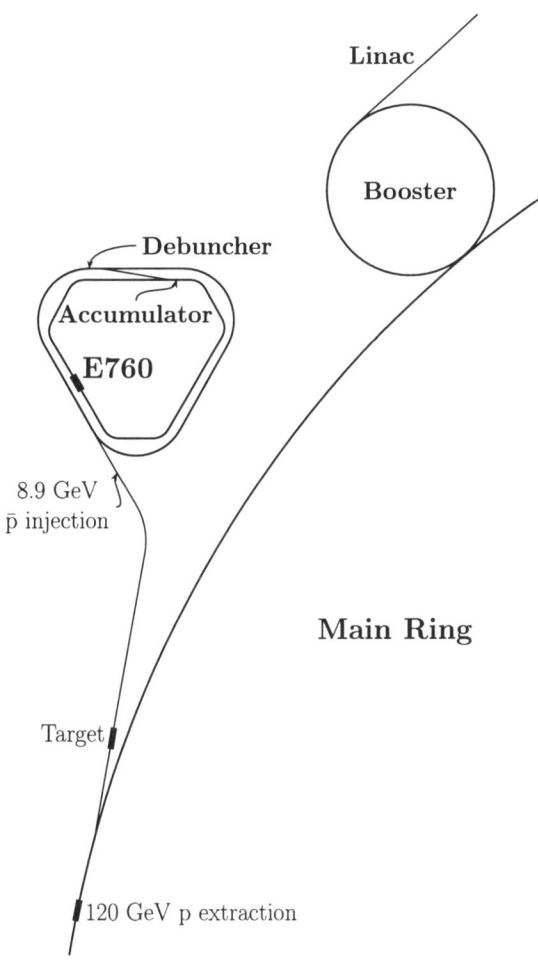

FIGURE 4: Fermilab Antiproton Accumulator.

4.2 The E760 Detector

A schematic of the E760 detector is shown in Figure 5. The non-magnetic E760 detector system is cylindrically symmetric about the beam axis. It covers the complete azimuth $0° < \phi < 360°$ and the laboratory polar angles $2° < \theta < 70°$. Two sets of cylindrical scintillators (H1, H2) provide fast triggering; a threshold Čerenkov counter provides electron/pion discrimination and fast triggering; and a straw chamber (SC) a radial projection chamber (RPC), a multiwire proportional chamber (MWPC), and a set of limited streamer tubes (LST) provide charged particle tracking. The lead-glass central calorimeter, which covers the full azimuth and the polar angles $11° < \theta < 70°$, provides electron and photon energy and angle measurement, and fast triggering. For the calorimeter, the average energy resolution is $\sigma_E/E = 1\% + 6\%/\sqrt{E(\text{GeV})}$, the average polar angle resolution $\sigma_\theta = 6$ mrad, and the average azimuthal angle resolution $\sigma_\phi = 12$ mrad. The tracking detectors improve the position resolution for charged particles, to $\sigma_\theta = 4$ mrad and $\sigma_\phi = 7$ mrad. A lead/scintillator sandwich forward calorimeter extends the polar angle coverage down to $2°$.

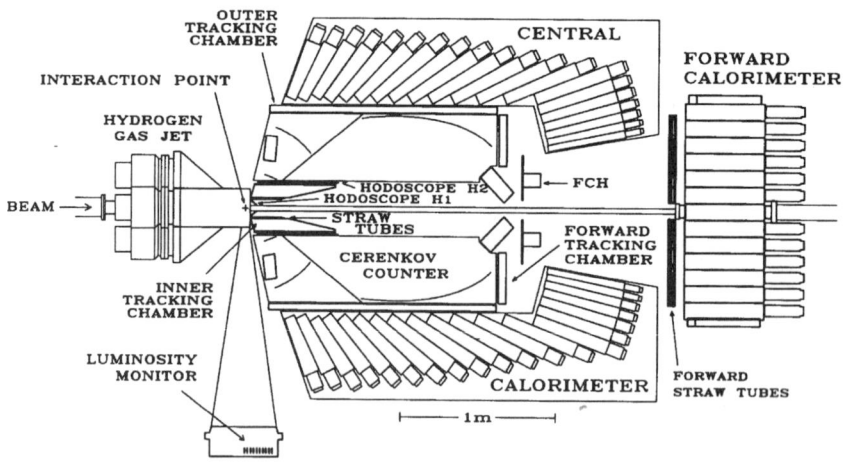

FIGURE 5: E760 Detector System.

4.3 The E760 Experimental Technique

The experimental technique entails decelerating the antiproton beam in small energy steps through charmonium resonances, and measuring the number of events of a given reaction at each step as a function of the beam energy. The measured events per integrated luminosity trace out an excitation curve, which is a convolution of the natural resonance line shape with the energy distribution of the antiproton beam. Analysis of these excitation curves yields precise measurements of the masses and widths of these states. Since all $\bar{c}c$ states are formed directly in $\bar{p}p$ annihilation, the mass resolution depends only on the resolution of the antiproton beam, not on detector resolution.

During E760, the accessible center-of-mass energy range was $2.9 - 4.3$ GeV. The beam momentum spread in the lab was $\sigma_p/p \sim 2 \times 10^{-4}$. The FWHM mass resolution was $\Gamma \sim 350 - 825$ keV. The mass error for a given energy point ranged from $\sim 0.06 - 0.2$ MeV. Typically 3.5×10^{11} stored antiprotons circulated at a frequency of ~ 0.62 MHz. The antiproton beam intersected an internal hydrogen gas jet target with a density 3.5×10^{13}. The resulting maximum instantaneous luminosity was $\sim 10^{31}$ cm^{-2}s^{-1}.

4.4 The E760 Beam Energy Measurement

Key to the precision measurement of narrow states such as the J/ψ and ψ' are the precise measurement of the average beam energy and the beam energy distribution. The average beam energy is determined from a measurement of the beam revolution frequency f, and the beam orbit length ℓ from the beam velocity

$$c\beta = \ell f.$$

Thus the error in the beam energy

$$\Delta E_{beam} = m_p c^2 \beta^2 \gamma^3 [(\Delta \ell/\ell)^2 + (\Delta f/f)^2]^{1/2},$$

where $\gamma = (1 - \beta^2)^{1/2}$ and m_p is the proton mass.

The revolution frequency distribution is measured from the beam longitudinal Schottky noise. Because the average frequency is accurate to $\Delta f/f \sim 10^{-7}$, the second term in the beam energy error equation is ignored.

The orbit length is determined by scanning the ψ' resonance, finding the resonance peak, and setting it equal to the known ψ' mass [4]. The 0.10 MeV error in the mass measurement translates to an uncertainty of 0.67 mm in the "reference" orbit length, which was found to be 474045.7 ± 0.7 mm.

Two errors arise in E760 charmonium mass measurements. One is a systematic error due to determination of the reference orbit length from the ψ' mass, and translates to a 0.033 MeV error in the determination of the J/ψ mass, for example. The other error is statistical and is due to the deviation of the antiproton beam from the reference orbit. Forty-eight beam position monitors (BPM's) determine this deviation during data taking. The error introduced by the BPM's results in an orbit length error of ± 1 mm, which corresponds to a mass error of 0.05 MeV at the J/ψ, for example.

The beam energy distribution is determined from the distribution of the revolution frequency distribution using

$$\frac{dp}{p} = \frac{1}{\eta} \frac{df}{f}.$$

The accelerator parameter η can be determined in several ways [5]. The most accurate method for determining η in E760 involves analysis of the J/ψ and ψ' excitation curves.

5. CHARMONIUM RESONANCES

As mentioned above, the small charmonium cross sections were extracted from the huge hadronic background by selecting states by their decay to electromagnetic final states. The E760 charmonium measurements reported herein result from analysis of the following decay channels

$$\bar{p}p \rightarrow (\bar{c}c) \rightarrow e^+e^-$$
$$\bar{p}p \rightarrow (\bar{c}c) \rightarrow \gamma\gamma$$
$$\bar{p}p \rightarrow (\bar{c}c) \rightarrow J/\psi X \rightarrow (e^+e^-)X.$$

5.1 Triplet-S and Triplet-P states

Shown in Figures 6(a) and (b) are excitation curves for the J/ψ and χ_2 resonances. These curves indicate the quality of the data and the excellent resolution of the antiproton beam. Note that even in the case of the J/ψ, where the beam width is larger than the resonance width, the narrow resonance width may be determined, if the beam resolution function is well known.

Results for the J/ψ, ψ', χ_1 and χ_2 widths and masses are shown in Table 1 [5, 6]. Using the ψ' as a calibration point, as described in Section 4.4, E760 determined the masses of the other three states. These mass measurements agree the previous results, but have about a factor three smaller errors.

The measurement of the J/ψ and ψ' widths is of theoretical importance to PQCD calculations. The E760 measurements of $\Gamma(J/\psi) = 91 \pm 13$ keV and $\Gamma(\psi') = 287 \pm 40$ keV are significantly larger than those of previous measurements which had been the accepted values for almost twenty years. The E760 result is the first direct measurement of these widths; unlike the previous measurements from the e^+e^- experiments, the E760 value does not depend on knowledge of absolute luminosity, detector acceptances or efficiencies. This measurement also requires a much smaller radiative correction. Mark III has recently reported a new branching ratio $B(\psi' \to e^+e^-) = 5.92 \pm 0.25$ [7] which is also independent of absolute luminosity, detector acceptances and efficiencies. This branching ratio result is smaller than previous results. Both the larger E760 J/ψ and ψ' widths and the Mark III result are consistent with the hadronic efficiency in each of the old e^+e^- experiments having been underestimated by about 25%. The hadronic efficiencies were quoted with large errors in each case.

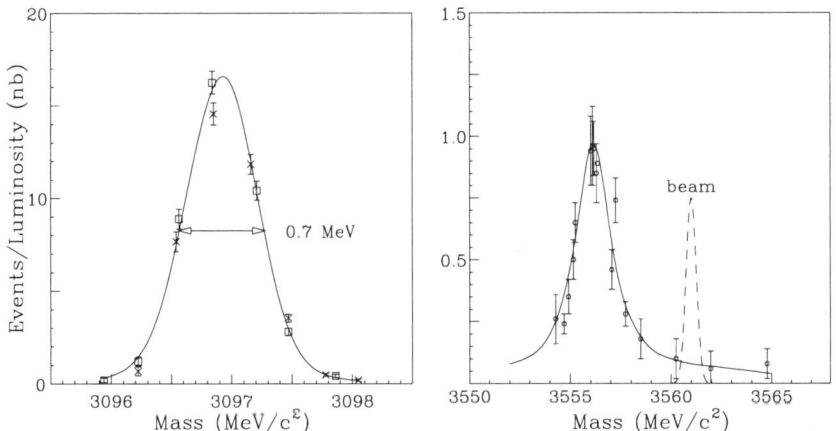

FIGURE 6: (a) $J/\psi \to e^+e^-$ excitation curve, and (b) $\chi_2 \to J/\psi\gamma \to (e^+e^-)\gamma$ excitation curve.

The BES experiment in Beijing has recently submitted a paper for publication in Physics Letters B which contains a new measurement of the J/ψ width of 84.4 ± 8.9 keV [8]. This width measurement is also significantly higher than measurements prior to E760.

The χ_1 and χ_2 widths were measured essentially for the first time in E760. Therefore branching ratio measurements may be converted to width measurements and compared to PQCD predictions for P-wave resonances for the first time.

Another interesting E760 measurement of χ_2 parameters is the decay width $\Gamma(\chi_2 \to \gamma\gamma)$ [9]. The two-photon results for χ_2 are shown in Table 2. Note

TABLE 1: E760 results compared to the world average prior to E760 for triplet-S and triplet-P states[††].

	Mass (MeV)		Γ (keV)	
	E760 [5, 6]	PDB [4]	E760 [5, 6]	PDB [4]
J/ψ	3096.87 ± 0.06	3096.93 ± 0.09	91 ± 13	68 ± 10
ψ'	–	3686.00 ± 0.10	287 ± 40	243 ± 43
χ_1	3510.53 ± 0.13	3510.6 ± 0.5	880 ± 136	< 1300
χ_2	3556.15 ± 0.14	3556.3 ± 0.4	1980 ± 184	2600^{+1200}_{-900}

TABLE 2: Comparison of E760 and CLEO χ_2 two-photon widths.

Experiment	$\Gamma(\chi_2 \to \gamma\gamma)$ (keV)	Channel
CLEO [10]	1.08 ± 0.40	$\gamma\gamma \to \chi_2 \to \gamma J/\psi$
E760 [9]	0.32 ± 0.09	$\bar{p}p \to \chi_2 \to \gamma\gamma$

that the E760 value is about a factor three smaller than the CLEO result [10]. Rumor has it that a new CLEO analysis decreases the result shown in Table 2 significantly.

5.2 Singlet-P state

Measurement of the $^1P_1$ state is vital for determination of parameters of the QCD potential. The spin-spin interaction splits triplet and singlet states of a given L. For the usual QCD potential, in which the confinement part is a Lorentz scalar, this splitting is only finite for $L = 0$, because it is proportional to the radial wave function at $r = 0$. If the mass splitting between the spin-weighted average of the ^3P_J and the $^1P_1$ is defined to be

$$\Delta M \equiv \langle M(^3P_J) \rangle - M(^1P_1),$$

the lowest order theoretical prediction is $\Delta M = 0$. Various second order effects have led to predictions in the range $-5 < \Delta M < 5$ MeV.

The $^1P_1$ is difficult to detect experimentally. The e^+e^- experiments were unlikely to find it because odd charge conjugation states are not populated in the radiative decays of ψ'. Even in $\bar{p}p$, the production is helicity-suppressed. Because of these difficulties and the importance of the $^1P_1$, E760 devoted half its luminosity ($\mathcal{L} \sim 16\,\text{pb}^{-1}$) to the $^1P_1$ search.

Three decay modes were examined for a $^1P_1$ signal

$$^1P_1 \rightarrow \eta_c\gamma \rightarrow (\gamma\gamma)\gamma,$$
$$^1P_1 \rightarrow J/\psi\pi\pi \rightarrow (e^+e^-)\pi\pi,$$
$$\text{and } ^1P_1 \rightarrow J/\psi\pi^0 \rightarrow (e^+e^-)(\gamma\gamma).$$

Of these, a signal was seen only in the third mode. The $^1P_1$ excitation curve [11] measured by E760 is shown in Figure 7. The resonance parameters are

$$M(^1P_1) = 3526.2 \pm 0.15 \pm 0.20 \,\text{MeV},$$
$$\Delta M = -0.9 \pm 0.3 \,\text{MeV},$$
$$\text{and } \Gamma \leq 1.1 \,\text{MeV (90\% C.L.)}.$$

A range for the product of the branching ratios may be found by assuming a reasonable range of values for the resonance width $100 \,\text{keV} \leq \Gamma(^1P_1) \leq 500 \,\text{keV}$ and assuming $B(J/\psi \rightarrow e^+e^-) = (6.3 \pm 0.2)\%$ [12]

$$1.7 \pm 0.4 \leq B(\bar{p}p \rightarrow ^1 P_1) \, B(^1P_1 \rightarrow J/\psi\pi^0) \times 10^7 \leq 2.3 \pm 0.6.$$

The statistics in the E760 measurement are rather poor, but the probability for statistical fluctuations giving rise to the observed structure is only one in four hundred. Confirmation of the $^1P_1$ signal, and attempted observation in alternate channels are key goals of E835.

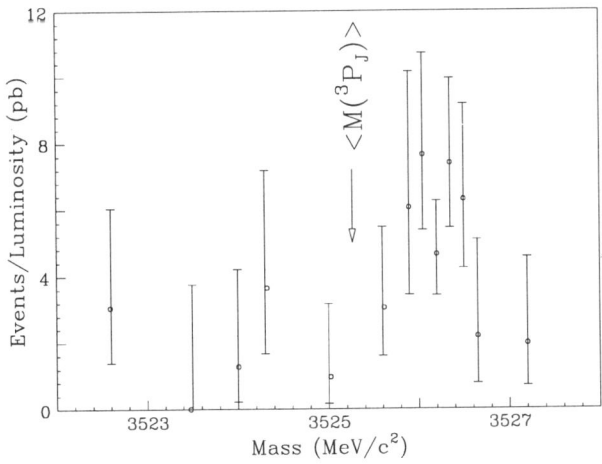

FIGURE 7: $^1P_1 \rightarrow J/\psi\pi^0 \rightarrow (e^+e^-)\gamma\gamma$ excitation curve.

The observed $^1P_1$ signal has some interesting features. Since ΔM is small, the confinement potential is probably scalar. It was rather surprising that it was seen in the isospin-violating channel $^1P_1 \to J/\psi\pi^0$, but not in the isospin-allowed channel $^1P_1 \to J/\psi\pi\pi$, which has a smaller available phase space. However, the ratio of branching ratios

$$\frac{B(^1P_1 \to J/\psi\pi\pi)}{B(^1P_1 \to J/\psi\pi^0)} \leq 0.18 \,(90\% \text{C.L.})$$

agrees with the prediction of Voloshin [13].

5.3 Singlet-S states

The previous η_c and η_c' width and mass results are compiled in Table 3. Note that the only measurement of the η_c' was made by Crystal Ball; this has not yet been confirmed.

TABLE 3: Existing η_c and η_c' resonance parameters, prior to E760.

Experiment	$M(\eta_c)$ (MeV)	$\Gamma(\eta_c)$ (MeV)
DM2 [15]	2974.4 ± 1.9	
Mark III [16]	2969 ± 6	
R704 [17]	$2982.6^{+2.7}_{-2.3}$	$7.0^{+7.5}_{-7.0}$
Mark III [18]	2980.2 ± 1.6	$10.1^{+33.0}_{-8.2}$
Crystal Ball [19]	2984 ± 6	11.5 ± 4.5
Mark II [20]	2982 ± 8	
PDB [12]	2978.8 ± 1.9	$10.3^{+3.8}_{-3.4}$

Experiment	$M(\eta_c')$ (MeV)	$\Gamma(\eta_c')$ (MeV)
Crystal Ball [21]	3594 ± 5.0	< 8.0 (95% C.L.)

For observation in E760, the most likely electromagnetic decay is $\eta_c \to \gamma\gamma$. The E760 η_c excitation curve [14] is shown in Figure 8; the resulting resonance parameters are shown in Table 4. Note that the E760 value of the η_c mass is 9 ± 3 MeV higher than the previously accepted value. This measurement is significant, because the $\eta_c - J/\psi$ mass splitting, along with the ψ' mass, provides the PQCD estimate for the η_c' mass, and in fact indicates that the η_c' mass should be about 30 MeV higher than where Crystal Ball claimed discovery. In addition, the η_c width is found to be larger than the world average, although the E760 error bars are large.

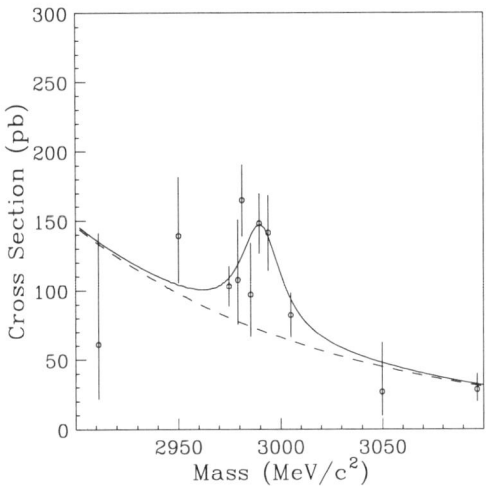

FIGURE 8: E760 $\eta_c \to \gamma\gamma$ excitation curve. Measured background contamination due to $\pi^0\pi^0$ and $\pi^0\gamma$ continuum production is shown by the dashed line.

TABLE 4: E760 results for η_c resonance parameters.

Experiment	$M(\eta_c)$ (MeV)	$\Gamma(\eta_c)$ (MeV)
E760 [14]	$2988.3^{+3.3}_{-3.1}$	$23.9^{+12.6}_{-7.1}$

The $\eta_c \to \gamma\gamma$ partial width is compared to other results in Table 5. It is interesting to note that the other experiments listed in Table 5 had to assume the existing value for the η_c width in their determinations. Another η_c width measurement with improved statistics would be very valuable and is planned for E835.

TABLE 5: η_c two-photon width summary.

Experiment	$\Gamma(\eta_c \to \gamma\gamma)$ (keV)	Channel(s)
CLEO [22]	5.7 ± 2.4	$\gamma\gamma \to \eta_c \to K^0_s K\pi$
L3 [23]	8.0 ± 3.3	$\gamma\gamma \to \eta_c \to 12$ channels
ARGUS [24]	11.3 ± 4.2	$\gamma\gamma \to \eta_c \to K^0_s K\pi$
E760 [14]	6.7 ± 3.1	$\bar{p}p \to \eta_c \to \gamma\gamma$

E760 also searched for the η'_c resonance, in the channel $\eta'_c \to \gamma\gamma$. With a small luminosity expenditure ($\mathcal{L} = 6\,\text{pb}^{-1}$), the results were inconclusive. A systematic search for the η'_c is part of the E835 agenda.

6. NON-CHARMONIUM RESULTS FROM E760

In conjunction with the E760 charmonium physics program, other physics measurements were made for "free." These are mentioned here only briefly.

6.1 Timelike proton electromagnetic form factor

The proton electromagnetic form factor [25] was measured by E760 in the timelike momentum transfer region $8.9 < q^2 < 13.0\,\text{GeV}^2$ for the first time. In the course of charmonium studies, $\bar{p}p \to e^+e^-$ data was taken at charmonium resonances which cannot decay directly to e^+e^-. This permitted analysis of three q^2 points: 9 GeV^2 (η_c), 12.5 GeV^2 ($^1P_1$) and 13.0 GeV^2 (η_c' search region). The timelike form factors derived from these cross sections are shown in Figure 9, along with timelike [26] and spacelike [27] form factor measurements from previous experiments. Like the spacelike form factor data, the E760 data combined with the other timelike data exhibit an α_s^2-type saturation for $|q^2| \gtrsim 5\,\text{GeV}^2$. The value for the timelike form factor at saturation is, however, about a factor two larger than that of the spacelike form factor.

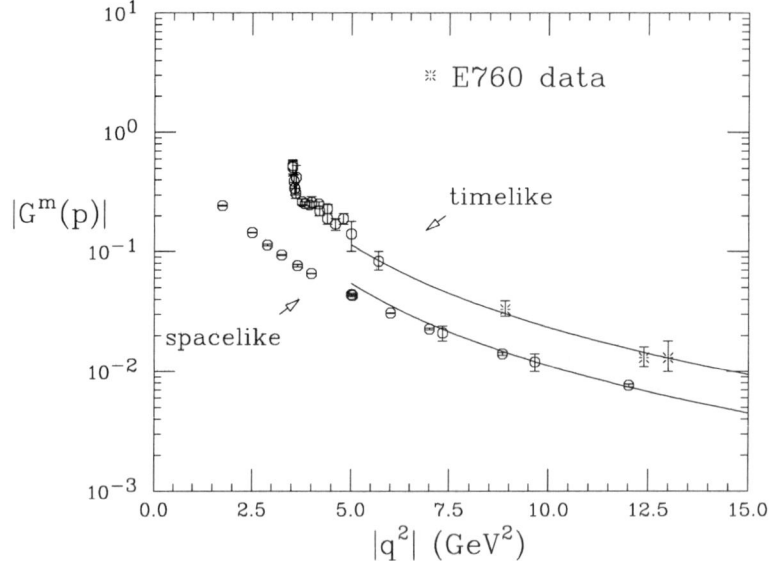

FIGURE 9: E760 timelike proton form factors (star symbols) and timelike and spacelike proton form factor measurements from previous experiments. The curves are fits to the data of the form $|G^m(p)|(|q^2|) \propto q^{-4}\alpha_s^2(|q^2|)$, where $\alpha_s^2(|q^2|) \propto 1/ln(|q^2|/\Lambda^2)$, with $\Lambda = 0.2$ GeV.

6.2 $\bar{p}p$ forward elastic scattering parameters

Measurements of antiproton-proton forward elastic scattering parameters in the $3.7 < p_{lab}(\bar{p}) < 6.2$ GeV/c region were made in E760 [28], with significantly smaller errors than previous measurements in this region. In contrast to pp scattering, there is little good-quality $\bar{p}p$ scattering data in this momentum region. The total cross section is related to the imaginary part of $\bar{p}p$ forward elastic scattering amplitude via the optical theorem

$$\sigma_T = (4\pi/p)\, Im(f(0))\,.$$

The imaginary part, therefore, has been fairly well determined. The real part, however, is found from the ρ parameter

$$\rho = \frac{Re(f(0))}{Im(f(0))}$$

and can only be determined by measuring differential cross sections very accurately at very small momentum transfers. In E760, recoil protons are measured in the momentum transfer region $0.001 \leq |t|(\text{GeV/c})^2 \leq 0.02$. The E760 technique permits measurement of the ρ parameter with very small errors.

6.3 Light-quark spectroscopy

E760 has also made several measurements of light-quark spectroscopy [29]. In the all-neutral data taken at various charmonium resonances, final states containing $\pi^0\pi^0\pi^0$, $\pi^0\pi^0\eta$, $\pi^0\eta\eta$, and $\eta\eta$ were studied. A structure with mass 1508 ± 10 MeV and width 103 ± 15 MeV was seen in the $\pi^0\pi^0$ decay channel in the reactions

$$\bar{p}p \rightarrow \pi^0\pi^0\pi^0 \rightarrow 6\gamma$$
$$\text{and } \bar{p}p \rightarrow \pi^0\pi^0\eta \rightarrow 6\gamma\,.$$

The second channel is suppressed relative to the first. This structure is identified with the $f(1520)$ glueball candidate because of the similar mass and width to the resonance found by the Crystal Barrel experiment [30]. J^{PC} assignments are difficult in E760 particularly because of the limited geometric acceptance available. In the $\eta\eta$ decay channel, three additional states have been observed, with masses 1488 ± 10 MeV, 1748 ± 10 MeV, and 2104 ± 20 MeV.

7. FERMILAB EXPERIMENT E835

E835 will significantly improve and expand on the results of E760. The equipment improvements include an increase in the instantaneous luminosity of about a factor seven, due to improvement of the accelerator complex which will increase the circulating $\bar{p}$ beam current by a factor three, and a reduction of the jet temperature which will increase the jet density by a factor two and a half.

The detector improvements include a replacement of most of the inner charged tracking detectors. The inner detectors for E760 and E835 are shown in Figures 10 and 11. For E835, there will be three sets of scintillators: H1, H2′ and H2. All three will be used for the charged trigger, and H2 will be used for a dE/dx measurement. Two new sets of straw chambers, SC1 and SC2, will be used for ϕ tracking. A silicon detector, SIL, will provide θ tracking information. Two layers of scintillating fibers, SF1 and SF2, will provide θ tracking and will be part of the trigger. The new detectors are expected to improve the charged particle position resolution to $\sigma_\theta \simeq 1$ mrad and $\sigma_\phi \simeq 3$ mrad, which is likely to permit a fruitful study of hadronic final states in E835.

The primary physics goals of E835 are spectroscopy of the $^1P_1$, η_c, η_c', and χ_0. As mentioned above, both the $^1P_1$ and η_c states have been studied in E760, but require additional data for improved statistics. In particular, it is hoped that the $^1P_1$ can be observed in additional decay modes. Since the η_c' was not seen in E760, a search for it is vital. In addition to a search for its $\gamma\gamma$ final state, a search for its M1 decay to J/ψ will be attempted. The χ_0 was not studied in E760, primarily because of its tiny branching ratio to $J/\psi\gamma$, which is twenty times smaller than the branching ratio $B(\chi_2 \to J/\psi\gamma)$. However, a measurement of the χ_0 width comparable to that of χ_2 would be very interesting because PQCD predictions relate several partial widths for the two states in a very straightforward manner. The expected luminosity budget for measurement of these states is shown in Table 6.

If there is time left over in the running period, some additional issues will be addressed. One possibility is the spectroscopy of $^3D_2(2^{--})$ and $^1D_2(2^{-+})$. These states are expected to be narrow, since they cannot decay to $D^0\bar{D}^0$ (parity violation), and are expected to lie below $D^0\bar{D}^*$ threshold (3875 MeV). Another possibility is the study of J/ψ and ψ' formation in nuclei.

8. CONCLUSIONS

E760 has been very successful, producing many new charmonium resonance parameter measurements with only 30 pb^{-1} in 10 months' running. E835 is expected to greatly improve the world data on charmonium with an expected

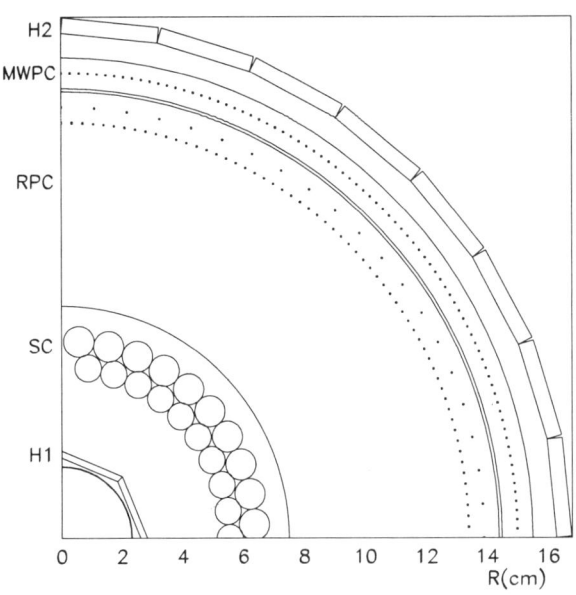

FIGURE 10: One quadrant of the azimuth for the E760 inner detector system.

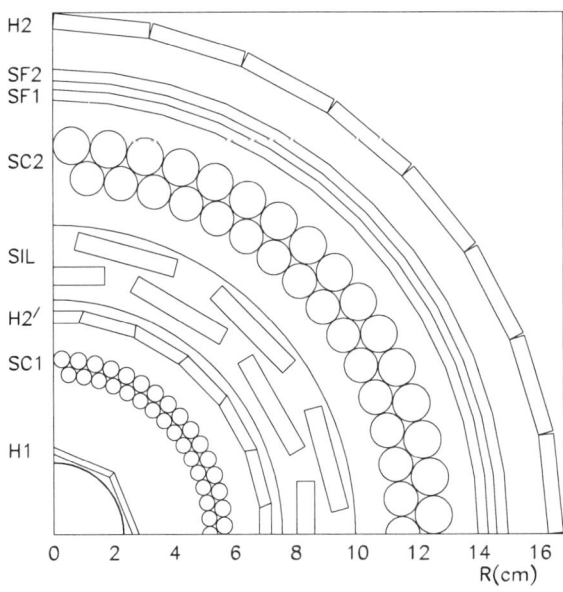

FIGURE 11: One quadrant of the azimuth for the E835 inner detector system.

TABLE 6: Anticipated E835 luminosity budget.

Channel	$\int \mathcal{L}$ (pb^{-1})	% error (Γ_{tot})	% error ($B_{in}B_{out}$)
$^1P_1 \to J/\psi \pi^0$	75	30	30
$\eta_c \to \gamma\gamma, \phi\phi$	20	25	15
$\eta_c' \to \gamma\gamma, J/\psi\gamma$	45	20	15
$\chi_0 \to J/\psi\gamma$	20	25	20
Total	160		

integrated luminosity of 200 pb^{-1} in 18 months' running. E835 is scheduled to begin running during the next Fermilab fixed-target run in April 1996.

9. ACKNOWLEDGEMENTS

This research has been supported in part by the U.S. Department of Energy, The U.S. National Science Foundation, and the Italian Istituto Nazionale di Fisica Nucleare. The author wishes to thank X. Fan, T. Pedlar, K.K. Seth, and G. Zioulas for valuable assistance in the preparation of this talk.

[†]The E760/E835 Collaboration consists of the following institutions:
Fermi National Accelerator Laboratory, Batavia, Illinois 60510, U.S.A.
I.N.F.N. and University of Ferrara, 44100 Ferrara, Italy
I.N.F.N. and University of Genoa, 16146 Genoa, Italy
University of California at Irvine, California 92717, U.S.A.
Northwestern University, Evanston, Illinois 60208, U.S.A.
Pennsylvania State University, University Park, Pennsylvania 16802, U.S.A.
I.N.F.N. and University of Turin, 10125 Turin, Italy

[††]The widths of the J/ψ and ψ' shown here differ from the published results of Reference [5] due to an improvement in the radiative correction calculation.

References

[1] J.J. Aubert et al., Phys. Rev. Lett. **33**, 1404 (1974).

[2] J.-E. Augustin et al., Phys. Rev. Lett. **33**, 1406 (1974).

[3] G.S. Abrams et al., Phys. Rev. Lett. **33**, 1453 (1974).

[4] Particle Data Group, *Phys. Lett.* **239**, 1 (1990).

[5] T.A. Armstrong *et al.*, *Phys. Rev.* **D47**, 772 (1993).

[6] T.A. Armstrong *et al.*, *Nucl. Phys.* **B373**, 35 (1992).

[7] D. Coffman *et al.*, *Phys. Rev. Lett.* **68**, 282 (1992).

[8] J.Z. Bai *et al.*, *submitted to Phys. Lett. B*, 1995.

[9] T.A. Armstrong *et al.*, *Phys. Rev. Lett.* **70**, 2988 (1993).

[10] J. Dominick *et al.*, *Phys. Rev.* **D50**, 4265 (1994).

[11] T.A. Armstrong *et al.*, *Phys. Rev. Lett.* **69**, 2337 (1992).

[12] Particle Data Group, *Phys. Rev.* **D45**, 1 (1992).

[13] M.B. Voloshin, *Yad. Fiz.* **43**, 1571 (1986) [*Sov. J. Nucl. Phys.* **43**, 1011 (1986)].

[14] T.A. Armstrong *et al.*, *accepted for publication in Phys. Rev. D*, 1995.

[15] D. Bisello *et al.*, *Nucl. Phys.* **B350**, 1 (1991).

[16] Z. Bai *et al.*, *Phys. Rev. Lett.* **65**, 686 (1990).

[17] C. Baglin *et al.*, *Phys. Lett.* **B187**, 191 (1987).

[18] R.M. Baltrusaitis *et al.*, *Phys. Rev.* **D33**, 629 (1986).

[19] J. Gaiser *et al.*, *Phys. Rev.* **D34**, 711 (1986).

[20] T. Himel *et al.*, *Phys. Rev. Lett.* **45**, 1146 (1980).

[21] C. Edwards *et al.*, *Phys. Rev. Lett.* **48**, 70 (1982).

[22] A. Bean *et al.*, contributed paper # 293, *Proc. 16th Int. Lepton-Photon Symposium*, Ithaca, NY 1993.

[23] O. Adriani *et al.*, *Phys. Lett.* **B318**, 575 (1993).

[24] H. Albrecht *et al.*, *Phys. Lett.* **B338**, 390 (1994).

[25] T.A. Armstrong *et al.*, *Phys. Rev. Lett.* **70**, 1212 (1993).

[26] D. Bisello *et al.*, *Nucl. Phys.* **B224**, 379 (1983); D. Bisello *et al.*, *Z. Phys.* **C48**, 23 (1990); G. Bassompierre *et al.*, *Phys. Lett.* **68B**, 477 (1977); G. Bassompierre *et al.*, *Nuovo Cimento* **73A**, 347 (1983); G. Bardin *et al.*, *Phys. Lett.* **255**, 149 (1991); G. Bardin *et al.*, *Phys. Lett.* **257**, 514 (1991).

[27] R.G. Arnold *et al.*, *Phys. Rev. Lett.* **57**, 174 (1986); P. Bosted *et al.*, *Phys. Rev. Lett.* **68**, 3841 (1992).

[28] T.A. Armstrong *et al.*, *in preparation*.

[29] T.A. Armstrong *et al.*, *Phys. Lett.* **B307**, 394 (1993); *ibid.*, **B307**, 399 (1993).

[30] C. Amsler *et al.*, *Phys. Lett.* **B340**, 259 (1994); *ibid.*, **B342**, 433 (1994); C. Amsler and F. Close, *Phys. Lett.* **B353**, 385 (1995).

Electroweak Radiative Corrections and Measurements of R_{had}

Morris L. Swartz

Stanford Linear Accelerator Center
Stanford University, Stanford, California, 94309

Abstract. The interpretation of the world's most precise electroweak data is limited by the precision of very old e^+e^- cross section measurements in the 1-5 GeV region. New measurements below charm threshold with a 5% systematic normalization uncertainty would eliminate this limitation. Additionally, the structure of the charm threshold region is not known well. New cross section measurements in the charm threshold region would clarify the structure of the charm threshold and would help to resolve differences in $\Delta\alpha_{had}$ analyses.

1. Introduction

At the current time, a large program of precise electroweak measurements is being conducted throughout the world. The object of this program is to test the electroweak Standard Model by comparing the measured values of a large set of electroweak observables with the predictions of the Minimal Standard Model (MSM). The Standard Model calculations have been performed to full one-loop accuracy and partial two-loop precision by a large community of researchers. The largest of the loop-level corrections are the vacuum polarization corrections to the gauge boson propagators which are characterized by the four one-particle-irreducible amplitudes shown in Figure 1. The largest effects are associated with the one-particle-irreducible contributions to the photon self-energy $\Pi_{\gamma\gamma}(q^2)$ or the related quantity $\Pi'_{\gamma\gamma}(q^2) \equiv (\Pi_{\gamma\gamma}(q^2) - \Pi_{\gamma\gamma}(0))/q^2$ evaluated at the Z mass scale $q^2 = M_Z^2$. These quantities are usually absorbed into the definition of the running electromagnetic coupling $\alpha(q^2)$,

$$\alpha(q^2) \equiv \frac{\alpha_0}{1 - \left[\Pi'_{\gamma\gamma}(q^2) - \Pi'_{\gamma\gamma}(0)\right]}, \tag{1}$$

where $\alpha_0 = 1/137.0359895(61)$ is the electromagnetic fine structure constant. This quantity is also represented as the fractional change in the electromagnetic coupling constant $\Delta\alpha$,

$$\Delta\alpha(q^2) = \frac{\alpha(q^2) - \alpha_0}{\alpha(q^2)} = \Pi'_{\gamma\gamma}(q^2) - \Pi'_{\gamma\gamma}(0). \tag{2}$$

At the Z-mass-scale, $\Delta\alpha$ is approximately 0.06 which leads to 100% corrections in the magnitudes of some Z-pole asymmetries! It is clear that $\Delta\alpha(M_Z^2)$ must be calculated very accurately.

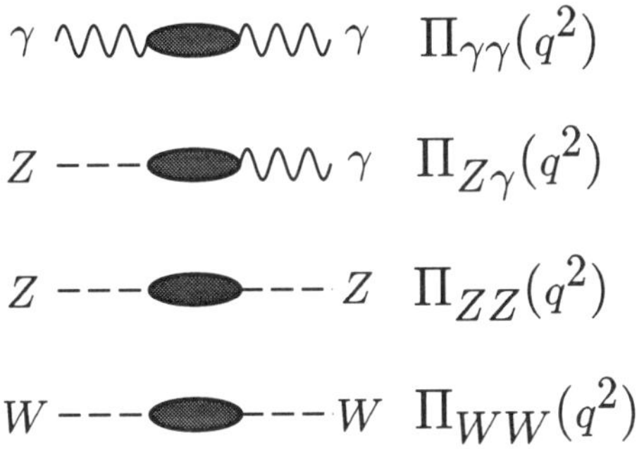

Figure 1. The one-particle-irreducible vacuum polarization amplitudes that modify the gauge boson propagators of the Minimal Standard Model.

Using analytic techniques and the optical theorem applied to the amplitude for s-channel Bhabha scattering, the quantity $\Delta\alpha$ has been related to the cross section for the process $e^+e^- \to \gamma^* \to$ all (σ_{tot}) as follows,[1]

$$\Delta\alpha(q^2) = \frac{1}{\pi} P \int_{4m_e^2}^{\infty} ds \frac{q^2}{s^2(s-q^2)} \mathrm{Im}\Pi_{\gamma\gamma}(s) = \frac{\alpha_0}{3\pi} P \int_{4m_e^2}^{\infty} ds \frac{q^2}{s(q^2-s)} R_{tot}(s), \tag{3}$$

where $R_{tot}(s)$ is the ratio of the total cross section to the (massless) muon pair cross section $\sigma_{\mu\mu}(s) = 4\pi\alpha^2(s)/3s$ at the center-of-mass energy $\sqrt{s}$. The cross section σ_{tot} is the physical cross section which has been corrected for initial state radiation.

It is straightforward to evaluate equation (3) for the continuum leptonic cross sections.[2] In the limit that the scale q^2 is much larger than the square of the lepton mass m_ℓ^2, the contribution of the continuum leptonic cross sections is given by the following expression,

$$\Delta\alpha_\ell(q^2) = \frac{\alpha_0}{3\pi} \sum_\ell \left[-\frac{5}{3} + \ln\frac{q^2}{m_\ell^2} \right]. \tag{4}$$

The remaining contributions to R_{tot} consist of the continuum hadronic cross section and the $J^P = 1^-$ resonances and are labelled R_{had}. Since the cross sections for the resonances and low energy continuum are not accurately calculable from first principles, experimental inputs are used to evaluate their contributions equation (3). The contribution of open top quark production to the integral is accurately calculable and since the top quark mass is not known precisely, only the five flavor hadronic cross section is included in R_{had}. The corresponding contribution to $\Delta\alpha(q^2)$ is therefore,

$$\Delta\alpha_{had}(q^2) = \frac{\alpha_0}{3\pi} P \int_{4m_\pi^2}^{\infty} ds \frac{q^2}{s(q^2-s)} R_{had}(s). \tag{5}$$

1.1 Note On Hadronic Vacuum Polarization Corrections

It is important to note that $\Delta\alpha_{had}(M_Z^2)$ is related to but not identical to the hadronic vacuum polarization correction that affects the anomalous magnetic moment of the muon, Δa_μ^{had}. These quantities are compared graphically in Figure 2. The quantity Δa_μ^{had} is also evaluated with dispersion integral techniques yielding the following expression,

$$\Delta a_\mu^{had} = \left(\frac{\alpha_0 m_\mu}{3\pi}\right)^2 \int_{4m_\pi^2}^{\infty} ds \frac{K(s)}{s^2} R_{had}(s), \tag{6}$$

where the kernal $K(s)$ varies from 0.63 at $s = 4m_\pi^2$ to 1.0 at $s = \infty$. Near threshold, the Δa_μ^{had} integrand is proportional to R_{had}/s^2 whereas the $\Delta\alpha_{had}(M_Z^2)$ integrand is proportional to R_{had}/s. This has the consequence that the uncertainty on Δa_μ^{had} is dominated by the precision of experimental measurements in the interval $4m_\pi^2 < s < 2$ GeV^2 whereas the uncertainty on $\Delta\alpha_{had}(M_Z^2)$ is dominated by the precision of measurements in the interval 4 GeV$^2 < s < 25$ GeV^2.

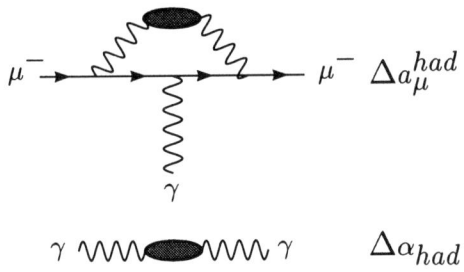

Figure 2. The hadronic vacuum polarization corrections to the anomalous magnetic moment of the muon Δa_μ^{had} and to the photon propagator $\Delta \alpha_{had}$.

2. Current Status of $\Delta \alpha_{had}(M_Z^2)$ Determinations

Equation (5) has been evaluated at the Z boson mass scale a number of times.[3-8] The most recent evaluations by Martin and Zeppenfeld,[6] Eidelman and Jegerlehner,[7] Burkhardt and Pietrzyk,[8] and this author[9] yield

$$\Delta\alpha_{had}(M_Z^2) = \begin{cases} 0.02739 \pm 0.00042, & \text{Reference 6} \\ 0.0280 \pm 0.0007, & \text{Reference 7} \\ 0.0280 \pm 0.0007, & \text{Reference 8} \\ 0.02752 \pm 0.00046, & \text{Reference 9.} \end{cases} \quad (7)$$

The authors of Reference 6 use perturbative QCD to parameterize the continuum $R_{had}(s)$ above $\sqrt{s} = 3$ GeV and linear interpolation of measured data below that point. The two-body final states $\pi^+\pi^-$ and K^+K^- are fit to parameterizations which include the ρ, ω, and ϕ resonances. The remaining resonance contributions are calculated from an analytic expression which results from integrating a Breit-Wigner lineshape and depends upon the masses, widths, and leptonic widths of each resonance. The authors of Reference 7 use linear interpolation (trapezoidal integration) of measured data points to evaluate the continuum, $\pi^+\pi^-$, and K^+K^- contributions. Above $\sqrt{s} = 40$ GeV, they use perturbative QCD to evaluate R_{had}. The contributions of the ω, ϕ, J/ψ-family, and Υ-family resonances are included by integrating a Breit-Wigner lineshape. The authors of Reference 8 use smoothed averages of data to evaluate the continuum contribution, a parameterization to evaluate the $\pi^+\pi^-$ contribution, and the analytic expression to evaluate the contribution of the remaining resonances. Finally, our result[9] is evaluated by fitting the continuum, $\pi^+\pi^-$, and K^+K^- contributions to parameterizations using a technique that automatically accounts for correlations. The resonance contributions are evaluated by numerically integrating appropriate lineshapes.

The values of $\Delta\alpha_{had}(M_Z^2)$ listed in equation (7) lead to the following values of $\alpha(M_Z^2)$,

$$\alpha^{-1}(M_Z^2) = \begin{cases} 128.98 \pm 0.06, & \text{Reference 6} \\ 128.90 \pm 0.09, & \text{References 7,8} \\ 128.96 \pm 0.06, & \text{Reference 9}, \end{cases} \quad (8)$$

which would lead to differences of 0.00020 in the predicted value of the effective weak mixing angle at the Z pole $\sin^2\theta_W^{\text{eff}}$. Since the precision of the current world average value[10] of this quantity is 0.00028, it is clearly quite important to determine $\Delta\alpha_{had}(M_Z^2)$ with improved accuracy.

3. Summary of the MS Analysis

The following is a summary of the recent MS analysis which can be used to indicate the sources of the limitations in our current knowledge of $\Delta\alpha(M_Z^2)$.

3.1 THE DATA

The approach to the evaluation of equation (5) is driven by the low-energy structure of the hadronic cross section. In the energy interval between pion-pair threshold and $W \equiv \sqrt{s} = 2$ GeV, the cross section for the $\pi^+\pi^-$ final state is large and dominated by the $\rho(770)$ and $\omega(782)$ resonances. The three-pion cross section is also large above its threshold and is dominated by the $\omega(782)$ and $\phi(1020)$ resonances in the interval $3m_\pi < W \leq 1$ GeV. At $W = 1$ GeV, the non-resonant three-pion cross section is quite small. The K^+K^- cross section in this region is dominated by the $\phi(1020)$ resonance and becomes small at $W = 2$ GeV. Finally, the hadronic continuum defined as at least three particles in the final state is small below 1 GeV. The hadronic cross section is therefore decomposed into four parts,

$$R_{had}(s) = R_{\text{cont}}(s) + R_{\pi^+\pi^-}(s) + R_{K^+K^-}(s) + R_{\text{res}}(s), \quad (9)$$

where: $R_{\text{cont}}(s)$ is the continuum contribution in the interval $1 \text{ GeV}^2 \leq s < \infty$; $R_{\pi^+\pi^-}(s)$ is the pion-pair contribution in the interval $4m_\pi^2 \leq s < 4 \text{ GeV}^2$; $R_{K^+K^-}(s)$ is the kaon-pair contribution in the interval $4m_K^2 \leq s < 3.24 \text{ GeV}^2$; and $R_{\text{res}}(s)$ is the contribution of the $\omega(782)$ [excluding $\pi^+\pi^-$ final states], $\phi(1020)$ [excluding K^+K^- final states], ψ-family, and Υ-family resonances.

3.2 ANALYSIS TECHNIQUE

The continuum, $\pi^+\pi^-$, and K^+K^- contributions to $\Delta\alpha_{had}(M_Z^2)$ are evaluated by numerically integrating parameterized functions which are fit to each set of data. The correlations induced by common normalization uncertainties are incorporated into the fitting procedure by defining χ^2 as follows,

$$\chi^2 = \sum_i \frac{\left[R_{had}^i - (1+\lambda_j\alpha_i)R_{fit}(s_i;a_k)\right]^2}{\sigma_i^2(\text{ptp})} + \sum_j \lambda_j^2, \quad (10)$$

where R_{had}^i is the value of R_{had} measured at energy s_i, α_i is the fractional normalization uncertainty associated with the ith measurement, $R_{fit}(s_i;a_k)$ is a parametric function chosen to suitably model the data, $\sigma_i(\text{ptp})$ is the point-to-point uncertainty associated with the ith measurement, and λ_j are fit parameters which are constrained to have zero mean and unit width. This form preserves shape information and propagates the normalization uncertainties into the parameters a_k of the function R_{fit}.

The uncertainty on $\Delta\alpha_{had}(M_Z^2)$ is estimated using two techniques. In the first, the parameter uncertainties are propagated to the calculated value of $\Delta\alpha_{had}(M_Z^2)$ using the following expression which is valid for any function of the parameters,

$$\delta^2(\Delta\alpha_{had})_{exp} = \sum_{k,l} \frac{\partial(\Delta\alpha_{had})}{\partial a_k} E_{kl} \frac{\partial(\Delta\alpha_{had})}{\partial a_l}, \quad (11)$$

where the derivatives are calculated numerically and $E_{kl} = \langle \delta a_k \delta a_l \rangle$ is the parameter error matrix that is extracted from the fitting procedure. The second error estimate is performed by constructing a large ensemble of data sets by shifting the measured data points $R_{had}^i(\text{meas})$ as follows,

$$R_{had}^i(\text{set } j) = R_{had}^i(\text{meas}) + f_{ij}^{\text{ptp}}\sigma_i(\text{ptp}) + f_{ij}^{\text{norm}}\sigma_i(\text{norm}), \quad (12)$$

where the factors f_{ij} are Gaussian-distributed random numbers of unit variance. The entire fitting and integration procedure is then applied to each member of the ensemble. The uncertainty on $\Delta\alpha_{had}(M_Z^2)$ is determined from the central 68.3% of the ensemble distribution.

The use of a fitting function has the problem that one may introduce bias through the choice of parameterization. This effect is evaluated by varying the parameterizations as much as ingenuity and computer time allow. The quoted contributions to $\Delta\alpha_{had}(M_Z^2)$ are those corresponding to the best fits. Each contribution is assigned a parameterization uncertainty $\delta(\Delta\alpha_{had})_{prm}$ based upon the spread of results corresponding to reasonable fits.

3.3 THE HADRONIC CONTINUUM

The contribution of the hadronic continuum $\Delta\alpha_{had}^{cont}$ is determined from a fit to experimental measurements in the interval 1 GeV$< W <$ 34 GeV. An additional pseudo-measurement at $W = M_Z$ is included by converting the suitably modifed* Particle Data Group average value[11] of $\alpha_s(M_Z^2)$ (0.116±0.005) into a cross section value using the third-order QCD expression for R_{had},[12]

$$R_{had}(M_Z) = 3.807 \pm 0.006. \tag{13}$$

The normalization uncertainties on the measurements below charm threshold are almost uniformly 15-20%. The latest version of this analysis incorporates a new and precise R_{had} measurement by the Crystal Ball Collaboration at charm threshold.[13] This measurement has a 7% normalization uncertainty and represents a major improvement in the determination of $\Delta\alpha_{had}^{cont}(M_Z^2)$.

None of the cross section measurements in the charm-threshold region separate the resonant contribution from the continuum contribution. Since this analysis incorporates the $\psi(4040)$, $\psi(4160)$, and $\psi(4415)$ resonances observed by the DASP Collaboration[14] into the resonance contribution, the cross section measurements in the charm-threshold region are not used. However, the shape of the non-resonant continuum measured in the DASP fit to the resonances is used in the construction of the fitting function.

The piecewise-continuous fitting function is constructed from polynomials and threshold functions in the region 1 GeV$< W <$ 15 GeV. Above $W =$ 15 GeV, the third-order QCD expression for R_{had} is used with $\alpha_s(M_Z^2)$ allowed to vary as a free parameter. The data and the result of the fit used to to evaluate the central value of $\Delta\alpha_{had}^{cont}$ are shown in Fig. 3. The error bars include the point-to-point and the normalization uncertainties. The fit quality is reasonable (χ^2/dof $= 110/100$). Note that the presence of the new Crystal Ball measurement pulls the fit to larger values of R_{had} at charm threshold. The large overlapping data sets of the $\gamma\gamma2$[15] and Mark I[16] Collaborations span the interval from 1.42 GeV to 3.65 GeV and are effectively renormalized upward by the Crystal Ball point which pulls the fit to larger values throughout the region.

---

* Since possible anomalies in the Z lineshape would bias the determination of $\alpha_s(M_Z^2)$ from the lineshape parameters, the Z lineshape information is excluded from the Particle Data Group average. Additionally, since the PEP/PETRA R_{had} measurements are included explicitly in the fit, they are also excluded from the PDG average.

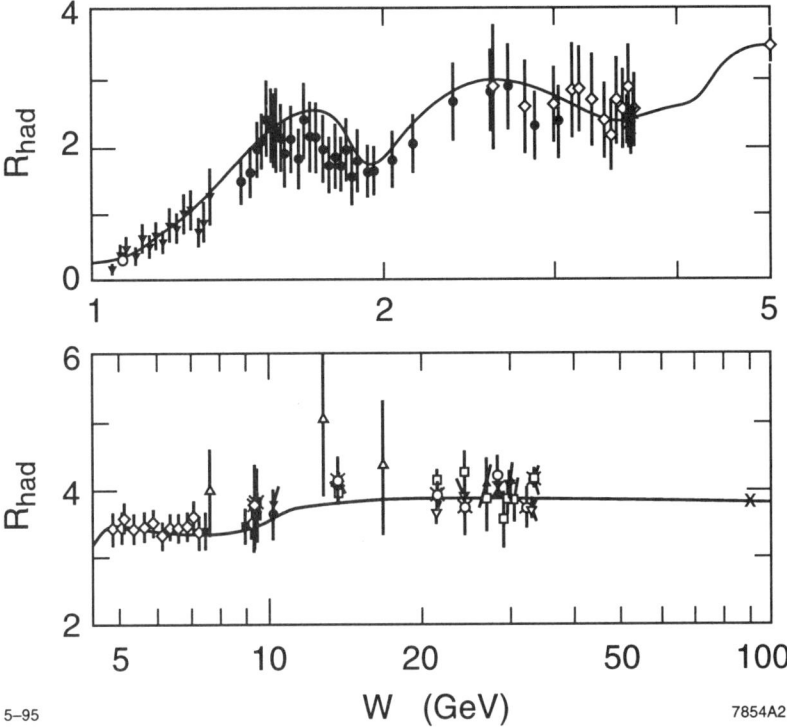

Figure 3. The continuum R_{had} measurements including normalization uncertainties. The fit used to to evaluate the central value of $\Delta\alpha_{had}^{cont}$ is shown as the solid curve.

3.4 THE $\pi^+\pi^-$ AND K^+K^- FINAL STATES

The processes $e^+e^- \to \pi^+\pi^-$ and $e^+e^- \to K^+K^-$ are described by the electromagnetic form factors, $F_\pi(s)$ and $F_K(s)$, which are related to the hadronic cross section ratio R_{had} for each process as follows,

$$R_{had}^{\pi^+\pi^-}(s) = \frac{1}{4}|F_\pi(s)|^2\beta_\pi^3, \qquad R_{had}^{K^+K^-}(s) = \frac{1}{4}|F_K(s)|^2\beta_K^3, \qquad (14)$$

where β_π and β_K are the velocities of the final state particles in the e^+e^- center-of-mass frame. It is clear that measurements of the form factors are equivalent to measurements of R_{had}.

Nine sets of $|F_\pi|^2$ measurements are fit to a function which is a sum of the Gounaris-Sakurai form[17] used by Kinoshita, Nizic, and Okamoto[18] to model the $\rho(770)$; and the Breit-Wigner amplitudes of the $\omega(782)$ and two additional resonances. The result of the fit is shown as a solid line in Fig. 4. The fit preferred a resonance of width 0.44 GeV at mass 1.15 GeV and a second resonance of width 0.18 GeV at mass 1.71 GeV. The fit quality is found to be good ($\chi^2/\text{dof} = 138.3/127$).

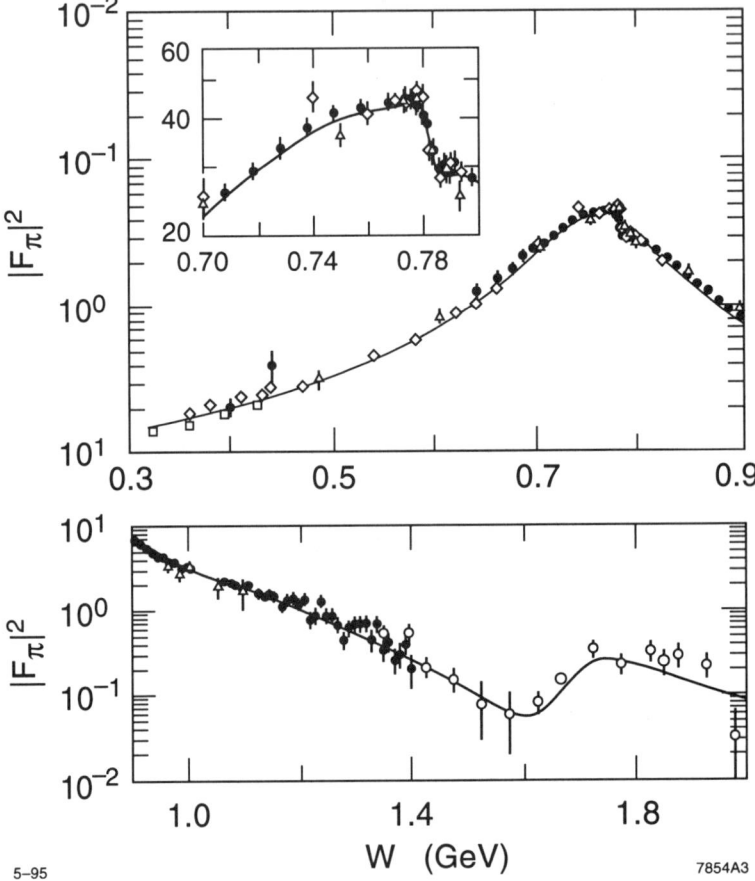

Figure 4. Measurements of $|F_\pi(W)|^2$ are compared with the best fit which is shown as a solid line. The error bars include normalization uncertainties.

Five sets of $|F_K|^2$ measurements are fit to a function which is a sum of a Breit-Wigner lineshape with an energy-dependent width for the $\phi(1020)$; and amplitudes for the $\rho(770)$, $\omega(782)$, and two additional resonances. The result of the fit is shown as a solid line in Fig. 5. The fit preferred a resonance of width 0.17 GeV at mass 1.35 GeV and a second resonance of width 0.24 GeV at mass 1.68 GeV. The fit quality is found to be good ($\chi^2/\text{dof} = 48.9/44$).

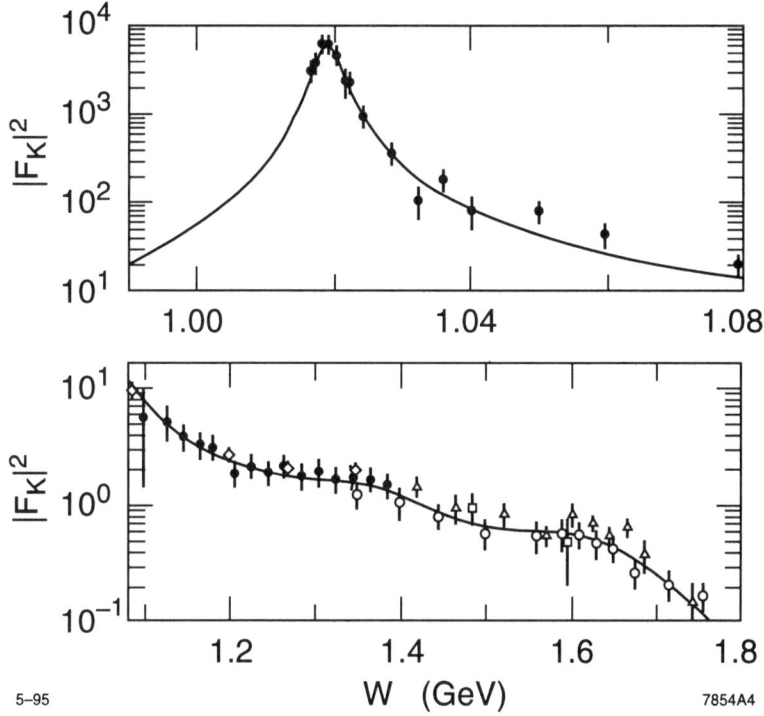

Figure 5. Measurements of $|F_K(W)|^2$ are compared with the best fit which is shown as a solid line. The error bars include normalization uncertainties.

3.5 THE RESONANCES

The resonances comprise the remaining portion of the total e^+e^- cross section. The total cross section for each resonance can be represented by a relativistic Breit-Wigner form with an energy-dependent total width,[19]

$$\sigma_{res}(s) = \frac{12\pi}{m} \frac{\sqrt{s}\Gamma_{ee}\Gamma_{fs}(s)}{(s-m^2)^2 + s\Gamma_{tot}^2(s)}, \qquad (15)$$

where: m, Γ_{ee}, and Γ_{tot} are the mass, electronic width, and energy-dependent total width of the resonance; and Γ_{fs} is the energy-dependent width corresponding to the final states considered in the analysis. Note that the electronic widths are physical widths (not corrected for vacuum polarization effects). In order to incorporate the Breit-Wigner cross section described by equation (15) into equation (5), it must be scaled to the electromagnetic point cross section,

$\sigma_{\mu\mu}(s) = 4\pi\alpha^2(s)/3s$, yielding the following expression,

$$\Delta\alpha_{had}^{res}(q^2) = \frac{\alpha_0 q^2}{4\pi^2} P \int_{4m_\pi^2}^{\infty} ds \frac{\sigma_{res}(s)}{\alpha^2(s)[q^2-s]}, \tag{16}$$

which has the slightly unpleasant feature that it incorporates $\alpha(s)$, the quantity that is being evaluated, into the integrand. To avoid this problem, the $\Delta\alpha_{had}(s)$ parameterization given in Ref. 4 is used to generate a first-order estimate of $\alpha(s)$ for use in equation (16).

Equation (16) is evaluated for the $\omega(782)$, $\phi(1020)$, ψ-family, and Υ-family resonances by performing a Simpson's rule integration over the interval $m - 60\Gamma_{tot}$ to $m + 60\Gamma_{tot}$ (the lower limit of the ω integration is the threshold for 3π decay). The masses and widths used to evaluate equation (16) are taken from the 1994 Review of Particle Properties.[11]

3.6 FINAL RESULT

The various contributions to $\Delta\alpha_{had}(M_Z^2)$ are summarized and summed in Table 1. The resulting value is

$$\Delta\alpha_{had}(M_Z^2) = 0.02752 \pm 0.00046. \tag{17}$$

Including the leptonic contribution, we find $\alpha^{-1}(M_Z^2)$ to be,

$$\alpha^{-1}(M_Z^2) = 128.96 \pm 0.06, \tag{18}$$

where the uncertainties on the lepton masses contribute negligibly to the total uncertainty. This result differs by one of its standard deviations from the (common) result given in References 7 and 8 and it differs by 0.3 standard deviations from the result given in Reference 6. However, since the different analyses make use of many of the same inputs, the results are not independent measurements of $\Delta\alpha_{had}(M_Z^2)$ but reflect differences in assumptions and technique.

Table 1: Summary of the various contributions to $\Delta\alpha_{had}$.

Contribution	W Region (GeV)	$\Delta\alpha_{had}(M_Z^2)$	$\delta(\Delta\alpha_{had})_{exp}$	$\delta(\Delta\alpha_{had})_{prm}$
Continuum	1.0-∞	0.022106	0.000366	0.000196
$\pi^+\pi^-$	0.280-2.0	0.003240	0.000057	0.000169
K^+K^-	0.987-1.8	0.000356	0.000032	0.000030
Resonances	$\omega^{(a)}$	0.000307	0.000010	0.000003
"	$\phi^{(b)}$	0.000296	0.000012	0.000004
"	ψ (6 states)	0.001101	0.000059	0.000023
"	Υ (6 states)	0.000118	0.000005	0.000003
Total		0.02752	0.00038	0.00026

$^{(a)}$Doesn't include $\pi^+\pi^-$ final states.
$^{(b)}$Doesn't include K^+K^- final states.

It is clear that the uncertainty on $\Delta\alpha_{had}(M_Z^2)$ is dominated by the experimental uncertainty on the continuum contribution. In order to understand the origin of this uncertainty, the uncertainty on the integrand of the continuum dispersion integral (integrated over W rather than s) is plotted as a function of W in Figure 6. The integrand uncertainty is calculated using equation (11) to estimate the uncertainty on $R_{fit}(W)$ at each energy point. The dashed curve shows the uncertainty before the Crystal Ball data point is included in the fit and the solid curve shows the uncertainty after its inclusion. Note that the overall uncertainty on $\Delta\alpha_{had}^{cont}$ is dominated by the poor precision of the data in the 1 GeV to 5 GeV region. A detailed comparison[9] of this analysis with the one described in Reference 7 indicates that much of the difference with that result stems from questions dealing with the optimal use of the rather poor quality data in this region. Improved data in this region would likely result in a convergence of these approaches.

The improvement that could be achieved with a modest set of new data is simulated by reducing the normalization uncertainty on the Mark I measurments in the region 2.6 GeV$< W <$ 3.65 GeV from 20% to 5%. The resulting uncertainty on $\Delta\alpha_{had}^{cont}$ is improved by a factor of two, which would eliminate $\Delta\alpha_{had}(M_Z^2)$ as a limitation for some time.

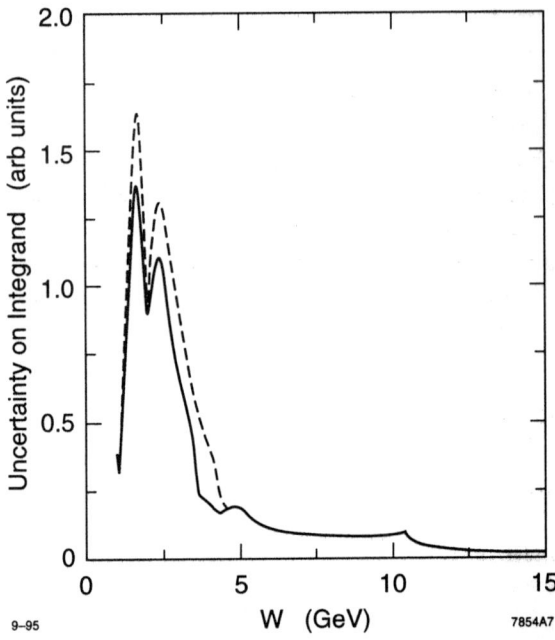

Figure 6. The uncertainty on the integrand of the dispersion integral (integrated over W rather than s) in arbitrary units. The dashed curve shows the uncertainty before the Crystal Ball data point is included in the fit and the solid curve shows the uncertainty after its inclusion.

4. Conclusions

The interpretation of the world's most precise electroweak data is limited by the precision of very old e^+e^- cross section measurements in the 1-5 GeV region. New measurements below charm threshold with a 5% systematic normalization uncertainty would eliminate this limitation for quite some time. Additionally, the structure of the charm threshold region is not known well. The three resonances observed by the DASP have not been confirmed by later experiments (the Crystal Ball Collaboration[13] observes only two resonances and a broad shoulder). New cross section measurements in the charm threshold region would clarify the structure of the charm threshold and would help to resolve differences in $\Delta\alpha_{had}$ analyses. A high luminosity Tau-Charm Factory could perform these measurements more quickly and efficiently than lower luminosity machines but is not crucial to perform the measurements. However, a Tau-Charm Factory quality detector is crucial. In order to control the systematic normalization uncertainty, the detector needs a high degree of redundancy in tracking, calorimetry, and triggering to measure the relevant efficiencies. And perhaps even more than redundancy, the detector needs large solid angle coverage. Most of the older experiments had limited coverage and found that their acceptance calculations were quite sensitive to event structure modelling. The Crystal Ball detector was hermetic in the region of polar angle $|\cos\theta| < 0.9$ and was much less sensitive to this problem leading to the 7% uncertainty mentioned earlier. An ideal detector would have an even larger acceptance to minimize reliance on event structure simulations (which are generally inadequate at low energies).

Acknowledgements:

This work was supported by Department of Energy Contract No. DE-AC03-76SF00515.

REFERENCES

1. N. Cabibbo and R. Gatto, *Phys. Rev.* **124**, 1577 (1961).
2. See G. Burgers and W. Hollik, CERN-TH-5131/88, August 1988, and CERN 88-06, September 1988.
3. F.A. Berends and G.J. Komen, *Phys. Lett.* **63B**, 432 (1976); F. Jegerlehner, *Z. Phys.* **C32**, 195 and 425 (1986); B.W. Lynn, G. Penso, and C. Verzegnassi, *Phys. Rev.* **D35**, 42 (1987).
4. H. Burkhardt, F. Jegerlehner, G. Penso, and C. Verzegnassi, *Z. Phys.* **C43**, 497 (1989).
5. F. Jegerlehner, *Progress in Particle and Nuclear Physics*, Vol 27, ed. A. Faessler, Pergamon Press, Oxford 1991, p. 32; F. Jegerlehner, *Proceedings of the Theoretical Advanced Study Institute in Elementary Particle Physics*, Boulder, 1990, ed. M. Cvetic and P. Langacker, World Scientific, Teaneck N.J. 1991, p. 476.
6. A.D. Martin and D. Zeppenfeld, *Phys. Lett.* **B345**, 558 (1995).
7. S. Eidelman and F. Jegerlehner, PSI-PR-95-1, BudkerINP 95-5, January 1995.
8. H. Burkhardt and B. Pietrzyk, PSI-PR-95-1, LAPP-EXP-95.05, June 1995.
9. M.L. Swartz, SLAC-PUB-95-7001, hep-ph/9509248, September 1995.
10. See the presentation of A. Olshevsky at the 1995 International Europhysics Conference on High Energy Physics, Brussels, Belgium.

11. Review of Particle Properties: L. Montanet, *et al.*, *Phys. Rev.* **D50**, 1173 (1994).

12. S.G. Gorishny, A.L. Kataev, and S.A. Larin, *Phys. Lett.* **B259**, 144 (1991); L.R. Surgladze, M.A. Samuel, *Phys. Rev. Lett.* **66**, 560 (1991); Erratum: ibid, 2416.

13. Crystal Ball Collaboration: A. Osterheld, *et al.*, SLAC-PUB-4160, December 1986.

14. DASP Collaboration: R. Brandelik, *et al.*, *Phys. Lett.* **76B**, 361 (1978); H. Albrecht, *et al.*, *Phys. Lett.* **116B**, 383 (1982).

15. $\gamma\gamma 2$ Collaboration: C. Bacci, *et al.*, *Phys. Lett.* **86B**, 234 (1979).

16. MARK I collaboration: J.L. Siegrist, *et al.*, *Phys. Rev.* **D26**, 969 (1982); J.L. Siegrist, SLAC-Report No. 225, October 1979.

17. G.J. Gounaris and J.J. Sakurai, *Phys. Rev. Lett.* **21**, 244 (1968).

18. T. Kinoshita, B. Nizic and Y. Okamoto, *Phys. Rev.* **D31**, 2108 (1985).

19. This form follows from the assumption that the resonance adds an imaginary part to the photon propagator $\mathrm{Im}\Pi_{\gamma\gamma}(s) = -\sqrt{s}\Gamma_{tot}(s)$ and from the inclusion of final state phase space factors into the cross section. This form should be a better approximation near threshold which is included in the range of integration for the ϕ and ω resonances.

Theoretical Predictions for Exotic Hadrons

T.Barnes

Computational and Theoretical Physics Group, Oak Ridge National Laboratory
Oak Ridge, TN 37831-6373, USA
and
Department of Physics and Astronomy, University of Tennessee
Knoxville, TN 37996-1200, USA

Abstract. In this contribution we discuss current theoretical expectations for the properties of light meson "exotica", which are meson resonances outside the $q\bar{q}$ quark model. Specifically we discuss expectations for gluonic hadrons (glueballs and hybrids) and multiquark systems (molecules). Experimental candidates for these states are summarized, and the relevance of a TCF to these studies is stressed.

I. INTRODUCTION

The most exciting developments in QCD spectroscopy involve searches for resonances which are external to the conventional $q\bar{q}$ quark model of mesons. There are two general classes of such states, which are those with dominant gluonic excitations "gluonic hadrons" and states with more quarks and antiquarks than the familiar $q\bar{q}$ states.

Since QCD is a theory which contains both quarks *and* gluons as dynamical degrees of freedom, we would expect to see evidence of both these building blocks in the spectrum of physical color-singlet hadrons. It is remarkable, however, that of the hundreds of hadronic states now known, most can be described as states made only of quarks and antiquarks in the nonrelativistic quark model, and none of the remaining problematic resonances have been established as having dominant gluonic valence components. The best evidence for the presence of gluons at low energies is indirect, for example in the Breit-Fermi one-gluon-exchange Hamiltonian used in potential models and in the $q\bar{q} \leftrightarrow s\bar{s}$ configuration mixing evident in the η and η'.

In addition to these gluonic states, one may also form color singlet combinations from multiquark systems of quarks and antiquarks, beginning with $q^2\bar{q}^2$.

Although these have been quite controversial, it now appears that light multiquark resonances do exist in nature, *albeit* as bound meson pairs "molecules" rather than single four-quark clusters.

Experimental studies now in progress may alter the status of hadronic exotica considerably, since there are now several resonances that, if confirmed, appear to be likely candidates for glueballs, hybrids and additional molecules. As we shall see, these states share several common features with theoretical expectations for these unusual hadronic states.

In this contribution we will review current theoretical expectations for gluonic hadrons and molecules, and briefly discuss some of the experimental candidates for these states.

II. GLUEBALLS

A. Introduction

A priori one would expect glueballs to be the most attractive gluonic hadrons experimentally, since they might be expected to differ most noticeably from $q\bar{q}$. In practice this naive expectation may not be realized; studies of the light glueball spectrum using lattice gauge theory have found that the lowest-lying glueball is a scalar, and its coupling to two-pseudoscalar final states suggests a typical hadronic width. The next glueballs encountered at higher masses are predicted to be 0^{-+} and 2^{++}, and states which couple to two transverse gluons (presumably the lightest glueballs) do not contain exotic J^{PC}.

Although there have been many studies of the spectrum and quantum numbers expected for glueballs [1], the results of lattice gauge theory should be treated as the most relevant to experiment, since they bear the closest resemblance to full QCD. The assumptions of quenched lattice gauge theory are that decay channels do not modify glueball masses significantly (since the neglect of quarks implies stable light glueballs) and that the extrapolations to small lattice spacing and large lattice volume do not introduce important biases. If glueballs are not very broad objects, the assumption of stable glueballs should not introduce large mass errors.

There are lattice predictions for the masses of glueballs with various J^{PC} [2]; the most reliable is presumably for the scalar glueball ground state, which is predicted to have a mass of

$$M(0^{++}) = \begin{cases} 1.550(50) & \text{GeV [3]} \\ 1.740(71) & \text{GeV [4]} \end{cases}. \tag{1}$$

The corresponding mass estimate for the tensor glueballs is in the 2.2-2.4 GeV range,

$$M(2^{++}) = \begin{cases} 2.270(100) & \text{GeV [3]} \\ 2.359(128) & \text{GeV [4]} \end{cases} ; \qquad (2)$$

with the pseudoscalar glueball at a similar mass.

There are obvious problems associated with the identification of a scalar state near 1.5 GeV. The f_0 sector is the most complicated of all meson sectors, with at least six problematical states, $f_0(980)$, $f_0(1300)$, $f_0(1365)$, $f_0(1500)$ $f_0(1590)$ and $f_0(1710)$. Since this sector contains broad and overlapping resonances, the problem of identifying unusual states against the $q\bar{q}$ and $s\bar{s}$ background, and the related problems of separating individual resonances from interference and threshold effects are daunting ones. If the scalar glueball does have a typical hadronic width, as suggested by the work of Sexton et al. [5], it may be quite difficult to identify this state convincingly. Amsler and Close [6] note that the near degeneracy of the pure (quenched) LGT glueball and the $L = 1$ $q\bar{q}$ and $s\bar{s}$ multiplets may lead to complicated mixing effects, so the physical states may be nontrivial combinations in flavor space, as in the η-η' sector.

The tensor glueball may be an easier experimental target, since the expected mass is far above the lowest-lying 2^{++} quarkonium states. Here the problem is that the mass region above 2 GeV is poorly explored, so it is not yet possible to distinguish a tensor glueball from the background of radial-$^3P_2$ and $^3F_2$ $q\bar{q}$ and $s\bar{s}$ states. This lack of adequate information regarding the higher mass quarkonium spectrum is even more of a problem in the 0^{-+} sector.

B. Expectations for glueball properties

Since we have no confirmed glueballs and the states predicted are in channels with a complicated or poorly explored resonance spectrum, it would be useful to have reliable theoretical predictions of glueball properties as a guide. The data we are likely to have on gluonic candidates in the near future are their masses, widths and strong decay amplitudes. Here a very characteristic naive glueball signature can be given, although it is easy to imagine ways in which this signature might be violated.

As gluons at the bare lagrangian level have equal strength couplings to quarks of all flavors, one can make the assumption that flavor-symmetric couplings to hadron final states are approximately valid for physical glueballs. This gives a characteristic flavor-singlet branching fraction to pseudoscalar pairs, which is (neglecting phase space differences)

$$\Gamma(G \to \pi\pi : K\bar{K} : \eta\eta : \eta\eta' : \eta'\eta')/(\text{phase space}) = 3 : 4 : 1 : 0 : 1 . \qquad (3)$$

Of course this simple pattern should at least incorporate the $|\vec{P}|$ from phase space for an S-wave decay, and there is in addition a decay form factor which depends on the unknown scalar glueball wavefunction and the decay mechanism. Experience with the $^3P_0$-model $f_0(q\bar{q})$ decay amplitude to $\pi\pi$, which has a node near the physical point [7], suggests that the naive pattern of flavor-singlet decay amplitudes may indeed be far from the physical couplings.

The accuracy of naive flavor-singlet couplings can be tested for a pure (quenched) scalar glueball in lattice gauge theory through a determination of the glueball-Ps-Ps three point function. Preliminary results for this coupling [5] indicate that flavor-singlet symmetry may indeed be badly violated at the amplitude level, and higher-mass Ps pairs are preferred in the decay. In view of the relatively large errors it is important to improve the statistics of this interesting lattice gauge theory measurement. An extension of this work to the decay amplitudes of tensor and pseudoscalar glueballs would also be a very useful contribution.

In future experimental work it may be possible to determine or limit electromagnetic couplings of glueball candidates. Measurements of one-photon ($R \to \gamma q\bar{q}$) and two-photon ($R \to \gamma\gamma$) transition rates of these resonances are extremely important because theorists can calculate these for $q\bar{q}$ states with reasonably accuracy [8]. The radiative transition rates of a relatively pure glueball would clearly be anomalous relative to expectations for the corresponding $f_J(q\bar{q})$ state. If physical glueballs are indeed strongly mixed linear combinations of gluonic, $q\bar{q}$ and $s\bar{s}$ basis states, a convincing way to identify the flavor components of these mixed states would be through a comparison of the relative rates

$$\Gamma(R \to \gamma\rho^0 : \gamma\omega : \gamma\phi)$$

since these act as flavor tags. Similarly, $\gamma\gamma$ couplings can be used to locate the scalar nonstrange f_0 $q\bar{q}$ signal, since this state should have a strong coupling to $\gamma\gamma$. Results on this reaction have already been obtained by the Crystal Ball in the reaction $\gamma\gamma \to \pi^\circ\pi^\circ$ [9]. Since a glueball should have suppressed couplings to $\gamma\gamma$, measurements of the $\gamma\gamma$ couplings of the various f_J states and other light resonances would be very important contributions to light meson spectroscopy at a TCF.

C. Summary of glueball candidates

At present the two most prominent experimental candidates for glueballs are the scalar $f_0(1500)$ and the $\xi(2230)$, which is probably a tensor. The scalar candidate has a mass and width (as reported by Crystal Barrel [10]) of

$$M(f_0) = 1520 \, ^{+20}_{-55} \, \text{MeV} \qquad (4)$$

and

$$\Gamma(f_0) = 148 \, ^{+20}_{-25} \text{ MeV} . \tag{5}$$

The $f_0(1500)$ seems rather too massive to be a nonstrange $^3P_0$ $q\bar{q}$ state, but is consistent with the lower mass estimates from LGT for a scalar glueball. The width is also quite narrow for a $^3P_0$ $q\bar{q}$ state at this mass. The decay pattern to pseudoscalar pairs is however inconsistent with flavor symmetry; the squared invariant couplings cited by Amsler [10] are

$$\Gamma(f_0(1500) \to \pi\pi : K\bar{K} : \eta\eta : \eta\eta')/(\text{p.s.}) =$$

$$1 \; : \; < 1/8.6 \; (95\%\text{c.l.}) \; : \; 0.24 \pm 0.12 \; : \; 0.35 \pm 0.15 \; . \tag{6}$$

A priori this argues against a pure glueball interpretation, and subsequent work by Amsler and Close [6] has investigated the possibility that these decays may be consistent with a scalar glueball that has important $q\bar{q}$ and $s\bar{s}$ components, leading to an $\eta\eta'$ mode and suppressing the $K\bar{K}$ mode. The limit on the coupling to $K\bar{K}$ is actually inferred from another experiment, and a more careful study of this coupling including interferences at the Crystal Barrel appears to find a much larger $K\bar{K}$ coupling [11]. This state has also been reported in a recent reanalysis of the MarkIII data on $\psi \to \gamma\pi^+\pi^-\pi^+\pi^-$ by Bugg et al. [12]; in this channel the $f_0(1500)$ appears dominantly in the "$\sigma\sigma$" mode of two S-wave $\pi\pi$ pairs.

The second glueball candidate, which might be the $\xi(2230)$ previously reported by MarkIII [13] in ψ radiative decays, is reported by BES [14] to have very anomalous properties for a tensor above 2 GeV. The mass and width BES cite for this state in $K_S K_S$ are

$$M(\xi) = 2232 \, ^{+8}_{-7} \pm 15 \text{ MeV} \tag{7}$$

and

$$\Gamma(\xi) = 20 \, ^{+25}_{-16} \pm 10 \text{ MeV} , \tag{8}$$

with similar results in $P\bar{P}$, K^+K^- and $\pi^+\pi^-$. If this narrow state is confirmed it is a remarkable discovery indeed. The mass is consistent with LGT expectations for the lightest tensor glueball (2), and the narrow width implies that this is certainly not a tensor quarkonium state. Since the couplings to $\pi\pi$ and $K\bar{K}$ appear to be approximately flavor symmetric [14], this appears to be a natural glueball candidate.

Although Godfrey, Kokoski and Isgur [15] previously suggested that the $^3F_2$ and $^3F_4$ $s\bar{s}$ states expected near this mass could be relatively narrow, subsequent work by Blundell and Godfrey [16] has shown that other modes such as $K_1(1270)K$ are large, so $\Gamma(f_2(s\bar{s})) \geq 400$ MeV. Similarly for the $^3F_4$ Blundell and Godfrey now find a broader state given these additional modes, $\Gamma(f_4(s\bar{s})) \geq 130$ MeV. Thus the $s\bar{s}$ assignments now appear implausible if the $\xi(2230)$ does indeed have an experimental width of < 50 MeV.

Several of the properties reported for this narrow $\xi(2230)$ are disturbing. It has surprisingly small branching fractions to pseudoscalar pairs in view of the available phase space [14]; branching fractions of only a few percent are implied by the PS185 limit on $P\bar{P} \to \xi \to K\bar{K}$. A more important concern is that the reported statistical significance in each of the four channels studied by BES is rather small, $\approx 3\sigma$. A caution is appropriate because some previously reported narrow effects were subsequently found to be artifacts (for example the $\zeta(8.3)$). In view of the remarkable properties reported for this state, measurement of these channels with higher statistics is an extremely important task for any e^+e^- facility operating at the ψ mass.

Although we have only discussed the $f_0(1500)$ and $\xi(2230)$ glueball candidates, this is largely because they have attracted considerable attention recently. Several other states with similar masses and the same quantum numbers, notably the $f_0(1710)$, should also be considered glueball candidates [5]. Measurements of strong branching fractions and electromagnetic decays of this and other glueball candidates should be considered high priorities at a TCF.

III. HYBRIDS

A. Introduction

Hybrid mesons may be defined as resonances in which the dominant valence basis state is $q\bar{q}$ combined with a gluonic excitation. Hybrids are attractive experimentally because, unlike glueballs, they span complete flavor nonets and hence provide many possibilities for experimental detection. In addition, the lightest hybrid multiplet is expected to include at least one J^{PC}-exotic (forbidden to $q\bar{q}$). In the bag model, for example, the lightest gluon mode has $J^P = 1^+$, so the lowest-lying $q\bar{q}g$ multiplet contains the quantum numbers

$$J^{PC_n}(q\bar{q}g) = \begin{cases} 0^{-+}, 1^{-+}, 2^{-+} & (S_{q\bar{q}} = 1), \\ 1^{--} & (S_{q\bar{q}} = 0) \end{cases}. \quad (9)$$

The flux tube model extends this bag model list by adding a degenerate set with reversed $\{P, C\}$ to the lowest hybrid multiplet. Constituent gluon models differ in that their lowest hybrid multiplet has P-wave $q\bar{q}$ quantum numbers [17] and so is nonexotic, although exotics appear in excited hybrid multiplets. An investigation of $q\bar{q}g$ interpolating fields [18] shows that hybrids can have any J^{PC}.

B. Hybrid masses.

Hybrids have been studied using a wide range of models and techniques. These are the MIT bag model [19], constituent gluon models [17,20,21], the flux tube model [22–31], an adiabatic heavy-quark bag model [32], heavy-quark lattice gauge theory [33] and QCD sum rules [34–38]. There have been no published Monte Carlo lattice gauge theory studies of hybrid masses; a study of exotic hybrid masses would be an interesting application of this technique. In all the theoretical approaches employed to date the lightest hybrids (H_q, involving u, d flavors) are predicted to have masses in the $\approx 1\frac{1}{2}$-2 GeV region. A summary of hybrid mass predictions for the especially interesting 1^{-+} exotic is given in the table below, taken from [28]. A more detailed discussion of these predictions and the literature on hybrids is given by Barnes, Close and Swanson [28]; for other recent reviews of hybrids see [39].

Much of the recent interest in hybrids has derived from the flux tube model, which gives rather precise predictions for masses and decay modes of hybrids. The original flux tube references [23–25] cited masses of ≈ 1.9 GeV for the lightest (u,d) hybrid multiplet, ≈ 4.3 GeV for $c\bar{c}$ hybrids and ≈ 10.8 GeV for $b\bar{b}$ hybrids. There is an overall variation of about 0.2-0.3 GeV in these predictions, as indicated in Table I. Multiplet splittings are usually neglected in the flux tube model. This approximation may not be justified; a large inverted spin-orbit term was found for hybrids by Merlin and Paton [25].

TABLE I. Predicted 1^{-+} Hybrid Masses.

state	mass (GeV)	model	Ref.
$H_{u,d}$	1.3-1.8	bag model	[19]
	1.8-2.0	flux tube model	[22–25,28]
	2.1-2.5	QCD sum rules (most after 1984)	[35–37]
	2.1	constituent gluon model	[21]
H_c	≈ 3.9	adiabatic bag model	[32]
	4.1-4.5	flux tube model	[23–25,28]
	4.1-5.3	QCD sum rules (most after 1984)	[35–37]
	4.19(3) $\pm$ sys.	HQLGT	[33]
H_b	10.49(20)	adiabatic bag model	[32]
	10.8-11.1	flux tube model	[23–25]
	10.6-11.2	QCD sum rules (most after 1984)	[35–37]
	10.81(3) $\pm$ sys.	HQLGT	[33]

A recent Hamiltonian Monte Carlo study [28] of the flux tube model determined hybrid masses without using the questionable approximations of the earlier flux tube model studies, such as an adiabatic separation of quark and flux-tube motion and a small oscillation approximation for the flux tube. This Monte Carlo study generally confirmed the accuracy of the earlier flux-tube model mass estimates, both for $q\bar{q}$ and $c\bar{c}$ mesons (compared to experiment) and for hybrids (compared to the earlier approximate analytical calculations). These flux tube predictions are shown in Fig.1 below for light quarks and in Fig.2 in the discussion of charmonium hybrids.

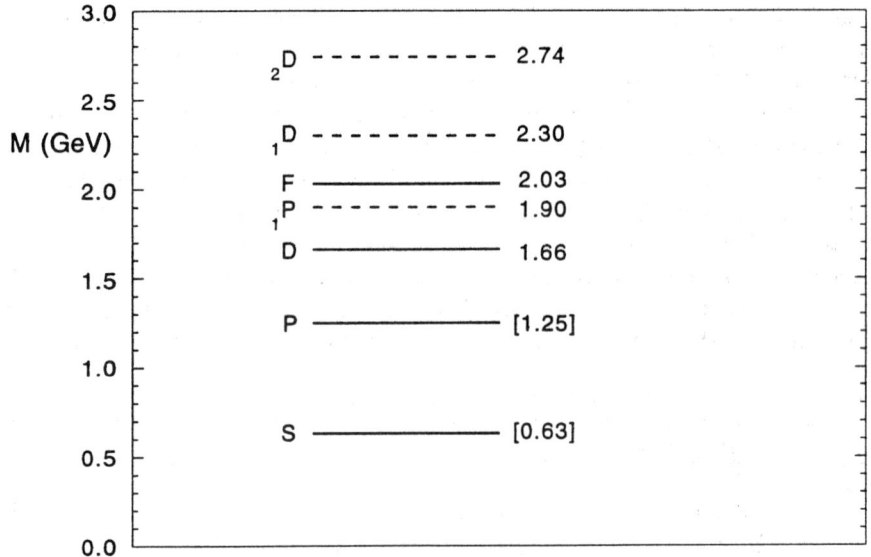

Fig.1. The light qqbar (q=u,d) and hybrid spectrum in the flux tube model [28].

By varying the model parameters over a plausible range, this study concluded that the lightest hybrid masses in the flux tube model were

$$M(H_{u,d}) = 1.8 - 1.9 \text{ GeV} \quad (10)$$

for light quark hybrids and

$$M(H_c) = 4.1 - 4.2 \text{ GeV} \quad (11)$$

for charmonium hybrids. Excited hybrids were also considered, and the first hybrid orbital excitation ($_\Lambda L = {_1}D$) was found at about 2.3 GeV, 400 MeV above the lightest ($_1P$) hybrids. The same numerical result was found earlier by Merlin [26] using the adiabatic approximation. This $_1D$ multiplet contains the J^{PC} states $(1,2,3)^{\pm\mp}$ and $2^{\pm\pm}$, which includes the exotics $1^{-+}, 2^{+-}$ and 3^{-+}. One way to test the experimental candidates for ground-state hybrids near 1.8 GeV [40] and 1.6-2.2 GeV [41] would be to search for members of this excited $_1D$ hybrid multiplet about 0.4 GeV higher in mass.

C. Light hybrid decay modes.

Theoretical models predict rather characteristic two-body decay modes for hybrids. Both constituent gluon [20] and flux tube [27] models find that the lightest hybrids decay preferentially to pairs of one $L_{q\bar{q}}=0$ and one $L_{q\bar{q}}=1$ meson "S+P", for example πf_1 and πb_1. These unusual modes previously received little experimental attention because they involve complicated final states, which may explain why hybrids were not discovered previously. The flux-tube decay predictions of Isgur, Kokoski and Paton [27] are quite interesting because they suggest that many hybrids are so broad that they will be effectively invisible, whereas a few hybrids should be narrow enough to be easily observable in certain channels. The $I=1$ $J^{PC}=1^{-+}$ exotic had already been cited as an attractive experimental candidate, and this work suggested that this state should be relatively narrow, $\Gamma_{tot} \approx 200$ MeV, and that the S+P modes πb_1 and πf_1 should be the dominant final states. These studies have motivated several experimental investigations of πb_1 and πf_1, which show possible indications of resonant amplitudes in 1^{-+}.

These original flux tube decay calculations were for the three exotic J^{PC} quantum numbers in the lowest flux-tube multiplet. Since this multiplet contains a total of eight J^{PC} assignments, $1^{\pm\pm}$ (for $S_{q\bar{q}}=0$) and $2^{\pm\mp}$; $1^{\pm\mp}$; $0^{\pm\mp}$ (for $S_{q\bar{q}}=1$), one might wonder whether any of the *nonexotic* hybrids are narrow enough to be observed. The decay amplitudes of these nonexotic hybrids were recently calculated by Close and Page [29], who also checked the exotic decay amplitudes and found reasonable numerical agreement with Isgur, Kokoski and Paton.

Close and Page predict that many of these nonexotic hybrids are also so broad as to be effectively unobservable. There are two striking exceptions. One is a 1^{--} ω-hybrid with a total width of only ≈ 100 MeV, which decays to $K_1(1270)K$ and $K_1(1400)K$; this should be searched for in $K_1 K$ final states, perhaps in photoproduction. A second interesting nonexotic hybrid is a π_2, with $\Gamma_{tot} \approx 170$ MeV. This may be the high-mass state which has been reported in several photoproduction experiments a mass near 1775 MeV [40]. Other notable conclusions are that 1) several other hybrids, including exotics, have total widths near 300 MeV and so should be observable, and 2) the $I=0$ 0^{+-}

exotic found by Isgur et al. to have $\Gamma_{b_1\pi} = 250$ MeV actually has very large $K_1 K$ modes and so should be unobservable.

In addition Close and Page investigate the "forbidden" decay modes such as $\hat{\rho}(1900) \to \rho\pi$, and find that, due to differences in the ρ and π spatial wavefunctions, these S+S modes are present with partial widths of typically ~ 10 MeV. An important $\rho\pi$ coupling was found earlier by deViron and Govaerts [38] using QCD sum rules. Thus it is interesting to search relatively straightforward modes such as $\rho\pi$ for hybrids, in addition to the favored but more difficult S+P modes such as $b_1\pi$, πf_1 and $K_1 K$.

D. Prospects for charmonium hybrids at a TCF.

The predictions of the recent flux tube model calculations ([28], shown below) and heavy-quark LGT [33] that hybrid charmonium states should appear beginning at 4.1-4.2 GeV are especially relevant for the physics program of a Tau-Charm Factory.

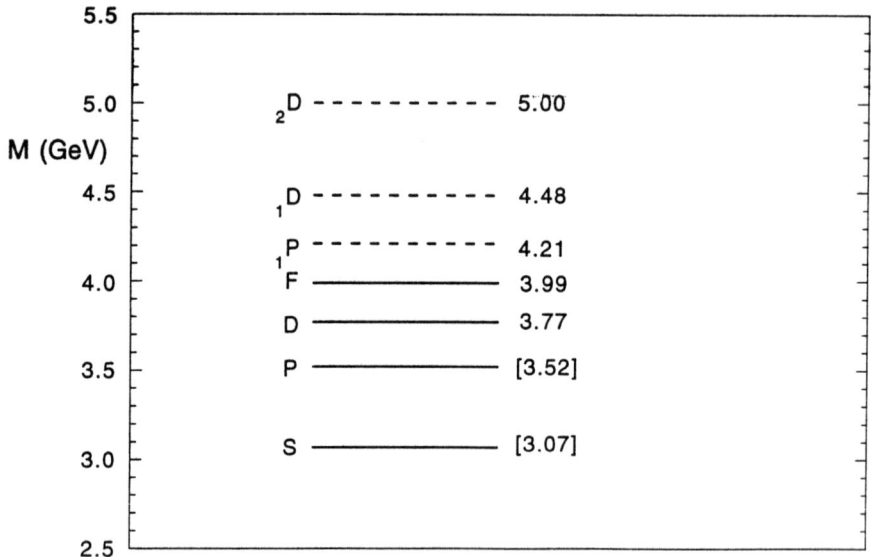

Fig.2. Charmonium and ccbar-hybrid masses in the flux tube model [28].

Charmonium spectroscopy is rather well understood up to about 3.8 GeV, so searches for unusual states should be straightforward near this mass. Since only a few open charm channels occur below 4.3 GeV, for a considerable range of masses one might anticipate rather narrow hybrid resonances. This possibility is supported by the theoretical preference of hybrids for S+P decay modes, which have thresholds of about 4.3 GeV for $c\bar{c}$ and 11.0 GeV for $b\bar{b}$. Calculations of the decay widths of charmonium hybrids have been carried out in the flux tube model by Close and Page [31], assuming masses of $\approx$ 4.1-4.2 GeV. The partial widths (to D^*D) are found to be quite small, typically only $\sim 1 - 10$ MeV. Thus if there are relatively unmixed charmonium hybrids, the 1^{--} vector hybrids should appear as narrow spikes in R in this mass range. For this reason a detailed scan of R starting near the open charm threshold would be a first priority at a Tau-Charm Factory.

Close and Page subsequently speculate about a more complicated possibility, which is that the $\psi(4040)$ and $\psi(4160)$ may be equal-weight linear combinations of $3S$ $|c\bar{c}\rangle$ and 1^{--} $c\bar{c}$-hybrid basis states. (The usual assignment is that the $\psi(4040)$ is a $3S$ $c\bar{c}$ and the $\psi(4160)$ is a $2D$ $c\bar{c}$ [43].) The Close-Page linear combinations would explain why the e^+e^- widths are approximately equal and relatively large for both states, which is surprising if one is a D-wave $c\bar{c}$. The assignments for the ψ states above open-charm thresholds can be tested by measurements of their branching fractions to $DD, D^*D, ..., D_s^*D_s^*$. The branching fractions predicted by these models are very sensitive to the initial state assignments [42]; unfortunately they have not yet been measured accurately. Determination of these branching fractions would be another high priority at a TCF.

Finally, we note that the non-vector hybrids can also be produced at a TCF through a "continuum cascade", as suggested by D.Bugg, and discussed in references [43,44]. In this approach one produces a high-mass $c\bar{c}$ system in the continuum, for example at 5 GeV; this may then decay hadronically to hybrid charmonium levels of various J^{PC} accompanied by a light hadron or hadrons. The $c\bar{c}$-hybrid in turn decays hadronically to a characteristic state such as the ψ. Thus one can search for example for the decay chain

$$e^+e^- \to c\bar{c} \to H_c\, \eta; \quad H_c \to \eta\psi; \quad \psi \to e^+e^-$$

in the final state $\eta\eta e^+e^-$, triggering on a lepton pair at the ψ mass and $\gamma\gamma$ pairs from the two ηs. The $\eta\psi$ invariant mass distribution can then be studied for evidence of hybrids or $c\bar{c}$ states. Other quantum numbers can be investigated by replacing η by other hadrons, for example $(\pi\pi)_S$, in the hadronic cascades.

E. Hybrid Experimental Candidates

There are several experimental candidates for hybrids, but just as for glueballs there are no generally accepted states at present.

In the exotic channels (which would provide the most convincing evidence for hybrids), previous claims by GAMS that a resonant signal had been detected in the 1^{-+} wave of $\pi\eta$ [45] have now been withdrawn. A KEK experiment [46] finds evidence for a resonant 1^{-+} $\pi\eta$ wave, but with the mass and width of the $a_2(1320)$; this surprising result obviously must be checked carefully for "feedthrough" of the a_2 amplitude. VES [47] has studied $\pi\eta$ and $\pi\eta'$ and report a broad, higher-mass effect in $\pi\eta$ and especially in $\pi\eta'$, near 1.6 GeV. The phase motion of the 1^{-+} component has not yet been determined. Studies of the πf_1 final state suggested by the flux tube model are underway [41,47], and preliminary evidence for a possible 1^{-+} signal has been reported by E818 at BNL [41].

There have been several observations of a photoproduced $I=1$ state in $\rho\pi$ and πf_2 at about 1775 MeV [40], which is too heavy to be the $\pi_2(1670)$ without complicated interference effects. Although the quantum numbers of this state have not been determined definitively, 1^{-+} is preferred over 2^{-+}. A possible narrow 1^{-+} state has been reported by GAMS in $\eta\eta'$ at a mass of 1910 MeV [48]; here there are rather few events, so it will be important to improve the statistics. Several experiments plan future studies of these channels, including E818 (to study $\pi^- f_1$) [49] and E852 (to study πf_1 and $\pi\eta$) [50] at BNL.

In addition to exotic hybrids there are several nonexotic candidates; recall for example the Close-Page result that a hybrid with π_2 quantum numbers is expected to be relatively narrow, and should be visible in πf_2. One way to distinguish hybrids from $q\bar{q}$ spin-singlet states is through their strong decay amplitudes; for example, in the π_2 sector the relative F/P and D/S amplitude ratios in $\pi_2(q\bar{q}) \to \rho\pi$ and πf_2 are reasonably well constrained in the $^3P_0$ and flux tube decay models [51]. These decay models provide an interesting selection rule for $q\bar{q}$ decays; they forbid the decay of a spin-singlet $q\bar{q}$ state to two final spin-singlet quarkonia,

$$(q\bar{q})_{S=0} \not\to (q\bar{q})_{S=0} + (q\bar{q})_{S=0} .$$

In the π_2 channel this selection rule forbids the decay of a $^1D_2$ $q\bar{q}$ π_2 to a $^1S_0$ π plus a $^1P_1$ b_1,

$$\pi_2(q\bar{q}) \not\to \pi b_1$$

but allows it for a hybrid π_2 which does not have the $q\bar{q}$ pair in an $S=0$ configuration. Close and Page find the πb_1 mode of a π_2 hybrid should be rather large, so it is especially important to search the πb_1 channel for evidence of a 2^{-+} signal.

Other nonexotic hybrid candidates which have been suggested recently are a $\pi(1800)$ reported by VES [52] and the nonstrange 1^{--} states near 1.4-1.7 GeV [53]. The $\pi(1800)$ is cited as a possible hybrid because it has unusual branching fractions, including a significant coupling to $\pi\eta\eta$, apparently through the glueball candidate $f_0(1500) \to \eta\eta$. This $\pi(1800)$ is also reported by VES in $\omega\rho$,

$\eta a_0(980)$, $\pi f_0(980)$ and $\pi f_0(1300)$. The decay mode $\pi(1800) \to \rho\pi$ is notably absent, and πf_2 is also weak or absent.

Although the weakness of the $\rho\pi$ S+S mode is indeed suggestive of a hybrid, a $\pi(1800)$ second radial excitation is expected in quark potential models (Godfrey and Isgur [54] predict 1.88 GeV), so one should consider this assignment as well. Radial quarkonia can have unusual branching fractions due to nodes in their decay amplitudes, and in the $^3P_0$ decay model with SHO wavefunctions the amplitude for $\pi(3S) \to \rho\pi$ has a node at $M = 1.88$ GeV for $\beta = 0.35$ GeV. The weakness of the $\rho\pi$ mode is therefore understandable for a 3S state. The same model however predicts a weak $\pi f_0(1300)$ mode, which disagrees with experiment. The decay amplitude for $\pi(3S) \to \pi\rho(2S)$ is predicted to be quite large [55], so a search for a $\pi\rho(1450)$ final state would be useful.

The unusual properties of the nonstrange $I = 0$ and $I = 1$ vectors near 1.5 GeV have led to suggestions that hybrid vector states may be present near this mass [53,56]. In $I = 1$, for example, the two states $\rho(1450)$ and $\rho(1700)$ are usually assigned to 2^3S_1 and $^3D_1$ respectively, but the very large $\rho(1450) \to 2(\pi^+\pi^-)$ mode [56] is in conflict with quark model expectations for a 2^3S_1 state [56–58]. A better understanding of these vector states may require a detailed isobar analysis of their quasi two-body strong decay modes.

These comparisons of strong decay modes illustrate the importance of having an accurate understanding of the decays of radially excited $q\bar{q}$ states. Careful studies of the strong decays of radially excited $q\bar{q}$ candidates such as the $\pi(1300)$, $\rho(1450)$, $\phi(1680)$, $\pi(1800)$ and so forth will be required if we are to distinguish $q\bar{q}$ from non-$q\bar{q}$ states with identical quantum numbers.

IV. MULTIQUARK SYSTEMS AND MOLECULES

A. Introduction

Multiquark systems have had a complicated history, and current theoretical expectations for these states now differ radically from the earliest suggestions. In the pre-QCD quark model era it was thought that multiquark hadrons should exist as resonances in the hadron spectrum. After the discovery of QCD and confinement it was still widely expected that multiquark hadrons should exist (in color singlet sectors), and models typically predicted a very rich spectrum of states. In the light $q^2\bar{q}^2$ sector these "baryonium" resonances were expected to appear beginning at about 1 GeV. It was clear however that there were problems with these predictions, because in the relatively uncomplicated flavor-exotic channels such as $I = 2$ $J^{PC} = 0^{++}$ no $q^2\bar{q}^2$ resonances were observed [59] whereas they were predicted to be relatively light (≈ 1.2 GeV in the MIT bag model). Similarly, the evidence for dilambda hypernuclei [60] makes the existence of an H six-quark resonance well below $\Lambda\Lambda$ threshold (another bag model prediction) appear very unlikely.

The problem with these predictions of multiquark resonances such as $q^2\bar{q}^2$ was that they were above $(q\bar{q})(q\bar{q})$ thresholds, and could spontaneously dissociate "fall-apart" into two mesons [61]. Thus the mass predictions in models which assumed *a priori* that the $q^2\bar{q}^2$ system existed as a single hadron were spurious, because the physical eigenstates were usually a continuum of scattering states [62]. Whether single multiquark clusters exist as resonances under any conditions is a detailed dynamical question, which should be investigated using models that allow the system itself freedom to choose between a single cluster or separate color singlets. At present it appears that single $q^2\bar{q}^2$ hadronic clusters may only exist as resonances in heavy-light systems such as $c^2\bar{q}^2$ [63].

More realistic models of multiquark systems were subsequently developed which gave the $q^2\bar{q}^2$ system freedom to choose dynamically between a bound system and a two-meson scattering state. The variational calculations of Weinstein and Isgur [64] are the best known of these studies; in this work it was found that most 0^+ sectors of the light $q^2\bar{q}^2$ system had two free mesons as the ground state, but that the $I = 0$ and $I = 1$ $qs\bar{q}\bar{s}$ sectors actually had a weakly bound, deuteronlike $K\bar{K}$ pair as the ground state. These states were obvious assignments for the problematical $f_0(980)$ and $a_0(980)$ resonances, which were difficult to explain as $^3P_0$ $q\bar{q}$ states but could easily be understood as $K\bar{K}$ systems with nuclear binding energies of 10s of MeV. These states have been the "prototypes" for hadron molecules, although they remain somewhat controversial. We note in passing that molecule states as a general category are not at all controversial, since the $> 10^4$ known nuclear levels are all examples of hadronic molecules. Here we will discuss meson molecules; candidates also exist in baryon sectors, for example the $\Lambda(1405)$, which may be a $\bar{K}N$ bound system [65].

Signatures for the *a priori* most likely molecular states [66] can be abstracted from our experience with short-ranged hadronic forces and the Weinstein-Isgur results:

1) *J^{PC} and flavor quantum numbers of an L=0 hadron pair.*

2) *A binding energy of at most about $50 - 100$ MeV.*

3) *Strong couplings to constituent channels.*

4) *Anomalous EM couplings relative to expectations for conventional quark model states.*

B. Experimental molecule candidates

1) $f_0(975)$ and $a_0(980)$: The "$K\bar{K}$-molecules".

Weinstein and Isgur [64] found an exception to the fall-apart phenomenon in the scalar sector, with parameters corresponding to the $qs\bar{q}\bar{s}$ system. Here weakly-bound deuteronlike states of kaon and antikaon were found to be the ground states of the four-quark system; Weinstein and Isgur refer to these as "$K\bar{K}$ molecules". The scalars $f_0(975)$ and $a_0(980)$ were obvious candidates for these states, having masses just below $K\bar{K}$ threshold and strong couplings to strange final states. Subsequently the $\gamma\gamma$ couplings of the $f_0(975)$ and $a_0(980)$ were found to be anomalously small relative to expectations for light $^3P_0$ $q\bar{q}$ states ($q = u, d$), as discussed in references [67,68]. The status of the $K\bar{K}$ molecule assignment and the many points of evidence in its favor have been discussed recently by Weinstein and Isgur [69,70].

Morgan and Pennington have argued against a molecule interpretation of the $f_0(975)$ [71]. Their criticism however applies to a $K\bar{K}$ potential model in which the $f_0(975)$ is a single pole in the scattering amplitude. The more recent work of Weinstein and Isgur [69,70] incorporates couplings to meson-meson channels and heavier $^3P_0$ $q\bar{q}$ states, so the physical resonances are not only $|K\bar{K}\rangle$. Since there has been much criticism of the idea of a pure $K\bar{K}$ bound state, a direct quote from Weinstein and Isgur [69] (regarding the $I = 0$ state) is appropriate:

"Despite its name and location, the "$K\bar{K}$ molecule" is not a simple $K\bar{K}$ bound state. Its stability is dependent on its couplings to the other $I = 0$ channels and at $E = M_{S^*}$ the coupled-channel wavefunction has substantial components of the other states."

Although the f_0 and a_0 states remain dominantly $K\bar{K}$, these modifications may answer the objections of Morgan and Pennington. Pennington suggests that the term "deuteronlike" may be a misnomer, if couplings to other states than $K\bar{K}$ play an important rôle in these states [67]. Thus it appears that the important question regarding the f_0 and a_0 may be one of detail, specifically how large the subdominant non-$K\bar{K}$ components are in these states and how they can be observed experimentally.

The experimental measurements which would be most useful for studies of these states at a TCF are 1) their $\gamma\gamma$ widths, which are as yet rather poorly known, and 2) their cross sections in ψ hadronic decays, in $\psi \to \omega f_0$ and ϕf_0. (The latter are flavor-tagging and in studies at BES have shown that the $f_0(980)$ does appear to be a mixed flavor state.) Other interesting measurements at low energies are the radiative transitions $\phi \to \gamma f_0$ and γa_0, which depend strongly on the scalar assignment [72] and may be measured at DAPHNE [73] and CEBAF [74].

2) $f_1(1420)$

Since the $f_1(1420)$ is above the K^*K threshold of 1390 MeV it is a candidate for a nonresonant threshold enhancement ($K^*\bar{K}+h.c.$) rather than a molecular bound state. This possibility was suggested by Caldwell [75], and satisfies the criteria of lying just above the K^*K threshold (antiparticle labels are implicit) and having quantum numbers allowed for that pair in S-wave. The apparent width of the enhancement should not be narrower than the intrinsic width of the K^*, and indeed the PDG values are similar, $\Gamma(f_1(1420)) = 56 \pm 3$ MeV and $\Gamma(K^*) = 50$ MeV. Longacre [76] found that a model with an S-wave nonresonant ($K^*\bar{K} + h.c.$) enhancement gives a good description of this state, and Isgur, Swanson and Weinstein [77] also favor this possibility. The (off-shell) $\gamma\gamma^*$ couplings of the $f_1(1420)$ relative to expectations for a 1^{++} $s\bar{s}$ state may provide a test of the hadron-pair model.

Another test of this K^*K-assignment is in radiative transitions; the dominant radiative mode of a K^*K system will arise from the radiative transition of the K^* constituent, $K^* \to K\gamma$, implying a partial width of

$$\Gamma(f_1(K^*K) \to \gamma K\bar{K}) \approx 80 \text{ KeV} , \qquad (12)$$

and a characteristic pattern of preference for $K^\circ \bar{K}^\circ$ over K^+K^- by about a factor of two. An $s\bar{s}$ state would give a similar radiative partial width, $\Gamma(f_1(1420)(s\bar{s}) \to \gamma\phi) \approx 50$ KeV if we scale from the $\Gamma(f_2' \to \gamma\phi) = 96$ KeV of Godfrey and Isgur [54]. Although the radiative rates are similar, there is a crucial difference in the two assignments: The $s\bar{s}$ decay is to $\gamma\phi$, so the final $K\bar{K}$ pair will clearly originate from a $\phi(1019)$ peak. The $K\bar{K}$ events from a K^*K system should instead have a broad distribution in invariant mass. Thus, the two $f_1(1420)$ assignments can easily be distinguished through the $K\bar{K}$ invariant mass distribution observed in $f_1(1420) \to \gamma K\bar{K}$.

3) Other possible molecules.

There are many other possible molecular states, which can only be mentioned briefly here. In the meson sector these include the $f_0(1710)$, which could be a vector-vector molecule involving $K^*\bar{K}^*$ [78-80]. This can be tested at a TCF by searches for a large $K\bar{K}\pi\pi$ mode. Similarly, the $f_0(1500)$ glueball candidate might be a nonstrange vector-vector system [78,81], which would explain the weakness of the $K\bar{K}$ mode. The $f_0(1365)$ should also be considered a possible vector-vector molecule, in view of its very large coupling to $\rho\rho$ despite the near absence of phase space. The 2^{++} state reported by VES [52] in $\rho^\circ\rho^\circ$ is another possible vector-vector molecule, although its appearance in the $\rho\rho$

D-wave may be a problem. The $\psi(4040)$, which shows a strong preference for $D\bar{D}$ over $D^*\bar{D}^*$ (opposite to expectations from phase space) was one of the earliest molecule candidates [82]. As the $c\bar{c}$ assignment for this state is a $3S$ radial excitation, this anomalous branching fraction may be due to a node in the decay amplitude near the $D^*\bar{D}^*$ momentum [42]. Finally, there are several molecule candidates in baryon sectors, such as the $\Lambda(1405)$ [65] (which as a possible $\bar{K}N$ is the earliest molecule candidate excluding nuclei), KN-flavor "Z^*" exotics (discussed in [83,84]) and dibaryons [85].

Since molecular bound states are a special aspect of $2 \to 2$ hadron scattering amplitudes, one might anticipate that an understanding of these scattering amplitudes will lead to reliable predictions of molecules. Theoretical work along these lines is in progress; at present there are different predictions for molecules depending on the scattering mechanism assumed. In one pion exchange models [78,80] many bound states are predicted which should be experimentally observable. In scattering calculations assuming quark-gluon forces (see [79,83,86-89] and references cited therein) few channels are found to have sufficiently strong attractions to form bound states; the vector-vector system [79,87] is one of the few. One of the principal limitations of hadron-hadron scattering calculations at the quark-gluon level is the absence of $q\bar{q}$ annihilation in most studies. Annihilation is known to be an important effect when allowed, for example in the $K\bar{K}$ molecules [69,70]. An extension of this work to include $q\bar{q}$ annihilation is in progress [90].

V. ACKNOWLEDGMENTS

I would like to thank the organisers of the Argonne Tau-Charm meeting, especially José Repond and Jasper Kirkby, for the opportunity to present these results and to discuss hadron physics with my fellow participants. I am grateful to E.S.Ackleh, C.Amsler, C.Brindle, D.Bugg, S.-U.Chung, F.E.Close, G.Condo, K.Danyo, A.Donnachie, A.Dzierba, P.Geiger, S.Godfrey, N.Isgur, P.R.Page, M.R.Pennington, E.S.Swanson and N.Törnqvist for discussions and material that contributed to this work. This research was sponsored in part by the United States Department of Energy under contract DE-AC05-84OR21400, managed by Lockheed Martin Energy Systems, Inc.

---

[1] F.E.Close, Rep. Prog. Phys. 51, 833 (1988).
[2] For the earlier LGT glueball literature see for example C.Michael and M.Teper, Nucl. Phys. B314, 347 (1989); P.deForcrand, G.Schierholz, H.Schneider and

M.Teper, Phys. Lett. B152, 107 (1985); G.Berg, Nucl. Phys. B221, 109 (1983); and references cited therein.

[3] G.Bali et al., Phys. Lett. B309, 378 (1993).

[4] H.Chen et al., IBM report IBM-HET-94-1 (contribution of A. Vaccarino to Lattice 93).

[5] J. Sexton et al., IBM report IBM-HET-94-5 (contribution to Lattice 94).

[6] C.Amsler and F.E.Close, "Evidence for Glueballs", Rutherford Laboratory and CERN report CCL-TR-95-003 (April 1995); ibid., Rutherford Laboratory reports RAL-95-036 (May 1995) and RAL-TR-95-003 (July 1995).

[7] E.S.Ackleh, T.Barnes and E.S.Swanson, Oak Ridge National Laboratory report ORNL-CTP-95-09.

[8] See for example E.S.Ackleh and T.Barnes, Phys. Rev. D45, 232 (1992); T.Barnes, in Proceedings of the IXth International Workshop on Photon-Photon Collisions (World Scientific, 1992) eds. D.O.Caldwell and H.P.Paar, pp.263-274; and references cited therein.

[9] J.K.Bienlein (Crystal Ball Collaboration), in Proceedings of the Ninth International Workshop on Photon-Photon Collisions (La Jolla, 22-26 March 1992), eds. D.O.Caldwell and H.P.Paar (World Scientific, 1992), pp.241-257.

[10] V.V.Anisovich et al., Phys. Lett. B323, 233 (1994); C.Amsler, in Proceedings of the XXVII Int. Conf. on High Energy Physics (Glasgow, 20-27 July 1994), Zürich report UZH-PH-50/94.

[11] C.Amsler, personal communication.

[12] D.Bugg et al., Phys. Lett. B353, 378 (1995).

[13] R.M. Baltrusaitis et al., Phys. Rev. Lett. 56, 107 (1986).

[14] T.Huang, contribution to the Argonne Workshop on a Tau-Charm Factory (June 1995); see also T.Huang et al., CCAST report BIHEP-TH-95-11.

[15] S.Godfrey, R.Kokoski and N.Isgur, Phys. Lett. B141, 439 (1984).

[16] S.Godfrey, personal communication; H.G.Blundell and S.Godfrey, "The $\xi(2220)$ Revisited: Strong Decays of the 1^3F_2 and 1^3F_4 $s\bar{s}$ Mesons.", Carleton University report OCIP/C 95-11.

[17] D.Horn and J.Mandula, Phys. Rev. D17, 898 (1978).

[18] T.Barnes, "The Bag Model and Hybrid Mesons", in Proceedings of the SIN Spring School on Strong Interactions (Zuoz, Switzerland, April 9-17, 1985). Also distributed as University of Toronto report UTPT-85-21 (April 1985).

[19] T.Barnes, Caltech Ph.D. thesis (1977), unpublished; T.Barnes, Nucl. Phys. B158, 171 (1979); T.Barnes and F.E.Close, Phys. Lett. 116B, 365 (1982); M.Chanowitz and S.R.Sharpe, Nucl. Phys. B222, 211 (1983); T.Barnes, F.E.Close and F.deViron, Nucl. Phys. B224, 241 (1983); M.Flensburg, C.Peterson and L.Sköld, Z. Phys. C22, 293 (1984).

[20] M.Tanimoto, Phys. Lett. 116B, 198 (1982); Phys. Rev. D27, 2648 (1983); A.LeYaouanc, L.Oliver, O.Pène, J.-C.Raynal and S.Ono, Z. Phys. C28, 309 (1985); F.Iddir, A.LeYaouanc, L.Oliver, O.Pène, J.-C.Raynal and S.Ono, Phys.

Lett. B205, 564 (1988); S.Ishida, H.Sawazaki, M.Oda and K.Yamada, Phys. Rev. D47, 179 (1992); Prog. Theor. Phys. 82, 119 (1989).

[21] J.M.Cornwall and S.F.Tuan, Phys. Lett. B136, 110 (1984).

[22] N.Isgur and J.Paton, Phys. Lett. 124B, 247 (1983).

[23] J.Merlin and J.Paton, J. Phys. G11, 439 (1985).

[24] N.Isgur and J.Paton, Phys. Rev. D31, 2910 (1985).

[25] J.Merlin and J.Paton, Phys. Rev. D35, 1668 (1987).

[26] J.Merlin, Oxford University Ph.D. thesis (unpublished); J.Paton, personal communication.

[27] N.Isgur, R.Kokoski and J.Paton, Phys. Rev. Lett. 54, 869 (1985).

[28] T.Barnes, F.E.Close and E.S.Swanson, Oak Ridge National Laboratory / Rutherford Laboratory Report ORNL-CTP-95-02 / RAL-94-106, hep-ph/9501405, Phys. Rev. D (to appear).

[29] F.E.Close and P.R.Page, Nucl. Phys. B443, 233 (1995).

[30] F.E.Close and P.R.Page, Rutherford Laboratory report RAL-94-122, hep-ph/9412301.

[31] F.E.Close and P.R.Page, Oxford University / Rutherford Laboratory report OUTP-95-13P / RAL-95-122, hep-ph/9507407.

[32] P.Hasenfratz, R.R.Horgan, J.Kuti and J.-M.Richard, Phys. Lett. 95B, 299 (1980).

[33] S.Perantonis and C.Michael, Nucl. Phys. B347, 854 (1990), and references cited therein.

[34] I.I.Balitsky, D.I.Dyakanov and A.V.Yung, Phys. Lett. 112B, 71 (1982); Sov. J. Nucl. Phys. 35, 761 (1982); Z. Phys. C33, 265 (1986).

[35] J.I.Latorre, S.Narison, P.Pascual and R.Tarrach, Phys. Lett. 147B, 169 (1984); J.I.Latorre, P.Pascual and S.Narison, Z. Phys. C34, 347 (1987); S.Narison, "QCD Spectral Sum Rules", Lecture Notes in Physics Vol.26, p.375 (World Scientific, 1989).

[36] J.Govaerts, F.deViron, D.Gusbin and J.Weyers, Phys. Lett. 128B, 262 (1983); (E) Phys. Lett. 136B, 445 (1983); J.Govaerts, L.J.Reinders, H.R.Rubinstein and J.Weyers, Nucl. Phys. B258, 215 (1985); J.Govaerts, L.J.Reinders and J.Weyers, Nucl. Phys. B262, 575 (1985); J.Govaerts, L.J.Reinders, P.Francken, X.Gonze and J.Weyers, Nucl. Phys. B284, 674 (1987).

[37] J.Govaerts, F.deViron, D.Gusbin and J.Weyers, Nucl. Phys. B248, 1 (1984).

[38] F.deViron and J.Govaerts, Phys. Rev. Lett. 53, 2207 (1984).

[39] See for example T.Barnes, ORNL-CCIP-93-11 / RAL-93-065 and F.E.Close, RAL-93-053, in Proceedings of the Third Workshop on the Tau-Charm Factory (Marbella, Spain, 1-6 June 1993); T.Barnes, ORNL-CCIP-93-14 / RAL-93-069 in Proceedings of the Conference on Exclusive Reactions at High Momentum Transfers (Marciana Marina, Elba, Italy, 24-26 June 1993); F.E.Close, Rep. Prog. Phys. 51, 833 (1988); C.Dover, in Proceedings of the Second Biennial Conference on Low Energy Antiproton Physics (Courmayeur, 14-19 Sept. 1992);

A.Dzierba, Indiana University report IUHEE-93-2, in Proceedings of the BNL meeting on Future Directions in Particle and Nuclear Physics at Multi-GeV Hadron Facilities (Brookhaven, N.Y. 4-6 March 1993); S.Godfrey, in Proceedings of the BNL Workshop on Glueballs, Hybrids and Exotic Hadrons (AIP, 1989), ed. S.-U. Chung; D.Hertzog, Nucl. Phys. A558, 499c (1993); N.Isgur, CEBAF-TH-92-31, in Proceedings of the XXVI International Conference on High Energy Physics (Dallas, August 1992); G.Karl, Nucl. Phys. A558, 113c (1993).

[40] G.Condo et al., Phys. Rev. D43, 2787 (1991); this state may have been seen earlier by D.Aston et al., Nucl. Phys. B189, 15 (1981).
[41] J.H.Lee et al., Phys. Lett. B323, 227 (1994).
[42] P.R.Page, Nucl. Phys. B446, 189 (1995); see also A.LeYaouanc et al., Phys. Lett. B71, 397 (1977); ibid., Phys. Lett. B72, 57 (1977).
[43] T.Barnes, in Proceedings of the 3rd Workshop on the Tau Charm Factory (Edition Frontieres 1994), eds. J.Kirkby and R.Kirkby, p.41.
[44] F.E.Close, in Proceedings of the 3rd Workshop on the Tau Charm Factory (Edition Frontieres 1994), eds. J.Kirkby and R.Kirkby, p.73.
[45] D.Alde et al., Phys. Lett. B205, 397 (1988).
[46] H.Aoyagi et al., Phys. Lett. B314, 246 (1993).
[47] G.M.Beladidze et al., Phys. Lett. B313, 276 (1993).
[48] Yu. Prokoshkin, presentation at HADRON95.
[49] S.U.Chung, personal communication.
[50] A.Dzierba, personal communication.
[51] P.Geiger and E.S.Swanson, Phys. Rev. D50, 6855 (1994), find that $\pi_2(q\bar{q}) \to \rho\pi$ has $F/P \approx 0.7$. The process $\pi_2(q\bar{q}) \to \pi f_2$ (not discussed in that reference) has $D/S \approx 0.2$ and $G/S \approx 0.01$. (P.Geiger, personal communication, and T.Barnes, unpublished.)
[52] D.Rybachikov, contribution to HADRON95.
[53] A.Donnachie and Yu. Kalashnikova, Z.Phys C59, 621 (1993).
[54] S.Godfrey and N.Isgur, Phys. Rev. D32, 189 (1985).
[55] T.Barnes and F.E.Close, in preparation.
[56] A.B.Clegg and A.Donnachie, Z. Phys. C62, 455 (1994).
[57] G.Busetto and L.Oliver, Z. Phys.C 20, 247 (1983).
[58] R.Kokoski and N.Isgur, Phys. Rev. D35, 907 (1987).
[59] W.Hoogland et al., Nucl. Phys. B126, 109 (1977).
[60] S.Aoki et al., Prog. Theor. Phys. 85, 1287 (1991), and references cited therein.
[61] R.L.Jaffe, Phys. Rev. Lett. 38, 195, 617E (1977).
[62] N.Isgur, Acta Physica Austraica, Suppl. XXVII, 177 (1985).
[63] J.P.Ader, J.M.Richard and P.Taxil, Phys. Rev. D25, 2370 (1982); G.Grondin, unpublished.
[64] J.Weinstein and N.Isgur, Phys. Rev. Lett. 48, 659 (1982); Phys. Rev. D27, 588 (1983); see also A.Astier et al., Phys. Lett. B25, 294 (1967); A.B.Wicklund et

al., Phys. Rev. Lett. 45, 1469 (1980).

[65] R.H.Dalitz and S.F.Tuan, Phys. Rev. Lett. 2, 425 (1959); *ibid.*, Ann. Phys. (NY) 3, 307 (1960); see also J.J.Sakurai, Ann. Phys. (NY) 11, 1 (1960).

[66] T.Barnes, "Signatures for Molecules", Invited contribution to the XXIX Recontres de Moriond, (Meribel, France, 19-26 March 1994); Oak Ridge National Laboratory report ORNL/CCIP/94-08; proceedings published as "QCD and High Energy Hadronic Interactions" (Editions Frontieres, Gif-sur-Yvette, 1994), pp.587-598.

[67] M.R.Pennington, University of Durham report DTP-94/26 (April 1994), Proceedings of the Meeting on Two-Photon Physics from DAΦNE to LEP200 and Beyond, eds. F.Kapusta and J.Parisi (Paris, February 1994).

[68] T.Barnes, Phys. Lett. 165B, 434 (1985); E.P.Shabalin, Yad. Fiz. 46, 852 (1987); T.N.Truong, in Proceedings of the HADRON '89 International Meeting on Hadron Spectroscopy (Ajaccio, 1989), pp.645; N.Brown and F.E.Close, Rutherford Laboratory report RAL-91-085.

[69] J.Weinstein and N.Isgur, Phys. Rev. D41, 2236 (1990).

[70] J.Weinstein, Phys. Rev. D47, 911 (1993).

[71] K.L.Au, D.Morgan and M.R.Pennington, Phys. Rev. D35, 1633 (1987); D.Morgan and M.R.Pennington, Phys. Lett. 258B, 444 (1991); *ibid.*, Rutherford Laboratory report RAL-92-070 (December 1992).

[72] F.E.Close, N.Isgur and S.Kumano, Nucl. Phys. B389, 513 (1993).

[73] N.Brown and F.E.Close, "Scalar Mesons and Kaons in Phi Radiative Decays and their Implications for Studies of CP Violation at DAPHNE." The DAPHNE physics handbook, vol. 2, pp.447-463 (ed. L.Maiani), also distributed as Rutherford Laboratory report RAL-91-085 (Dec 1991).

[74] A.Dzierba *et al.*, CEBAF proposal E-94-016; A.Dzierba, "Measuring Rare Radiative Decays of the Φ Meson at CEBAF", in Proceedings of the Second Workshop on Physics and Detectors for DAPHNE (DAPHNE95).

[75] D.O.Caldwell, Mod. Phys. Lett. A2, 771 (1987); Proceedings of the BNL Workshop on Glueballs, Hybrids and Exotic Mesons (Upton, N.Y. 29 August - 1 September 1988), ed. S.U.Chung (AIP 1989), pp.465-471.

[76] R.S.Longacre, Phys. Rev. D42, 874 (1990).

[77] N.Isgur, E.S.Swanson and J.Weinstein, work in progress.

[78] N.Törnqvist, in Proceedings of the International Conference on Hadron Spectroscopy "HADRON '91", (World Scientific, 1992; eds. S.Oneda and D.C.Peaslee), pp.795-798; Phys. Rev. Lett. 67, 556 (1991).

[79] K.Dooley, E.S.Swanson, and T.Barnes, Phys. Lett. 275B, 478 (1992); K.Dooley, in Proceedings of the 4th International Conference on Hadron Spectroscopy "Hadron '91" (College Park, Md. 12-16 August 1991), (World Scientific, 1992), pp.789-794.

[80] T.E.O.Ericson and G.Karl, Phys. Lett. B309, 426 (1993); G.Karl, Nucl. Phys. A558, 113c (1993).

[81] Yu.S.Kalashnikova, in Proceedings of the International Conference on Hadron Spectroscopy "HADRON '91", (World Scientific, 1992; eds. S.Oneda and D.C.Peaslee), pp.777-782.

[82] V.A.Novikov *et al.*, Phys. Rep. C41, 1 (1978); M.B.Voloshin and L.B.Okun, JETP Lett. 23, 333 (1976); A.DeRújula, H.Georgi and S.L.Glashow, Phys. Rev. Lett. 38, 317 (1977); S.Iwao, Lett. Nuovo Cimento 28, 305 (1980).

[83] T.Barnes and E.S.Swanson, Phys. Rev. C49, 1166 (1994); see also K.Maltman and S.Godfrey, Nucl. Phys. A452, 669 (1986), who reach rather different conclusions regarding attractive channels.

[84] J.S.Hyslop, R.A.Arndt, L.D.Roper and R.L.Workman, Phys. Rev. D46, 961 (1992).

[85] R.A.Arndt, L.D.Roper, R.L.Workman and M.W.McNaughton, Phys. Rev. D45, 3995 (1992).

[86] T.Barnes and E.S.Swanson, Phys. Rev. D46, 131 (1992); for closely related work on meson-meson scattering see B.Masud, J.Paton, A.M.Green and G.Q.Liu, Nucl. Phys. A528, 477 (1991); D.Blaschke and G.Röpke, Phys. Lett. B299, 332 (1993); K. Martins, D. Blaschke and E. Quack, Phys. Rev. C51, 2723 (1995).
A.LeYaouanc, L.Oliver, O.Péne and J.-C.Raynal, Phys. Rev. D42, 3123 (1990).

[87] E.S.Swanson, Ann. Phys. (NY) 220, 73 (1992).

[88] T.Barnes, E.S.Swanson and J.Weinstein, Phys. Rev. D46, 4868 (1992).

[89] T.Barnes, S.Capstick, M.D.Kovarik and E.S.Swanson, Phys. Rev. C48, 539 (1993).

[90] T.Barnes and E.S.Swanson, in preparation.

Status of Hadron Spectroscopy at LEAR

Curtis A. Meyer

Carnegie Mellon University, Pittsburgh, PA 15217

Abstract

Recent observations by the Crystal Barrel experiment of two scalar resonances, $f_0(1370)$ and $a_0(1450)$ have allowed us to clarify the members of the scalar nonet. In addition, a third scalar, $f_0(1500)$, appears to be supernumerary, and is a candidate for the scalar glueball expected near 1550 MeV.

Light Quark Meson Spectroscopy

Light quark meson spectroscopy studies the mesons which are made up of u, d, and s quarks. Within the quark model, these mesons come in nonets of definite spin, parity and C–parity, J^{PC}, where these quantum numbers are produced by the relative state of the quark–antiquark pair. The spin $\frac{1}{2}$ quarks and antiquarks can combine into either a spin singlet, $S = 0$ state, or a spin triplet, $S = 1$ state. In addition, there can be a relative orbital angular momentum L between the quark and the antiquark, $L = 0, 1, 2, \cdots$. Finally, the $q\bar{q}$ system can have radial excitations where the principal quantum number N takes on values $1, 2, \cdots$. The total angular momentum J is then formed by combining L and S. Using these quantum numbers, we can describe a given nonet via spectroscopic notation as $N^{2S+1}L_J$, or in terms of the conserved quantum numbers of the strong interaction $I^G J^{PC}$. Where the system has isospin I, G–parity, $G = (-1)^{L+S+I}$, total spin J, parity $P = -(-1)^L$ and C–parity, $C = (-1)^{L+S}$.

Fig. 1 shows the known light–quark mesons classified in this manner, as well as a few additional states which do not seem to fit into the scheme. It is actually the states which do not fit into the $q\bar{q}$ classification, the so called *exotics*, which are the main interest in light quark meson spectroscopy. These exotics include *glueballs*, bound states of gluons, *hybrid mesons*, meson states containing a valance gluon, and so called 4-quark states, $q\bar{q}q\bar{q}$, all of which are predicted by models of low energy QCD. There are several tools available to identify these states:

- Quantum Numbers, $(I^G)J^{PC}$ identify if the state can be assigned to a nonet, or if the nonet is full, they identify the state as supernumerary.

© 1996 American Institute of Physics

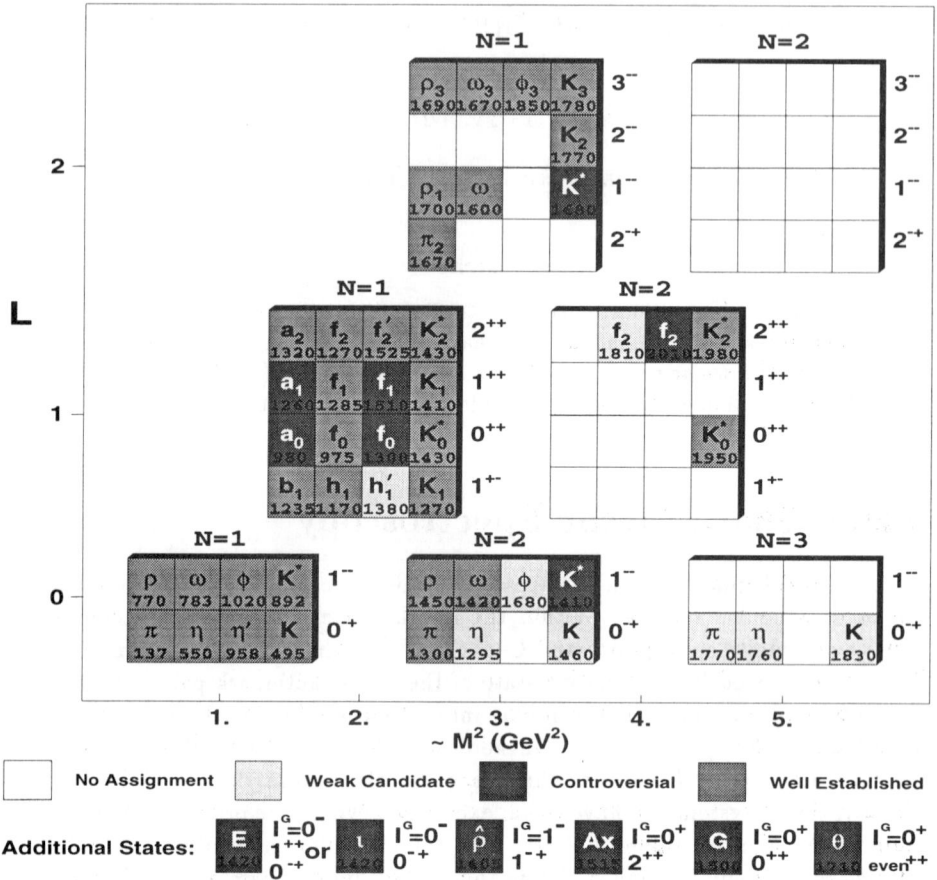

Figure 1: The expected spectrum of light-quark mesons.

- Decay patterns, $Br(x \to ab) : Br(x \to cd) : \cdots$ measure the quark content of the state. In particular the mixing of $\bar{u}u$, $\bar{d}d$ and $\bar{s}s$.

- Production Mechanisms, particularly gluon rich mechanisms provide likely hunting grounds for exotics.

    - Radiative J/Ψ decay, (gluon rich) — $J/\Psi \to \gamma x$
    - Proton antiproton annihilation, (gluon rich) — $\bar{p}p \to xa$
    - Central Production (gluon rich) — $pp \to p_f x p_s$.
    - Photoproduction, (gluon poor) — $\gamma p \to xp$
    - Two Photon Production (gluon poor) — $\gamma\gamma \to x$

As an example of an exotic state, consider the *glueball* – a bound state of 2 or 3 gluons. In most models, the ground state glueball is expected to be a scalar object, $I = 0$, $J^{PC} = 0^{++}$. In addition, lattice calculations, and the fluxtube model both predict that the ground state glueball should have a mass of 1500 to 1600 MeV/c^2. Unfortunately, these are exactly the quantum numbers of the scalar mesons, f_0 and f_0'. The mass is also more or less where one would expect to find the scalar mesons. In this case, the quantum numbers alone will be insufficient to identify the glueball. However, we can also examine the decay pattern of a glueball. Assuming flavor democracy, or equal coupling to u, d and s quarks, one would expect the following decay patterns:

$$\pi\pi : \eta\eta : \eta\eta' : K\bar{K} = 3 : 1 : 0 : 4 \qquad (1)$$

For production, one would expect to see the glueball in the gluon–rich production mechanisms: radiative J/Ψ, $\bar{p}p$ annihilation, and central production, while one would expect this to not be present in channels like photoproduction, and $\gamma\gamma$. Finally, if the glueball were to lie near in mass to mesons of the same quantum numbers, one would not be surprised to see mixing of these states, which could distort the decay pattern.

The Analysis Method

Most of the results presented here come from the Crystal Barrel Detector at LEAR, [1] a nearly 4π apparatus for both charged particles and photons. The detector is used to study $\bar{p}p$ annihilation at rest and in flight, (up to 1.9GeV/c), and has been very productive in the field of meson spectroscopy. The detector is shown in Fig. 2. The antiprotons enter along the axis of the 1.5 T solenoidal field, and stop in a liquid hydrogen target. They annihilate, and the resulting $\pi^\pm$, $K^\pm$ and γ's are tracked outward through the detector.

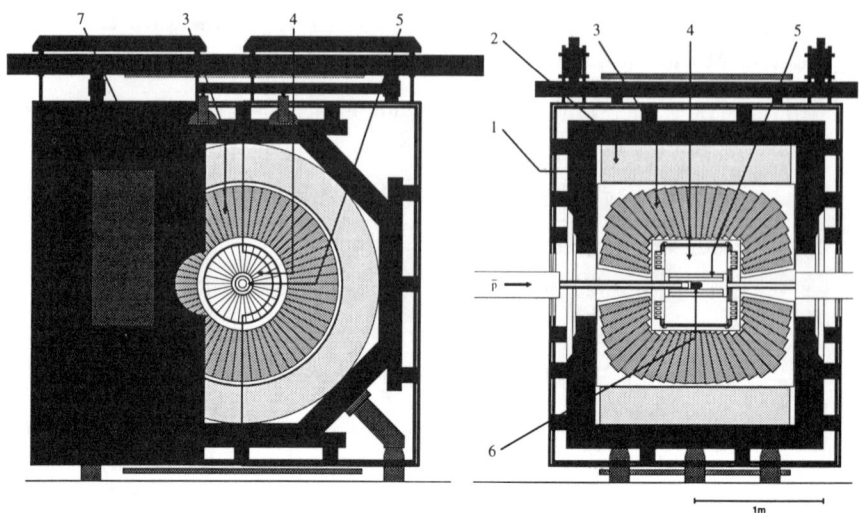

Figure 2: The Crystal Barrel detector. Magnet yoke (1), magnet coils (2), CsI–calorimeter (3), JDC (4), PWC (5), target (6), one half of the endplate (7).

Immediately outside the hydrogen target are two cylindrical multiwire proportional chambers used to either trigger on or veto events with charged particles. Next is a 23 layer cylindrical drift chamber, (JDC). The chamber is divided axially into 30 sectors, with an 8mm wire spacing between adjacent radial layers. The wires are read out at each end to provide a z–measurement through charge division. Outside the JDC is the CsI(Tl) barrel built from 1380 16 radiation length crystals. The crystals are arranged in a center–pointing geometry, and the calorimeter has achieved an energy resolution of 2.5% at 1GeV.

The data presented here arise from $\bar{p}p$ annihilations at rest. For annihilations in a liquid hydrogen target, it is known that most of the $\bar{p}p$ atoms annihilate from an initial S-state, $^1S_0$ or $^3S_1$. In this situation, the quantum numbers of the the initial state are known as 0^{-+} an 1^{--} respectively. These data are analyzed within the framework of the isobar model. Assuming that we have three mesons in the final state, a, b and c, then the transition from the initial to final state is assumed to proceed through a series of quasi two–body intermediate states, A, B and C. The decay momentum $\vec{p}$ describes the two particles from the initial $\bar{p}p$ system, while the decay momentum $\vec{q}$ is measured in the rest frame of the intermediate two–body state. As an example, the following processes might be possible.

$$\bar{p}p \xrightarrow{J^{PC}} Ac \xrightarrow{L,\vec{p}} \overbrace{(ab)c}^{l,\vec{q}}$$
$$\bar{p}p \to aB \to a(bc)$$
$$\bar{p}p \to bC \to b(ac)$$

The transition amplitude $\mathcal{A}$ can be written for a particular partial wave as:

$$\mathcal{A}_{J^{PC}}(\vec{p},\vec{q}) = \sum_{sym} \underbrace{Z_{J^{PC},l,L}(\vec{p},\vec{q})}_{spin-parity\ fcn.} \overbrace{D_L(\vec{p})}^{barrier\ fcn.} \underbrace{F_l(\vec{q})}_{dynamical\ fcn.} \quad (2)$$

where Z is a spin–parity function, D is a centrifugal barrier function, and F is a dynamical function. In the case of a single resonance, F would be written as a Breit–Wigner function. However, in the case of the scalar meson sector, where one has many broad, overlapping resonances, a more detailed parametrization of F is needed. In this case, the K–matrix approach is used to describe the decay of a resonance into several different final states [2]. The real, symmetric K–matrix can be written as:

$$K_{ij} = \sum_{alpha} \frac{g_{\alpha i} g_{\alpha j} D^l_{\alpha i}(q,q_\alpha) D^l_{\alpha j}(q,q_\alpha)}{m_\alpha^2 - m^2} + c_{ij}$$

where m_α are the K–matrix poles, $g_{\alpha i}$ describes the coupling of the α'th pole to the i'th decay channel, $D^l_{\alpha i}(q,q_\alpha)$ is a centrifugal barrier term and c_{ij} is a real background constant. The coupling g can itself be written as

$$g_{\alpha i} = \sqrt{\frac{m_\alpha \Gamma_{\alpha i}}{\rho_i(m_\alpha)}}$$

where $\Gamma_{\alpha i}$ is the partial decay width, and ρ_i describes the two–body phase space, $\rho_i = \frac{2q_i}{m_i}$. The T–matrix can then be written in terms of the K–matrix as:

$$T = (1 - iK\rho)^{-1} K.$$

It is then the poles of the T–matrix which describe the resonance parameters. In addition, the production is described using the P–vector approximation of Aitchison.

$$P_i = \sum \frac{\beta_\alpha g_{\alpha i}}{m_{\alpha^2} - m^2},$$

which then leads to the F–vector,

$$\vec{F} = (1 - iK\rho)^{-1} \vec{P} \quad (3)$$

This parametrization for F is used in equation 2 to fit the data, and extract the resonance parameters.

The Scalar Mesons

The first three data sets discussed in this paper come from an all-neutral trigger, (no charged particles in the PWC). The 6γ final state is obtained from $16.8 \cdot 10^6$ all neutral annihilations at rest, where we reconstruct $\bar{p}p \to \pi°\pi°\pi°$ (712k events), $\bar{p}p \to \pi°\pi°\eta$ (374k events) and $\bar{p}p \to \pi°\eta\eta$ (198k events) where $\pi°$ and η are detected in their 2γ decay modes. These three data sets have been analyzed using a coupled channel technique in the K–matrix formalism. This formalism allows for example a $\pi°\pi°$ resonance to be observed in both the $\pi°\pi°\pi°$ and $\pi°\pi°\eta$ final states. The same decays are involved in both, but a different production mechanism is allowed. In addition, a resonance decaying to both $\pi°\pi°$ and $\eta\eta$ could be observed via the same production mechanism in both $\pi°\pi°\pi°$ and $\pi°\eta\eta$, but a different decay is involved in each. A very important element of these analysis is the correct treatment of the isospin 0, $\pi\pi$ S–wave scattering which we refer to as σ. We have taken the parametrization of Au, Morgan and Pennington [3], (AMP) to describe the $\pi\pi$ scattering as a function of the $\pi\pi$ invariant mass up to the $f_0(980)$. This must then smoothly connect to our solution for higher $\pi\pi$ masses. The following summarizes the results in this analysis.

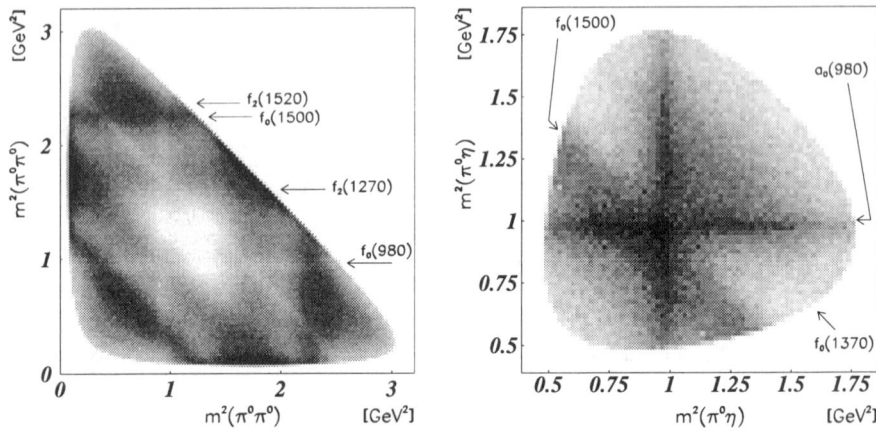

Figure 3: The Dalitz plots for $\bar{p}p \to \pi°\pi°\pi°$ (left) and $\eta\eta\pi°$ (right).

The $\bar{p}p \to \pi°\pi°\pi°$ Dalitz plot (Fig. 3–left) has 3–fold symmetry, meaning that resonances are seen simultaneously as three bands, vertical, horizontal and diagonal. In the vertical, one sees a band near 0.9 GeV^2 corresponding to the $f_0(980)$. The two blobs at the upper and lower edge of the plot near 1.7 GeV^2 corresponds to the $f_2(1270)$, and the vertical band near 2.25 GeV^2 corresponds to a new scalar resonance, the $f_0(1500)$. The blob near 2.4 GeV^2 is caused by reflections from the low–mass $\pi\pi$ S–wave system, and a tensor state, ($J^{PC} = 2^{++}$) near 1550 MeV. In addition, the fit to this Dalitz plot requires a broad, inelastic scalar state at a mass of 1370 MeV, which is probably the $f_0(1300)$. There is a very broad background pole at a mass 1000 MeV/c². The mass and width of this latter pole tend to move around. [4],[5],[6]

The $\eta\eta\pi°$ Dalitz plot (Fig. 3-right) has a 2-fold symmetry. Resonances in the $\eta\pi°$ system are seen as horizontal and vertical bands, while resonances in the $\eta\eta$ system are seen as diagonal bands. The Dalitz plot shows a clear signal for the $a_0(980) \to \eta\pi°$ as the horizontal and vertical crossing bands near 0.96 GeV². In addition, there are two diagonal bands corresponding to the $f_0(1370) \to \eta\eta$ and the $f_0(1500) \to \eta\eta$. [4],[7],[8]

The Dalitz plot for $\bar{p}p \to \eta\pi°\pi°$ (Fig. 4-left), exhibits a very complicated interference pattern which is evidence for dominance of a single initial state. The $\eta\pi°\pi°$ final state arises almost entirely from the $^1S_0$ state of the $\bar{p}p$ atom. The analysis explains the data using the following amplitudes. [$\pi\pi$ S-wave]

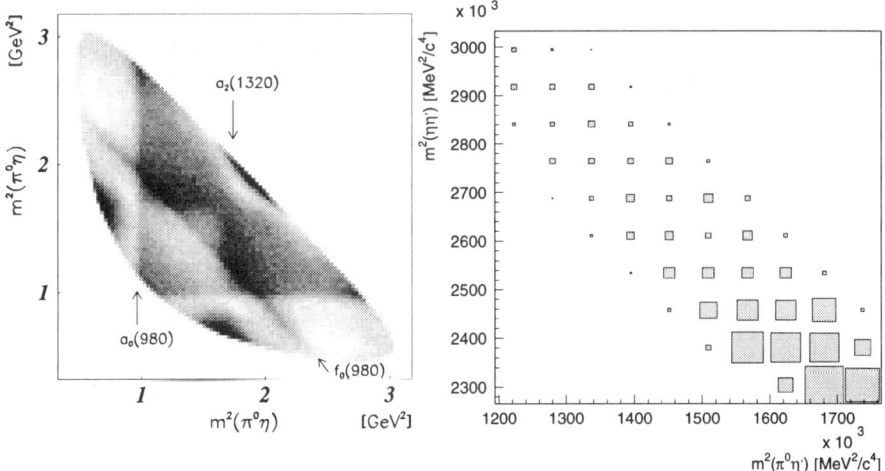

Figure 4: The Dalitz plots for $\bar{p}p \to \pi°\pi°\eta$ (left) and $\eta'\eta\pi°$ (right).

Including the AMP parametrization [3], $f_0(980)$ and $f_0(1370)$ explain 50 to 55% of the data. [$\pi\eta$ S-wave] The $a_0(980)$ explains 13 to 15% of the data. In addition the data demand a second resonance, the $a_0(1450)$, $m = (1470 \pm 25)$, $\Gamma = (265 \pm 30)$, which explains 5% of the data. Ignoring the second resonance causes the χ^2/N_{dof} to increase from 1.2 to 2.5 [9]. [$\pi\eta$ P-wave] A broad non-resonant background wave is needed to explain about 1.5 to 4.5% of the data. We exclude a resonance interpretation with the mass and width reported by GAMS for the $\hat{\rho}(1405)$ [10]. [$\eta\pi$ d-wave] The $a_2(1320)$ describes 18 to 28% of the data. In addition, a second a_2 is needed to account for about 2% of the data. However, its mass is at the edge of phase space, and we do not claim that this is a resonance. [4]

313

A simultaneous spin–parity analysis [4] of $\pi^\circ\pi^\circ\pi^\circ$, $\eta\pi^\circ\pi^\circ$ and $\eta\eta\pi^\circ$ yields the following parameters for the two isoscalar scalar resonances and the new isovector scalar resonance:

$$f_0(1370) \quad m = 1390 \pm 30 \quad \Gamma = 390 \pm 80$$
$$f_0(1500) \quad m = 1500 \pm 10 \quad \Gamma = 154 \pm 30$$
$$a_0(1450) \quad m = 1470 \pm 25 \quad \Gamma = 265 \pm 30$$

and the product branching ratios for $f_0(1500)$ are measured to be:

$$BR(\bar{p}p \to f_0(1500)\pi^\circ, f_0(1500) \to \pi^\circ\pi^\circ) = (1.27 \pm 0.33) \cdot 10^{-3}$$
$$BR(\bar{p}p \to f_0(1500)\pi^\circ, f_0(1500) \to \eta\eta) = (0.60 \pm 0.17) \cdot 10^{-3}.$$

In addition to the previous three final states, an independent spin–parity analysis of $\bar{p}p \to \eta'\eta\pi^\circ$ has been made using 977 $\eta'\eta\pi^\circ$ events reconstructed from the 6γ final state [11]. The Dalitz plot from these events is shown in Fig. 4–right. The enhancement in the lower right corner of the Dalitz plot is identified as a $J^{PC} = 0^{++}$ object in the $\eta'\eta$ system. This isospin 0 object is identified as the $f_0(1500)$ decaying into $\eta'\eta$. We measure its mass, width and branching fraction to be:

$$m = (1545 \pm 25) \quad \Gamma = (100 \pm 40)$$

$$BR(\bar{p}p \to f_0(1500)\pi^\circ, f_0(1500) \to \eta'\eta) = (1.70 \pm 0.40) \cdot 10^{-4}.$$

A second analysis of these data using a mass independent width to fit the f_0 finds an equally good solution with a mass of 1500 MeV/c^2 and a width of 120 MeV/c^2. This solution has a branching ratio that is 20% higher than quoted above.

The $K\bar{K}\pi^\circ$ final state is very important in classifying mesons. In Fig. 5–left is shown a very preliminary Dalitz plot for the $K_L K_L \pi^\circ$ final state. The $K^*(890)$ can be seen as crossing horizontal and vertical bands near 0.8 GeV^2. In addition, there are at least three diagonal bands ($K_L K_L$ resonances). Preliminary results from the spin parity analysis need the $f_0(1500)$ decaying into $K_L K_L$. However at present, we only find an upper limit which is consistent with an older bubble chamber limit [12] of

$$BR(\bar{p}p \to f_0(1500)\pi^\circ, f_0(1500) \to K_L K_L) < (3.6) \cdot 10^{-4}.$$

The system $\bar{p}p \to \pi^+\pi^-\pi^\circ\pi^\circ\pi^\circ$ has been examined to search for resonances decaying into $\rho^+\rho^-$ [13]. The mass–mass plot, (Fig. 5–right) shows an enhancement in the $\rho^+\rho^-$ data over the background. The analysis of these data show the presence of a scalar resonance X decaying into $\rho\rho$ and $\sigma\sigma$, where σ is the $\pi\pi$ S-wave. We find $m_X = 1374 \pm 38$ MeV and $\Gamma_X = 375 \pm 61$ MeV. Under the assumption that this is isoscalar, it is probably the inelastic $f_0(1370)$ seen above, and we estimate that its $4\pi:2\pi$ decay ratio is approximately 5:1.

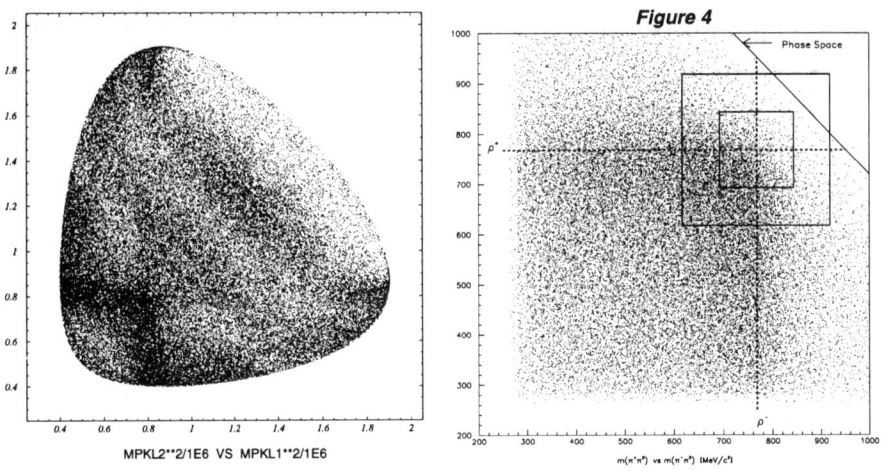

Figure 5: The Dalitz plots for $\bar{p}p \to \pi^0 K_L K_L$ (left) and the mass–mass plot from $\pi^+\pi^-\pi^0\pi^0\pi^0$.

The Scalar Nonet

In Table 1 are shown the ground state $L = 1$ nonets in the quark model as assigned by the Particle Data Group [14]. Normally, the octet and singlet $I = 0$ members of the nonets are mixed. In the case of an ideally mixed nonet, (complete separation between the $u\bar{u} + d\bar{d}$ and the $s\bar{s}$ components), $\Theta = 35.3°$. In table 1 are given the two mixing angles for the PDG assignments for all except the scalars. Most of these are fairly close to ideally mixed.

Looking at the 0^{++} assignment, it is fairly clear that the $a_0(980)$ and the $f_0(980)$ do not fit well into the nonet. Other evidence from both two-photon couplings and J/ψ decays indicate that these states are not simple $q\bar{q}$ systems. If these are not members of the scalar nonet, what are the members? The $f_0(1370)$, (probably the $f_0(1300)$), is a good candidate for the lightest $I = 0$ member. In addition the $a_0(1450)$ fits naturally in as the $I = 1$ member. Using the mass of the $K_0^*(1430)$, we can compute a scalar nonet mixing angle of either $62°$ or $118°$. The question then arises as to where is the f_0'? There are several additional 0^{++} states that could possibly be this state. The $f_0(1500)$ from Crystal Barrel, the $f_0(1520)$ from LASS [15], the $f_0(1590)$ or G–meson from GAMS [16], and possibly the $f_J(1710)$ or Θ. Using the mixing angle for the scalar nonet, we would expect the f_0' to have a mass of 1500–1800 MeV and relative decay widths as given in table 2. Correcting the measured branching fractions for the $f_0(1500)$ by its decay momentum yields the relative decay strengths given in the $f_0(1500)$ row of table 2. The most obvious conclusion

$^{2S+1}L_J$ J^{PC}	$I=1$ $ud, u\bar{u}, d\bar{d}$	$I=0$ $u\bar{u}, d\bar{d},$		$I=0$ $s\bar{s}$		$I=1/2$ $\bar{s}u, \bar{s}d$	Θ
$^1P_1$ 1^{+-}	$b_1(1235)$ $\Gamma=155$	$h_1(1170),$ $\Gamma=360,$		$h'_1(1380)$ $\Gamma=80$	$K\bar{K}^*$	$K_1(1270)$ $K_1(1400)$	41°
$^3P_0$ 0^{++}	$a_0(980)$ $\Gamma=57$	$f_0(975),$ $\Gamma=47,$	$20\% K\bar{K}$	$f_0(1300)$ $\Gamma=400$	$93\% \pi\pi$	$K_0^*(1430)$ $\Gamma=287$	
$^3P_1$ 1^{++}	$a_1(1260)$ $\Gamma=400$	$f_1(1285),$ $\Gamma=25,$	$K\bar{K}^*$	$f_1(1510)$ $\Gamma=35$	$10\% K\bar{K}$	$K_1^*(1400)$ $K_1(1270)$	38°
$^3P_2$ 2^{++}	$a_2(1320)$ $\Gamma=110$	$f_2(1270),$ $\Gamma=185,$	$5\% K\bar{K}$	$f'_2(1525)$ $\Gamma=85$	$71\% K\bar{K}$	$K_2(1430)$ $\Gamma=105$	31°
New 0^{++} Assignment							
$^3P_0$ 0^{++}	$a_0(1450)$ $\Gamma=250$	$f_0(1370),$ $\Gamma=267,$	$\approx 70\% 4\pi$	$f'_0(????)$		$K_0^*(1430)$ $\Gamma=287$	62°

Table 1: The $L=1$ mesons shown with their widths and some decay fractions. The top part of the table shows the PDG assignments, while the bottom row shows this assignment for the scalar mesons.

from this is it is not possible to identify the $f_0(1500)$ as the f'_0. An identification as the first radial excitaion of the the $f_0(1370)$ also seems rather unlikely. The mass difference is to small, and the relatively narrow width of the $f_0(1500)$ make it improbable that it is a radial excitation. The obvious question is *What is the $f_0(1500)$?*.

θ	$\pi\pi$	$\eta\eta$	$\eta'\eta$	$K\bar{K}$
62°	3	0.2	1.7	14.2
118°	3	0.3	0.3	1.4
100°	3	0.15	0.5	3.5
$f_0(1500)$	3	0.70 ± 0.27	1.00 ± 0.46	≤ 0.36
Pure Glueball	3	1	0	4

Table 2: The expected decay fractions for the f'_0 for three mixing angles. In addition, the measured values for the $f_0(1500)$ and a pure scalar glueball are given.

Lattice calculations predict that the lowest mass scalar glueball should have a mass around 1550 MeV. However, for a pure glueball, one expects the flavor blind decays rates as given in equation 1. These are clearly in disagreement with the measured values of the $f_0(1500)$. The $f_0(1500)$ is not a *pure* glueball. However, the $f_0(1500)$ sits more or less in the middle of the isoscalar members of the scalar nonet, $f_0(1370)$, and an expected $f'_0(\sim 1500 - 1800)$. It would be quite surprising if the $f_0(1500)$ were not mixed into the scalar nonet. Amsler and Close [17] argue that this is exactly what has occurred. In this paper, they are able to explain the decay pattern of the $f_0(1500)$ as a glueball mixed into the scalar nonet.

If the $f_0(1500)$ is the scalar glueball, then one would expect to see it in other production mechanisms. Both radiative J/Ψ decays and central production are expected to be gluon rich channels. Recent reanalysis of Mark III data [18] on $J/\Psi \to \gamma(\pi^+\pi^-\pi^+\pi^-)$ show evidence for a state at 1505 MeV/c² decaying to two S-wave di-pion systems, $(\sigma\sigma)$. In addition, recent analysis by the WA91 collaboration [19] on centrally produced $\pi^+\pi^-$ and $\pi^+\pi^-\pi^+\pi^-$ final states,

$$pp \to p_f(\pi^+\pi^-)p_s \text{ and } pp \to p_f(\pi^+\pi^-\pi^+\pi^-)p_s$$

show evidence for the $f_0(1500)$. These data can be described as an interference between the $f_0(1370)$ and the $f_0(1500)$, with both resonances being observed in both the $\pi^+\pi^-$ and $\pi^+\pi^-\pi^+\pi^-$ final states. In addition, the GAMS experiment has observed a $J^{PC} = 0^{++}$ state decaying to $4\pi^\circ$, $\eta\eta$ and $\eta\eta'$ [16]. This centrally produced state is quoted with a mass of 1590, but may be the $f_0(1500)$.

Summary

There is currently good evidence that the scalar gluball has been observed. This object, the $f_0(1500)$ has a decay pattern which is most simply described as the scalar glueball mixed into the scalar nonet. In addition, it is observed in three different gluon rich production mechanisms. However, there are several missing measurements which are needed to clarify this. In particular, a measurement of $f_0(1500) \to \bar{K}K$ is quite important. In addition, the $f_0'(1500-1800)$ needs to be found. We hope to see both of these completed at LEAR in the next couple of years.

References

[1] E. Aker et al., Nucl. Instrum. Methods A**321**, 69, (1992).

[2] S. U. Chung, et al., submitted to **Z. Phys. C**, (1994).

[3] K. L. Au, D. Morgan and M. R. Pennington, Phys. Rev. D**35**, 1633, (1987).

[4] C. Amsler, et al., **Coupled Channel Analysis of Antiproton Proton Annihilation into $\pi^\circ\pi^\circ\pi^\circ$, $\pi^\circ\eta\eta$ and $\pi^\circ\pi^\circ\eta$**. Accepted for publication in Phys. Lett. B., (1995).

[5] C. Amsler, et al., Phys. Lett. B**342**, 433, (1995).

[6] V. V. Anisovich, *et al.*, Phys. Lett. B**323**, 233, (1994).

[7] C. Amelser *et al.*, **High Statistics Study of the** $f_0(1500)$ **decay into** $\eta\eta$. Accepted for publication in Phys. Lett. B, (1995).

[8] C. Amsler *et al.*, Phys. Lett. B**291**, 347, (1992).

[9] C. Amsler, *et al.*, Phys. Lett. B**333**, 277, (1994).

[10] M. Boutemeur and M. Poulet, Hadron 89, ed. F. Binon, J. M. Frère, J. P. Peigneux – (Ed. Frontiérs 1989), p.119.

[11] C. Amsler *et al.*, Phys. Lett B**340**, 259, (1994).

[12] L. Gray *et al.*, Phys. Rev. D**27**, 307, (1983).

[13] C. Amsler, *et al.*, Phys. Lett. B**322**, 431, (1994).

[14] K. Hikasa, *et al.*, Phys. Rev. D**50 II**, (1994).

[15] D. Aston *et al.*, Nucl. Phys. B**301**, 525, (1988).

[16] F. Binon *et al.*, Nuovo Cimento **78A**, 313, (1983), and **80A**, 303, (1984).

[17] C. Amsler, F. E. Close, Phys. Lett. B**353**, 385, (1995).

[18] D. V. Bugg, I. Scott, B. S. Zou, V. V. Anisovich, A. V. Sarantsev, T. H. Burnett, S. Sutlief, Phys. Lett. B**353**, 378, (1995).

[19] F. Antinori *et al.*, Phys. Lett. B**353**, 589, (1995).

Precision Charmonium Physics

Walter Toki*

*Department of Physics, Colorado State University
Ft. Collins, CO. 80523

ABSTRACT. Charmonium physics at a tau charm factory is discussed. The current status of several topics in J/ψ, ψ′ and gluonium physics is reviewed and future physics results from a high luminosity tau charm factory are considered.

I. INTRODUCTION

The physics of the charmonium states, J/ψ and ψ′, produced at a tau charm factory offers an excellent opportunity for important and unique contributions to physics and a large potential for new discoveries. This is mostly due to the ultra high increase in data from a tau charm machine. With a monochromater the beam spread can be reduced to produce extremely high instantaneous data rates for the narrow width J/ψ and ψ′ resonances. The event rates and the events produced per day in the tau charm factory with a beam spread, ~1 MeV, and for a monochromater[1] with a spread of 0.14 MeV are listed in Table I.

Table I. Event rates at a tau charm factory			
Resonance	peak σ	instantaneous rate at 10^{33} $cm^{-2} sec^{-1}$	#events/day at 50% efficiency
J/ψ	~2600 nb	2.6 khz	112×10^6
ψ′	~800 nb	0.8 khz	34×10^6
J/ψ monochronometer	~16000 nb	16 khz	688×10^6
ψ′ monochronometer	~5000 nb	5 khz	215×10^6

In Table II, the current sample of J/ψ and ψ′ events are listed.

Table II. Existing J/ψ and ψ′ data sets		
Experiment	J/ψ data	ψ′ data
MarkII	1.3M	1M
Crystal Ball	2M	1.3M
Mark III	5.8M	.3M
DM2	8M	
BES	9M	4M

As can be seen from the above Tables, the potential increase in number of events from the existing data to a tau charm factory machine is staggering. In a single day, we are expecting a hundred fold increase of the entire data available today. In a short run, factors of a 1000 should be easily achieveable. In a hundred days of

© 1996 American Institute of Physics

running we would expect to collect ~6×10^{10} and ~2×10^{10} of produced J/ψ and ψ' events.

In this paper we explore various physics topics possible with a large increase in J/ψ and ψ' data. Much of this material has been taken from the previous tau charm workshops at SLAC[2], Marbella[3], Dubna[4], and from the literature. We begin with a brief discussion of detector requirements and procede with an examination of the physics topics.

II. REQUIREMENTS FOR A TAU CHARM DETECTOR

The detector requirements for charmonium physics at a tau charm factory have been discussed elsewhere[5]. The main improvements compared to previous detectors are good particle identification to separate pions from kaons, good photon detection efficiency and resolution, good forward acceptance, and very fast data aquisition. These are very similar to the requirements for B factory detectors.[6] Here we discuss issues important to charmonium physics.

For particle identification, pion/kaon separation will be important to identify decays such as J/$\psi \rightarrow \gamma KK\pi$ and to separate backgrounds such as $\gamma 3\pi$ and 3π. Previous detectors, Mark III and BES used time of flight detectors with ~200-300 picosecond resolution that allow 3σ separation up to 800 MeV. Newer systems such as the DIRC, planned for the Babar detector, would dramatically improve the particle separation up to and beyond 1.5 GeV which is more than adequate for tracks from charmonium decays.

For photon detection, a crystal calorimeter similar to the CLEO II detector would make an important improvement over previous detectors for tau charm physics. The old detectors (except Crystal Ball) used gaseous cells in a Pb sandwich or liquid Argon detectors which both had very poor energy resolution. A crystal detector would dramatically improve the energy resolution and low energy photon detection efficiency. This would help to detect low energy photons, reconstruct or reject π^{o}'s and improve the mass resolution in studies of radiative J/ψ decays and of radiative transitions of charmonium states in ψ' decays.

The forward acceptance of the detector is important for spin parity tests of states produced in radiative decays. In many spin parity analyses of decays such as J/$\psi \rightarrow \gamma + X$, $X \rightarrow KK$, $KK\pi$, $\pi\pi$, etc., the polar angle is sensitive to the spin of X. Typically, the decays follow $1 + \cos^2\theta$ or $\sin^2\theta$. Previous detectors had a flat acceptance to $|\cos\theta| \sim 0.7$. This limited range allowed only marginal discrimination between spin assignments. A next generation tcf detector should achieve forward acceptances down to $|\cos\theta| \sim 0.95$.

The data aquisition for the highest rate J/ψ decays at a tau charm factory is required to cope with a signal trigger rate of ~16Khz. At Mark III and BES, a typical event required ~10Kbytes. Assuming a similar event size would require a data aquisition rate of ~160 Mbytes/sec. Currently, cheap off the shelf technology uses 8mm tapes which have a rate of 5 Gbytes per 2 hours or 0.69 Mbytes/sec which corresponds to a requirement of 230 conventional 8mm tape drives for a tcf. These requirements are similar to existing fixed target charm experiments.[7] This technology should improve by the time a tcf detector is operational.

The data reconstruction rate may be estimated from the BES experiment which is about 10 events/sec reconstruction on a medium size UNIX workstation with about 100 MIPS cpu power. Per running day, the rate would be about ~600M events and projecting a 100 day run is projected, this would correspond to ~60x10^9 events. These requirements are comparable to future fixed target charm experiments. To finish one pass of reconstruction would require an equivalent of 200 medium size UNIX workstations for one year. In the future, the UNIX workstations would probably be replaced by much cheaper farms of PC's (future generations of PENTIUM like microprocessor chips) running LINUX.

III. PHYSICS TOPICS

In this section various physics topics are reviewed. The current status of existing measurements and the possible improvements by a tcf are discussed.

IIIa. Tests of the Charmonium Model

The measurements of the single most precise value of charmonium states are listed in Table III.

Table III. Status of Charmonium Measurements (single most precise value)					
State	Mass MeV	Width	Spin	BR % (ee or $\gamma\gamma$)	hadronic decays
η_c	2974.4±1.9	>15 MeV	0^{-+}	7±3 KeV	few decays
J/ψ	3096.9±.09	85.5±6KeV	1^-	5.92±.15±.2	many decays
χ_0	3417±.8	13.5±3	consistent	.04±.02	few decays
χ_1	3510±.04	.88±.11	consistent		few decays
χ_2	3556±.07	1.98±.17	consistent		few decays
$^1P_1$	3526.2±.2	.9±.44	?	?	pp,J/ψ+π
$\eta_c(2s)$	no signal E760	?	?	?	?
ψ'	3686±0.1	.308±.036	1^-	.88±.13	many decays
$\psi(3.77)$	3764±5	24±5	1^-		DD
$\psi(4.03)$	4040±10	52±10	1^-		D*D*,D*D, DD,DsDs
$\psi(4.14)$	4159±20	78±20	1^-		D*D*,D*D, DD,DsDs*
$\psi(4.42)$	4414±7	33±10	1^-		

As is seen in the table, many branching ratio measurements are still quite poor. Most of the precise mass and width measurements have been made by the antiproton gas jet experiment at Fermilab.[8]

The important electromagnetic decays that can be well measured in a tcf include $\eta_c \to \gamma\gamma$ (currently measured to ~8% precision and with large disagreements between experiments) and $J/\psi \to \gamma\gamma\gamma$ (not yet seen). A Monte Carlo study[9] has been performed using 10^9 J/ψ events. The total number of produced events for the two and three photon measurements were 10^4 and 2×10^4, respectively. These would lead to 1% precision measurements at a tcf.

Among the interesting new measurements feasible with a high statistic ψ' data set, is a study of the hadronic decays of the χ states. So far only about ten decays modes of each of the three χ states have been seen. At a tcf, a level of knowledge comparable to what is now known in J/ψ decays would be achieveable for χ decays. In a possible discovery scenario, the scalar χ decays into the scalar glueball (G) plus two pions such as $\chi_0 \to G\pi\pi$ or GS*, with $G \to \pi\pi, KK$. This could be the simplest decay into all s-wave states and may be an ideal place to search for glueballs.

IIIb. Gluonium Searches

Recently Gluonium searches have become a hot topic. QCD Lattice Gauge theories are now confidently predicting[10] the scalar glueball ($J^{pc}=0^{++}$) mass to be ~1600 MeV and the tensor glueball (2^{++}) mass to be ~2200 MeV. Unfortunately the widths are not well predicted and as a result experimentally pinning down their existence has been problematic. A favored area for glueball searches has been in radiative J/ψ decays where two gluon formation is expected to enhance the production of tensor and scalar glueballs.

A candidate for the scalar glueball observed in radiative J/ψ decays has been the resonance previously called the θ which is observed in K^+K^- decays with a mass ~1700 MeV.[11] In a simplified analysis this state[12] was thought to be a tensor. Another glueball candidate is the $\xi(2.2)$ seen in $J/\psi \to \gamma\xi$, $\xi \to KK$ and more recently in $\xi \to \pi\pi, p\bar{p}$. This is experimentally consistent as a tensor.[13]

Recently[14], in the Crystal Barrel experiment observed a new scalar resonance (called $f_0(1500)$) at a mass of 1500 MeV with a width ~120 MeV and decaying into $\pi^+\pi^-$, $\pi^\circ\pi^\circ$, $\eta\eta$ and $\eta\eta'$, but not into KK. In the $p\bar{p} \to 3\pi$ decay mode, the analysis was performed on a +700K data sample. Previously this resonance was thought to be a tensor resonance called the AX, but a new spin parity analysis including interference of $(\pi\pi)_s$ waves, determined the spin to be zero. Also, an independent analysis[15] of old Mark III data on the decay of $J/\psi \to 4\pi$, reveals a scalar resonance near 1500 MeV which might be identified as the same resonance.

In a tau-charm factory the search for glueballs can be significantly improved by

1. large amounts of data (factor 1000) that would enable reliable spin parity analyses of J/ψ→γKK, γππ, γKKπ, γηππ, γηη, γηη'. The existing data has poor statistics which was marginal for spin parity tests.
2. a tcf detector with larger forward acceptance to help discriminate between $\sin^2\theta$ and $1+\cos^2\theta$ production of the resonances as observed in radiative J/ψ decays.
3. a tcf detector with a crystal calorimeter to improve the detection of radiative decays and reject backgrounds with π°'s.

The high statistics data and improved tcf detector will have a major impact in the study of the θ, ξ, and the $f_0(1500)$ resonances. It will be possible to make definitive spin-parity analysis. This may be the only successful approach towards conclusively detecting gluonium states predicted by QCD.

IIIc. Rare Decays of the J/ψ

Very high statistics J/ψ data provides the opportunity to search for very rare decays such as weak J/ψ decays.[16] Based on a comparison[17] of the width of the J/ψ to the lifetimes of charm mesons, we expect a branching ratio of a few $\times 10^{-7}$. A sample of 6×10^{10} J/ψ decays would contain ~450, (D_s or D_s^*)+(π,eν,μν,ρ) decays into φπ, with φ→K^+K^-, assuming the weak decays are 100% into D_s or D_s^*. These decays[18] are particularly interesting as a check of heavy quark effective theory which can provide accurate predictions for these rates.

IIId. Engineering Measurements

Very high statistics J/ψ and ψ' data will enable the measurement of many charmonium and light quark mesons and baryons with very high precision. These measurements, although not leading directly to new physics, are very important for interpreting results from other experiments. In some cases the lack of precision in the existing charmonium data is limiting or will soon be limiting the precision of several measurements from the hadron colliders and the b factories.

<u>Charmonium Measurements</u>

Many experimental measurements need precise charmonium branching ratios. These include measurements of CP eigenstates of charm-anticharm mesons which are secondaries from B meson CP eigenstates. Furthermore, the antiproton gas jet target experiments require precise branching ratios of decays into proton-antiproton pairs in order to normalize their measurements. Table IV lists the decay modes needed by other experiments.

As can be seen, all measurements, except the J/ψ leptonic decay, have large errors and many have been made ~20 years ago. The measurements with the J/ψ, η_c and χ states can be normalized by the clever use of tagged reactions where the produced charmonium rate is normalizable. The ψ' is more difficult as it cannot be produced in the recoil of some other reaction. This requires a careful energy scan at a tcf and measurements of backgrounds below threshold.

Table IV.		
mode	no. of events	branching ratio (%)
$J/\psi \rightarrow ee, \mu\mu$	large	5.92±.15±.2 (MK3)
$J/\psi \rightarrow p\bar{p}$		1.19±.04±.3 (DM2)
$\psi' \rightarrow \pi\pi J/\psi$		32±4 (MK1)
$\psi' \rightarrow ee, \mu\mu$		.88±.13 (MK1)
$\chi_0 \rightarrow 4\pi$		3.7±.7 (MK1)
$\chi_1 \rightarrow 4\pi$		2.2±.8 (MK1)
$\chi_2 \rightarrow 4\pi$		1.6±.5 (MK1)
$\eta_c \rightarrow p\bar{p}$	18±6	.1±.3±.4 (DM2)

Light Quark Mesons and Baryons

High Statistics J/ψ decays ($\sim 10^{10}$) can provide a large sample of decays of $J/\psi \rightarrow p\bar{p} + (\omega, \eta, \eta')$ of $\sim 1.3 \times 10^7$, 2×10^7, and $.9 \times 10^7$ events. These decays may be a clean source of mesons suitable for precision measurements of their branching ratios. For each mode, events can be easily tagged by fitting the recoil spectrum of the $p\bar{p}$ system. It should be possible to achieve absolute precisions of $\sim 0.5\%$. Currently the particle data tables have absolute precisions of 2-5% for the larger decay modes of the ω, η, and η'.

Systematic Errors and Detector Calibration

The tau charm factory will provide precision measurements of many charm, tau and charmonium branching ratios. Currently, many measurement in charm decays from the CLEO group are approaching errors of ~1% and are limited not by statistics, but rather by systematics. At a tcf many of the J/ψ and ψ' decays will permit a careful check of the detector performance and detector systematics which will be necessary to have reliable and well understood precision tau and charm measurements. This is possible because;

1. High rate J/ψ and ψ' decays will produce copious amounts of virtually all hadronic and leptonic tracks. This includes photons, kaons, pions, protons, antiprotons, neutrons, antineutrons, K_L's, electrons, and muons in a large range of momenta and angle. These can be used to determine the tracking efficiency, particle identification efficiency, particle misidentification rates, and the hadronic backsplash in the drift chamber and shower counters. In particular the nonphysics backgrounds such as the hadronic backsplash could be measured for different particle species and be saved in a data dictionary used to simulate backgrounds in the Monte Carlo programs.

2. High rate J/ψ and ψ′ decays will produce reactions with known production rate and production matrix element. This includes for example the decays J/ψ→ρπ and ψ′→ J/ψ ππ. These reactions can be used to test the detector response such as the tracking as a function of absolute momentum and angle. This will be very important to validate the monte carlo simulation of the detector performance which is a stringent requirement for precision measurements. Monte Carlo simulation of backgrounds must include production rates of all charm, tau and continuum backgrounds and a means of simulating all non-physics backgrounds.

IIIe. ψ′ Puzzle

The study of ψ′ production and decay has led to puzzling findings. The decay[19], ψ′→ρπ, is suppressed whereas there is a large rate in J/ψ decays. In addition the rate of ψ′ production as observed by the CDF group is much larger (factor ~30) than predicted by QCD models. Perhaps there are new aspects of the ψ′ yet to be discovered. Since the suppression of the ψ′ decays into vector-pseudoscalars is experimentally well established, the essential next step is a complete measurement of all hadronic ψ′ decays. Basically the all ψ′ decays should be catalogued in a similar fashion to what has been done for the J/ψ. With the complete measurement of all modes, a pattern should emerge to distinguish the decay mechanism at work in the ψ′ and the J/ψ.

IIIf. Other Measurements

High Statistics Production of Strange Particles for Weak Decay studies

The J/ψ has large branching ratios into pairs of strange baryons such as $\Lambda \bar{\Lambda}$ and $\Xi \bar{\Xi}$. A study[20] of these decays for a measurement of CP violations was done. In the Standard Model, nonleptonic hyperon decays can have CP violation through Penguin diagrams. A large sample of hyperon decays contained in one year of J/ψ running could be sensitive to CP violation if longtitudinal polarization[21] is provided in the beam. CP odd parameters can be constructed from sums and differences of helicity decay amplitudes of the hyperon decays and, in the case of the Cascades, the tau charm factory is sensitive to levels predicted by the Standard Model.[22]

Phase Shifts Studies (chiral symmetries)

With high statistics J/ψ and ψ′ data it will be possible to perform careful studies of the phase shifts. An interesting possibility is to check the Chiral symmetry predictions in J/ψ→ππ ψ′. The Chiral Lagrangians[23] constrain the matrix elements and recent data[24] appears to fit s-wave terms. An interesting future measurement will be to test the ππ system for higher order waves to determine if the Chiral Lagrangian includes other terms besides s-wave. Such studies may be an important step towards testing model predictions[25] of phase shifts which are needed for the measurement of CP phases in direct CP violation.

Study of the 4 GeV center of mass energy regions

The e^+e^- energy regions above charm threshold offer several interesting opportunities for studies of D and D* states.

Threshold production of DD, D*D and D*D* might be measureable at a tcf with a monochromater. This could result in very precise measurements of the D and D* masses. The mass error is $\sim \delta m = \delta p(p/m)$, where p is the momentum of the D or D*. If the momentum of the D or D* is very small, then in principle the error in the mass due to the tracking momentum errors could be reduced arbitrarily to zero and the mass errors would be only limited by the beam energy error (~ 0.1 MeV). Of course, this depends on the excitation production cross section of ee$\rightarrow$ DD, D*D and D*D*. As this cross section becomes smaller as threshold is approached, there will be a trade off between the number of events produced and the resolution. This needs to be experimentally optimized.

A large D*D* cross section at 4.03 GeV has been observed. The rate is larger than D*D which has more phase space. It has been argued[26] that this is due to the formation of a D*D* molecule. Therefore, it is important to scan carefully the 4.03 GeV region with a narrow beam spread and to measure the production rates of DD, D*D and D*D* (as well as $D_s D_s$) as a function of center of mass energy.

IV. SUMMARY

Precision Charmonium physics is uniquely possible at a tau charm factory. The increase in data is a 1000 fold over previous experiments. The possibilities for new physics are;

1. Many charmonium measurements are either very poor or nonexistent. Important measurements include the two and three photon widths of the J/ψ and the η_c.
2. Search for gluonium states will be significantly improved with high statistics spin parity studies of the resonances observed in radiative J/ψ decays. Definitive spin parity determinations should be possible.
3. Further progress in understanding the ψ' puzzles requires a large increase in statistics to measure the complete pattern of hadronic decays.
4. The ultra high rate of J/ψ decays enable a high statistics study of light quark mesons. A tau charm factory could contribute many accurate branching ratio measurements.
5. The ultra high rate of J/ψ and ψ' decays permits a careful study of detector acceptances and backgrounds which will be necessary to reduce the systematic errors which are expected to limit precision in charm and tau measurements.

ACKNOWLEDGEMENTS

The author wishes to thank Dr. Repond for an excellent workshop which gathered many experts on tau, charm and charmonium physics.

REFERENCES

[1] P. Yennie, Phys. Rev. Lett., 34, 239(1975).

[2] "Proceedings of the Tau-Charm Factory Workshop", May 23-27, 1989, ed. L. Beers, SLAC Report 343 and "Proceedings of the Tau-Charm Factory in the ERA of B factories and CESR", August 15-16, 1994, ed. L. Beers and M. Perl, SLAC Report 451.

[3] "Third workshop on the Tau-Charm Factory", Jun 1-6, 1993, Marbella, Spain, Ed. J. Kirkby and R. Kirkby.

[4] "Proceedings of the Workshop on the JINR c-tau Factory", May 29-31, 1991, JINR, Dubna.

[5] J. Kirkby, these proceedings.

[6] see the design reports for CLEO II(Cornell), Babar(SLAC), and BELLE(KEK).

[7] The E791 experiment has a 5kHz trigger rate, 0.5x109 data events, and is expected to reconstruct the data in one year using a farm of 20 MIP UNIX workstations.

[8] C. Ginsberg, these proceedings.

[9] R. Mir, "Proceedings of the Tau-Charm Factory Workshop", May 23-27, 1989, ed. L. Beers, SLAC Report 343.

[10] T. Barnes, these proceedings.

[11] L. P. Chen, PhD thesis, University of Vanderbilt, 1990.

[12] R. Baltrusaitus et. al., Phys. Rev. D35, 2077 (1987)

[13] T. Huang, these proceedings.

[14] C. Meyer, these proceedings.

[15] D. V. Bugg, submitted to Phys. Lett.

[16] F. Gilman, these proceedings; R.Verma and A. Kamal, Phys. Lett. B252,690(1990); M.A. Sanchis-Lozano, "Third workshop on the Tau-Charm Factory", Jun 1-6, 1993, Marbella, Spain, Ed. J. Kirkby and R. Kirkby.

[17] W. Toki, "Proceedings of the Tau-Charm Factory Workshop", May 23-27, 1989, ed. L. Beers, SLAC Report 343.

[18] F. Gilman, private communication.

[19] For a review see W. Toki, "Proceedings of the 1989 International Symposium on Heavy Quark Physics", Cornell, June 1989, ed. P. Drell and D. Rubin.

[20] E. Gonsalez and J.I. Illana, "Third workshop on the Tau-Charm Factory", Jun 1-6, 1993, Marbella, Spain, Ed. J. Kirkby and R. Kirkby.

[21] M. Tigner, these proceedings.

[22] J. Donoghue et al., Int. Jour. of Mod. Phys. A2(1987) 147.

[23] For a recent tcf related discussion see N. di Bartolomeo, "Third workshop on the Tau-Charm Factory", Jun 1-6, 1993, Marbella, Spain, Ed. J. Kirkby and R. Kirkby; for a general review see J. Donoghue, E. Golowich, B. Holstein, "Dynamics of the Standard Model", Cambridge University Press, 1994.

[24] D. Coffman et. al., Phys. Rev. Lett. 68, 282 (1992).

[25] M. Lu, M. Wise, M. Savage, Phys. Lett. 337, 133 (1994); A.Datta, S. Pakvasa, Phys. Lett. B344, 430 (1992).

[26] For a recent discussion see T. Barnes, "Third workshop on the Tau-Charm Factory", Jun 1-6, 1993, Marbella, Spain, Ed. J. Kirkby and R. Kirkby.

CHARM PHYSICS:
TESTS OF THE
STANDARD MODEL

QUO VADIS, FASCINUM?

I.I. Bigi

Physics Dept., University of Notre Dame du Lac
Notre Dame, IN 46556, U.S.A.
e-mail address: BIGI@UNDHEP.HEP.ND.EDU

Abstract

The recent progress in our understanding of heavy-flavour decays allows us to define more reliably which future measurements on charm decays are needed to further advance our understanding of QCD and to get our theoretical tools ready for treating beauty decays. After sketching the theoretical landscape I list those required measurements. I argue that some – in particular those concerning inclusive semileptonic charm decays and their lepton spectra – can presumably be performed only at a tau-charm factory.

1 Overview

There are four questions I would like to address in reviewing the long-term goals of charm decay studies: *why, how, where and when?*

Why?

There exists a triple motivation for a detailed analysis of charm decays: (i) it provides novel probes of QCD, (ii) it sharpens our tools for dealing with beauty decays and (iii) it represents one of the more promising avenues to finding the hoped-for 'unexpected', namely the intervention of New Physics; this would most clearly be realized through the observation of CP asymmetries [1, 2].

How?

A comprehensive program is required with its cornerstones being the measurements of (a) lifetimes, (b) semileptonic branching ratios, (c) nonleptonic branching ratios and (d) lepton spectra in exclusive as well as inclusive semileptonic charm decays.

Where?

Three different experimental environments have been considered for high-statistics charm decay studies, namely (α) B factories, (β) photo- and hadro-production and (γ) a τ-charm factory.

When?

There are two benchmark dates in evaluating the impact of a τ-charm factory, namely

(A) 1998 - 2000: the next round of fixed target experiments at FNAL will be finished by then; the data on charm decays from CLEO will have reached a new dimension statistically and systematically; BABAR and BELLE will commence data taking;

(B) $\sim$ 2005: the analysis of charm and beauty decays at the B factories should have reached a fully mature level; new initiatives like CHARM2000 will have had a run; the LHC with its dedicated program on beauty physics will hopefully start up.

My talk will be organized as follows: in Sect.2 I will sketch the relevant theoretical framework for charm decays; in Sect.3 I list the database that is required – or at least desired – for a deeper understanding to come about; in Sect.4 I review various experimental stages before presenting an outlook in Sect.5.

2 Theoretical Framework

2.1 The Landscape

Four second-generation theoretical technologies have emerged for treating the impact of the strong interactions on heavy-flavour decays:

• *QCD Sum Rules* are employed for describing exclusive semileptonic and non-leptonic decays of hadrons with strangeness, charm and beauty.

• *Lattice Simulations of QCD* have worked their way down in distance scale to deal with exclusive semileptonic and non-leptonic charm decays, albeit only in the quenched approximation; quite significant jumps are required before beauty decays can be tackled and one can go beyond the quenched approximation.

- *Heavy Quark Effective Theory* (HQET) on the other hand deals with exclusive semileptonic decays of beauty hadrons; its applicability to charm decays is dubious.

- $1/m_Q$ *Expansions* are distinct from the other technologies in that they deal with inclusive transitions, semileptonic as well as non-leptonic ones. Similarly to HQET they apply best to beauty decays while their numerical usefulness in charm decays is a priori uncertain.

There is one point of particular relevance for our discussion here. Charm decays occupy a special place in the theoretical landscape: they represent common ground for all the QCD methods listed above, albeit sometimes only at the extreme range of their applicability [1]. Comparing the predictions from the various theoretical technologies, which emphasize complementary aspects of QCD, with detailed and comprehensive data has a two-fold benefit:

- It provides valuable insights into the inner workings of QCD in general, and on the interplay between perturbative and non-perturbative effects in particular.

- These findings can be extrapolated to beauty decays and applied there with *tested* confidence.

2.2 Inclusive Transitions

2.2.1 Total Rates

Total rates for a heavy-flavour hadron H_Q to decay into an inclusive final state f can be expressed through an expansion in powers of $1/m_Q$ [3]; through order $1/m_Q^3$ one obtains the following master equation[4]:

$$\Gamma(H_Q \to f) = \frac{G_F^2 m_Q^5}{192\pi^3}|KM|^2 \left[c_3^f \langle H_Q|\bar{Q}Q|H_Q\rangle + c_5^f \frac{\langle H_Q|\bar{Q}i\sigma\cdot GQ|H_Q\rangle}{m_Q^2} + \right.$$

$$\left. + \sum_i c_{6,i}^f \frac{\langle H_Q|(\bar{Q}\Gamma_i q)(\bar{q}\Gamma_i Q)|H_Q\rangle}{m_Q^3} + \mathcal{O}(1/m_Q^4) \right] \quad (1)$$

where the dimensionless coefficients c_i^f depend on the parton level characteristics of f (such as the ratios of the final-state quark masses to m_Q); KM denotes the appropriate combination of KM parameters, and $\sigma\cdot G = \sigma_{\mu\nu}G_{\mu\nu}$

---

[1]The non-perturbative contributions can be expressed through an expansion in powers of μ_{had}/m_Q; for charm one has $\mu_{had}/m_Q \sim 0.4$. It is at least smaller than unity, but not by a large factor.

with $G_{\mu\nu}$ being the gluonic field strength tensor. The last term in eq.(1) implies also the summation over the four-fermion operators with different light flavours q.

It is through the quantities $\langle H_Q|O_i|H_Q\rangle$ that the dependence on the *decaying hadron* H_Q, and on non-perturbative forces in general, enters; they reflect the fact that the weak decay of the heavy quark Q does not proceed in empty space, but within a cloud of light degrees of freedom – (anti)quarks and gluons – with which Q and its decay products can interact strongly. The practical usefulness of the $1/m_Q$ expansion rests on our ability to determine the size of these matrix elements. Through order $1/m_Q^3$ there are three types of expectation values that determine the non-perturbative corrections:

(i) The leading term can be expanded further:

$$\langle H_Q|\bar{Q}Q|H_Q\rangle = 1 - \frac{\langle(\vec{p}_Q)^2\rangle_{H_Q}}{2m_Q^2} + \frac{\langle\mu_G^2\rangle_{H_Q}}{2m_Q^2} + \mathcal{O}(1/m_Q^3) \quad (2)$$

where $\langle(\vec{p}_Q)^2\rangle_{H_Q} \equiv \langle H_Q|\bar{Q}(i\vec{D})^2 Q|H_Q\rangle$ denotes the average kinetic energy of the quark Q moving inside the hadron H_Q and $\langle\mu_G^2\rangle_{H_Q} \equiv \langle H_Q|\bar{Q}\frac{i}{2}\sigma\cdot G Q|H_Q\rangle$. The first term on the right-hand-side of eq.(2) represents the naive spectator ansatz.

(ii) The quantity $\langle\mu_G^2\rangle_{P_Q}$ is known from the meson hyperfine splitting:

$$\langle\mu_G^2\rangle_{P_Q} \simeq \frac{3}{4}(M_{V_Q}^2 - M_{P_Q}^2) , \quad (3)$$

where P_Q and V_Q denote the pseudoscalar and vector mesons, respectively. Therefore

$$\langle\mu_G^2\rangle_B \simeq 0.37\,GeV \ , \quad \langle\mu_G^2\rangle_D \simeq 0.41\,GeV \quad (4a)$$

For Λ_Q and Ξ_Q baryons one has instead

$$\langle\mu_G^2\rangle_{\Lambda_Q,\Xi_Q} \simeq 0 \quad (4b)$$

since the light diquark system inside Λ_Q and Ξ_Q carries no spin, whereas

$$\langle\mu_G^2\rangle_{\Omega_Q} \neq 0 \quad (4c)$$

For $\langle(\vec{p}_Q)^2\rangle_{H_Q}$ there exists an estimate from a QCD sum rules analysis[5]

$$\langle(\vec{p}_b)^2\rangle_B \simeq 0.5 \pm 0.1\,GeV \quad (5a)$$

and one can expect one from lattice QCD in the foreseeable future. We do have a model-independant lower bound on it[6]:

$$\langle(\vec{p}_b)^2\rangle_B \geq 0.37 \pm 0.1\,GeV . \quad (5b)$$

The *difference* in the kinetic energy of Q inside baryons and mesons can be related to the masses of charm and beauty hadrons:

$$\langle(\vec{p}_Q)^2\rangle_{\Lambda_Q} - \langle(\vec{p}_Q)^2\rangle_{P_Q} \simeq \frac{2m_b m_c}{m_b - m_c} \cdot \{[\langle M_B\rangle - M_{\Lambda_b}] - [\langle M_D\rangle - M_{\Lambda_c}]\} \quad (5c)$$

where $\langle M_{B,D}\rangle$ denote the 'spin averaged' meson masses: $\langle M_B\rangle \equiv \frac{1}{4}(M_B + 3M_{B^*})$ and likewise for $\langle M_D\rangle$. Using data one finds: $\langle(\vec{p}_Q)^2\rangle_{\Lambda_Q} - \langle(\vec{p}_Q)^2\rangle_{P_Q} = -(0.07 \pm 0.20)(GeV)^2$; i.e., the present measurement of M_{Λ_b} is not yet sufficiently accurate, but this will change in the foreseeable future.

(iii) The expectation values for the four-quark operators taken between *meson* states can be expressed in terms of a single quantity, namely the decay constant:

$$\langle H_Q(p)|\bar{Q}_L\gamma_\mu q_L)(\bar{q}_L\gamma_\nu Q_L)|H_Q(p)\rangle \simeq \frac{1}{4}f_{H_Q}^2 p_\mu p_\nu \quad (6)$$

where factorization has been assumed. The theoretical expectations for the decay constants are [7]

$$f_D \simeq 200 \pm 30\,MeV \quad , \quad f_B \simeq 180 \pm 30\,MeV \quad (7a)$$

$$f_{D_s}/f_D \simeq 1.15 - 1.2 \quad , \quad f_{B_s}/f_B \simeq 1.15 - 1.2 \quad (7b)$$

The size of the expectation values taken between *baryonic* states are quite uncertain at present. There exists more than one relevant contraction, and for the time being quark model estimates provide us with the only guidance! I will return to this point when discussing predictions of baryon lifetimes.

To illustrate the method I give the semileptonic and non-leptonic widths for charm hadrons through order $1/m_c^2$:

$$\Gamma_{SL}(H_c) \simeq \Gamma_0 \langle H_c|\bar{c}c|H_c\rangle \cdot \left(1 - \frac{2\langle\mu_G^2\rangle_{H_c}}{m_c^2} + \mathcal{O}(1/m_c^3)\right) \quad (8a)$$

$$\Gamma_{NL}(H_c) \simeq \Gamma_0 N_C \langle H_c|\bar{c}c|H_c\rangle \cdot \left[A_0\left(1 - \frac{2\langle\mu_G^2\rangle_{H_c}}{m_c^2}\right) - 4A_2 \frac{2\langle\mu_G^2\rangle_{H_c}}{m_c^2} + \mathcal{O}(1/m_c^3)\right] \quad (8b)$$

$$\Gamma_0 = \frac{G_F^2 m_c^5}{192\pi^3}|V(cs)|^2 \quad , \quad (8c)$$

where $A_{0,2}$ denote perturbative QCD corrections; I have ignored here the small phase space correction due to $m_s^2/m_c^2 \neq 0$. The following results can be read off from eqs.(8):

- For $m_c \to \infty$ the parton model spectator expression is recovered.

- Γ_{SL} as well as Γ_{NL} receive non-perturbative corrections, as does $BR_{SL}(H_c)$; the latter quantity is lowered since the last term in eq.(8b) enhances Γ_{NL} ($A_2 < 0$!)[4].

- Γ_{SL} is *not* universal for all charm hadrons H_c once $\mathcal{O}(1/m_c^2)$ contributions are included. One actually finds

$$\Gamma_{SL}(D) \,/\, \Gamma_{SL}(\Lambda_c) \,/\, \Gamma_{SL}(\Omega_c) \sim 1 \,/\, 1.5 \,/\, 1.2 \,, \tag{9}$$

where the difference in the Λ_c and Ω_c widths to this order are due to eqs.(4b,c); i.e., the ratio of semileptonic branching ratios does *not* coincide with the ratio of lifetimes when comparing mesons and baryons.

- The widths for D and Λ_c decays differ already in order $1/m_c^2$.

- The differentiation between the widths for D^0, D^+ and D_s mesons occurs at order $1/m_c^3$ as expressed in eq.(1), but left out in eq.(8). [2]

The underlying pattern can be expressed as follows:

$$\Gamma(\Lambda_Q) = \Gamma(P_Q^0) = \Gamma(P_Q^+) + \mathcal{O}(1/m_Q^2) \tag{10a}$$

$$\Gamma(\Lambda_Q) > \Gamma(P_Q^0) \simeq \Gamma(P_Q^+) + \mathcal{O}(1/m_Q^3) \tag{10b}$$

$$\Gamma(\Lambda_Q) > \Gamma(P_Q^0) > \Gamma(P_Q^+) + \mathcal{O}(1/m_Q^4) \tag{10c}$$

i.e., one predicts $\tau(\Lambda_b) < \tau(B_d) < \tau(B^-)$ as well as $\tau(\Lambda_c) < \tau(D^0) < \tau(D^+)$. Yet keeping in mind that the $1/m_Q$ expansion is at best a semi-quantitative tool for $m_Q = m_c$ ($\mu_{had}/m_c \sim 0.4$!) one cannot expect to make precise predictions for the charm lifetime ratios. On the other hand – and that is a central point of my message – measurements of the lifetime ratios for all weakly decaying charm hadrons can be used with great profit to disentangle the various contributions in charm decays. Such an anatomy will then pave the way for more reliable predictions on beauty decays. To say it differently: a comprehensive study of charm decays can be harnessed as Nature's microscope onto the numerically smaller effects in beauty decays.

In Tables 1 and 2 I juxtapose the theoretical expectations and predictions on charm and beauty lifetime ratios [8, 9, 10, 11, 12] with present data [13, 14].

As mentioned before, at present one has to rely on quark models to estimate the size of the relevant *baryonic* expectation values. Thus there is a model dependance in the predictions on *baryon* lifetimes – in contrast to the case with meson lifetimes. This is indicated in the tables by an asterisk.

---

[2] $SU(3)_{Fl}$ symmetry is very much observed in the expectation values $\langle \mu_G^2 \rangle_{D,D_s}$ and $\langle (\vec{p}_c^2) \rangle_{D,D_s}$.

Observable	QCD ($1/m_c$ expansion)	Data
$\tau(D^+)/\tau(D^0)$	~ 2 [for $f_D \simeq 200$ MeV] (mainly due to *destructive* interference)	2.547 ± 0.043
$\tau(D_s)/\tau(D^0)$	$1 \pm$ few$\times 0.01$	1.125 ± 0.042
$\tau(\Lambda_c)/\tau(D^0)$	~ 0.5 *	0.51 ± 0.05
$\tau(\Xi_c^+)/\tau(\Lambda_c)$	~ 1.3 *	1.75 ± 0.36
$\tau(\Xi_c^+)/\tau(\Xi_c^0)$	~ 2.8 *	3.57 ± 0.91
$\tau(\Xi_c^+)/\tau(\Omega_c)$	~ 4 *	3.9 ± 1.7

Table 1: QCD Predictions for Charm Lifetimes

Observable	QCD ($1/m_b$ expansion)	Data
$\tau(B^-)/\tau(B_d)$	$1 + 0.05(f_B/200 \text{ MeV})^2[1 \pm \mathcal{O}(10\%)] > 1$ (mainly due to *destructive* interference)	1.04 ± 0.05
$\bar{\tau}(B_s)/\tau(B_d)$	$1 \pm \mathcal{O}(0.01)$	0.98 ± 0.08
$\tau(\Lambda_b)/\tau(B_d)$	~ 0.9 *	0.76 ± 0.06

Table 2: QCD Predictions for Beauty Lifetimes

The general agreement with the data is remarkable, in particular for the charm system, where the expansion parameter is not much smaller than unity. A few more detailed comments are in order:

• The D^+-D^0 lifetime difference is driven mainly by a destructive interference [15] with 'Weak Annihilation' (WA) contributing not more than 10 - 20%. Within the accuracy of the expansion, the data are reproduced.

• The D_s-D^0 lifetime ratio can be treated with better theoretical accuracy, namely of order a few percent. The observed near equality of $\tau(D_s)$ and $\tau(D^0)$ represents rather direct evidence for the reduced weight of WA in charm meson decays[9].

• Even the expectations on the charm baryon lifetimes reproduce the data which is quite remarkable since there are constructive as well as destructive contributions to baryon lifetimes. It has to be noted though that the present measurements suffer from large uncertainties.

• However the prediction on $\tau(\Lambda_b)/\tau(B_d)$ appears to be in serious (though not yet conclusive) disagreement with the data. The details of what went into that prediction can be found in ref.[12]; here I want to state only the following conclusion. If $\tau(B_d)$ indeed exceeds $\tau(\Lambda_b)$ by 25 - 30 %, then a 'theoretical price' has to be paid. It strongly suggests that the present agreement between theoretical expectations and data on charm baryon lifetimes is largely accidental and most likely would not survive in the face of more precise measurements! To state it in a more constructive manner: more precise measurements on charm baryon lifetimes would then allow to isolate the source of the discrepancy between prediction and observation.

As already said, on general grounds one does not predict the semileptonic widths to be the same for all charm hadrons – apart from $\Gamma_{SL}(D^0) \simeq \Gamma_{SL}(D^+)$, which is protected by isospin invariance and Cabibbo suppression. A priori there could be a sizeable difference in $\Gamma_{SL}(D^0)$ vs. $\Gamma_{SL}(D_s)$ due to a WA contribution to the latter observable. It is then a non-trivial prediction that those two quantities largely coincide:

$$1\pm \sim \text{few }\% = \frac{\tau(D_s)}{\tau(D^0)} \simeq \frac{BR_{SL}(D_s)}{BR_{SL}(D^0)} \simeq 1\pm \sim 10\% \tag{11}$$

On the other hand a sizeable difference is expected in the semileptonic widths of baryons and mesons which is expressed as follows:

$$BR_{SL}(\Lambda_c) > BR_{SL}(D^0) \cdot \frac{\tau(\Lambda_c)}{\tau(D^0)} \simeq 0.5 \cdot BR_{SL}(D^0) \tag{12}$$

2.2.2 Lepton Spectra

A detailed study of the lepton spectra in inclusive semileptonic decays of D^0, D^+ and D_s mesons is highly desirable. One expects [16] sizeable differences between the energy spectra in D^0 and D_s and to a lesser degree also in D^+ decays. For there is a WA process that is Cabibbo allowed [forbidden] for D_s [D^+] mesons where the hadrons in the final state emerge from (double) gluon emission of the initial anti-quark line. These differences will show up mainly in the endpoint region. An analogous complication is expected for semileptonic B decays: hadronization affects the spectra in the endpoint region differently in B_d and B^- transitions. This creates a systematic uncertainty in the value extracted from inclusive decays that cannot be evaluated reliably unless

- one separates B_d and B^- decays or

- measures the corresponding effects for D mesons and extrapolates to B mesons through a $1/m_Q$ expansion.

2.3 Exclusive Charm Decays

2.3.1 Leptonic and Semileptonic Channels

Measuring $BR(D^+, D_s \to \mu\nu, \tau\nu)$ with *good* accuracy represents a high priority goal since it allows to extract the decay constants f_D and f_{D_s}. There exists considerable intrinsic interest in the value of these hadronic parameters; in addition – and maybe more urgently from a phenomenological perspective – once f_D and f_{D_s} have been well measured, one can confidently extrapolate to the beauty sector and predict f_B and f_{B_s}.

As discussed in detail in El'Khadra's talk [17] a host of theoretical tools can be brought to bear on exclusive semileptonic charm decays: QCD sum rules, Lattice QCD and HQET in addition to quark models. Confronting their predictions with comprehensive measurements of the relevant hadronic form factors and their dependance on the momentum transfer will provide us with valuable insights into the inner workings of QCD; it also will be of great benefit in quantitatively understanding exclusive semileptonic B decays.

2.3.2 Nonleptonic Modes

Most important is a general caveat: the relationship between the pattern in exclusive modes and in inclusive transitions is quite tenuous. The former in contrast to the latter are very sensitive to the dramatic behaviour of QCD in the infrared regime; there exist relatively straightforward examples [8] showing that while *individual* exclusive rates get enhanced or decreased significantly by the strong interactions, these effects average out to a large degree in the sum. No theoretical tools have been developed yet that can master these complexities and at present one can employ only models of uncertain reliability. Nevertheless there is a valid motivation behind such 'phenomenological engineering', in particular when applied to two-body modes. For it allows us – once sufficiently many branching ratios have been well measured – to extract the size of the contributing isospin amplitudes and their phase shifts [18]. This provides information that is essential for designing a strategy for CP studies and for interpreting its outcome.

2.3.3 Radiative Decays

While in the Standard Model there is no (short-distance) penguin operator generating $D \to \gamma + X$, γK^*, $\gamma \rho/\omega$ transitions, long distance dynamics can. One should note that even the inclusive rate receives contributions from a non-local (though higher-dimensional) operator. Thus the radiative branching ratios cannot be predicted in a reliable fashion. Yet measuring $BR(D \to \gamma K^*, \gamma \rho/\omega)$ helps us in two ways: On the one hand one can again extrapolate to the beauty system and obtain a reliable estimate for the impact of long-distance dynamics on the corresponding modes $B \to \gamma K^*$, $\gamma \rho/\omega$. This is important for any attempt to extract $|V(ub)|$ from these radiative B decays.

On the other hand one has opened up a new window onto New Physics; for it can manifest itself by producing a significant deviation of the ratio $BR(D \to \gamma \rho/\omega)/BR(D \to \gamma K^*)$ from $\tan^2(\theta_c)$.

3 Required/Desired Database

The preceding discussion should have made it clear that even without aiming at possible manifestations of New Physics the need for further data on charm decays has not diminished; the advances in our theoretical understanding actually allow us to define more precisely the kind of future measurement one needs for further progress. I will briefly sketch them.

While there is no need from theory to measure the D^+, D^0 and Λ_c lifetimes more precisely, it would be quite useful to determine $\tau(D_s)$ to within 1%. Clearly the greatest need for improvement exists for the Ξ_c and Ω_c lifetimes. A 5-10% accuracy in $\tau(\Xi_c^0)$, $\tau(\Xi_c^+)$ and $\tau(\Omega_c)$ would enable us to extract the size of the relevant baryonic matrix elements.

The benchmark to aim for in $BR(D^+, D_s \to \mu\nu, \tau\nu)$ is a 10% accuracy allowing to extract the decay constants to within 5%.

For practical reasons one wants to know the *absolute* branching ratios for charm hadrons to within a few percent. Such information which is sorely missing for D_s, Λ_c and Ξ_c decays [19] is needed, among other things, for proper charm counting in B decays and for determining the absolute values of $BR(B_s \to l\nu D_s^{(*)})$, $BR(B \to D\bar{D}_s^{(*)})$ and $BR(B_s \to D_s^{(*)}\bar{D}_s^{(*)})$. Likewise one wants to know the inclusive rates for $\Lambda_c \to \Xi + X_s$, $\Xi_c \to \Xi + X$ etc. Our ignorance here constitutes a major bottle neck in many studies, like using $l\Xi$ correlations to distinguish between Λ_b and Ξ_b decays.

It is also important to know the absolute branching ratios for the inclusive transitions $D_s \to l + X$, $\Lambda_c \to l + X$, $\Xi_c \to l + X$ to complement the information obtained from the lifetimes. A detailed analysis of the lepton spectra in $D, D_s \to l + X$ (and also in $\Lambda_c \to l + X$) would be of great theoretical help when extracting $|V(ub)|$ from the endpoint region in $B \to l + X$ decays.

The dependance of the hadronic form factors in exclusive semileptonic charm decays on the momentum transfer has to be measured directly and the analysis has to be extended to include also channels like $D^+, D_s \to l\nu\eta/\eta'$.

The data base for the program of 'theoretical engineering' in two-body modes referred to above has to be completed by analysing final states containing (multi)neutrals.

Enough statistics has to be accumulated to study doubly Cabibbo suppressed decays in detail.

The radiative channels $D, D_s \to \gamma K^*, \gamma \rho/\omega$ have to be searched for in a dedicated manner.

Finally it would be quite useful to remeasure the reaction $e^+e^- \to D\bar{D}+X$ for $E_{c.m.} \sim 5-6$ GeV. Old SPEAR data suggest an enhancement there; if true, it would point to rather virulent final state interactions in that interval. That region happpens to be the one that is also probed in $B \to D\bar{D}_{(s)}$; such effects would have an obvious impact on the CP phenomenology in those B decays.

4 Experimental Stage

The measurements listed above fall into three categories:

(A) The lifetimes can be measured by fixed target experiments (or at B factories).

(B) Some of the measurements might not be impossible in hadronic collisions or at a B factory, but certainly represent a stiff challenge there [20]. Determining $BR(D^+, D_s \to \mu\nu)$ with a 20% accuracy presumably belongs into that category, as do observing non-leptonic decays with multi-neutrals in the final state and extracting *absolute* branching ratios for D_s mesons and the charmed baryons. A τ-charm factory on the other hand offers the cleanest measurements [19].

(C) There are finally measurements that presumably will remain in the sole domain of a τ-charm factory. Among them are: studies of the lepton spectra in $D/D_s/\Lambda_c \to l+X$; the semileptonic branching ratios for the various charm hadrons and the identification of genuine radiative D decays which requires the efficient rejection of nonleptonic modes like $D \to K^*\pi^0 \to K^*\gamma[\gamma]$, i.e. where one photon escapes detection; this can probably be achieved only by making use of the excellent energy resolution available due to beam-energy constraints at a threshold machine.

5 Outlook

Let me start out with some very general statements which I then relate to the purpose of our meeting. A comprehensive program on Heavy Flavour Physics is essential in any serious quest to unveil Nature's Grand Design. Detailed studies of charm decays have to form an integral part of such a program. In an 'ideal' or at least 'optimal' world a τ-charm factory plays a central role in such studies. This factory is justified by its unique capabilities to advance our understanding of QCD, and it would run with good luminosity in the energy range 3 GeV $\leq \sqrt{s} \leq$ 6 GeV; its discovery potential for New Physics would serve as the 'icing on the cake'. The 'ideal' world is defined as one where a τ-charm factory would be running by now; in an 'optimal' world it would start delivering data by the end of the millenium, like the asymmetric B factories. Not surprisingly, our world is not ideal and probably not optimal. There is still an excellent motivation for a first-class τ-charm factory starting up later, yet the emphasize will shift somewhat. High precision studies of τ and charmonium physics would represent the superb primary justification. On the other hand, the battle lines for open charm physics might be re-drawn. I expect that our experimental colleagues will devise some ingenious new methods for obtaining at least decent measurements of absolute branching ratios for charmed baryons etc. For cost and time (also running time) reasons it might make more sense then to limit the effective energy range of the machine to 3 GeV $\leq \sqrt{s} \leq$ 4.4 GeV and concentrate on the core part of open charm physics, namely the weak decays of D and D_s mesons with a two-fold purpose: to perform measurements of absolute branching ratios, semileptonic decays and their lepton spectra that cannot be made in other experimental environments – and to follow up on tantalizing hints for the intervention of New Physics that might have surfaced in the meantime!

Acknowledgements: I am grateful to T.D. Lee for sharing his insights and his enthusiasm with us. It was a stimulating meeting nicely organized by J. Repond. This work was supported by the National Science Foundation under grant number PHY 92-13313. I also thank the Institute for Nuclear Theory at the University of Washington for its hospitality and the Department of Energy for partial support during the write-up of this manuscript.

References

[1] G. Burdman, these Proceedings.

[2] I.I. Bigi, Invited talk given at HQ94, Charlottesville, Virginia, Oct. 1994, preprint UND-HEP-94-BIG11, to appear in the Proceedings.

[3] for the first suggestion, see: M. Shifman, M. Voloshin, 1982, in: V. Khoze, M. Shifman, *Uspekhi Fiz. Nauk* **140** (1983) 3 [*Sov. Phys. Uspekhi* (1983) 387]; *Sov. Journ. Nucl. Phys.* **41** (1985) 120.

[4] I.I. Bigi, N.G. Uraltsev, A. Vainshtein, *Phys. Lett.* **B293** (1992) 430; (E) **B297** (1993) 477; B. Blok, M. Shifman, *Nucl. Phys.* **B399** (1993) 441; 459.

[5] P. Ball, V. Braun, *Phys. Rev.* **D49** (1994) 2472.

[6] I.I. Bigi, M. Shifman, N.G. Uraltsev, A. Vainshtein, *Phys. Rev.* **D52** (1995) 196, with references to earlier work.

[7] G. Martinelli, Invited talk given at the 6th Intern. Symposium on Heavy Flavour Physics, Pisa, Italy, June 1995, to appear in the Proceedings.

[8] I.I. Bigi, N.G. Uraltsev, *Phys. Lett.* **B280** (1992) 120; B. Blok, M. Shifman, in: Proceedings of the Third Workshop on the Physics at a Tau-Charm Factory, Marbella, Spain, June 1993, R. & J. Kirkby (eds.), Editions Frontieres, 1994.

[9] I.I. Bigi, N.G. Uraltsev, *Z. Physik* **C 62** (1994) 623.

[10] I. I. Bigi, B. Blok, M. Shifman, N. Uraltsev, A. Vainshtein, in: '*B* Decays', ed. by S. Stone, World Scientific, Revised Second Edition, 1994, p.132.

[11] note also the earlier predictions: N. Bilic, B. Guberina, J. Trampetic, *Nucl. Phys.* **B248** (1984) 261; M. Shifman, M. Voloshin, *Sov. J. Nucl. Phys.* **41** (1985) 120; *JETP* **64** (1986) 698; B. Guberina, R. Rückl, J. Trampetic, *Z. Phys.* **C33** (1986) 297; R. Rückl,

in: Proc. International School of Physics "Enrico Fermi", Course XCII, Varenna, 1984, ed. by N. Cabibbo, North-Holland, Amsterdam, 1987, p. 43.

[12] for an update, see: I. Bigi, Invited talk given at the 6th Intern. Symposium on Heavy Flavour Physics, Pisa, Italy, June 1995, preprint UND-HEP-95-BIG06, to appear in the Proceedings.

[13] S. Malvezzi, Invited talk given at the 6th Intern. Symposium on Heavy Flavour Physics, Pisa, Italy, June 1995, to appear in the Proceedings.

[14] V. Sharma, Invited talk given at the 6th Intern. Symposium on Heavy Flavour Physics, Pisa, Italy, June 1995, to appear in the Proceedings.

[15] B. Guberina, S. Nussinov, R. Peccei, R. Rückl, *Phys. Lett.* **B89** (1979) 111.

[16] I.I. Bigi, N.G. Uraltsev, *Nucl. Phys.* **B423** (1994) 33.

[17] A. El'Khadra, these Proceedings.

[18] For a recent analysis, see: F. Bucella et al., *Phys. Lett.* **B302** (1993) 319.

[19] P. Roudeau, these Proceedings.

[20] J. Wiss, these Proceedings.

Charm Decay in Fixed Target: Present and Future

Jim Wiss

Department of Physics
University of Illinois at Urbana-Champaign
1110 W. Green St.
Urbana, IL 61801-3080

Abstract

I summarize recent data on the decay of charmed mesons into hadronic and semileptonic final states obtained primarily in fixed target experiments at Fermilab. In the semileptonic sector, these data provide tests of lattice gauge theories and quark models; while in the hadronic sector, they can be used to test factorization models and help elucidate the important role of final state interactions. I will also make a few guesses at the charm particle decay physics at fixed target facilities in the future.

PACS numbers: 13.25.+m, 14.40.Jz

1 Semileptonic Decays

The emphasis in this talk is primarily on measurement physics for processes with reasonably large available data sets, rather than on limit physics or very rarely observed decays. We begin with semileptonic decays:

Table 1: Semileptonic Physics

$e^\pm$, ft	$D^o \to K^- \ell \nu$	$f_+(q^2)$
ft	$D^+ \to K^* \ell \nu$	$A_1(0), V_1(0), A_2(0)$
$e^\pm$, ft	$D_s^+ \to \phi \ell \nu$	$\mathcal{B}$ ff
$e^\pm$	$D^+ \to \pi \ell \nu$	V_{cd}/V_{cs} $\{Q^2\}$
$e^\pm$, ft	$D_s^+ \to \eta$ & $\eta' \ell \nu$	{Vec/PS}
$e^\pm$	$\Lambda_c^+ \to \Lambda \ell \nu$	$\mathcal{B}$, Polar
$e^\pm$	$\Xi_c \to \Xi \ell \nu$	$\mathcal{B}$
$e^\pm$	$\Omega_c \to \Xi \ell \nu$	$\mathcal{B}$

Above is a highly schematic table which summarizes the states which have been studied, how they have been studied, and either the realized or future {} physics potential of such measurements. Results from fixed target experiments (ft) continue to complement those from e^+e^- annihilation ($e^\pm$). Charm semileptonic studies provide a wealth of information, including important probes of quark dynamics through measurements of form factors ($f_+(q^2)$) and ($A_1(0), V(0), A_2(0)$), model dependent information on the absolute branching ratios ($\mathcal{B}$) for the D_s^+, Λ_c^+, Ξ_c and Ω_c, information on CKM matrix elements (V_{cd}/V_{cs}), and tests of HQET. Determination of CKM matrix elements (V_{bu}) and more stringent tests of HQET will be possible through the interplay of studies of both charm and beauty semileptonic decay results.

There has been a long standing theoretical problem with the observed ratio of vector to pseudoscalar decay widths for $D^+ \to K^* \ell \nu$ relative to $D^+ \to K \ell \nu$ [1]. Recent work [2] by Scora and Isgur may have solved this problem. Because of the undetected ν, one only partially reconstructs the final state leaving the important experimental challenge of proving exclusivity of the final state; ie one must establish that one is observing the claimed final state without additional, undetected neutrals. A variety of experimental techniques can be brought to bear on the problem of isolating semileptonic decays from both non-charm and charm backgrounds. Frequent use is made of D^* tagging. One often has the ability to exploit the charge correlations between leptons and kaons or D^* decay pions and thus eliminate backgrounds through a wrong sign subtraction.

1.1 $D \to$ pseudoscalar $\ell^+\nu$

These decays are particularly interesting since they can provide detailed information on the q^2 dependence of the charm semileptonic form factors, $f_+(q^2)$. The decay rate expression for PS $\ell\nu$ is:

$$\frac{d\Gamma}{dq^2} = \frac{G_F^2 |V_{cq}|^2 P^3}{24\pi^3} \{|f_+(q^2)|^2 + m_\ell^2 \, |f_-(q^2)|^2 ...\} \qquad (1)$$

where P is momentum of the pseudoscalar particle in the D frame. One of the two possible form factors, $f_-(q^2)$, becomes unimportant in the limit of zero lepton mass. Two parameterizations are used for $f_+(q^2)$:

$$f_+(q^2) = \frac{f_+(0)}{(1 - q^2/m_{pole}^2)} \quad \text{or} \quad f_+(q^2) = f_+(0)e^{\alpha q^2} \qquad (2)$$

The first form is motivated by the belief that the coupling of the $c\bar{q}$ quarks to the virtual $W^\pm$ should be dominated by bound states of the $c\bar{q}$ system. For the case of $D \to K\ell\nu$ decay, one expects that m_{pole} should be set to the mass of the vector $D_s^*(2110)$ since it has the same spin-parity as the $c\bar{s}$ current of the form factor. The second form is motivated [3] by the ISGW model. Figure 1(a) illustrates the difference between $f_+^2(q^2)$ for a pole form, an exponential form, and a linear form. Over the restricted q^2 range available for the presently studied $D^\circ \to K^-\ell^+\nu$ decay, one is primarily measuring just the slope ratio, $(df_+/dq^2(0))/f_+(0)$. To go futher, we will probably have to wait for measurements of $D \to \pi e^+\nu$ so that the q^2 domain can extend much closer to the location of the anticipated D^{+*} (rather than D_s^{+*}) pole.

Figure 1(b) illustrates that $f_+^2(q^2)$ has a rather subtle, asymmetric, and difficult to measure influence on $d\Gamma/dq^2$. Most of one's ability to measure beyond $f_+(0)$ and $df_+/dq^2(0)$ occurs at large q^2 where the rate is low. One can measure the product of $f_+^2(0)$ and the appropriate CKM matrix element by integrating the $d\Gamma/dq^2$ expression given by Eqn. (1) using the measured

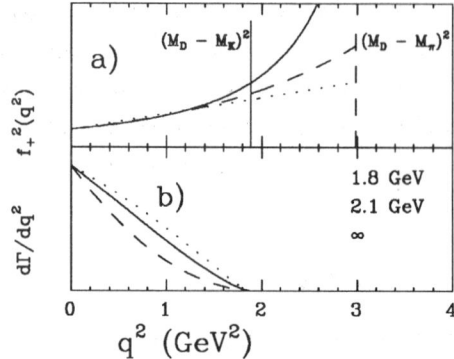

Figure 1: (a) Various parameterizations of $f_+^2(q^2)$ over the kinematic range for $D^o \to Ke^+\nu$ (to the left of the vertical solid) and $D \to \pi e^+\nu$ (to the left of the vertical dashed). The pole form (solid), exponential form (dashed) and a linear form (dotted) are displayed. (b) $d\Gamma/dq^2$ for $m_{pole} = 2.1\ GeV$ (solid), $m_{pole} = \infty$ (dashed) , $m_{pole} = 1.8\ GeV$ (dotted)

$f_+(q^2)/f_+(0)$ shape and setting the integrated width to the measured total width for $\Gamma(PS\ \ell\nu)$. Experiments generally measure a given final state semileptonic width $\Gamma(f\ \ell\nu)$ by first determining the absolute branching fraction of the decay $\mathcal{B}(f\ \ell\nu)$ from the yield of the $X\ell\nu$ state relative to the yield of a decay with a known absolute branching ratio. Γ then follows from $\mathcal{B}$ and the lifetime of the particular charm species (C):

$$\Gamma(C \to f\ell\nu) = \hbar \mathcal{B}(f\ \ell\nu)\ /\ \tau_C \tag{3}$$

1.1.1 $D \to K\ell^+\nu$

Much of the information on the detailed decay shapes originally came from fixed target experiments which exploit their generally excellent vertexing capability in order to "close" the decay kinematics and measure q^2. The momentum of the unobserved neutrino can be measured to within a two fold ambiguity by balancing p_t about the line between the primary and secondary vertex. The kinematics is most easily done by boosting along the D direction until the sum of the longitudinal kaon and lepton momentum vanishes. q^2 smearing is considerable which creates significant complications in the fitting procedure. Traditionally[3, 4, 5] one exploits the reaction $D^* \to (K\ell\nu)$ as a method to eliminate backgrounds from other charm semileptonic sources. The most recent results employing a D^* tag come from CLEO[3] (Fig. 3)

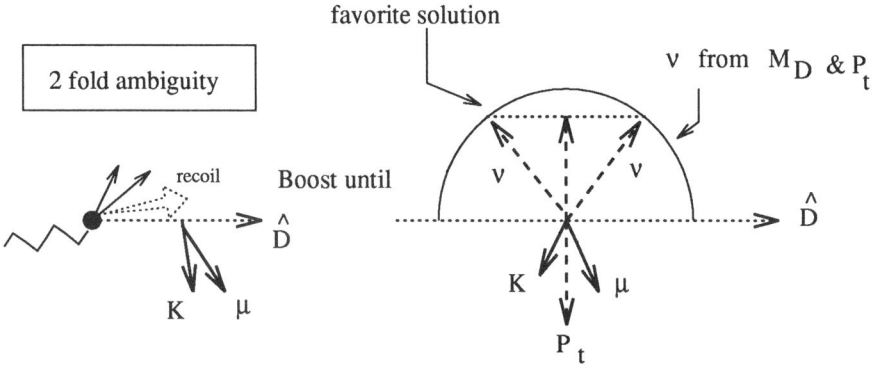

Figure 2: The kaon and lepton are boosted along the $\hat{D}$ direction into a frame where their momentum sum lies transverse to the $\hat{D}$ direction. The neutrino P_t balances the P_t of the kaon + lepton and its energy can be computed from the P_t and mass of the D. The neutrino momentum is then known to a 2-fold ambiguity corresponding to the intersection of the P_t line and the energy circle. The favorite solutions used by most groups is marked.

and E687[6] (Fig. 4). Recently E687 [6] greatly increased their statistics for this mode by including a sample of $\approx$ 1850 inclusive $D^o \to K^-\mu^+\nu$ events where a D^* tag was not required. Of course without the cleansing power of a D^* tag there will be an inevitable increase in backgrounds from both misidentified muons and other semileptonic decay processes such as $D^+ \to (K^-\pi^+)\mu^+\nu$, $D^0 \to (K^-\pi^o)\mu^+\nu$, and $D_s^+ \to \phi\mu^+\nu$. However, the background was somewhat reduced by a series of kinematic cuts. The Dalitz plot for the surviving sample was fit to a combination of signal, and semileptonic and misidentification backgrounds. The Dalitz projections are shown in Fig. 5. Table 2 summarizes information on the $f_+(q^2)$ form factor which describes the $D^o \to K^-\ell^+\nu$ decay.

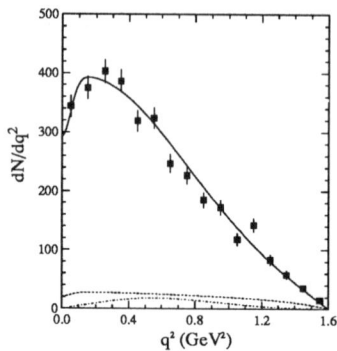

Figure 3: The uncorrected q^2 dependence from CLEO's 2700 event sample of D^* tagged events where the modes $D^0 \to K^-\ell^+\nu$ and $D^+ \to K_s\ell^+\nu$ are combined.

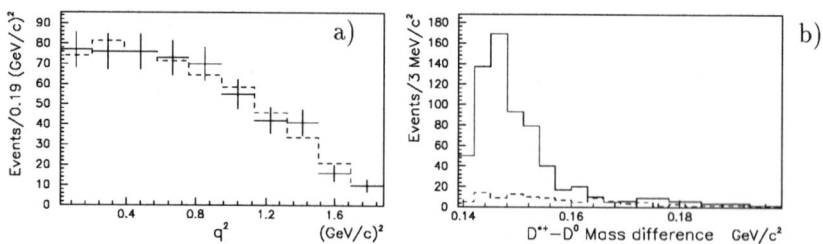

Figure 4: a) The uncorrected q^2 distribution for $K^-\mu^+\nu$ events from ≈ 500 event, D^* tagged sample of E687. b) The $D^{*+} - D^0$ mass difference for right sign (solid) and wrong sign (dashed) events.

Table 2: $K\ell\nu$ form factor results

Exp.	Mode	m_{pole}	$\mid f_+(0) \mid$
E691[4]	$K^-e^+\nu_e$	$2.1^{+0.4}_{-0.2} \pm 0.2$	$0.79 \pm 0.05 \pm 0.06$
CLEO(91)[5]	$K^-e^+\nu_e$	$2.1^{+0.4+0.3}_{-0.2-0.2}$	$0.81 \pm 0.03 \pm 0.06$
CLEO(93)[3]	$K^-l^+\nu_l$	$2.00 \pm 0.12 \pm 0.18$	$0.77 \pm 0.01 \pm 0.04$
MKIII[7]	$K^-e^+\nu_e$	$1.8^{+0.5+0.3}_{-0.2-0.2}$	$\mid V_{cs} \mid (0.72 \pm 0.05 \pm 0.04)$
E687 (prlm)	$K^-\mu^+\nu_\mu$ tag	$1.97^{+0.43+0.07}_{-0.22-0.06}$	$0.71 \pm 0.05 \pm .03$
E687 (prlm)	$K^-\mu^+\nu_\mu$ inc	$1.87^{+0.11+0.07}_{-0.08-0.06}$	$0.71 \pm 0.03 \pm 0.02$

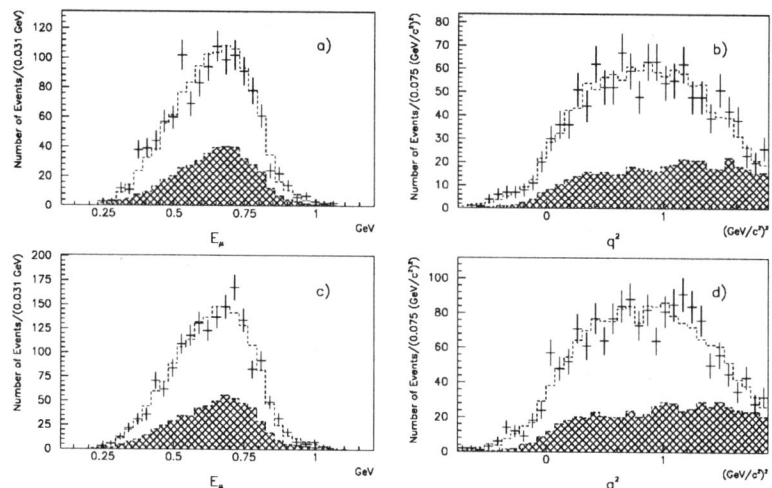

Figure 5: a) and c) Projection of the E_μ (muon energy in the D rest frame) Dalitz variable. The fit shown by the histogram includes background contributions shown by the shaded histogram. b) and d) Projection of the q^2 Dalitz variable. The upper row is for 1990 data and the lower is for 1991 data.

The average of all results appear consistent with the expected D_s^{*+} pole mass of 2.1 GeV; however the inclusive E687 results are about 1.7 σ lower than the D_s^{*+} mass. At present this is little more than a measurement of the slope of $f_+(q^2)$ near $q^2 = 0$. CLEO obtains an exponential fit to the alternative form: $f_+(q^2) \propto \exp\left[(0.29 \pm 0.04 \pm 0.06)\, q^2\right]$, which, as illustrated in Figure 1 (a), is nearly indistinguishable from the pole form but has more symmetrical error bars. The $f_+(0)$ values are also consistent with theoretical estimates: $f_+(0) \approx 0.7 - 0.9$.

1.1.2 Are non-parametric $f_+^2(q^2)$ measurements possible?

This is likely to be of considerable help to theorists in the future. Predictions tend to be most reliable at high q^2 where there is the maximum overlap of the initial and final state quark wave functions. In the context of a quark model [2], the form factors at low q^2 require knowlege of the high momentum tails of these wave functions. Unfortunately, as shown in Fig. 1, measurements tend to be dominated by the low q^2 behavior.

FOCUS (E831) is a follow-up photoproduction experiment to E687, with an anticipated sample of 45,000 $D^\circ \to K^- \ell \nu$ events. An important component of this measurement in a fixed target environment is the inevitable q^2 smearing due to ambiguity and vertex errors. Figure 6 illustrates the q^2 resolution using a simulation performed by my graduate student student, Amir Rahimi. The

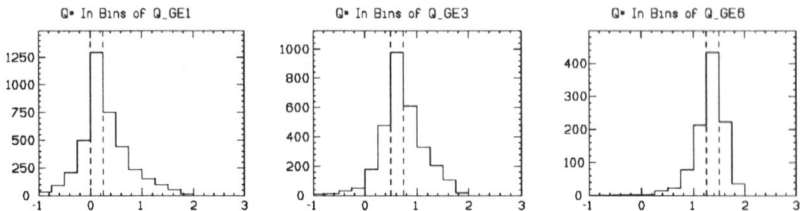

Figure 6: The reconstructed q^2 in three bins of real q^2. The true q^2 lies between the dashed lines.

observed q^2 distribution can then be de-convolved with the resolution to give parameterization free measurements of $f_+(q^2)$ which are shown in Fig. 7. The errors in an 8 bin deconvolution are significantly worse that $\sqrt{2}$ times the errors for a 4 bin deconvolution indicating the important role of q^2 smearing in degrading the measurement. A τc factory operating near $D\overline{D}$ threshold

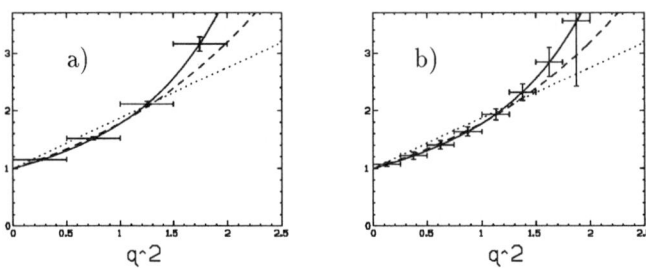

Figure 7: Anticipated errors in an 4 bin a) and 8 bin b) $f_+(q^2)$ deconvolution. The curves give the $f_+(q^2)$ expected for the pole (solid), exponential (dash), linear (dot) dependence.

would have significant advantages in this measurement since it could much more accurately determine the momentum of the missing neutrino through energy-momentum conservation.

1.2 $D \to$ vector $\ell^+ \nu$

The vector $\ell\nu$ decay process involves a hadronic current describing the overlap of the D and vector meson wave functions which (in the limit of zero lepton mass) can be described by two axial and one vector form factor. A variety of

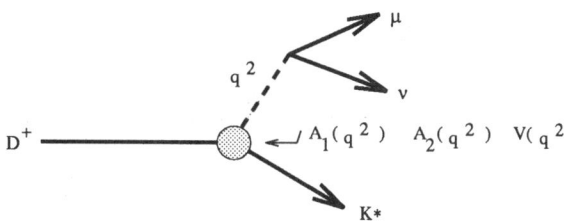

Figure 8: Two axial and one vector form factor are required to describe the current which connects the D to the vector meson in $D \to$ vector $\ell\nu$ decay.

theoretical methods including QCD sum rules[11] , quark models[12, 13, 14], and lattice gauge theory[15][16] have been brought to bear on the prediction of these three form factors. Although the full expression for the decay width is rather lengthy, a clear exposition can be found in the seminal reference [17]. It has become customary to assume that the q^2 dependence of the form factors is dominated by the poles of the cq system with the same spin-parity as the form factor. Hence for $D \to K^*\ell\nu$ decays one expects the $D_s^{*(*)}$ spectrum of poles (2.1 GeV for the vector and 2.5 GeV for the axial). Given the narrow q^2 domain, this is tantamount to assuming values for the form factor q^2 slope near 0. This leaves one with three measurements $A_1(0)$, $A_2(0)$, and $V(0)$. It has become traditional to factor out $A_1^2(0)$ from the decay width, leaving two ratios: $R_v = V(0)/A_1(0)$ and $R_2 = A_2(0)/A_1(0)$ which serve to describe the shape of the decay distribution (See Fig. 9). The R_v and R_2 values can be used to obtain Γ_ℓ/Γ_t which is the ratio of the q^2 integrated widths for the W^+ to be longitudinally polarized ($|1,0>$) as opposed to transversely polarized ($|1,\pm 1>$) with respect to its D frame momentum axis.

The value of $A_1(0)$ then follows from the decay width which can be estimated by measuring the branching ratio of the semileptonic decay with respect to a reference state and then using the absolute branching fraction of the reference state and D lifetime to compute a total decay width (eq. 3).

Form factor measurements have been made for both the $D^+ \to K^{*+}\ell\nu$ decay[17, 18, 19] and the $D_s^+ \to \phi\ell^+\nu$ decay[20, 21, 24]. Figure 10 shows two representative vector $\mu^+\nu$ signals from E687.

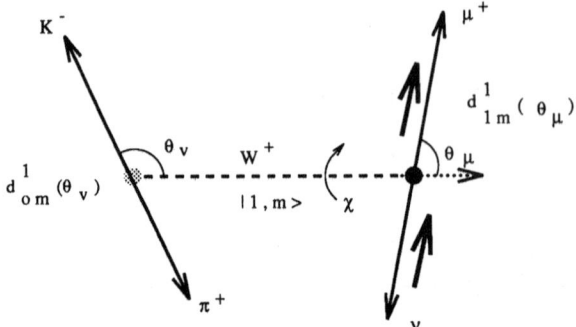

Figure 9: Illustration of the three decay angles used in describing the decay of eg $D \to \overline{K}{}^* \ell \nu$. The form factors determine a q^2 dependent spin of the virtual W^+ which then dictates the decay distribution in terms of θ_v – the polar angle describing the vector → two pseudo-scalar decay, θ_μ – the polar angle describing the decay of the virtual $W \to \ell \nu$, and χ – the azimuthal acoplanarity angle between the vector meson and virtual W decay planes.

1.2.1 $D^+ \to \overline{K}{}^{*0} \ell^+ \nu$

At present, information on the form factors comes primarily from E691 [17], E653 [18], and E687 [19]. These fixed target experiments use vertexing methods to estimate the ν momentum and thus measure considerably smeared values of q^2 and the three decay angles. All three published measurements are consistent. Table 3 compares the average[8] along with a recent, representative Lattice Gauge Theory calculation[15].

Table 3: $\overline{K}{}^* \ell \nu$ form factor ratios

	R_v	R_2	Γ_l / Γ_t
< E687/E691/E653 >	1.86 ± 0.20	$0.74 \pm .14$	1.21 ± 0.10
LGT[15]	$1.99 \pm .22 {}^{+0.31}_{-0.35}$	$0.70 \pm .16 {}^{+0.20}_{-0.15}$	$1.21 \pm .12 {}^{+0.15}_{-0.13}$

When these shape parameters are fed back into the decay rate expression to obtain values for $A_1(0)$, the agreement with the older theory predictions is not good.

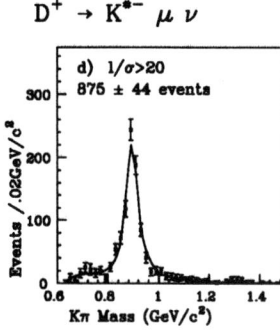
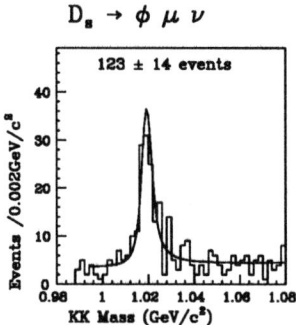

Figure 10: Two vector $\mu\nu$ signals from E687 where the vector daughters and a μ are in a common detached vertex. Left: Wrong sign subtracted $K^-\pi^+$ mass distribution showing a prominent $\overline{K}^{*0}$ (896). Right: K^+K^- mass distribution showing a prominent ϕ (1020).

Table 4: $\overline{K}^*\ell\nu$ form factors

Exp	$A_1(0)$	$A_2(0)$	$V(0)$
E691	0.49 ± .07	0.0 ± 0.2	1.2 ± 0.3
E653	0.57 ± 0.08	0.47 ± 0.16	1.2 ± 0.3
E687	0.59 ± 0.05	0.46 ± 0.11	1.0 ± 0.30
Average	0.56 ± 0.04	0.40 ± 0.08	1.1 ± 0.2
LGT (BES)	0.83 ± .14 ± 0.28	0.59 ± 0.14 $^{+0.24}_{-0.23}$	1.43 ± 0.45 $^{+0.48}_{-0.49}$

Other theoretical estimates for $A_1(0)$ tend to cluster around 0.8.[8] This indication of a $K^*\ell\nu$ shortfall is borne out by directly comparing the ratio of $K^*\ell\nu/K\ell\nu$. The recent, preliminary result from E687 is $\Gamma(K^*\ell\nu)/\Gamma(K\ell\nu) = 0.62 \pm 0.11 \pm 0.02$ which is consistent with the average of value [8] from CLEO-I , CLEO-II and E653: 0.55 ± 0.053. Earlier theoretical estimates had $K^*\ell\nu/K\ell\nu \approx 1$ which is almost a factor of two larger than the experimental values. A similar miss-match between theoretical expectation[10] and the measured [9] widths is observed for the pseudoscalar $D_s^+ \to \eta\ell^+\nu + \eta'\ell^+\nu$ relative to the vector decay $D_s^+ \to \phi\ell\nu$. Recently Scora and Isgur [2] updated their quark model for semileptonic decay by improving, for example, the quark wave functions used to compute form factors. They have re-examined their pseudoscalar-vector predictions with these improvements and find $K^*\ell\nu/K\ell\nu =$

0.54 which is in much closer agreement with the data.

1.2.2 $D_s^+ \to \phi\ell^+\nu$

A recent, unresolved experimental controversy has arisen concerning the relationship between the form factors for the decays $D_s^+ \to \phi\mu^+\nu$ and $D^+ \to \overline{K}^{*0}\mu^+\nu$ which are expected [15] [25] to be very close.

Below are three recent measurements:

Table 5: $D_s^+ \to \phi\ell^+\nu$ form factor ratios

	R_v	R_2	Γ_ℓ/Γ_t
E653[20]	$2.3^{+1.1}_{-0.9} \pm 0.4$	$2.1^{+0.6}_{-0.5} \pm 0.2$	$.54 \pm .21 \pm .10$
E687[21]	$1.8 \pm 0.9 \pm 0.2$	$1.1 \pm 0.8 \pm 0.1$	$1.0 \pm .5 \pm .1$
CLEO[24]	$0.9 \pm 0.6 \pm 0.3$	$1.4 \pm 0.5 \pm 0.3$	$1.1 \pm 0.3 \pm 0.2$
<informal>	1.43 ± 0.5	1.63 ± 0.37	0.74 ± 0.18
PDG D^+[8]	1.89 ± 0.25	0.73 ± 0.15	$1.23 \pm .13$

To my mind the situation remains rather murky. Given the large size of the errors, the measurements from all three experiments are consistent but the form factors measured for $D_s^+ \to \phi\mu^+\nu$ are inconsistent with those measured for $D^+ \to \overline{K}^{*0}\mu^+\nu$ at about the 2σ level. The new CLEO numbers have brought the R_v average for $\phi\ell\nu$ below the value for $\overline{K}^{*0}\ell\nu$; while E653 has brought the R_2 average for $\phi\ell\nu$ above the value for $\overline{K}^{*0}\ell\nu$.

1.2.3 Future prospects for $A_1(0), A_2(0), V(0)$

FOCUS, the follow on experiment to E687, is anticipated to have data sets of from ten to thirty times the present E687 data set. One can estimate the anticipated statistical errors in vector $\ell\nu$ physics using the *a priori* error matrix formula [26]:

$$E^{-1}_{\alpha\beta} = N \int \frac{d\vec{x}}{I(\vec{x})} \frac{\partial I(\vec{x})}{\partial t_\alpha} \frac{\partial I(\vec{x})}{\partial t_\beta} \tag{4}$$

For an experiment with perfect resolution on the q^2 and the decay angles, employing the standard likelihood fit which uses all four decay quantities, this formula predicts statistical errors of $\sigma(R_v) = 4.7/\sqrt{N}$ and $\sigma(R_2) = 4.1/\sqrt{N}$

for form factors near the E691/E653/E687 average. Given the substantial resolution smearing, my experience is that $\sigma(R_V) = 7/\sqrt{N}$ and $\sigma(R_2) = 5/\sqrt{N}$ are probably more realistic error estimates. [1] If the vector and axial pole masses are included as fit parameters as opposed to being assumed, the error on the vector pole mass is expected to be $24/\sqrt{N}$; while the error on the axial pole mass is expected to be $56/\sqrt{N}$. The statistical error in R_v essentially doubles when the pole mass is unconstrained.

FOCUS expects to collect a fully reconstructed, clean sample of 20000 $\overline{K^{*o}}\mu\nu$ decays, 2250 $\phi\mu\nu$ decays, and 500 $\rho\mu\nu$ decays. In pole constrained fits, these yields suggest that in the absence of systematic errors R_2 could be measured to 5 %, 14%, and 29 % for the K^*, ϕ, and ρ decay.

At present the theoretical error for R_2 is ±22% in typical Lattice Gauge Theory calculations[15] which is nearly identical to the present experimental error on $\overline{K^{*o}}\mu\nu$. Much of this present theoretical error can be reduced with increased computing, but it is estimated that there is an intrinsic theoretical error of ±5% in part due to the approximation schemes used to extract the number, such as the "quenched approximation". This theoretical error matches anticipated error in FOCUS for $\overline{K^{*o}}\mu\nu$. Of course, it is always possible that theoretical progress will greatly improve or eliminate the need for such assumptions. These calculations also show that statistical errors on the vector pole mass in $\overline{K^{*o}}\ell\nu$ decays would be 160 MeV which is roughly the present precision of the vector pole mass for the $f_+(0)$ form factor as determined from $K\ell\nu$ decay.

Systematic uncertainties are, of course, much harder to predict. I believe instrumental systematics for the $\overline{K^{*o}}\mu\nu$ decay are likely to be very small. As a way of illustrating this, consider the 18 bin fit employed by E687 [19] where one fits for R_v and R_2 by measuring the fractions of decays which are observed in 18 bins of $\cos\theta_v \times \cos\theta_\ell \times q^2$ (See Figure 11). Consider possible systematic problems that arise from a scenario where the center 10 mrad of a muon detector system has suffered a loss in efficiency. Lack of knowledge of this efficiency will create a systematic error to the extent that the fraction of events with a muon produced in the central 10 mrad varies from bin to bin. Figure 12 a), however, shows that this fraction is remarkably constant over the 18 bins. Often, as in E687, one triggers events on an hadronic energy threshold.

---

[1] These are the errors expected in the 18 bin likelihood fit used by E687 [19] where the $\cos\theta_v \times \cos\theta_v \times q^2$ space is divided into 3 bins, 3 bins, and 2 bins respectively.

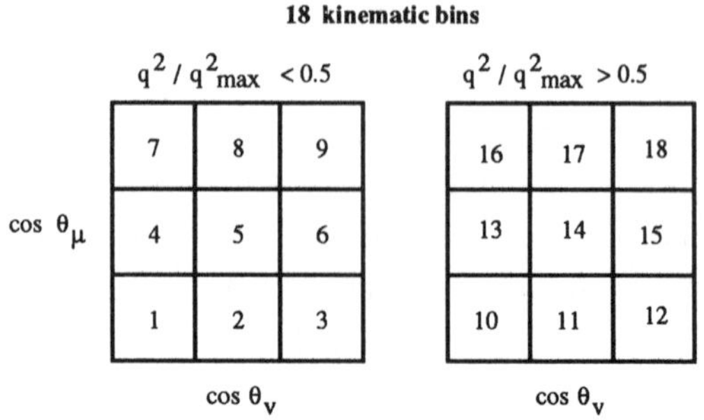

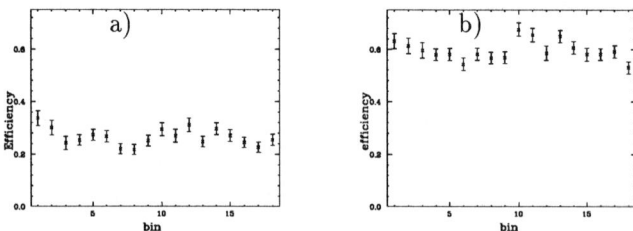

Figure 11: 18 kinematic bins used by E687 to fit the $\overline{K}^*\mu\nu$ form factor ratios.

Figure 12: a) The fraction of $\overline{K}^{*0}\mu\nu$ decays with a muon detected in the central 10 mrad as a function of the $\cos\theta_v \times \cos\theta_\ell \times q^2$ bin number. b) The fraction of decays with more than 70 GeV of energy deposited in a hadron calorimeter.

The exact threshold is often difficult to properly model and correct for. Figure 12 b) shows that the fraction of photoproduced events with a hadronic energy deposition exceeding 70 GeV is very uniform over the 18 bins.

2 Hadronic Charm Decays

There are many ways of studying hadronic charm decay which can be organized according to how closely one looks at the details of the decay. On the most inclusive level, the roughly one order of magnitude difference in the lifetimes of ground state charmed particles (as shown in Table 6) is due to differences in their hadronic decay rate, since one expects only slight variations in the rate of semileptonic decay.

Table 6: E687 lifetimes (ps)

D^+	D^o	D_s^+
$1.048 \pm .015 \pm .011$	$0.413 \pm .004 \pm .003$	$0.475 \pm 0.020 \pm 0.007$
Λ_c^+ (udc)	Ξ_c^+ (csu)	Ξ_c^o (csd)
$0.215\pm0.016\pm0.008$	$0.41^{+0.11}_{-0.08} \pm 0.02$	$0.101^{+0.025}_{-0.017} \pm 0.005$
	Ω_c^0 (css)	
	$0.089^{+0.027}_{-0.020} \pm 0.028$	

At the next level of detail, one can measure the inclusive decay width for specific final states. Factorization models make predictions for the width for spectator charmed meson decays into exclusive, two pseudoscalar final states. Comparison of experimental data to these predictions underscores the important role of final state interactions (FSI) where the two decay daughters undergo a strong rescattering after their initial formation. Such FSI effects can affect the interference between the various isospin amplitudes contributing to the decay as well as grossly enhancing the rate for suppressed, non-spectator processes such as $D^o \to \overline{K}^o K^o$.

Finally, at the most detailed level, one can study the kinematical dependence of decay amplitudes into exclusive three or four body final states. For the most part, multibody charm decays are strongly dominated by resonant substructure and often have sizeable quasi-two-body contributions which allows one to extend factorization tests to vector-pseudoscalar and vector-vector final states. Such analyses provide unique new probes of final state interactions by measurement of the phase shifts between the interfering amplitudes for the various resonant channels. In addition they provide additional examples of suppressed processes such as $D^o \to \overline{K}^{*o} K^{*o}$.

2.1 Two pseudoscalar decays: Factorization models

In factorization models, such as the influential Bauer, Stech, and Wirbel (BSW) model[27], the Cabibbo allowed quark interaction is represented by a product of 4 local operators which are organized into an effective charged current interaction (amplitude a_1) and neutral current interaction (amplitude a_2). The spectator hadronic decays of charmed mesons can be organized according to whether they can only proceed via the $(\bar{s}c) \times (\bar{u}d)$ charged current process

(Class 1), the $(\bar{u}c) \times (\bar{s}d)$ neutral current process (Class 2), or an interfering mixture (Class 3). Fig. 13 illustrates the type of destructive interference for a typical Class 3 decay of D^+ which will help explain its lower decay rate relative to the D^o. In the charged current a_1 process of Fig. 13, one of the

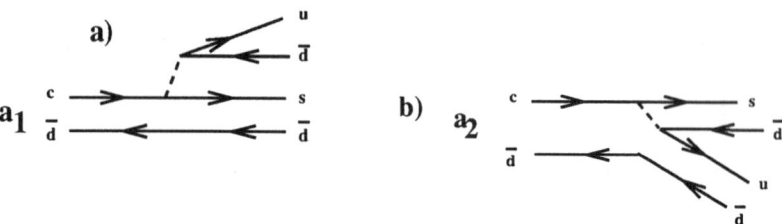

Figure 13: Interference between the effective charged current a) and effective neutral current b) process for the Class 3 decay $D^+ \to \overline{K^o}\pi^+$.

effective charged currents, $< \overline{K^o}|(\bar{s}c)|D^+ >$, is governed by the same form factors probed in semileptonic decay; while the other, $< \pi^+|(\bar{u}d)|0 >$, which describes how the pion is ejected from the external W^+, is related to the pion decay constant. The neutral current a_2 process is characterized by an internal rather than external W^+.

2.2 Complications of final state interactions

The weak amplitudes computed in a factorization model will be real. The fact that the observed amplitudes from underlying weak process pick up complex relative phases due to the final state interactions (FSI) is a consequence of the Migdal-Watson theorem which states that the observed amplitudes are related to the bare amplitudes by the square root of a strong interaction S matrix [27, 28]:

$$\begin{pmatrix} a_1 \\ a_2 \end{pmatrix} = \begin{pmatrix} \eta\, e^{2i\delta_1} & i\sqrt{1-\eta^2}e^{i(\delta_1+\delta_2)} \\ i\sqrt{1-\eta^2}e^{i(\delta_1+\delta_2)} & \eta\, e^{2i\delta_2} \end{pmatrix}^{1/2} \begin{pmatrix} a_1 \\ a_2 \end{pmatrix}_{bare} \quad (5)$$

We can think of the amplitudes in Eqn. 5 as being two interfering isospin amplitudes such as the $A_{3/2}$ and $A_{1/2}$ amplitudes which combine with Clebsch-Gordon coefficients to form the amplitude for charge variants of the $D \to K\pi$ decay. The 2×2, unitary S matrix[28] includes the strong interaction phase shifts, $\delta_{1,2}$, as well as, an an elasticity parameter, η. The presence of off diagonal terms ($\eta < 1$) can easily move the data away from factorization predictions by eg boosting $A_{3/2}$ at the expense of $A_{1/2}$. Even elastic ($\eta = 1$) FSI, which result in a diagonal S matrix, can still change the observed width of a particular two body D decay by affecting the relative interference between two isospin amplitudes.

One of my colleagues on E687, Dr. Daniele Pedrini, compiled a list of phase shifts and isospin amplitudes for a forthcoming review article[34] which are summarized below in Table 7 using the decay widths from the 1994 Review of Particle Properties. Table 7 demonstrates that the phase shifts between different isospin amplitudes are often nearly $90°$ indicating the general importance of FSI effects. [8].

Table 7: Isospin Decompositions

Mode	Ratio of amplitudes	$\delta = \delta_I - \delta_{I'}$
$K\pi$	$|A_{1/2}|/|A_{3/2}| = 3.99 \pm 0.25$	$86° \pm 8°$
$K^\star\pi$	$|A_{1/2}|/|A_{3/2}| = 5.14 \pm 0.54$	$101° \pm 14°$
$K\rho$	$|A_{1/2}|/|A_{3/2}| = 3.51 \pm 0.75$	$0° \pm 40°$
$K^\star\rho$	$|A_{1/2}|/|A_{3/2}| = 5.13 \pm 1.97$	$42° \pm 48°$
KK	$|A_1|/|A_0| = 0.61 \pm 0.10$	$47° \pm 10°$
$\pi\pi$	$|A_2|/|A_0| = 0.72 \pm 0.14$	$82° \pm 9°$

2.3 The Decays $D^o \to K^+K^-$, $\pi^+\pi^-$

Both decays, $D^o \to K^+K^-$ and $D^o \to \pi^+\pi^-$, proceed entirely through the effective charged currents in the BSW model which therefore predicts [27] (independent of the values for a_1 and a_2) a width ratio of $\Gamma(K^+K^-)/\Gamma(\pi^+\pi^-) = 1.4$. The experimental values for this ratio continue to be considerably larger than this prediction as shown below.

Table 8: Recent measurements of $\Gamma(D^\circ \to K^+K^-)/\Gamma(D^\circ \to \pi^+\pi^-)$

E687[29]	WA82[30]
$2.53 \pm 0.46 \pm 0.19$	$2.23 \pm 0.81 \pm 0.46$
E691[31]	CLEO[32]
$1.95 \pm 0.34 \pm 0.22$	$2.35 \pm 0.37 \pm 0.28$

This discrepancy has been frequently attributed to inelastic FSI [33]. An alternative explanation based on the relative interference of Penguin diagrams has also been proposed[35].

2.4 Non-spectator decays

An interesting example is the decay $D^\circ \to K^\circ \overline{K^\circ}$ which proceeds by the nearly cancelling diagrams (because of cancelling CKM vertex factors) depicted in Fig. 14 c). Based on the signal in Fig. 14, E687[36] quotes relative branching ratios of $K^\circ \overline{K^\circ}/K^+K^- = 0.51 \pm 0.18 \pm 0.19$ and $K^\circ \overline{K^\circ}/\overline{K^\circ}\pi^+\pi^- = 0.039 \pm 0.013 \pm 0.013$, which are somewhat higher than the earliest numbers form E400 [37] and than a more recent value from CLEO [38], $K^\circ \overline{K^\circ}/\overline{K^\circ}\pi^+\pi^- = 0.021^{+0.011+0.002}_{-0.008-0.002}$. Presumably, the relatively large branching fraction for this suppressed, non-spectator, Cabibbo suppressed decay is due to FSI effects.

3 Resonant substructure

The $D_s^+, D^+ \to K^-K^+\pi^+$ Dalitz plots, obtained in E687 [40] are particularly instructive. We note that both the ϕ and $\overline{K}^*$ bands have a node due to angular momentum conservation in the center of the band. The D_s^+ Dalitz plot is very highly dominated by the $\phi\pi^+$ and $\overline{K}^{\circ*}K^+$ channels; while the D^+ also has a significant contribution from a broad resonance. We further note the pronounced asymmetry between the two $\overline{K}^*$ lobes for the $D^+ \to K^-K^+\pi^+$. We believe this lobe asymmetry is due to interference of the $\overline{K}^{*\circ}K^+$ channel with a broad, spinless resonance channel with a nearly constant amplitude, which we (temporarily) write as $\cos\delta + i\sin\delta$. The interference term is then proportional to:

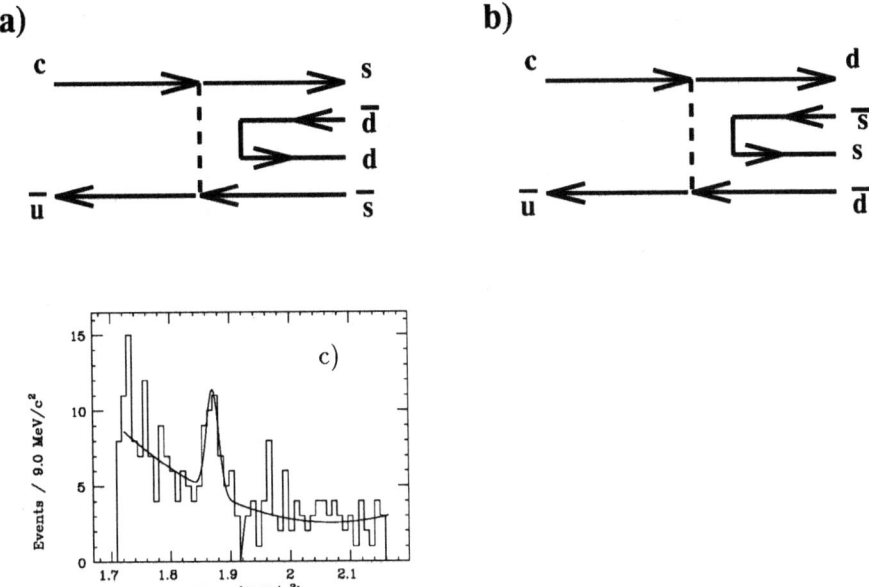

Figure 14: a) and b) Two exchange diagrams for $K^o \overline{K^o}$ decay and c) the recent E687 signal for $D^o \to K_s^o K_s^o$.

$$Re\left\{(\cos\delta + i\sin\delta)^* \frac{\cos\theta}{M_r^2 - M_{K\pi}^2 - i\Gamma M_r}\right\} =$$

$$\frac{(M_r^2 - M_{K\pi}^2)\cos\theta\cos\delta}{(M_r^2 - M_{K\pi}^2)^2 + \Gamma^2 M_r^2} + \frac{\Gamma M_r \cos\theta \sin\delta}{(M_r^2 - M_{K\pi}^2)^2 + \Gamma^2 M_r^2}$$

The asymmetry comes about because the $\overline{K^{o*}}K^+$ decay amplitude contains an angular factor ($\cos\theta$, where θ is the angle between the two kaons in the $\overline{K^{o*}}$ rest frame) which causes the interference term to change sign from the left lobe to the right lobe. We get an interference term from both the real part of the Breit-Wigner as well as the imaginary part. Since the real part of a Breit-Wigner reverses sign as one passes through the resonance (thus cancelling the interference), any net interference is due to the second term which is proportional to $\sin\delta$. The net lobe asymmetry provides visible evidence for final state interactions since all bare amplitudes must be real ($\delta = 0$ or $180°$).

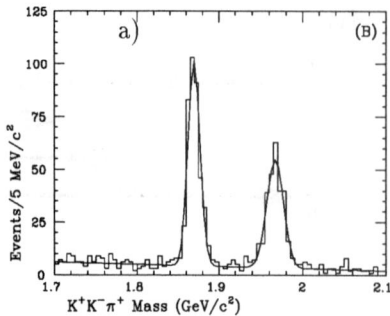
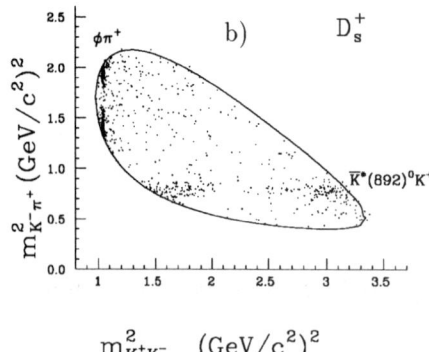
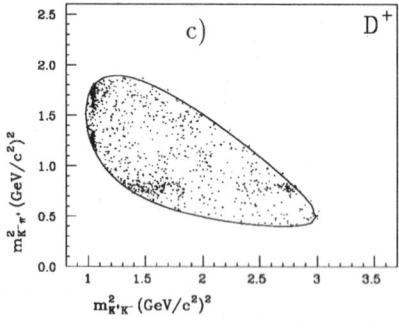

Figure 15: a) The mass spectrum for out of target $K^+K^-\pi^+$ decays. b) and c) The full sample $D_s^+, D^+ \to K^-K^+\pi^+$ Dalitz plots.

After exploring many possibilities, E687 settled on the $\overline{K}_0^*(1430)^0\,K^+$ channel as the most likely channel interfering with the $\overline{K^{*0}}K^+$.

We turn next to a discussion of the $D_s^+ \to K^+K^-\pi^+$ Dalitz plot. A close examination of the Dalitz plot for the nearly background free, out of target data shows an accumulation of events in (what should be) the ϕ band angular node. E691 [41] discovered that the $D_s^+ \to \pi^+\pi^-\pi^+$ Dalitz plot is strongly dominated by the channel $D_s^+ \to f(980)\pi^+ \to (\pi^+\pi^-)\pi^+$. We therefore expect a contribution from the known dikaon decay of the $f(980)$ which should populate the ϕ node region as shown in Fig. 16. The inclusion of additional contributions from the $f_J(1710)\pi^+$ and $\overline{K}_0^*(1430)^0\,K^+$ significantly improved the quality of the fits in the vicinity of the K^* peak (R_1) and just above it

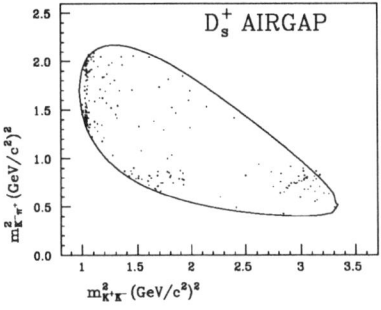
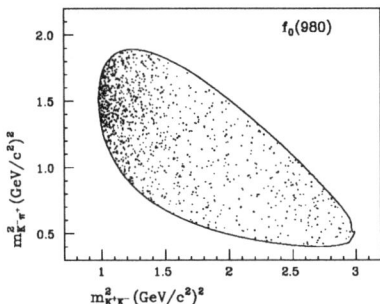

Figure 16: The E687, D_s^+ Dalitz plot and simulated $D_s^+ \to f(980)\pi^+ \to (K^+K^-)\pi^+$ decays.

(R_2) as shown in Fig. 17.

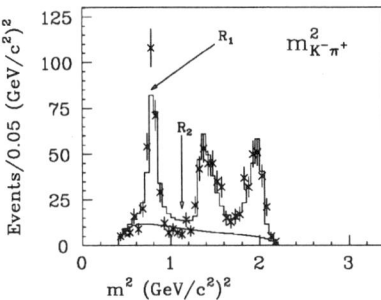
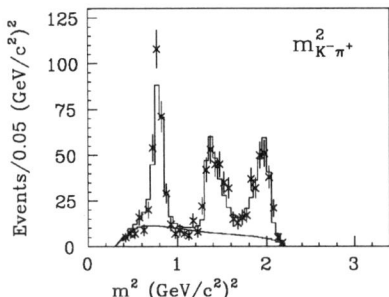

Figure 17: Comparing the $m^2(K^-\pi^+)$ 3 amplitude (left) and 5 amplitude (right) D_s^+ fit to the data (error bars). The background, deduced from the mass sidebands, is shown in the lower histogram.

3.1 Amplitude Formalism

Following the original analysis by Mark III [39], the experiments fit their amplitudes to a coherent sum of resonance channel contributions. The decay of a given quasi-two-body channel of the form $(ab)_r c$ where particles a and b are the daughters of a spin J resonance r is described by a decay function $\mathcal{M}_r$ which is the product of two (unimportant) vertex form factors ($F_D \times F_r$), an order J Legendre polynomial representing the decay angular wave function, and a relativistic Breit Wigner: $\mathcal{M}_r = F_D\, F_r \times |\vec{c}|^J |\vec{a}|^J P_J(\cos\theta_{ac}^r) \times BW(m_{ab})$.

Each decay function is multiplied by a constant, complex amplitude factor $a_r = |a_r| \exp(i\delta_r)$, with a squared modulus related to the relative importance of the channel, and a strong phase shift δ_r.

Neglecting CP violations, the δ_r phases for $\overline{D}$ decays are exactly the same as for D decays. One simply changes all particles to antiparticles in the $\mathcal{M}_r$ decay function. One can always factor out the phase of one resonance channel, and hence only the phase shift differences, $\Delta\delta$, relative to a reference phase can be measured by fitting the Dalitz intensity. According to Watson's Theorem, the presence of a relatively imaginary phase difference ($\sin(\Delta\delta) \neq 0$) can only be due to FSI. In two body decays, FSI effects can be measured through the interference of isospin amplitudes; while in three body decays, FSI effects are measured through the interference of resonant channels.

Since the phase factors in the D_s^+ and D^+ amplitudes both originate from FSI, one is tempted to speculate that they might be the same. If the D_s^+ and D^+ were degenerate in mass, the S-matrix describing rescattering $\phi\pi$ and $\overline{K}^*K$ would be the same. For elastic FSI ($\eta = 1$ in Eqn. 5), it is easy to see the observed amplitudes pick up a relative phase shift (given by $\delta_1 - \delta_2$) which is common to both D_s^+ and D^+ in the limit of degenerate masses. If $\eta < 1$ on the other hand, the observed phases will depend on the ratio of the ϕ and $\overline{K}^*$ amplitudes which are different for the D_s^+ and D^+. The universality of Dalitz amplitude phases might provide a clue to the elasticity of FSI.

3.2 Dalitz Results

Table 9 compares the E687 [40] results on the channels common to both the $D^+, D_s^+ \to K^+K^-\pi$ decays. The fractions f_r are known as decay fractions and represent the ratio of the integrated Dalitz intensity for a single resonance r divided by the intensity with all contributions present. The D^+ amplitude consists of nearly equal contributions of $\overline{K}^*K$, $\phi\pi$ and $\overline{K}^*(1430)^0 K^+$; while the D_s^+ is strongly dominated by just the $\overline{K}^*K$, and $\phi\pi$. The decay fraction for the $f_o(980)$ and $f_J(1710)$ contributions to the D_s^+ sum to 14 %.

Table 9: Comparison of $D^+, D_s^+ \to K^+K^-\pi$ amplitudes

Parameter	D^+	D_s^+
$\delta_{\overline{K}^*(892)^0 K^+}$	$0°$ (fixed)	$0°$ (fixed)
$\delta_{\phi\pi^+}$	$-159 \pm 8 \pm 11°$	$178 \pm 20 \pm 24°$
$\delta_{\overline{K}^*(1430)^0 K^+}$	$70 \pm 7 \pm 4°$	$152 \pm 40 \pm 39°$
$f_{\overline{K}^*(892)^0 K^+}$	$0.301 \pm 0.020 \pm 0.025$	$0.478 \pm 0.046 \pm 0.040$
$f_{\phi\pi^+}$	$0.292 \pm 0.031 \pm 0.030$	$0.396 \pm 0.033 \pm 0.047$
$f_{\overline{K}^*(1430)^0 K^+}$	$0.370 \pm 0.035 \pm 0.018$	$0.093 \pm 0.032 \pm 0.032$

Both the D^+ and D_s^+ could be fit entirely by quasi-two-body processes without the inclusion of a non-resonant contribution. It is interesting to note that both charm states have a real relative phase between the dominant $\overline{K}^*K$ and $\phi\pi$ channels which indicates absent or cancelling FSI phase shifts. The large error on the $\overline{K}^*(1430)^0 K^+$ relative phase for the D_s^+ precludes a meaningful test of my conjecture of a universal phase shift for that channel.

Measurements of the branching ratio $D_s^+ \to \overline{K}^{*0}K^+/\phi\pi^+$ from this first fully coherent fit $(0.92 \pm 0.11 \pm 0.09)$ are quite consistent with previous values from CLEO $(1.00 \pm 0.17 \pm 0.12)$, E691 $(0.87 \pm 0.13 \pm 0.05)$, and NA14 $(0.85 \pm 0.34 \pm 0.20)$ and theoretical estimates which range from 0.6 to 0.8 in the BSW model[27] and 1.02 in a model of Bedaque et al [42].

Our amplitude fit for $D^+ \to K^-K^+\pi^+$ enables us to correct for acceptance variations across the Dalitz plot as well as the effects of resonant channel interference to obtain a much improved inclusive branching ratio $D^+ \to K^+K^-\pi^+/K^-\pi^+\pi^+ = 0.0976 \pm 0.0042 \pm 0.0046$. It has become common practice to quote "branching ratios" for resonant decay amplitudes such as $D^+ \to \phi\pi^+/K^-\pi^+\pi^+$, but only such inclusive ratios as $D^+ \to K^+K^-\pi^+/K^-\pi^+\pi^+$ are legitimate.

Several groups have performed Dalitz analyses for the $D^o, D^+ \to K\pi\pi$ final state. Although these fits provide a good qualitative match to the data, often discrepancies are apparent in comparisons between the data and mass projections as shown in Fig. 18. The fit fractions obtained by the various experiments [44, 45, 46] are in excellent agreement; while there is little agreement concerning the relative phases, as illustrated in Fig. 19 for $D^o \to K_s^o\pi^+\pi^-$. There is excellent agreement, however, between phases obtained by E687 [44]

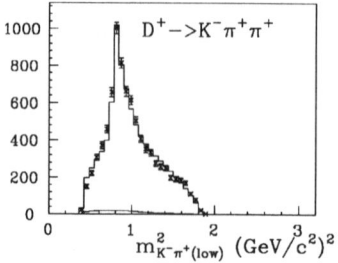 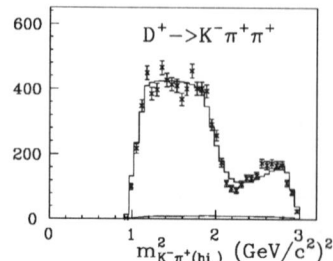

Figure 18: Comparison of the lower and higher $K^-\pi^+$ mass projection in E687 data (error bars) and our fit (histogram). The high projection does not match near 2.5 GeV^2.

and Argus [46]. These analyses provide a wealth of information on new decay modes which can be compared to models based on factorization, QCD sum rules, and $1/N_c$ expansions. Generally agreement of the models with the data is only at about the ±60 % level. The isospin amplitudes show nearly imaginary relative phases as reported in Table 7. It is interesting to note that all experimental groups report a sizeable non-resonant contribution to the $D^+ \to K^-\pi^+\pi^+$ Dalitz plot which makes it unique among the plots discussed here.

3.3 Four body decays

Fig. 20 shows invariant mass plots for the decays $D^\circ \to \pi^-\pi^+\pi^+\pi^-$, $K^+K^-\pi^+\pi^-$, $K^+K^-K^-\pi^+$, and $K^-\pi^+\pi^+\pi^-$. E687 [29] recently measured the relative ratio $4\pi/K3\pi = 0.095\pm0.007\pm0.002$. E687 finds that $KK2\pi/4\pi = 0.37$ whereas $KK/\pi\pi = 2.5$ which is interesting but not totally unexpected[43]. Both results are in good agreement with earlier measurements by ARGUS , CLEO , E691.

E687 also has a first measurement of $3K\pi/K3\pi = 0.0028 \pm 0.0007 \pm 0.0001$. They find, using an incoherent amplitude analysis that the dominant resonant contributions are $K^+K^-\rho^\circ$ (36 ± 9%) and $\overline{K}^{*\circ}K^{*\circ}$ (21% ± 9%). The process $\overline{K}^{*\circ}K^{*\circ}$ should be suppressed in a manner analogous to $D^\circ \to K^\circ_s K^\circ_s$ as discussed in Fig. 14. A previous analysis by CLEO concluded that the $KK\pi\pi$ final state was dominated by $\phi\rho$. However E687 finds only a small (7 ± 4%) $\phi\rho$ contribution.

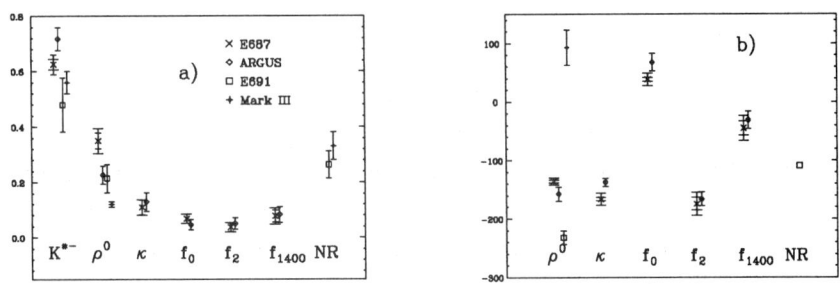

Figure 19: Comparison of the decay fractions a) and phases b) obtained by different experiments.

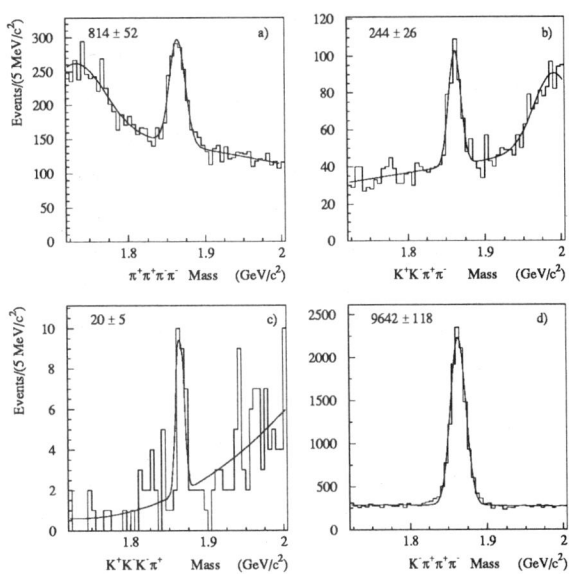

Figure 20: Four 4-body decays of the D^o obtained by E687

4 Summary

The recent data on $D \to K\ell\nu$ agrees reasonably well among experiments and with theory. Over the q^2 domain probed by experiments, the data are consistent with a form factor dominated by poles of the D_s^* spectrum, but the q^2 domain is insufficient to establish a pole rather than an exponential form. Data on $D \to \overline{K}^*\ell\nu$ are consistent among experiments and the R_2 and R_v are in good agreement with theoretical expection. However the experimental $\Gamma(\overline{K}^*\ell\nu)/\Gamma(\overline{K}\ell\nu)$ ratio remains nearly a factor of two lower than the original theoretical expectation. Recently [2] an updated version of the ISGW model is able to predict a $\overline{K}^*\ell\nu/\overline{K}\ell\nu$ ratio consistent with experimental data. At present, the situation on the R_2 and R_v form factors for $D_s^+ \to \phi\ell\nu$ is rather murky. The results are consistent between experiments, but are inconsistent by about 2σ with the form factor ratios measured for $D^+ \to \overline{K}^*\ell\nu$ which runs counter to theoretical expectations.

Data on the two pseudoscalar decay of charmed mesons underscores the importance of final state interactions, both in creating a nearly imaginary relative phase between isospin amplitudes and in modifying the expected width ratios for spectator processes such as $KK/\pi\pi$ and non-spectator processes such as $D \to \overline{K}^\circ K^\circ$. Dalitz analyses of 3 pseudoscalar decays provides new handles on FSI effects which create phase shifts between the interfering resonant channels. The Dalitz plots for $KK\pi$ and $K3\pi$ are dominated by quasi-two body contributions. The one exception to this pattern is the decay $D^\circ \to K^-\pi^+\pi^+$. Amplitude fits qualitatively reproduce many of the features of the data, although they have unacceptably large χ^2 indicating that the description is oversimplified. There is general agreement on the ratios of the $K3\pi$, $KK\pi\pi$, $KKK\pi$, and 4π decays, but experimental controversy on the ϕ content of $KK\pi\pi$. E687 observes an unexpectedly large $D^\circ \to \overline{K}^{*\circ}K^{*\circ}$ contribution.

4.1 Wish lists

I believe that charm semileptonic physics will continue to provide interesting and exciting physics for many years to come. In the future, one could extend form factor measurements to $\pi\ell\nu$, $\rho\ell\nu$, $\eta\ell\nu$ and charmed baryons. A non-parametric measurement of $f_+(q^2)$ for the decay $D \to \pi\ell\nu$ could significantly discriminate between models and provide guidance to phenomenologists. High data samples might allow for precise measurements of the D decay constant

(F_D) from processes such as $D, D_s^+ \to \mu\nu$.

There is also quite a bit of intriguing charm hadronic decay physics for the future. It would be very useful to obtain more precise values for $\tau(D_s^+/D^o)$, $\tau(\Xi_c)$, and $\tau(\Omega_c^o)$. Such precise lifetime measurements should provide insight into the relative importance of various charm decay processes.

Present data provides several intriguing mysteries which include the large level for supposedly greatly suppressed processes such as $K^o\overline{K}^o$ and $K^{*o}\overline{K}^{*o}$, the large apparent non-resonant contribution for $D^+ \to K^-\pi^+\pi^+$, and the presence of unusual daughters (with uncertain quark contents) such as $f_0(980)$ and $f_J(1710)$ in D_s^+ decay. More data would be required to put most these effects statistically "over the top". A very sophisticated, high statistics analysis may be required to ultimately understand the $D^+ \to K^-\pi^+\pi^+$ decay, which was the first charm particle Dalitz plot to be studied[47], still cannot be fit satisfactory [29]. One can look forward to the application of factorization to doubly Cabibbo suppressed decays, additional 3 and 4 body amplitude analyses, and spin dependent amplitude analyses to charm baryons.

4.2 Fixed Target Charm Experiments

Table 10 summarizes some fixed target charm experiments, which are actively publishing, which will run in the next FNAL fixed target run, or which are being discussed for the future. In the near future, we anticipate a flood of publications from E791 which amassed an enormous charm sample. The WA89 experiment in the CERN hyperon beam, has recently observed the Ω_c^o in several new decay modes and presented a preliminary lifetime for this state at the Heavy Flavor 95 Conference in Pisa.

Table 10: Fixed Target Charm Experiments

Exp.	Beam	Reconstructed Sample
E687	γ	80 K
E791	π	250 K
WA89	Σ	10 K $25\ fs < \tau(\Omega_c) < 75\ fs$
FOCUS	γ	10^6
E781	Σ	10^6 1/2 baryons
Charm2000	p	10^8

The experiments E781 (SELEX) and E831 (FOCUS) will take data in the next FNAL fixed target run. Based on our experience with E687, we anticipate an order of magnitude more photoproduced charm in FOCUS than in the present E687 sample. E781 is very encouraged by the hyperon production experiences of WA89 and anticipates a very large sample of charm strange baryons. Last summer there was a very well attended, enthusiastic workshop at Fermilab to explore the possibilities of acquiring a truly enormous charm sample at a future FNAL fixed target run.[48] It is clear that there is a lot of interesting charm physics for the future, and a large number of people eager to study it!

It is a pleasure to acknowledge the efforts of my colleagues on E687 with special thanks to Rob Gardner, Daniele Pedrini, Matteo Boschini, and Will Johns.

1. J.P. Cumalat , in **The Fermilab Meeting DPF92**, 10-14 November 1992, published by World Scientific Publishing Co, edited by Carl H. Albright, Peter H. Kasper , Rajendran Raja, John Yoh.

2. Daryl Scora and Nathan Isgur, Semileptonic Meson decays in the Quark Model; An Update, CEBAF-TH-94-14 (1994)

3. CLEO Collab., Phys. Lett. B317 (1993) 647.

4. E691 Collab., J.C. Anjos et al., Phys. Rev. Lett. 62 (1989) 1587.

5. CLEO Collab.,Phys. Rev. D44 (1991) 3394.

6. W. Johns et al., *The Albuquerque Meeting - Particles and Fields '94*, Sally Seidel Editor, World Scientific (1995) 590
7. Mark III Collab., Phys. Rev. Lett. 66 (1991) 1011.
8. Particle Data Group, M. Aguilar-Benitez *et al.*, Review of Particle Properties, Phys. Rev. D 50 (1994) 1173
9. CLEO Collab., (M. Battle, et al.), CLEO-CONF-94-18, Jul 1994. 17pp. Submitted to Int. Conf. on High Energy Physics, Glasgow, Scotland, Jul 20-27, 1994.
10. D. Scora, Ph.D. thesis, University of Toronto, 1993
11. P. Ball, V.M. Braun, H.G. Dosch, M. Neubert, Phys. Lett. B 259 (1991) 481;
P. Ball, V.M. Braun, H.G. Dosch, Phys. Rev. **D** 44 (1991) 3567.
12. J.G. Korner and G.A. Schuler, Z. Phys. **C46** (1990) 93.
13. M. Bauer, B. Stech, M. Wirbel, Z. Phys. **C29** (1985) 637;
M. Bauer and M. Wirbel, Z. Phys. **C42** (1989) 671.
14. F.J. Gilman and R.L. Singleton, Phys. Rev. **D41** (1990) 142.
15. C.W. Bernard, A.X. El-Khadra, and A. Soni, Phys. Rev. D 45 (1992) 869.
16. V. Lubicz, G. Martinelli, M.S. McCarthy, C.T. Sachrajda, Nucl. Phys. **B356** (1991) 301.
17. E691 Collab., J.C. Anjos *et al.*, Phys. Rev. Lett. 65 (1990) 2630.
18. E653 Collab., K. Kodama *et al.*, Phys. Lett. B 274 (1992) 246.
19. E687 Collab., P.L. Frabetti *et al.*, Phys. Lett. B 307 (1993) 262.
20. E653 Collab., K. Kodama *et al.*, Phys. Lett. B 309 (1993) 483.
21. E687 Collab., P.L. Frabetti, *et al.*, Physics Letters B 328 (1994) 187
22. CLEO Collab., J. Alexander, *et al.*, Phys. Lett. B337:(1994) 405
23. E687 Collab., P.L. Frabetti *et al.*, Phys. Lett. B 313 (1993) 253.
24. CLEO Collab., F. Butler *et al.*, Phys. Lett. B 324 (1994) 255
25. V. Lubicz, G. Martinelli, M.S. McCarthy, and C.T. Sachrajda, Phys. Lett. B 274 (1992) 415
26. See W.T. Eadie *et al.*, *Statistical Methods in Experimental Physics*, Elsevier, New York, 1988.

27. M. Bauer, B. Stech, M. Wirbel, Z. Phys. C34 (1987) 103.
28. Hanquing Zheng, P. Scherrer Ins. Preprint PSI-PR-95-06 (95)
29. P. Frabetti *et al.*, Phys. Lett. B321 (1994) 295
30. M.Adamovich *et al.*, Phys. Lett.B280 (1992) 163
31. J. C. Anjos *et al.*, Phys. Rev. D44,(1991)R3371
32. J. Alexander *et al.*, Phys. Rev. Lett. B65 (1990) 1184
33. Czarnecki,et al.,Z.Phys.C54(1992)411
    Kamal,Pham,Phys.Rev.D50(1994)R1832
34. Thomas E. Browder , Klaus Honscheid , and Daniele Pedrini, to be submitted to Ann. Rev. Nucl. Part. Sci.
35. Sanda,A.,Phys.Rev.D22(1980)2814
36. P. Frabetti *et al.*, Phys. Lett. B340 (1994)254
37. J.P. Cumalat *et al.*,Phys. Lett. B210 (1988)253
38. R. Ammar *et al.*, Phys Rev. D44 (1991)3383
39. J. Adler *et al.*, Phys. Lett. B196(1987)107
40. P. Frabetti *et al.*, to be published in Physics Letters.
41. J.C. Anjos *et al.*, Phys. Rev. Lett. 62 (1989) 125
42. Bedaque,Das, Mathur, Phys. Rev.D49 (1994) 269
43. I. Bigi,Phys. Lett.90B(1980)177
44. P. Frabetti *et al.*,Phys. Lett.B331(1994)217
45. J.C. Anjos *et al.*,Phys. Rev. D48(1993)56
46. H. Albrecht *et al.*,Phys. Lett.B308(1993)435
47. James Ernest Wiss, Charm Meson Production by e^+e^- Annihilation, Ph.D. Thesis, University of California, Berkeley (1977)
48. The Future of High-Sensitivity Charm Experiment: Proceedings of the CHARM2000 Workshop, Fermilab June 7-9 , 1994, Editors Daniel M. Kaplan , Simon Kwan, FERMILAB-CONF-94/190

Charm Physics at CLEO and Future Prospects at a B-Factory

Don Fujino

Department of Physics
Ohio State University, Columbus, OH 43210
(Representing the CLEO Collaboration)

Abstract

I will review three recent results in charm physics from CLEO and speculate on the future of the charm physics program at a B-factory.

I. INTRODUCTION

In this talk I will present three recent results in charm physics from the CLEO Collaboration. These topics are observations of the semileptonic decays $D_s^+ \to \eta \ell \nu$ and $D_s^+ \to \eta' \ell \nu$, observation of the isospin forbidden decay $D_s^{*+} \to D_s^+ \pi^0$, and the discovery of the Ξ_c^{*0} charmed baryon. I will next discuss the prospects for charm physics at a future B-Factory.

The CLEO II detector at the Cornell e^+e^- storage ring CESR operates on and just below the $\Upsilon(4S)$ resonance (~ 10.6 GeV), as shown in Figure 1. CLEO has a rich program in B physics since the $\Upsilon(4S)$ decays exclusively into a pair of B mesons, B^+B^- or $B^0\bar{B}^0$, that are nearly at rest. However, since the $\Upsilon(4S)$ comprises only about 25% of the total hadronic cross section, CLEO takes about a third of its data 60 MeV below the resonance to better understand the continuum backgrounds underneath the $\Upsilon(4S)$ peak. For those of us working in charm physics, these $B\bar{B}$ events are considered the background, but luckily they can be suppressed.

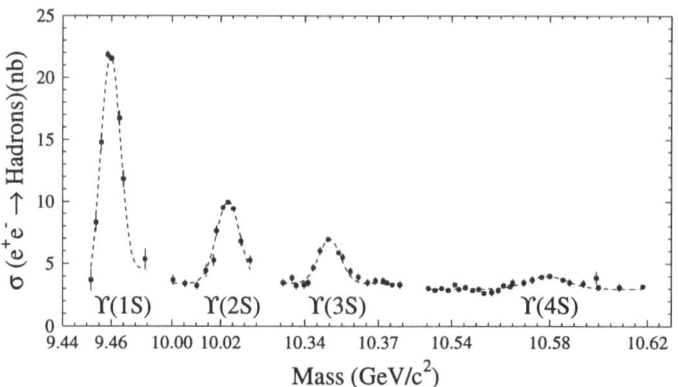

FIG. 1. The e^+e^- hadronic cross section showing the first four Υ resonances. CESR operates on and just below the $\Upsilon(4S)$ resonance.

CESR is currently the highest luminosity e^+e^- machine in the world and has delivered to date a total of ~ 5 fb^{-1} of data, which corresponds to roughly 3 million $B\bar{B}$ and 5 million $c\bar{c}$ events. The CLEO II detector [1] is a large solenoidal detector with 67 tracking layers and a CsI electromagnetic calorimeter that provides efficient π^0 and η reconstruction. Pions, kaons, and protons are identified using time-of-flight (TOF) and dE/dx information.

At CESR energies all charmed hadrons are produced. These include the D^0, D^+, and D_s^+ mesons and their excited states and the ground state charmed baryons (Λ_c, Σ_c, Ξ_c, and Ω_c) and their excited states. The $e^+e^- \to c\bar{c}$ events at CESR energies have a very different event topology than $e^+e^- \to c\bar{c}$ events at a Tau-Charm Factory, where the charmed hadrons would be produced nearly at rest. The continuum events at CLEO are relatively jetty and the charmed hadrons are produced with a hard fragmentation spectrum. We can discriminate between charmed hadrons from the continuum and from B decays since the $\Upsilon(4S) \to B\bar{B}$ events will be spherical (not jetty), and the charmed hadrons from B's are relatively soft with a maximum momentum of $x_p < 0.5$, where $x_p \equiv P/\sqrt{E_{beam}^2 - M^2}$ is the scaled momentum. A cut of $x_p > 0.5$ greatly reduces the combinatoric backgrounds with little loss in signal efficiency for charmed hadrons from the continuum.

A common technique to obtain clean samples of D mesons is to require that they come from a D^*, namely $D^{*+} \to D^0\pi^+$, $D^{*+} \to D^+\pi^0$, and $D_s^{*+} \to D_s^+\gamma$. This D^* trick leads to a loss of the inclusive D meson signal by a factor of 4 – 10, depending on the species, but reduces backgrounds typically by as much as two orders of magnitude.

II. $D_s^+ \to \eta\ell\nu$ AND $D_s^+ \to \eta'\ell\nu$ SEMILEPTONIC DECAYS

CLEO has recently observed both the $D_s^+ \to \eta\ell\nu$ and $D_s^+ \to \eta'\ell\nu$ decay modes. These two modes complete the list of Cabibbo allowed semileptonic decay modes for charmed mesons, as $D \to K^{(*)}\ell\nu$ and $D_s^+ \to \phi\ell\nu$ have been studied extensively in the past. The Fermilab E653 experiment [2] has seen evidence for $D_s^+ \to (\eta + \eta')\ell\nu$, but not each mode individually.

We cannot fully reconstruct semileptonic decay modes due to the missing neutrino, but we can nevertheless obtain clean samples of $D_s^+ \to \eta\ell\nu$ and $\eta'\ell\nu$ by requiring an energetic $\eta^{(\prime)}$ and lepton to be in the same jet, where $\eta^{(\prime)}$ denotes an η or η'. We identify η and η' mesons via the decay channels $\eta \to \gamma\gamma$ and $\eta' \to \eta\pi^+\pi^-$. Electrons are identified using dE/dx and energy/momentum information; muons candidates must penetrate five interaction lengths of iron.

To reduce combinatorics we place minimum momentum cuts for the ℓ, $\eta^{(\prime)}$, and $\eta^{(\prime)}\ell$ combinations. We also require the $\eta^{(\prime)}\ell$ invariant mass to lie in the range 1.2 – 1.9 GeV/c^2. The invariant mass distributions for $\gamma\gamma$ and $\eta\pi^+\pi^-$ after all selection cuts, Figures 2a and 2b, show signals of 578 ± 30 η events and 43 ± 8 η' events. The major source of background is due to fake leptons. Spurious $\eta^{(\prime)}\ell$ correlations can also arise due to leptons from semileptonic decays (D or B mesons) and $\eta^{(\prime)}$ mesons from the fragmentation process in $c\bar{c}$ events or from the other B meson in the event. The Cabibbo suppressed decays $D^+ \to \eta^{(\prime)}\ell\nu$ also contribute $\sim 10\%$ to the signal. The $\eta\ell$ and $\eta'\ell$ invariant mass spectra are shown in Figures 2c and 2d, respectively. Here the estimated backgrounds are plotted as the dashed lines.

We can also observe the $D_s^+ \to \eta\ell\nu$ decay by tagging the D_s^+ that come from $D_s^{*+} \to D_s^+\gamma$, i.e. by performing the D^* trick. The pseudo mass difference $\Delta M \equiv M_{\eta\ell\gamma} - M_{\eta\ell}$ will still be relatively narrow even with the missing neutrino, as shown in Figure 3. The backgrounds in the D^* tag method (dashed

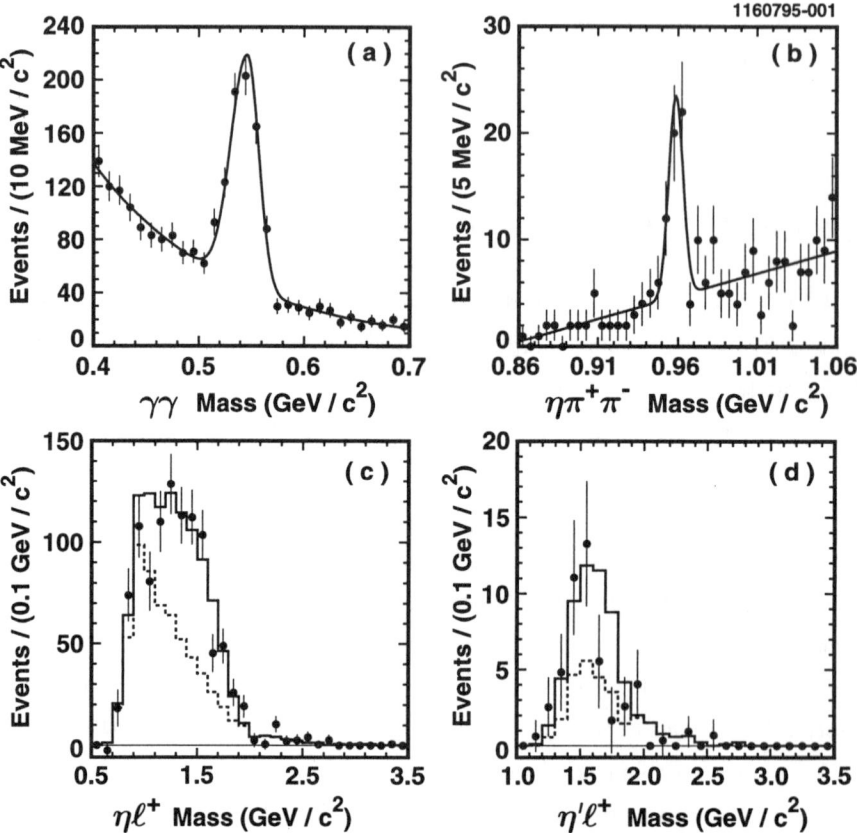

FIG. 2. The a) $\gamma\gamma$ and b) $\eta\pi^+\pi^-$ invariant mass distribution for $D_s^+ \to \eta\ell\nu$ and $D_s^+ \to \eta'\ell\nu$ candidates in the non D_s^{*+} tag analysis. The c) $\eta\ell^+$ and d) $\eta'\ell^+$ invariant mass distribution for the candidates. The dashed lines denote the predicted backgrounds.

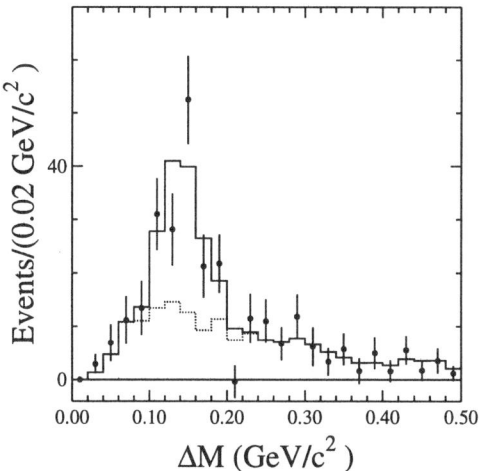

FIG. 3. The pseudo mass difference $\Delta M \equiv M_{\eta\ell\gamma} - M_{\eta\ell}$ for $D_s^+ \to \eta\ell\nu$ candidates. The points are data, the solid histogram is a fit to the data, and the dashed histogram is the predicted background.

histogram) are primarily due to 1) fake leptons and 2) real $D_s^+ \to \eta\ell\nu$ decays with random photons. We obtain the $D_s^+ \to \eta\ell\nu$ signal by requiring ΔM to lie in the range $0.1 - 0.2$ GeV/c^2, obtaining the η yield in the $\gamma\gamma$ mass plot, and subtracting the background contributions.

The branching fractions for $D_s^+ \to \eta\ell\nu$ and $D_s^+ \to \eta'\ell\nu$ are measured relative to that of $D_s^+ \to \phi\ell\nu$ (see Table I). For $D_s^+ \to \eta\ell\nu$, the nontag and the D_s^{*+} tag methods yield consistent results, and we quote the weighted

TABLE I. Summary of the $D_s^+ \to \eta\ell\nu$ and $D_s^+ \to \eta'\ell\nu$ measurements and the theoretical predictions of ISGW2 and Kamal. The numbers in (parentheses)[brackets] are the ISGW2 predictions using an $\eta - \eta'$ mixing angle of $(-10°)$ and $[-20°]$.

Decay Mode	Method	$\mathcal{B}/\mathcal{B}(D_s^+ \to \phi\ell\nu)$	ISGW2 [3]	Kamal [4]
	nontag	$1.21 \pm 0.12 \pm 0.16$		
$D_s^+ \to \eta\ell\nu$	tag	$1.32 \pm 0.22 \pm 0.15$	$(1.17)[0.77]$	1.9 ± 0.6
	combined	$1.24 \pm 0.12 \pm 0.15$		
$D_s^+ \to \eta'\ell\nu$	nontag	$0.43 \pm 0.11 \pm 0.07$	$(0.50)[0.67]$	2.2 ± 0.7

average. The ISGW2 predictions [3] agree well with our measurements for an $\eta - \eta'$ mixing angle of $-10°$. However, the predictions by Kamal et al. [4] are in marked disagreement. This is somewhat surprising because Kamal assumes factorization, which implies $\mathcal{B}(D_s^+ \to \eta'\ell\nu)/\mathcal{B}(D_s^+ \to \eta\ell\nu) \approx \mathcal{B}(D_s^+ \to \eta'\pi^+)/\mathcal{B}(D_s^+ \to \eta\pi^+)$. CLEO has measured the right-hand side of the equation to be 1.22 ± 0.33 [5], whereas the left-hand side is only 0.35 ± 0.12.

An outstanding issue in charm semileptonic decays is the disagreement on the vector to pseudoscalar ratio, $\mathcal{B}(D \to \bar{K}^*\ell\nu)/\mathcal{B}(D \to \bar{K}\ell\nu)$. The experimental world average of this ratio is 0.56 ± 0.06 [6], while the theoretical predictions lie in the range $0.5 - 1.2$ [7]. We can now measure the V/P ratio for D_s^+ semileptonic decays, which should not differ from D semileptonic decays. From our results above, we obtain $\mathcal{B}(D_s^+ \to \phi\ell\nu)/\mathcal{B}(D_s^+ \to (\eta+\eta')\ell\nu) = 0.60 \pm 0.06 \pm 0.06$, in agreement with the D meson case and the ISGW2 prediction. E653 measures a somewhat lower value of $0.26^{+0.18}_{-0.07}$ [2].

III. ISOSPIN-VIOLATING DECAY $D_s^{*+} \to D_s^+\pi^0$

Up to now, the D_s^{*+} has only been observed in the electromagnetic decay $D_s^{*+} \to D_s^+\gamma$. The process $D_s^{*+} \to D_s^+\pi^0$ violates isospin since the D_s^{*+} has isospin-0 whereas the $D_s^+\pi^0$ system has isospin-1. However, isospin is not an exact symmetry. For example, the decay $\psi' \to \psi\pi^0$ occurs, albeit at a branching fraction of $\sim 10^{-3}$. Cho and Wise [8] have predicted the decay rate for $D_s^{*+} \to D_s^+\pi^0$; the decay proceeds first through the isospin-conserving decay involving a virtual η meson, $D_s^{*+} \to D_s^+\eta$, followed by the η mixing into a π^0. The $D_s^{*+} \to D_s^+\eta$ process is not suppressed by the Okubo-Zweig-Iizuka (OZI) rule because the η couples via its $s\bar{s}$ component. The mixing amplitude is proportional to $(m_d-m_u)/[m_s-(m_d+m_u)/2]$, and hence vanishes if the light quark masses are degenerate. Furthermore, the radiative decay $D_s^{*+} \to D_s^+\gamma$ is suppressed due to a partial cancellation of the charm and strange quark magnetic dipole moment. Cho and Wise have predicted that the ratio of partial widths $\Gamma(D_s^{*+} \to D_s^+\pi^0)/\Gamma(D_s^{*+} \to D_s^+\gamma)$ lies in the range $0.01 - 0.10$. Unfortunately, there are additional uncalculable corrections to their prediction that could be nonnegligible.

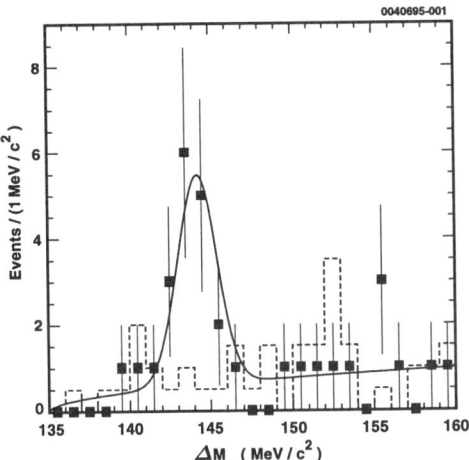

FIG. 4. The mass difference ΔM distribution for the isospin-violating decay $D_s^{*+} \to D_s^+ \pi^0$. The points are the data, the solid line is a fit to the data, and the dashed line is the estimated backgrounds.

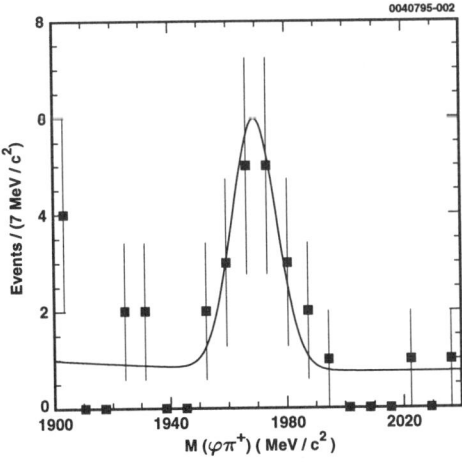

FIG. 5. The $\phi\pi^+$ invariant mass distribution for events in the ΔM signal region.

CLEO has looked for the isospin-violating decay $D_s^{*+} \to D_s^+ \pi^0$, where $D_s^+ \to \phi \pi^+$ and $\phi \to K^+ K^-$. We select $\phi \pi^+$ candidates that are within 2σ of the nominal D_s^+ mass, require the D_s^{*+} momentum to satisfy $x_p > 0.6$, and the π^0 momentum to exceed 0.25 GeV/c. The mass difference distribution, $\Delta M \equiv M_{D_s^+ \pi^0} - M_{D_s^+}$, is shown in Figure 4 and has a peak at $\Delta M = 144.2$ MeV/c^2 with $14.7^{+4.6}_{-4.0}$ signal events. Backgrounds from the D_s^+ and π^0 mass sidebands (dashed line) show no evidence of peaking in the signal region. We can also cut on ΔM and fit the $\phi \pi^+$ invariant mass distribution to obtain a consistent signal of $13.9^{+4.8}_{-4.1}$ events, shown in Figure 5.

We observe 944 ± 57 events in the radiative decay $D_s^{*+} \to D_s^+ \gamma$ with $D_s^+ \to \phi \pi^+$. Hence, the ratio of decay rates is measured to be

$$\frac{\Gamma(D_s^{*+} \to D_s^+ \pi^0)}{\Gamma(D_s^{*+} \to D_s^+ \gamma)} = 0.062^{+0.020}_{-0.018} \pm 0.022, \tag{1}$$

in agreement with the predictions of Cho and Wise, assuming the uncalculable corrections are small. In addition, the existence of the $D_s^{*+} \to D_s^+ \pi^0$ decay mode implies that the D_s^{*+} has natural spin-parity (0^+, 1^-, 2^+, ...). The radiative decay precludes the $0^\pm$ states, so the most likely J^P assignment is 1^-, i.e. a vector meson.

IV. OBSERVATION OF THE Ξ_c^{*0} EXCITED CHARMED BARYON

Last year CLEO reported on new decay modes of the Λ_c^+ charmed baryon [9] and the observation of orbital excited states of the Λ_c^+, namely, the $\Lambda_c^{*+}(2593)$ and the $\Lambda_c^{*+}(2625)$ states [10]. This year we have observed new decay modes of the Ξ_c^+ [11,12] and discovered a narrow state decaying into $\Xi_c^+ \pi^-$, which we believe to be the $J^P = \frac{3}{2}^+$ spin excitation of the Ξ_c^0 charmed baryon [13].

The Ξ_c baryons are composed of a charm, a strange, and a light (u or d) quark. The ground state isodoublet, Ξ_c^+ and Ξ_c^0, are $J^P = \frac{1}{2}^+$ states in which the lighter quarks are in a spin-0 configuration. The Ξ_c' baryon ($J^P = \frac{1}{2}^+$) and the Ξ_c^* baryon ($J^P = \frac{3}{2}^+$) are the spin excitations in which the light quarks form a spin-1 configuration. The Ξ_c' mass is expected to be below the $\Xi_c \pi$ mass threshold and hence must decay electromagnetically into $\Xi_c \gamma$. WA89 [14] has

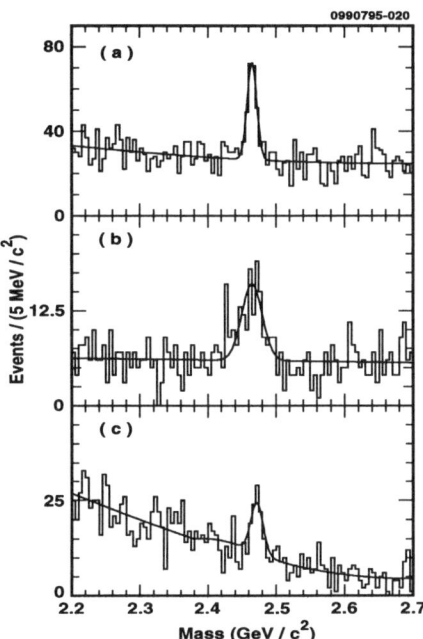

FIG. 6. The invariant mass distribution for a) $\Xi_c^+ \to \Xi^- \pi^+ \pi^+$ with $x_p > 0.4$, b) $\Xi_c^+ \to \Xi^0 \pi^+ \pi^0$ with $x_p > 0.6$, and c) $\Xi_c^+ \to \Sigma^+ K^{*0}$ with $x_p > 0.5$.

seen evidence for $\Xi_c'^+ \to \Xi_c^+ \gamma$, with a mass difference of around 95 MeV/c². The Ξ_c^* states are expected to be heavy enough to decay via a pion emission into $\Xi_c \pi$. We present evidence for $\Xi_c^{*0} \to \Xi_c^+ \pi^-$.

In addition to the previously reported decay modes $\Xi_c^+ \to \Xi^- \pi^+ \pi^+$ and $\Xi_c^+ \to \Lambda K^- \pi^+ \pi^+$, CLEO has also observed five new decay modes of the Ξ_c^+. These modes are $\Xi_c^+ \to \Xi^0 \pi^+$, $\Xi^0 \pi^+ \pi^0$, $\Xi^0 \pi^+ \pi^- \pi^+$, $\Sigma^+ K^- \pi^+$, and $\Sigma^+ K^{*0}$. The hyperons are detected through the decay channels $\Lambda \to p\pi^-$, $\Sigma^+ \to p\pi^0$, $\Xi^- \to \Lambda \pi^-$, and $\Xi^0 \to \Lambda \pi^0$, by searching for candidates with a decay vertex displaced from the primary interaction point.

The three largest and cleanest modes, shown in Figure 6, are chosen for the Ξ_c^{*0} search. We observe 160 ± 18, 76 ± 12, and 59 ± 12 events in the $\Xi_c^+ \to \Xi^- \pi^+ \pi^+$, $\Xi^0 \pi^+ \pi^0$, and $\Sigma^+ K^{*0}$ decay modes, respectively. The Ξ_c^+ candidates must have an invariant mass within 2.5σ of the nominal Ξ_c^+ mass. They are combined with each π^- track and the mass difference $\Delta M = M(\Xi_c^+ \pi^-) - M(\Xi_c^+)$ is computed. Figure 7 clearly shows a narrow peak of 55 ± 12 events at

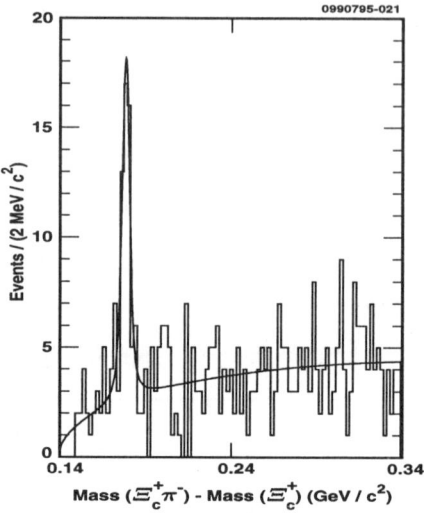

FIG. 7. The mass difference distribution, $M_{\Xi_c^+\pi^+} - M_{\Xi_c^+}$, for the decay mode $\Xi_c^{*0} \to \Xi_c^+ \pi^-$.

a mass difference of 178 MeV/c^2. The peak is fit to a Breit-Wigner convoluted with a Gaussian resolution function. The width is consistent with the detector resolution, and an upper limit of $\Gamma < 5.5$ MeV/c^2 is placed at the 90% C.L. This is consistent with the theoretical prediction of about 2.5 MeV/c^2 [15]. The mass difference is measured to be $178.2 \pm 0.5 \pm 1.0$ MeV/c^2, which is in the predicted range of $155 - 225$ MeV/c^2 [16]. Signals are also seen in each of the three Ξ_c^+ decay modes separately. A surprisingly large fraction of Ξ_c^+ charmed baryons, $(27 \pm 6 \pm 6)\%$, comes from Ξ_c^{*0} decays.

It is unlikely that the new state is an $L = 1$ orbital excitation of the Ξ_c^0 charmed baryon since we would expect a much greater mass splitting and a harder x_p distribution than what is observed. Although a spin-parity analysis was not performed, the most likely explanation for this narrow state is the Ξ_c^{*0} charmed baryon. This represents the first $J^P = \frac{3}{2}^+$ excited charmed baryon to be discovered, as the Σ_c^* and Ω_c^* states are yet unobserved.

V. FUTURE PROSPECTS AT A B-FACTORY

What charm physics can we expect at a future B-factory? Three are currently being constructed: asymmetric B-factories at SLAC and KEK, and a symmetric B-factory at Cornell. It is important to realize that the charm physics opportunities, for the most part, do not depend on whether the machine is symmetric or asymmetric, since charmed hadrons from the continuum are already heavily boosted. Of course, the quest for CP violation in $B^0\bar{B}^0$ events due to mixing depends critically on the boost of the center-of-mass frame.

The charm physics program at a B-factory will be greatly enhanced over the current CLEO II program due to three principal reasons:

- High luminosity

- Silicon vertex detectors

- Particle identification

The luminosity at a B-factory is expected to reach 2×10^{33} cm^{-2}s^{-1}, which translates to an integrated luminosity of $15-30$ fb^{-1}/year. This is a factor of 10 higher than what is presently attainable at CLEO.

Silicon vertex detectors can be used to provide a clean tag for D mesons, in addition to the commonly used D^* tag. The mean decay lengths at B-factory energies are quite long: 250 μm for D^0 and 600 μm for D^+ mesons. Since the decay length resolution will be around 70 μm, the vertex tag should be very efficient. The largest improvements over the D^* tag will occur for 3-prong D^+ decays, since the $D^{*+} \to D^+\pi^0$ tag is relatively inefficient and not as clean as the $D^{*+} \to D^0\pi^+$ tag. In $e^+e^- \to c\bar{c}$ events, only $\sim 15\%$ of the D^+ mesons come from D^{*+} decays. Also, the π^0 reconstruction efficiency is only 50%. Finally, the mass difference spectrum, $\Delta M \equiv M_{D^+\pi^0} - M_{D^+}$, still has a substantial combinatoric background due to the abundance of fake π^0's at low momenta. CLEO is currently installing a three layer silicon detector and will soon learn the benefits of vertex tagging.

In addition, a silicon vertex detector can provide information on the D^+ flight direction in semileptonic decays. This can be used to calculate the missing neutrino momentum, up to a two-fold ambiguity, a technique that is commonly employed at charm fixed target experiments. Up to now, CLEO has only been able to infer the D momentum from the thrust axis of the event and from the slow pion kinematics in D^* decays.

In the current CLEO II detector, the main weakness is particle identification which uses dE/dx and TOF information. The dE/dx crossover for π/K occurs at a momentum of 1.1 GeV/c, which results in a less than 2σ separation in the range 0.8 – 2.5 GeV/c. TOF provides 2σ separation for π's and K's up to 1 GeV/c. The majority of kaons from D decays unfortunately fall in the region of poor particle ID.

All three B-factories will have excellent $\pi/K/p$ separation over the full kinematic range (0 – 3 GeV/c). SLAC will use a DIRC detector in BABAR, Japan's BELLE detector will employ aerogel, and CLEO III has chosen a fast RICH to identify particles. Combinatoric backgrounds in D decays will be greatly reduced, once the kaon can be clearly identified. This will be especially beneficial for Cabibbo and doubly Cabibbo suppressed decays, whose signals are often masked by the corresponding Cabibbo allowed decay mode. Also, rare decays such as $B \to \pi\pi$ and $B \to K\pi$ will be easily distinguishable.

With these vast improvements many rare D processes will be explored, including $D^0 \bar{D}^0$ mixing, doubly Cabibbo suppressed decays, and possibly CP violation in charm decays. These topics are discussed in a later talk, and so I will conclude with a few speculations on the future of charm semileptonic decays and charm hadron spectroscopy.

We are just now becoming sensitive to Cabibbo suppressed semileptonic decays. CLEO has measured the $D^+ \to \pi^0 \ell \nu$ branching fraction with 53 observed events [17]. Evidence for $D^0 \to \pi^- \ell \nu$ is much weaker, from a recent 2.5σ measurement by CLEO [18] and a 1988 measurement by Mark III [19] based on 7 events. The B-factories with their excellent particle ID and high luminosity should easily be able to separate $D^0 \to \pi^- \ell \nu$ decays from the more prolific $D^0 \to K^- \ell \nu$ decays. Searches for the Cabibbo suppressed $D^+ \to \omega \ell \nu$

will benefit greatly from both a 3-prong D^+ vertex tag and from the narrow ω resonance. The $D^+ \to \rho^0 \ell \nu$ mode may also be observable if we can limit the contamination from $D^+ \to K^{*0} \ell \nu$.

Another goal in semileptonic decays is to measure the form factors, with the hope of relating the dynamics of $D \to \pi \ell \nu$ decays to $B \to \pi \ell \nu$ decays. The differential decay rate for $D^0 \to \pi^- \ell \nu$ is

$$\frac{d\Gamma}{dq^2} = \frac{G_F^2}{24\pi^3} |V_{cs}|^2 P_\pi^3 \left| f_+^\pi(q^2) \right|^2 \qquad (2)$$

where $f_+^\pi(q^2)$ is the q^2 dependent form factor. CLEO has measured the shape of the differential decay rate for $D^0 \to K^- \ell \nu$ [20], but was unable to distinguish between the pole form factor [21], $f_+(q^2) = f_+(0)/(1 - q^2/m_p^2)$, and the exponential form [22], $f_+(q^2) = f_+(0) \exp(\alpha q^2)$. The $D^0 \to \pi^- \ell \nu$ decay mode can provide more sensitivity to the shape of the form factor because it has a greater q^2 range than the $D^0 \to K^- \ell \nu$ decay mode.

Charmed meson spectroscopy is very rich due to the abundance of spin-spin and spin-orbit excitations. The D^* mesons ($S = 1$ spin excitations) and the D^{**} mesons ($L = 1$ orbital excitations) have been firmly established. In fact, all 6 narrow D^{**} states have been observed, namely the $J^P = 1^+$ and 2^+ states for D^0, D^+, and D_s^+ mesons. The mass splittings predicted by Heavy Quark Effective Theory (HQET) [23] are in good agreement with the observed D^{**} masses.

The spectroscopy of charmed baryons is equally rich, but experimentally not as well known. Most of the ground state $J^P = \frac{1}{2}^+$ charmed baryons (Λ_c, Σ_c, Ξ_c, Ξ'_c, and Ω_c) have been firmly established. Only the Ξ'^0_c state has not yet been observed, and the Ξ'^+_c has been observed only by WA89. We are just now at the threshold of discovering the lowest lying spin and spatial excitations. With a factor of 10 more data, most of the missing states will no doubt be discovered.

As stated in the previous section, the Ξ_c^{*0} is the only $J^P = \frac{3}{2}^+$ state that has been observed. Discovery of the Ξ_c^{*+} state is imminent. The Σ_c^* baryons decay into $\Lambda_c^+ \pi$ and are experimentally difficult (but not impossible) to measure because their total widths will be somewhat broad, about 20 − 30 MeV. The

Ω_c^* will be narrow since it is expected to decay radiatively via $\Omega_c^* \to \Omega_c \gamma$. Its detection is limited only by its small production rate.

The lowest lying spatial excitations in the charmed baryon system can be pictured as the diquark orbiting the charm quark with one unit of angular momentum. Only the Λ_c^{*+} states have been observed [10]. There are two excited Λ_c^+ states, the $\Lambda_c^{*+}(2593)$ and the $\Lambda_c^{*+}(2625)$, both which decay into $\Lambda_c^+\pi\pi$. They are predicted by HQET to be the $J^P = \frac{1}{2}^-$ and $\frac{3}{2}^-$ states, respectively. The Λ_c^{*+} baryons are relatively narrow resonances due to the phase space suppression, since it cannot decay into $\Lambda_c^+\pi^0$ due to isospin violation.

The $L = 1$ excitations for the Σ_c and Ξ_c baryons will decay via $\Sigma_c^{**} \to \Sigma_c\pi$ and $\Xi_c^{**} \to \Xi_c\pi$. Since the mass difference should be similar to the $\Lambda_c^{*+} - \Lambda_c^+$ mass difference of $310 - 340$ MeV/c^2, these states will most likely be broad and difficult to observe. Interestingly, the Ω_c^{**} should be narrow because it cannot decay into $\Omega_c\pi^0$ due to isospin violation, nor can it decay into $\Omega_c\pi\pi$ due to OZI suppression. If kinematically allowed, it will decay strongly into $\Xi_c K$, and hence will be narrow due to phase space suppression.

VI. CONCLUSIONS

CLEO continues to have a very active charm physics program. Recent results include the first measurement of the $D_s^+ \to \eta\ell\nu$ and $D_s^+ \to \eta'\ell\nu$ decay modes, the observation of the isospin-forbidden decay $D_s^{*+} \to D_s^+\pi^0$, and the discovery of the Ξ_c^{*0} ($J^P = \frac{3}{2}^+$) charmed baryon. With the completion of the B-factories by the turn of the century, there will undoubtably be a wealth of new discoveries in the charm sector.

REFERENCES

[1] Y. Kubota *et al.* (CLEO), Nucl. Inst. and Meth. **A320**, 66 (1992).

[2] K. Kodama *et al.* (E653), Phys. Lett. B **41**, 1581 (1993).

[3] D. Scora and N. Isgur, preprint CEBAF-TH-94-14.

[4] A.N. Kamal *et al.*, Phys. Rev. D **49**, 1330 (1994).

[5] P. Avery *et al.* (CLEO), Phys. Rev. Lett. **68**, 1279 (1992).

[6] Particle Data Group, L. Montanet *et al.*, Phys. Rev. D **50**, 1173 (1994).

[7] J.D. Richman and P.R. Burchat, preprints UCSB-HEP-95-08 and Stanford-HEP-95-01, to be submitted to Rev. of Mod. Phys.

[8] P. Cho and M.B. Wise, Phys. Rev. D **49**, 6228 (1944).

[9] R. Ammar *et al.* (CLEO), Phys. Rev. Lett. **74**, 3534 (1995).

[10] K.W. Edwards *et al.* (CLEO), Phys. Rev. Lett. **74**, 3331 (1995); H. Albrecht *et al.* (Argus), Phys. Lett. B **317**, 227 (1993); and P.L. Frabetti *et al.* (E687), Phys. Rev. Lett. **72**, 961 (1994).

[11] T. Bergfeld *et al.* (CLEO), CLNS 95/1349, CLEO 95-12, submitted to Phys. Rev. Letters.

[12] K.W. Edwards *et al.* (CLEO), CLNS 95/1353, CLEO 95-15, to be submitted to Phys. Rev. Letters.

[13] P. Avery *et al.* (CLEO), CLNS 95/1352, CLEO 95-14, submitted to Phys. Rev. Letters.

[14] E. Chudakov (WA89), "Charmed Baryon Production in the CERN Hyperon Beam," talk presented at Heavy Quark 94 Conf., Virginia, Oct. 1994.

[15] J. Rosner, DOE/ER/40561-21X-INT95-17-02, to be submitted to Phys. Rev. D. The expected Ξ_c^{*0} width can be related to the non-charmed Ξ^{*0} width by $\Gamma(\Xi_c^{*0})/\Gamma(\Xi^{*0}) = 0.75 p_1^3/p_2^3$, where p_1 and p_2 are the decay momenta for the two processes and 0.75 is the appropriate ratio of the overlap of the spin wave-functions.

[16] K. Maltman and N. Isgur, Phys. Rev. D **22**, 1701 (1980); J. Richard and P. Taxil, Phys. Lett. B **128**, 453 (1983); J. Korner and H. Siebert, Ann. Rev. of Nucl. and Part. Science **41**, 511 (1991).

[17] M.S. Alam *et al.* (CLEO), Phys. Rev. Lett. **71**, 1311 (1993).

[18] F. Butler et al. (CLEO), CLNS 95/1324, CLEO 95-3, submitted to Phys. Rev. Letters.
[19] J. Adler et al. (Mark III), Phys. Rev. Lett. **62**, 1821 (1989).
[20] A. Bean et al. (CLEO), Phys. Lett. B **317**, 647 (1993).
[21] M. Wirbel et al., Z. Phys. C **29**, 627 (1985); and J.G. Körner and G.A. Schuler, Z. Phys. C **38**, 511 (1988).
[22] N. Isgur et al., Phys. Rev. D **39**, 799 (1989).
[23] N. Isgur and M.B. Wise, Phys. Rev. Lett. **66**, 1130 (1991).

Prospects for High Precision Measurements of Charmed Hadron Properties at the τCF

P. Roudeau

Laboratoire de l'Accélérateur Linéaire

Bat. 200

91405 Orsay, France.

Abstract

The energy region between 3.7 and 5.7 GeV is the only place where charmed hadrons of a given flavour can be produced in pairs without accompanying particles. Kinematical constraints, related to the production of charmed hadrons close to threshold allow, in addition, to isolate unbiased and pure samples of events. Precision tests of the Standard Model and precise understanding of B decay properties require accurate and reliable measurements of D hadron decay properties. The τCF is ideally suited to complete this program.

Introduction

More accurate measurements of charmed particle properties are absolutely needed. At present the very poor knowledge of charmed hadron branching fractions and their decay dynamics limits the possible studies of several phenomena related directly to the charm flavour, as the hadronization of c quark jets or the measurement of the BR($Z^0 \to c\bar{c}$) at high energy e^+e^- colliders, or indirectly, as the decay properties of B hadrons, the hadronization of b quark jets and the measurement of the BR($Z^0 \to b\bar{b}$) at the Z^0 pole.

In addition, charmed hadrons are bound states of an heavy and a light quark and the mass of the charm quark is such that these systems are an extremely clean laboratory to study strong interactions at the frontier between the perturbative and non-perturbative regimes of QCD. A comparison of accurate data obtained at the τCF with precise predictions from lattice QCD will be very important. These studies have not only a fundamental character related to a better understanding of strong interactions, but they will allow also a better understanding of B decay properties and will play an important role in the evaluation of the B decay constant (f_B) and of several CKM matrix elements, such as V_{bu}, V_{td} and V_{ts}.

I. The Poor Knowledge of Charmed Hadrons

Throughout this section, the values quoted in PDG94 [1] will be used as reference values. Most of the lifetimes of weakly decaying charmed hadrons have been measured with high accuracy (better than 5%) which is still expected to be improved in future fixed target experiments. Present accuracies on absolute charmed hadron branching fractions have been summarized in Table 1.

Apart for the D^0, the accuracy of the branching factions is only at the 10% level and even worse for the D_s^+ and the Λ_c. Significant progress was achieved in the last years in measuring τ decays giving precision measurements at the 1% level and a consistent overall picture. In contrast to the situation in τ decays, the absolute scales for the decays of charmed particles remain poorly determined. Absolute branching fractions for charmed hadrons into a given final state can be measured if one is able to determine the number of produced hadrons. At the τCF, the energy of the machine can be selected such that only the D and its antiparticle (a $\bar{D}$) are produced in the final state. The inclusive reconstruction of one of these particles is then used to determine the complete final state: $D^0\overline{D^0}, D^+D^-, D_s^+D_s^-, \Lambda_c^+\Lambda_c^-$ or $\Xi_c\overline{\Xi_c}$. This approach has been used by the MARKIII (for D^0 and D^+) and BES (for the D_s^+) collaborations.

Decay channel	Present accuracy (PDG 94)	Comments
$D^0 \to K^-\pi^+$	3.5%	ALEPH, ARGUS, CLEO, MARKIII
$D^+ \to K^-\pi^+\pi^+$	7%(?)	CLEO, MARKIII, (CLEO has the best measurement with $\sigma_{syst.}=12\%$)
$D_s^+ \to \phi\pi^+$	11%(no !)	a more realistic value is 25%
$\Lambda_c^+ \to pK^-\pi^+$	14%	CLEO, from B $\to$ baryons + X, seems realistic
$\Xi_c^{0,+}$	nothing	
D^{**}, D_N	nothing	

Table 1: *Experimental accuracies on the absolute values of charmed branching fractions.*

However, their results are limited by the collected statistics; at the τCF the luminosity of the machine is expected to be more than two orders of magnitude higher.

Charged D^{*+} decays can be used to tag the production of D^0 mesons in jets at high energy e^+e^- colliders (this approach has been initially proposed by the HRS collaboration at PEP). The pion emitted in the transition $D^{*+} \to D^0\pi^+$ has characteristic properties:

- its charge defines the flavour of the charmed meson, D^0 or a $\overline{D^0}$.

- it is emitted with a very small transverse momentum relative to the D^* momentum ($\Delta m = m_{D^{*+}} - m_{D^0} - m_\pi \simeq 6$ MeV).

These pions can be observed as an excess of charged particles produced at small transverse momentum relative to a jet axis and their rate gives the number of produced D^0's. It has been shown by CLEO and ARGUS that, using data only, it is possible to count these pions. At the Z^0 pole, the results have to rely more heavily on the simulation. Furthermore, B decays which contribute also to D^{*+} production have to be carefully considered. In any case, due to the limited statistics collected by MARKIII, the most accurate measurement on the D^0 absolute branching fraction ($D^0 \to K^-\pi^+$) to date is obtained by CLEO and ALEPH using the D^{*+} tag.

A similar approach can be used to measure absolute branching fractions for the D^+ using the transition $D^{*+} \to D^+\pi^0$. However, this is less direct as for the D^0 because low p_t π^0s are also produced through the cascade $D^{*0} \to D^0\pi^0$ and one has to assume similar production rates in jets for D^{*0} and D^{*+} and a value for the absolute branching fractions for D^* decays into π^0s. Experimentally, it is also more difficult to reconstruct a low energy π^0 than a $\pi^\pm$.

Absolute Measurements of D_s^+ Branching Fractions

The cascade $D_s^* \to D_s^+ \gamma$ does not produce a γ with low p_t, as in the decays $D^* \to D\pi$, and there is no clear way to determine inclusively the production rate of D_s^+ mesons. At present, **there is no direct measurement** of any absolute branching fraction of the D_s^+ meson and all quoted values are based on the use of **hadronic models**. The present quoted uncertainty in PDG94 is 11% for the $D_s^+ \to \phi\pi^+$ branching fraction. This is, in my view, very optimistic and I shall explain why:

The value quoted for $BR(D_s^+ \to \phi\pi^+) = (3.5\pm0.4)\%$ has been obtained by averaging the results from three rather different measurements and hypotheses:

- the ratio $BR(D_s^+ \to \phi\ell\nu_\ell) / BR(D_s^+ \to \phi\pi^+)$ and the partial width $\Gamma(D^+ \to \bar{K}^{*0}\ell\nu_\ell)$ have been measured and a model has been used to relate $\Gamma(D_s^+ \to \phi\ell\nu_\ell)$ and $\Gamma(D^+ \to \bar{K}^{*0}\ell\nu_\ell)$.

- the measurement of the ratio $\Gamma(D_s^+ \to \mu^+\nu_\mu) / \Gamma(D_s^+ \to \phi\pi^+)$ and the factorization hypothesis applied to B meson decays which, as an example, relates decay channels like $\bar{B}^0 \to D^{*+}D_s^-$ to the semileptonic decays $\bar{B}^0 \to D^{*+}\ell\nu_\ell$ give two different expressions for f_{D_s} (the D_s^+ decay constant) versus $BR(D_s^+ \to \phi\pi^+)$. Their overlap provides values for the two quantities.

- the probability to produce a D_s^+ meson in a jet has been inferred from inclusive measurements of strange particles in jets and from an evaluation of the D^{**} rate (this is because non-strange D^{**}'s are usually too light to decay into D_s^+K and preferentially decay into non-strange D mesons and K).

Using recent measurements from ARGUS, CLEO and E687, each of these techniques gives a result with an accuracy of about 30%. This uncertainty mainly depends on the a priori quoted uncertainty for the dynamical hypothesis. At present it is difficult to quote any reliable uncertainty on the D_s^+ production rate in jets. Furthermore, the second approach should only be used as a test for factorization models containing D_s^+ mesons in the final state.

It is a priori unjustified to take the mean of these values and of their errors (some words of caution are in fact included in the long version of PDG94). The first approach seems to be the most promising since more detailed measurements of semileptonic decays of D^0, D^+ and D_s^+ mesons may add constraints on models in the future. In this approach, the main uncertainty comes from the theoretical evaluation of the ratio:

$$\Gamma(D_s^+ \to \phi \ell \nu_\ell) / \Gamma(D^+ \to \bar{K}^{*0} \ell \nu_\ell) = 1.02 - 0.84,$$

where the last value is derived using the ISGW2 model[2]. The two extreme values give

$$BR(D_s^+ \to \phi \pi^+) = (4.00 \pm 0.55)\% \text{ and } (3.30 \pm 0.45)\%,$$

respectively. Considering that these two values, and all those in between, have the same probability I get:

$$BR(D_s^+ \to \phi \pi^+) = (3.65^{+0.90}_{-0.80})\%$$

(outside the quoted uncertainty interval there is still a 31.7% probability to find the correct value for $BR(D_s^+ \to \phi \pi^+)$).

Absolute Branching Fractions of Charmed Baryons

The results on charmed baryons are less uncertain than the results on the D_s^+ when using B meson decays with baryons emitted in the final state. In this approach one assumes that one Λ_c^+ has been produced in each decay. In future, one can take into account the fraction of decays with Ξ_c baryons produced in place of the Λ_c^+. In this way an accuracy of 10% or better can be achieved on absolute branching fractions of the Λ_c^+. In a jet environment, it may be possible to use the decays $\Sigma_c^{++} \to \Lambda_c^+ \pi^+$ as a source of pions emitted at low p_t relative to the jet axis. But, a priori, a large background is expected from pions of similar properties coming from D^* decays. There will be also a contamination from $\Xi_c^{*+} \to \Xi_c^0 \pi^+$ and the problem of separating Λ_c^+ and Ξ_c states at the production level will remain.

At the τCF it is possible to measure accurately absolute charmed baryon branching fractions. In 1980, the MARKII collaboration [3] measured the production of charmed baryons at $\sqrt{s} = 5.2$ GeV and obtained:

$$\sigma(\Lambda_c^+ + \Lambda_c^-) \times BR(\Lambda_c^+ \to pK\pi) = 0.037 \pm 0.012 \ nb.$$

Assuming that charmed baryons are pointlike spin 1/2 particles, the energy dependence of the cross section is given by:

$$\sigma = \sigma_0 \beta (3 - \beta^2), \text{ where } \beta = \sqrt{1 - \frac{m^2}{E^2}}$$

which exhibits a steep increase of the cross section just above threshold. This approach is too naive, since several production processes have to be considered, such as:

$$e^+e^- \to \Lambda_c^+ \Lambda_c^-$$

$$e^+e^- \to \Sigma_c \bar{\Sigma}_c^*$$

$$e^+e^- \to \Sigma_c^* \bar{\Sigma}_c^*$$

Independently of phase space differences, it can be expected that these reactions have production rates given by the ratios 3 : 1 : 16 : 10 [4]. Therefore, the reaction $e^+e^- \to \Lambda_c^+ \Lambda_c^-$ could turn out to be only a very small fraction of the total rate measured by MARKII. Assuming all excited charmed baryons like Σ_c and $\bar{\Sigma}_c^*$ decay into $\Lambda_c^+ \pi$, it will not be necessary to base the analysis only on the first process to measure absolute branching fractions of the Λ_c^+. The machine energy can be set at $\sqrt{s} \simeq 5$ GeV (below the $e^+e^- \to \Lambda_c^+ \bar{N} \bar{D}$ threshold) and events will be tagged by reconstructing exclusive decay channels of the Λ_c^+. The analysis is expected to be more complicated than for $e^+e^- \to D\bar{D}$ since in applying the beam energy constraint on the Λ_c^+ candidate, not only one production process has to be considered. As a consequence the majority of the baryon production cross section at this energy can be used to study the properties of the Λ_c^+.

The measurement of absolute branching fractions for Ξ_c baryons can be envisaged in a similar way, running the machine at an energy below $\Xi_c \Lambda_c^- K$ production (5.244 GeV). Considering that the following ratios, $\frac{\Xi^-}{\Lambda^0} \simeq 7\%$ and $\frac{\Omega^-}{\Xi^-} \simeq 8\%$, have been measured in jet production at LEP, the production rate of these baryons is consistent with 5 to 10% of the Λ_c^+ rate. The production mechanisms of Ξ_c states have to be measured, at the working point energy of the machine, so that precise values for Ξ_c^+ and Ξ_c^0 production rates can be extracted. **The entire field of charmed baryons at the τCF needs more attention both from theorists and experimentalists to validate the proposed approach using realistic simulations of the physics and of the detector.**

In addition, it would be interesting to study a similar approach for the D_s^+ meson. Running the machine below the $e^+e^- \to D_s^+ \bar{D} K$ threshold (4.33 GeV) but much above the $D_s^+ D_s^-$ threshold, one can benefit from an higher production cross-section enhanced by the processes $e^+e^- \to D_s^* D_s^-$ and $e^+e^- \to D_s^* \bar{D}_s^*$

Absolute Branching Fractions of Excited Charmed Hadrons

Very little is known on $c\bar{q}$ excited states. The first orbital excitations (L=1), usually named D**, contain 4 states. Two states are narrow and two are expected to be broad. Masses and widths of the narrow states have been measured but no absolute value of any branching fraction is known. Broad states have not yet been seen, neither in fixed target experiments nor at high energy e^+e^- colliders, because of the high level of combinatorial background. As a consequence, production rates of these states in the fragmentation of charm quarks remain uncertain. At the τCF, following an old proposal by A. de Rujula, H. Georgi and S.L. Glashow (1976) [5], the following reactions can be studied:

$$e^+e^- \rightarrow (1P)\bar{D}^*, \ (1P)\bar{D}, \ (0P)\bar{D}^*$$

Running the machine at a fixed energy, well above the thresholds for these reactions, the mass distribution of the hadronic system, recoiling against a D or a D* meson, can be measured and the corresponding states identified in an unbiased way.

II. Charmed Hadron Production Rates at the τCF and B Factories

An update of the event rates expected at the τCF is given in Table 2. Compared to results quoted during the SLAC(1989) [6] and Marbella(1993)[7] workshops, the main differences are:

- the rates are now given for a integrated luminosity of 10 fb^{-1} which corresponds to the amount of data accumulated during a "normal year" (10^7s) running at a luminosity of 10^{33}cm^{-2}s^{-1},

- the cross section for D_s^+ production is taken from the recent BES measurement for the reaction $e^+e^- \rightarrow D_s^+ D_s^-$,

- additional decay channels for the D_s^+ meson have been considered and some recent measurements have changed significantly compared to previously used values,

- in semileptonic decays, the ratio for the production of vector and pseudo-scalar particles is about half the value assumed before.

The comparison between the expected event rates for charmed particles at the τCF and the B factories (BFs) can be done in several ways:

- by comparing production cross sections,

- by considering single tagged event rates. At the τCF the tagging efficiency is given by the reconstruction efficiency of some well established exclusive decay modes of charmed hadrons. To reduce the combinatorial background, only track combinations having a total energy compatible with the beam energy are used. At the BFs the tagging efficiency is given by the cuts applied to reduce the $B\bar{B}$ contribution and to isolate D^* candidates.

D^0	τCF	BF	τCF rate / BF rate
$\sigma(D^0 + \bar{D}^0)$	5.8 nb	1.3 nb	4.5
ϵ(tag.)	12 %	11 %	$\simeq 20$
$\sharp$(sing. tag.)	7×10^6		
D^+	τCF	BF	τCF rate / BF rate
$\sigma(D^+ + D^-)$	4.2 nb	0.5 nb	8.
ϵ(tag.)	7 %	7 %	$\simeq 40$
$\sharp$(sing. tag.)	3×10^6		
D_s^+	τCF	BF	τCF rate / BF rate
$\sigma(D_s^+ + D_s^-)$	0.6 nb(BES)	0.24 nb(?)	2.5
ϵ(tag.)	8 %	? no tag	?
$\sharp$(sing. tag.)	5×10^5		
Λ_c^+	τCF	BF	τCF rate / BF rate
$\sigma(\Lambda_c^+ + \Lambda_c^-)$	1.0 nb(MARKII)	0.2 nb(?)	5.
ϵ(tag.)	5 %(?)	? no tag	?
$\sharp$(sing. tag.)	$5. \times 10^5$		

Table 2: *Cross sections and events rates expected for a 10 fb^{-1} integrated luminosity.*

- doubly tagged event rates. They correspond to the exclusive reconstruction of the other D hadron in singly tagged events and provide essentially **background free samples** which can be used for several purposes: absolute branching fraction measurements and for the study of leptonic, semileptonic and some rare decay channels. **This class of events is unique to the τCF.**

It appears that the tagging efficiencies at the two facilities are similar, of the order of 10%, and the numbers of doubly tagged events at the τCF and of single tagged events at the BFs remain in the ratio of the production cross-sections. From results obtained by MARKIII and CLEOII it appears also that the signal/background ratio obtained with single tagged events at the τCF and BFs are rather similar. Thus if one compares the rates for a given reconstructed D signal, **at the same level of purity, there is an additional factor 5 increase in favour of the τCF**.

III. Goals in Charmed Hadron Spectroscopy

Absolute Branching Fractions

As explained in section II, major contributions to the understanding of the D_s^+, Λ_c^+, Ξ_c, and excited states are expected to emerge from the τCF. These data are important building blocks for studies of hadronization mechanisms of c and b quark jets and to measure B hadron decay branching fractions.

The τCF is also the best place to measure inclusive branching fractions of charmed hadrons, such as D → K X, Λ_c^+ → p X, ...

Systematic Measurement of D Hadron Decay Channels

This is a topic where a lot of information can be obtained at other facilities, like fixed target experiments or higher energy e^+e^- colliders. For example, for D^0 decays, the use of the cascade D^{*+} → $D^0\pi^+$ helps to reduce the combinatorial background and gives access to small branching fractions. Fixed target experiments are limited, at present, to final states with no or at most one neutral hadron in the final state. At the τCF, all decay modes of charmed hadrons can be measured with high statistical accuracy and good control of systematics. The reconstruction efficiency for charged and neutral hadrons can be monitored continuously using J/Ψ decays.

A reliable and complete measurement of these branching fractions can be used to measure the contributions from the different quark diagrams (external W emission, internal W emission, W exchange, annihilation, penguin). This information can be used in turn to understand B decays for which non-spectator contributions are expected to be reduced by $(\frac{m_c}{m_b})^\alpha$ with $\alpha \simeq 2$.

IV. D Meson Leptonic Decays

Leptonic branching fractions of charged D mesons are given by the following expression:

$$BR(D_q^+ \to \bar{\ell}\nu_\ell) = \frac{G_F^2}{8\pi} M_D\, m_\ell^2\, |V_{cq}|^2\, (1 - \frac{m_\ell^2}{M_D^2})^2\, f_{D_q}^2\, \tau_{D_q}.$$

The Cabibbo matrix elements $|V_{cq}|$ and the D_q lifetimes have been measured with high accuracy, to the % level, and leptonic decay channels can be used to measure the D decay constants:

$$f_{D_q}^2 = \frac{|\psi(0)|^2}{M_D}.$$

Lattice QCD calculations give the following results: $f_{D_s}^{th} = 230 \pm 30$ MeV and $\frac{f_{D^+}}{f_{D_s}} = 0.9 \pm 0.1$. Corresponding branching fractions and expected events rates, for an integrated luminosity of 10 fb^{-1} at the τCF are given in Table 3.

D^+ and D_s^+ leptonic branching fractions can be measured with an accuracy ranging between 2 and 3% (which corresponds to an error of 1 - 1.5% on f_D and f_{D_s}).

Decay channel	Branching fraction	♯ events (10 fb^{-1})
$D_s^+ \to \tau^+ \nu_\tau$	4%	5000
$D_s^+ \to \mu^+ \nu_\mu$	0.4%	1300
$D_s^+ \to e^+ \nu_e$	0.1%	
$D^+ \to \tau^+ \nu_\tau$	0.1%	
$D^+ \to \mu^+ \nu_\mu$	0.04%	900

Table 3: *Branching fractions and events rates corresponding to an integrated luminosity of 10 fb^{-1} at the τCF.*

The Best Method to Determine f_B

This may seem provocative, but I believe that **the best way to know accurately the B meson decay constant, f_B, lets's say to better than 5 %, is to measure f_D with high precision and to use lattice QCD** to extrapolate from the D to the B region.

A direct measurement of f_B using the channel $B^+ \to \tau^+ \nu_\tau$ is extremely difficult. At BFs, this requires the reconstruction of the accompanying B meson to reduce the background contributions. The expected branching fraction is of the order of 7×10^{-5} (using $\frac{|V_{bu}|}{|V_{bc}|} = 0.08$ and $f_B = 200$ MeV). This value has to be multiplied by the branching fraction of the τ into leptons. To have a distinct signature, very large samples of exclusive tagged events have to be collected. But even so, the large backgrounds coming from channels with an escaping K_L^0 might prevent an accurate measurement. Furthermore, this measurement gives the product $|V_{bu}| \times f_B$ and one has to rely on a precise value for $|V_{bu}|$ which must be determined separately. However, this CKM parameter is not expected to be known with an accuracy better than 10% in the near future. As we will argue, in the following, this is another place where precise measurements in the D sector (semileptonic decays) are essential.

Expected Accuracy on f_D in the Next Years

Within the next few years BES expects to collect enough $e^+e^- \to D_s^+D_s^-$ events and reach an accuracy of $\simeq 10\%$ on f_{D_s} (this may depend on the defined priorities for the physics program of the next years).

Recently, CLEO measured the ratio BR($D_s^+ \to \mu^+\nu_\mu$) / BR($D_s^+ \to \phi\pi^+$) using the cascade $D_s^* \to D_s^+\gamma$. The accuracy of this measurement seems to be limited by the control backgrounds from photons produced during the hadronization of jets. It appears difficult (at least to me) to reach $\frac{\sigma_{f_{D_s}}}{f_{D_s}} < 10\%$ based on this approach (even if ignoring the lack of a precise measurement of BR($D_s^+ \to \phi\pi^+$)).

V. D Meson Semileptonic Decays

Semileptonic decays of D mesons are another laboratory to study in a quantitative and precise way the nonperturbative regime of QCD.

D Meson Semileptonic Decays in 1994.

In the following, results quoted in PDG94 will be shortly summarized and commented on.

$D \to K\ell\nu_\ell$

The measured partial widths for the D^0 and the D^+ are rather different, $(8.9 \pm 0.5) \times 10^{10} \ s^{-1}$ and $(6.3 \pm 0.8) \times 10^{10} \ s^{-1}$, whereas they are expected to be equal. This is possibly another illustration of the difficulty to control experimental efficiencies for quite different final states, $D^0 \to K^-\ell^+\nu_\ell$, $D^+ \to \bar{K}^0\ell^+\nu_\ell$, and of the resulting underestimation of measurement errors when combining different experiments. The mean quoted value for this partial width is $(8.2 \pm 0.4) \times 10^{10} \ s^{-1}$.

$D \to K^*\ell\nu_\ell$

Here, the agreement between different experiments appears to be better. The mean partial width is $(4.6 \pm 0.4) \times 10^{10} \ s^{-1}$.

Inclusive Semileptonic Branching Fraction

In PDG94 this measurement is still dominated by the MARKIII result $(16.9 \pm 1.5) \times 10^{10}$ s^{-1}. It agrees with a new measurement from CLEO obtained for D^0 decays using the D^* tag: $(16.8 \pm 0.8) \times 10^{10}$ s^{-1}.

Other Exclusive Decays

Adding the $D \to K\ell\nu_\ell$ and $D \to K^*\ell\nu_\ell$ partial widths and comparing with the inclusive measurement shows that about $20 \pm 5\%$ or $(4.0 \pm 1.0) \times 10^{10}$ s^{-1} remain unaccounted for.

According to the recent ISGW2 model, the transition $c \to d\ell\nu_\ell$ may correspond to 5% of the total rate and the $D \to K^{**}\ell\nu_\ell$ to about 2%: $(13 \pm 5)\%$ remain to be explained. Experimentally, there is no evidence for nonresonant contributions. Therefore, the $K\ell\nu_\ell$ and the $K^*\ell\nu_\ell$ decay rates have been, presumably, underestimated by about 10% and the total present quoted uncertainty of 5% is too optimistic. **A consistent picture for D^0 and D^+ semileptonic decays can be only obtained assuming that at present branching fractions are known with an accuracy of 10 to 15 % only.**

Physics Goals and the τCF Contributions

As for other D decays, the τCF can provide accurate measurements for all semileptonic branching fractions of all charmed hadrons. In addition, the Q^2 dependence of the form factors can be precisely measured for all weakly decaying charmed hadrons.

As D hadrons are produced almost at rest, the inclusive lepton energy spectrum and also the inclusive mass distribution of the produced hadronic system accompanying the leptons can be measured directly. These measurements are important tests for lattice QCD computations (branching fractions and Q^2 dependence of form factors) and also of $1/m_Q$ expansion models (inclusive lepton distributions). As argued by several authors, e.g. ISGW2 [2], the precise understanding of D semileptonic decays can be used to extract $\mid V_{bu} \mid$ from $B \to \pi\ell\nu_\ell$ and $B \to \rho\ell\nu_\ell$ transitions using quark models.

Expected Rates at the τCF

Table 4 is an update of similar projections of event rates presented during previous τCF workshops. An integrated luminosity of 10 fb^{-1} has been assumed and the

Charmed hadron	Decay channel	♯ events reconst.
D^0	$K^-\ell\nu_\ell$	1.4×10^5
	$K^{*-}\ell\nu_\ell$	2.2×10^4
	$\pi^-\ell\nu_\ell$	$9. \times 10^3$
	$\rho^-\ell\nu_\ell$	1.6×10^3
D^+	$\bar{K}^0\ell\nu_\ell$	$5. \times 10^4$
	$\bar{K}^{*0}\ell\nu_\ell$	$5. \times 10^4$
	$\pi^0\ell\nu_\ell$	2.3×10^3
	$\rho^0\ell\nu_\ell$	2.0×10^3
	$\omega^0\ell\nu_\ell$	0.7×10^3
	$\eta\ell\nu_\ell$	0.8×10^3
	$\eta'\ell\nu_\ell$	0.2×10^3
D_s^+	$\eta\ell\nu_\ell$	1.5×10^3
	$\eta'\ell\nu_\ell$	1.7×10^3
	$\phi\ell\nu_\ell$	2.7×10^3
	$\bar{K}^0\ell\nu_\ell$	2.0×10^2
	$\bar{K}^{*0}\ell\nu_\ell$	1.8×10^2

Table 4: *Double tag events rates expected for 10 fb^{-1} integrated luminosity.*

values for the semileptonic branching fractions have been taken from PDG94 or from the ISGW2 model if they are as yet unmeasured. Quoted rates are the average over e and μ contributions and not the sum of the two channels. Only double tagged and, thus, background free events have been considered.

Form Factors of D → P$\ell\nu_\ell$

Many results on semileptonic decays of D^0 and D^+ mesons into P$\ell\nu_\ell$ (P stands for pseudoscalar particle) have already been obtained.

At CLEO these decay channels have been selected using the D^* tag, by counting the excess of events in the $\delta(M) = M(P\ell\pi) - M(P\ell)$ peak which appears over a significant background. The width of the peak and the level of the background depend on the value of the (Pℓ) mass. Since the K$\ell\nu_\ell$ branching fraction is relatively large (a few %) and this channel depends mainly on a single form factor, rather precise measurements are expected in the future.

At the τCF the main concern will be to improve the accuracy of these measurements and to obtain precise results on Cabibbo suppressed decay modes.

The latter can be isolated without any background as has already been demonstrated by MARKIII for the D → $K\ell\nu_\ell$ and D → $\pi\ell\nu_\ell$ channels. The technique relies on simple kinematical constraints.

It has been shown at this workshop that it is possible to measure the two helicity form factors at the τCF. The decay partial width can be written [8]:

$$\frac{d\Gamma}{d\cos\theta} \propto \sin^2\theta \frac{d\Gamma_L}{dQ^2} + 0(\frac{m_\ell^2}{Q^2}),$$

where $\frac{d\Gamma_L}{dQ^2} \propto P_{CM}^3 \mid F_+^V \mid^2$ and where the other contribution depends on the $\mid F_+^V \mid$ and $\mid F_-^V \mid$ invariant form factors. P_{CM} is the momentum of the pseudoscalar particle in the D rest frame and θ is the angle between the directions of the charged lepton and of the pseudoscalar particle in the $\ell\nu_\ell$ center of mass system. Q^2 is the square of the mass of the $\ell\nu_\ell$ system. The contribution from the second term can be measured in decays involving muons and it contributes mainly at small Q^2. Its effect can be seen in the angular distribution of the lepton which departs from the simple $\sin^2\theta$ behaviour. At large values of $\cos\theta$ an excess of events is expected but it corresponds to a small value for the muon momentum and is thus usually outside the acceptance of the detector. As the neutrino momentum is maximum in this region, it will be possible, when working with double tagged events at the τCF, to select these events using the reconstruction of the missing momentum.

In any case, it has to be noted that even if one is able to measure two helicity form factors for this decay at the τCF, one can not expect to be sensitive to the invariant form factor $\mid F_-^V \mid$ which, in the model of [8] is expected to have an effect at the % level.

Form Factors of D → $V\ell\nu_\ell$

These decays have already been studied in previous workshops [6], [7]. The dynamics is more complex as compared to D → $P\ell\nu_\ell$ transitions and the Q^2 dependence of three form factors has to be measured. The uniform experimental acceptance of the τCF detector allows to do these measurements with high accuracy. As an example, the forward/backward asymmetry in the charged lepton angular distribution gives a measurement of the invariant form factor $V(Q^2)$ but such an asymmetry can also be generated by a non uniform experimental acceptance versus the charged lepton energy.

VI. Conclusions

The high statistics and the unique production mechanisms close to threshold will allow the τCF in all domains of charm physics to access the $\simeq 1\%$ accuracy region, gaining about an order of magnitude in precision compared to other facilities. Such a precision is needed for accurate tests of the Standard Model and to understand the dynamics of heavy-light quark systems. The situation for charm is now very similar to the one for τ decays a few years ago. There exist many not very accurate measurements which, when combined, give the illusion of a reliable measurement. As the main source of uncertainties comes from systematics, we do not expect a large improvement in the next years, in contrast to the recent progress in τ physics. The reason lies in a basic difference between the τ and charm production mechanisms. Independently of the collider energy τ pairs are always produced without additional particles in the event. On the other hand, there is only one energy region where charm particles of the same flavour are produced in pairs without additional particles: the energy range of the τCF. At higher energies the flavour correlation between the two charmed hadrons is significantly lower.

Acknowledgments. I would like to thank all the organizers of this Workshop especially J. Repond for his help before, during, and after the meeting.

References

[1] Phys. Rev. D 50(1994) 1173.

[2] D. Scora and N. Isgur, CEBAF-TH-94-14.

[3] Abrams et al., Phys. Rev. Lett. 44(1980)10.

[4] J.G. Korner and M. Kuroda, Phys. Rev. D16(1977)2165,
A.C.D. Wright, Phys. Lett. 71B(1977)425,
K.O. Mikaelian and R.J. Oakes, Phys. Rev. D19(1979)1613,
J.G. Korner and M. Kuroda, Phys. Lett. 67B(1977)455.

[5] A. De Rujula, H. Georgi and S.L. Glashow, Phys. Rev. Lett. 3

[6] Proceedings of the Tau-Charm factory workshop, May 23-27, 1989

[7] Third Workshop on the Tau-Charm factory, 1-6 June 1993, Marb Éditions Frontières.

[8] J.G. Korner and G.A. Schuler, Z. Phys. C- Particles and Fie

CHARM PHYSICS: BEYOND THE STANDARD MODEL

Potential for Discoveries in Charm Meson Physics

Gustavo Burdman

Fermi National Accelerator Laboratory, P. O. Box 500, Batavia, IL 60510, USA.

Abstract. The possibility of using charm meson physics to test the Standard Model (SM) is reviewed. In the case of $D^0 - \bar{D}^0$ mixing, the SM contributions are carefully considered and the existence of a window for the observation of new physics is discussed. Some examples of extensions of the SM giving large mixing signals are presented. Finally, some distinctive aspects of CP violation and rare decays in charm mesons are discussed.

The D meson has been largely overlooked as a testing ground for the SM. The reason for this might be traced back to the fact that the most important effects in Flavor Changing Neutral Currents (FCNC) and in flavor mixing are brought about, in the SM, by the top quark. This has important effects in loops that couple to external down quarks (B mixing, radiative and rare B and K decays). In the SM, top quark loops do not couple to external up quarks and thus the SM loop effects in charm physics are expected to be very small. This is the case, for instance, for $D^0 - \bar{D}^0$ mixing and for rare decays: the SM predicts very small rates. However this can be viewed as a window of opportunity for observing effects coming from new physics at higher energy scales. Of these, the most interesting ones are those that are relatively small or even negligible in B and K physics but become observable contributions when looking at D physics. Here we will discuss where these opportunities are likely to be and some of the new physics scenarios giving interesting signals. We will first discuss $D^0 - \bar{D}^0$ mixing in the SM, focusing on our current understanding of the long distance contributions. This is a crucial point in order to establish whether or not there is a window to observe new physics in the measurement of this effect. Then we will go on to show some examples of extensions of the SM that could fill this experimental window. Among these are multi-Higgs doublet models, fourth-generation effects, supersymmetry, and tree-level FCNC effects induced by dynamical symmetry breaking scenarios. We will then move to briefly review the prospects of CP violation effects in D physics, both direct and associated with mixing, in the SM and beyond. Finally, we will take a look at radiative and rare decays and point out the relevance of some modes as tests of the SM.

$D^0 - \bar{D}^0$ MIXING IN THE STANDARD MODEL

The current experimental knowledge of $D^0 - \bar{D}^0$ mixing comes from the upper bound on the wrong-sign to right-sign ratio

$$r_D \equiv \frac{\Gamma(D^0 \to \ell^- X)}{\Gamma(D^0 \to \ell^+ X)} \simeq \frac{1}{2}\left[\left(\frac{\Delta m_D}{\Gamma}\right)^2 + \left(\frac{\Delta \Gamma_D}{\Gamma}\right)^2\right], \tag{1}$$

with the approximation in (1) valid for $\Delta m_D/\Gamma$, $\Delta\Gamma_D/\Gamma \ll 1$. From the latest E691 data [1] we know $r_D < 3.7 \times 10^{-3}$. If $\Delta\Gamma_D/\Gamma$ is neglected, this translates into an upper limit for the mass difference giving

$$\Delta m_D^{\text{exp.}} < 1.3 \times 10^{-13} \text{GeV} \tag{2}$$

Short Distance

In the SM the short distance $\Delta C = 2$ transition occurs via the box diagrams. The effective interactions at the m_c scale are described by the hamiltonian [2]:

$$\mathcal{H}_{\text{eff.}}^{\Delta C=2} = \frac{G_F}{\sqrt{2}} \frac{\alpha |V_{cs}^* V_{us}|^2}{8\pi \sin^2\theta_W} \frac{(m_s^2 - m_d^2)^2}{m_W^2 m_c^2} (\mathcal{O} + \mathcal{O}') \tag{3}$$

with $\mathcal{O} \equiv \bar{u}\gamma_\mu(1-\gamma_5)c\ \bar{u}\gamma^\mu(1-\gamma_5)c$ and $\mathcal{O}' \equiv \bar{u}(1+\gamma_5)c\ \bar{u}(1+\gamma_5)c$. The presence of the additional operator $\mathcal{O}'$ is due to the non negligible external momentum. The matrix elements of the operators can be parametrized by

$$\langle D^0|\mathcal{O}|\bar{D}^0\rangle = \frac{8}{3}m_D f_D^2 B_D \quad ; \quad \langle D^0|\mathcal{O}'|\bar{D}^0\rangle = -\frac{5}{3}\left(\frac{m_D}{m_c}\right)^2 m_D f_D^2 B_D' \tag{4}$$

In the vacuum insertion approximation one has $B_D = B_D' = 1$ and the short distance contribution to the mass difference is

$$\Delta m_D^{\text{SD}} \simeq 2.5 \times 10^{-17} \text{ GeV} \left(\frac{m_s}{0.3\text{GeV}}\right)^4 \left(\frac{f_D}{f_\pi}\right)^2 \tag{5}$$

Thus for typical values of f_D and m_s the short distance contributes to r_D with a value not above 10^{-8}, perhaps as small as 10^{-10}.

Long Distance

The contributions of the short distance box diagrams are not the only ones. The fact that light quarks with rather large CKM couplings to the charm quark can propagate between the D^0 and $\bar{D}^0$, hints the possibility of relatively important long distance contributions to mixing. The propagating degrees of freedom are hadrons rather than quarks. The situation is very different in K and B mixing, where there is always a very important effect of a heavy quark inside the box diagram loop, with large CKM couplings: the charm quark in $K^0 - \bar{K}^0$ and the top quark in $B^0 - \bar{B}^0$. In the latter, the effect of the top quark completely dominates and long distance contributions are expected to be negligibly small due to the small CKM couplings of the B meson to light hadrons. In the case of K mixing, the coupling to light-hadron intermediate states is still large as a consequence of which sizeable long distance contributions -of the same order of magnitude of the short distance ones- are expected [3]. The long distance contributions to $D^0 - \bar{D}^0$ mixing are inherently nonperturbative and cannot be calculated from first principles. It is however of paramount importance to estimate their size in order to understand to origin of a possible observation of the effect in future experiments.

A first observation is that the mass difference is a $SU(3)$ breaking effect. On the other hand, Δm_D is doubly Cabibbo suppressed whereas Γ is not. Therefore, a naive estimate of the effect would be given by

$$\frac{\Delta m_D}{\Gamma} \sim \lambda^2 \times (SU(3) \text{ breaking}) \simeq (10^{-3} - 10^{-2}) \qquad (6)$$

where $\lambda \simeq \sin\theta_c$. Specific calculations tend to give smaller results. There are basically two ways to attempt estimating the long distance effects: a dispersive approach and Heavy Quark Effective Theory (HQET).

Dispersive Approach

An estimate of the long distance contributions can be obtained by assuming they come from the propagation of hadronic states to which both D^0 and $\bar{D}^0$ can decay. There will be one, two, three, etc. particle intermediate states. Each of these groups can be further separated into sets whose contributions vanish separately in the $SU(3)$ limit. One of these sets is formed by the two-charged-pseudoscalar intermediate states $\pi^+\pi^-$, K^+K^-, $K^-\pi^+$ and $K^+\pi^-$. Thus computing their contribution to the mass difference, as shown schematically in Fig. 1, gives a concrete realization of the estimate in (6) for an $SU(3)$ set for which data is available. This was first done in [4].
Although these "self energy" diagrams will depend on the interaction chosen for the vertices, they have a universal imaginary part which typically comes

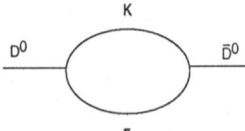

Figure 1: Long distance contribution from two charged pseudo-scalar intermediate states.

from a logarithm. This leads to the expression [4]

$$\frac{\Delta m_D^{l.d.}}{\Gamma} \simeq \frac{1}{2\pi} \ln \frac{m_D^2}{\mu^2} \left[B(\pi^+\pi^-) + B(K^+K^-) - 2\sqrt{B(K^-\pi^+)\,B(K^+\pi^-)} \right] \quad (7)$$

where all the branching ratios correspond to D^0 decays. The scale μ is a typical hadronic scale, $\mathcal{O}(1\ \text{GeV})$. Inserting the current available experimental values [5] in (7) we obtain

$$\frac{\Delta m_D}{\Gamma} \simeq 8.4 \times 10^{-4} \times \left(1.46 - \sqrt{b}\right) \quad (8)$$

where we have defined b by

$$\frac{B(K^+\pi^-)}{B(K^-\pi^+)} = b\,\tan^4\theta_c \quad (9)$$

The experimental measurement of the Doubly Cabibbo Suppressed Decays (DCSD) gives $b = 2.8 \pm 1.3$. Then, for any value of b in a 1σ interval, it is clear that the cancellation in this $SU(3)$ set is even better than expected in (6), giving $|\Delta m_D^{l.d.}/\Gamma| \simeq \mathcal{O}(10^{-4})$.

Several assumptions are implied in (7). First, the minus sign is obtained assuming there is no strong relative phase between the two $K\pi$ amplitudes. This is only true in the $SU(3)$ limit [6]. However, the effect of this phase is expected to be small [7]. More importantly, equation (7) is valid for massless particles in the loop and it assumes a constant coupling at the vertices which allows one to relate the product of the two $\Delta C = 1$ interactions in Fig. 1 to the actual decay amplitudes. The approximation regarding the internal masses is rather safe in the two-pseudoscalar (PP) case. However masses should be kept in the calculation when sets including vector mesons (PV and VV) are considered. On the other hand, large momenta in the loop should not contribute given that the coupling is expected to develop a momentum suppression at a typical hadronic scale. This suggests the existence of a physical cutoff for the integrals involved. It is instructive to see how this cutoff and the effect of the internal masses come about. The contribution of Fig. 1 to the mass difference obeys a dispersion relation of the form

$$\Sigma(p^2) = \frac{1}{\pi} \int_{s_0}^{\infty} \frac{Im[\Sigma(s)] \, ds}{(s - p^2 - i\epsilon)} \tag{10}$$

where $s_0 \equiv (m_1 + m_2)^2$, and m_1 and m_2 are the masses in the loop. Taking into account a subtraction forcing the condition $\Sigma(0) = 0$ and keeping the masses, the implementation of a cutoff Λ in the dispersive integral (10) gives [9]

$$\frac{\Delta m_D^{l.d.}}{\Gamma} \simeq \frac{m_D}{4\pi} \left\{ \frac{B(\pi^+\pi^-)}{p_{\pi\pi}} I(m_\pi, m_\pi, \Lambda) + \frac{B(K^+K^-)}{p_{KK}} I(m_K, m_K, \Lambda) \right.$$
$$\left. -2 \frac{\sqrt{B(K^-\pi^+) B(K^+\pi^-)}}{p_{K\pi}} I(m_\pi, m_K, \Lambda) \right\} \tag{11}$$

where p_{ij} is the magnitude of the three-momentum in the actual decay and

$$I(m_1, m_2, \Lambda) = -\int_{s_0}^{\Lambda^2} \frac{\sqrt{1 - \frac{s_0}{s}} \, ds}{s - m_D^2} \tag{12}$$

If the massless limit is taken in (11) one recovers (7) with the identification $\mu^2 = 2 m_D (\Lambda - m_D)$. Although the result depends strongly on the cutoff Λ, this can be interpreted as the value of s for which the internal momentum reaches its maximum. Not surprisingly, the value of Λ giving an internal momentum of ~ 1 GeV is the same giving $\mu \sim 1$ GeV. This is $\Lambda \simeq (2 - 2.2)$ GeV, not too far above m_D. Using this cutoff results in a contribution to the mass difference of

$$\frac{\Delta m_D^{l.d.}}{\Gamma} \simeq 6.5 \times 10^{-4} \times (1 - \sqrt{b})) \tag{13}$$

As mentioned earlier, this result does not differ drastically from what is obtained by using (7).

On the other hand, it is clear that the use of (7) to estimate the Pseudoscalar-Vector (PV) and Vector-Vector (VV) contributions is dangerous and it could result in an overestimate of these contributions [8]. We can repeat the same procedure carried out with the PP modes. However data is even scarcer in these cases. Measurements for all the PV and VV decays rates as well as a more precise determination of $\Gamma(D^0 \to K^+\pi^-)$ is needed in order to complete the picture of Δm_D in this approach. However, and as it was already pointed out in [4], it is very likely that the charged pseudoscalar contribution gives a good approximation to the order of magnitude of the effect. This is because, although there are many contributions, their relative signs are not fixed and some degree of cancellation is expected. A hint of these cancellations is already present in (11), which includes the interaction between the scale of softening of the effective vertex and the masses. It can be shown that in (11) the contributions of small and large internal masses will tend to have have different signs. This is merely an argument that makes cancellations plausible, but by no means a rigorous proof. It could be argued that several contributions might conspire to give a total long distance contribution to Δm_D orders of magnitude larger than (13). This scenario cannot be completely excluded in this approach until more data on PV and VV modes is available.

Heavy Quark Effective Theory

Yet another possible theoretical approach to D^0-$\bar{D}^0$ mixing is the application of the heavy quark effective theory (HQET). It was first noted in [10] that if one considers the charm quark mass to be much larger than the typical scale of the strong interactions, there would be no nonleptonic transitions to leading order in the resulting effective theory. They would require large momenta to be exchanged between the heavy quark and the light degrees of freedom, a subleading effect in inverse powers of the charm mass. As a consequence there are no new available operators in the low energy theory to produce $\Delta C = 2$ transitions. At scales below m_c these occur only due to operators present at the matching scale m_c plus the action of the renormalization group, which in this picture constitute the only "long distance" effects. Therefore, to leading order in HQET, Δm_D can be computed from quark operators. The nonperturbative physics enters in the matrix elements of these operators, and in [10, 11] are estimated using naive dimensional analysis. There are three groups of operators in the HQET. The first corresponds to four-quark operators, which are the HQET version of the box diagram. The second group, the six-quark operators, gives a modest enhancement over the first one [10, 11]. Finally, the eight-quark operators give a large enhancement in the matrix elements over the four-quark operators (a factor of ≈ 20), but they are suppressed by an overall factor of $\alpha_s/4\pi$. This seems to suggest that the large enhancement in Δm_D over the short distance box diagrams coming from individual contributions and accounted for in the dispersive approach above, is cancelled when all the contributions are summed over in order to make up for the $\alpha_s/4\pi$ suppression [10]. This is the HQET version of the cancellations among different $SU(3)$ sets (e.g. PP with VV, etc.). The size of the effect is estimated in [11], where QCD corrections and running are properly accounted for. The result is

$$\frac{\Delta m_D}{\Gamma} \approx (1-2) \times 10^{-5} \qquad (14)$$

where the uncertainty comes from the unknown relative signs of the various operators. Thus HQET predicts a value of Δm_D in the SM that is roughly an order of magnitude smaller than the dispersive estimates (8) and (13).

Of course the validity of the central HQET assumption, $m_c \gg \Lambda_{QCD}$, can be questioned. After all here Λ_{QCD} is actually a typical hadronic scale, not far below 1 GeV. It is not clear what is the size of the corrections. However, the most interesting conclusion is the suggestion that there is a cancellation among the sets of $SU(3)$-related intermediate states. On the other hand, the HQET and the dispersive results are consistent within the experimental errors in the determination of b in DCSD modes.

NEW PHYSICS AND $D^0 - \bar{D}^0$ MIXING

Proposed high statistics charm experiments are likely to probe $D^0 - \bar{D}^0$ mixing down to $r_D \sim 10^{-5}$ [12]. If the long distance contributions in the SM are below this sensitivity, then the question is: are there extensions of the SM that can fill this window and be compatible with all other low energy phenomenology? In this section we review a few examples of new physics scenarios that could produce a signal in these experiments.

1. Two-Higgs Doublet Models

As a first example of an extension of the SM we take a Two-Higgs doublet model with natural flavor conservation. That is, there are no tree level FCNC [13]. There will be two neutral scalars, a neutral pseudoscalar and a pair of charged scalars. In what is called Model II in the literature, the latter couples to the fermions as

$$\mathcal{L} = \frac{g}{\sqrt{2}m_W} H^+ \left\{ \cot\beta \, \bar{\mathcal{U}}_R \, M_u \, V_{ckm} \, \mathcal{D}_L + \tan\beta \, \bar{\mathcal{U}}_L \, V_{ckm} \, M_d \, \mathcal{D}_R + \text{h.c.} \right\} \quad (15)$$

where $\mathcal{U} \equiv (u, c, t)$, $\mathcal{D} \equiv (d, s, b)$, M_u and M_d are the diagonal quark mass matrices and $\tan\beta = v_2/v_1$; with v_1, v_2 the vacuum expectation values of the doublets. Incidentally, this type of couplings to fermions is the same as in the Higgs sector of the Minimal Supersymmetric Standard Model (MSSM), which is addressed separately below. The couplings in (15) induce an additional set of box diagrams where the $W^{\pm}$ is replaced by the charged Higgs. For large values of $\tan\beta$ the b quark is the dominant contribution, giving

$$\Delta m_D^{\text{2HDM}} \simeq \frac{G_F^2}{6\pi^2} \, m_D \, B_D \, f_D^2 \, \eta_{\text{QCD}} \, |V_{cb} V_{ub}^*|^2 \, m_b^2 \, \tan^2\beta \, F\left(\frac{m_b^2}{m_H^2}\right) \quad (16)$$

where η_{QCD} is a QCD correction and $F(x)$ is a known function of the mass ratios, resulting from the loop integrals. As it is obvious from (16), the effect can be important for large $\tan\beta$. This shows once again how charm meson physics can be complementary with B physics. The radiative decay $b \to s\gamma$ largely constrains the low $\tan\beta$ region given that, as it can be seen in (15), the top quark mass term amplifies that region of parameter space. As seen in Fig. 2, in the large $\tan\beta$ limit, charged Higgs masses below $m_H = 250$ GeV will give a contribution to which future experiments will be sensitive.

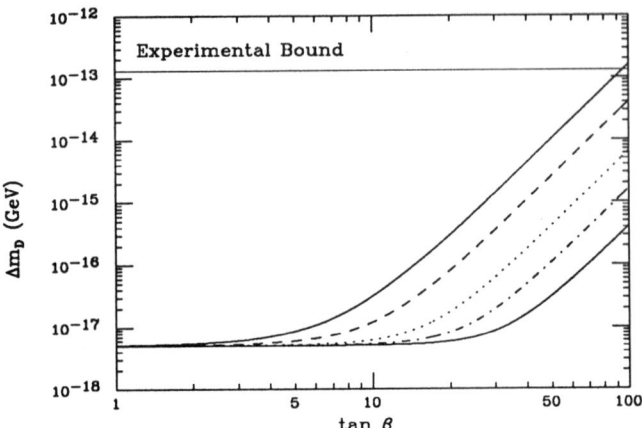

Figure 2: Δm_D in the two-Higgs doublet model II as a function of $\tan\beta$ and for several charged Higgs masses. From top to bottom $m_{H^\pm} = 50, 100, 250, 500, 1000$ GeV.

2. Heavy Down Quark

The reason why the SM box diagram contributions are so small is a very efficient GIM mechanism: the heaviest quark has very small couplings with c and u and is not so heavy anyway. The obvious question is then: what would happen if there was a fourth down quark in the loop with $Q = -1/3$ and a large mass. This could belong to a fourth generation or be just a singlet. Its contribution to the mass difference is

$$\Delta m_D^{b'} \simeq \frac{G_F^2 m_W^2}{6\pi^2}\, m_D\, B_D\, f_D^2\, \eta_{\mathrm{QCD}}\, |V_{cb'} V_{ub'}^*|^2\, F\left(\frac{m_{b'}^2}{m_W^2}\right) \qquad (17)$$

From direct searches it is known that $m_{b'} > 85$ GeV [5]. In the case of b' belonging to a fourth generation, the couplings to the second and first generations are constrained by the possible "leakage" of the CKM matrix from unitarity. The mixing factors must satisfy

$$|V_{ub'}| < 0.08\ ,\quad |V_{cb'}| < 0.6 \qquad (18)$$

although smaller mixing factors are expected. The effect of the b' quark is shown in Fig. 3. If the mixing factors are not too small a heavy b' could give a large effect, saturating the current experimental limit.

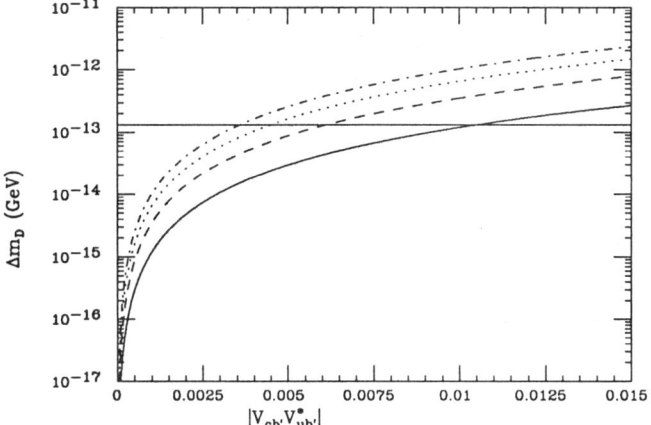

Figure 3: The contributions to Δm_D of a heavy $Q = -1/3$ quark, as a function of the mixing factor in a fourth-generation model. From top to bottom $m_{b'} = 400, 300, 200, 100$ GeV. The horizontal line shows the current experimental limit.

3. Tree Level FCNC

Tree-level FCNC are severely constrained by $K^0 - \bar{K}^0$ and $B^0 - \bar{B}^0$ mixings and decays like $K^+ \to \pi^+ \nu \bar{\nu}$ and $B \to X \ell^+ \ell^-$ [5]. However, it is possible to imagine scenarios where the up-quark sector is treated differently. Some of these new physics scenarios predict large $D^0 - \bar{D}^0$ mixing effects.

A first example is to relax the requirement of flavor conservation in multi-Higgs models. There will be tree-level flavor changing couplings of the neutral Higgses [14]. The form of the flavor conserving couplings to fermions suggests the parametrization of the flavor changing couplings given by

$$C_{ij} \sim \frac{\sqrt{m_i m_j}}{v_w} \delta_{ij} \qquad (19)$$

where i, j are the quark labels (e.g. u, c, t); and $v_w \simeq 246$ GeV. The $c - u$ coupling would then induce a contribution to the mass difference of the form

$$\Delta m_D^{h^0} = \frac{5}{12} f_D^2 \, B_D \, m_D \, \frac{m_c m_u}{v_w^2} \frac{\delta_{cu}^2}{m_h^2} \qquad (20)$$

where m_u is the current quark mass. The experimental limit is saturated for reasonable values of $\delta_{cu} \sim (0.2 - 1)$ and $m_h \sim (50 - 250)$ GeV.

Another interesting example is that of theories of dynamical symmetry breaking with large GIM violations, such as TopColor models [15]. Here the top quark gets a large mass after the breaking down to QCD of a group strongly coupled to the third generation. This generates a set of pseudo-Goldstone bosons (top-pions) that couple to the up-quark sector. After the quark weak eigenstates are rotated to the mass eigenstates, there will be flavor changing neutral couplings mediated by the top-pions of the form

$$C_{ij} \sim \frac{m_t}{\sqrt{2} f_{\tilde{\pi}}} U_L^{ti} U_R^{tj} \qquad (21)$$

where $f_{\tilde{\pi}} \simeq 50$ GeV is the top-pion decay constant and U_L and U_R are the mass matrices of the left and right up-quark sector respectively. These couplings lead to a mass difference given by

$$\Delta m_D^{t.c.} \simeq \frac{5}{12} f_D^2 B_D m_D \frac{m_t^2}{2 f_{\tilde{\pi}}^2} \frac{U_L^{tc} U_R^{tu} U_L^{tu} U_R^{tc}}{m_{\tilde{\pi}}^2} \qquad (22)$$

The mass matrices are not determined in general by the model but must obey, as in the SM, $V_{\text{ckm}} = U_L^\dagger D_R$. This suggests that a possible anzatz for $U_{L,R}$ is to take the "squared root" of V_{ckm}. This prescription gives, for instance, $U_L^{tc} \simeq (1/2) V_{cb}$. Following this in (22) and taking $m_{\tilde{\pi}} \simeq 200$ GeV [15], gives $\Delta m_D^{t.c.} \simeq 8 \times 10^{-14}$ GeV, right below the current experimental limit. In principle, other textures can be chosen that would not give such a large effect.

4. Supersymmetry

In addition to the box diagrams involving charged Higgses, there will be contributions from squarks + (gluinos, charginos or neutralinos) box diagrams. These vanish if the down squarks are degenerate. This is just a statement derived from GIM cancellations. However, there could be flavor changing, radiatively generated mass insertions [16]. They are thought to be small in the MSSM, but could be large in non minimal models [17]. For instance, the action of these mass insertions would allow for squark-gluino box diagrams. The resulting $\Delta C = 2$ hamiltonian is

$$\mathcal{H}_{\Delta C=2}^{\text{SUSY}} = \frac{\alpha_s^2}{216 \tilde{m}_0} \left\{ \frac{\delta \tilde{m}_{u_L c_L}^2}{\tilde{m}_0^2} G\left(\frac{m_{\tilde{g}}^2}{\tilde{m}_0^2}\right) (\bar{u}_L \gamma_\mu c_L)(\bar{u}_L \gamma^\mu c_L) \right.$$
$$\left. + (RR)^2 + (LL)(RR) + (LR)^2 + (LR)(LR) \right\} \qquad (23)$$

where $\tilde{m}_0$ is the universal scalar mass, $m_{\tilde{g}}$ is the gluino mass, $\delta \tilde{m}_{u_L c_L}$ characterizes the mass insertion and the $G(x)$'s are known functions. This way, the experimental upper limit on Δm_D can be translated into limits for the various terms entering in (23). These are

$$\frac{\delta \tilde{m}_{u_{ACB}}^2}{\tilde{m}_0^2} < \begin{cases} (0.2)^2 & \text{, for } (LL)^2, (RR)^2 \\ (3.6 \times 10^{-2})^2 & \text{, for } (LL)(RR) \\ (5 \times 10^{-2})^2 & \text{, for } (LR)^2, (RL)^2 \\ (0.1)^2 & \text{, for } (LR)(LR) \end{cases} \qquad (24)$$

Thus, the current experimental limit is already sensitive to non minimal SUSY effects.

There are several other extensions of the SM that would saturate the current experimental limit or at least give a signal in high statistics charm experiments. The above list is by no means exhaustive and is mainly intended to illustrate how the effect comes about in a variety of theories.

CP VIOLATION

The D system is not particularly sensitive to CP violation in the SM to the extent the K and B mesons are. Once again, this could imply there is a window of observation of new physics effects. In what follows we discuss some of the general features of CP violation in D mesons rather than going into specific calculations, both in the SM and beyond.

Direct CP Violation

The occurrence of direct CP violation requires the concurrence of both weak and strong relative phases between two or more amplitudes contributing to a given final state. In the SM, relative weak phases can only be obtained in Cabibbo suppressed decays, for instance, via the interference between spectator and penguin amplitudes. To estimate the size of the CP asymmetries this would generate, we write

$$a_{CP} \sim \frac{Im[V_{cd} V_{ud}^* V_{cs} V_{us}^*]}{\lambda^2} \sin\delta_{st} \frac{P}{S} \qquad (25)$$
$$\sim A^2 \eta \lambda^4 \sin\delta_{st} \frac{P}{S} \leq 10^{-3}$$

where δ_{st} is the strong relative phase between the penguin and the spectator amplitudes, and $A \sim 1$ and η are CKM parameters in Wolfenstein's parametrization. Specific model calculations for $D \to KK, \pi\pi, K^*K$, three-body modes, etc. yield this order of magnitude for the effect. New physics could enter, for instance, through large phases in the penguin diagram. This could give very large asymmetries of the order of one percent or larger. On the other hand, an even cleaner window are the Cabibbo allowed decays. These modes do not have two amplitudes with different weak phases and therefore the CP asymmetry is zero in the SM. There are new physics scenarios that provide extra phases and could give asymmetries as large as one percent. This is for instance the case in some left-right symmetric models [18]. The current experimental sensitivities for various modes is in the vicinity if 10% [19].

Indirect CP Violation

The interaction between mixing and CP violation in D mesons has recently received a lot of attention in the literature [20, 6, 7]. Here we shall focus only on one aspect, which can be condensed in the following question: if mixing is large (e.g. right below the current experimental limit) should CP violation in D decays be large ? The question is motivated by the fact that in B decays the large $B^0 - \bar{B}^0$ mixing is known to give large CP asymmetries. We first define the time-evolved states in the usual way, as

$$|D^0(t)\rangle = g_+(t)|D^0\rangle + \frac{q}{p}g_-(t)|\bar{D}^0\rangle$$

$$|\bar{D}^0(t)\rangle = \frac{p}{q}g_-(t)|D^0\rangle + g_+(t)|\bar{D}^0\rangle$$

with

$$\frac{p}{q} \equiv \sqrt{\frac{M_{12} - i\Gamma_{12}}{M_{12}^* - i\Gamma_{12}^*}} \qquad (26)$$

and the time evolution given by

$$g_\pm(t) = \frac{1}{2}e^{(-\frac{\Gamma_L}{2}t + im_L t)}\left[1 \pm e^{(-\frac{\Delta\Gamma_D t}{2} + i\Delta m_D t)}\right] \qquad (27)$$

We also need to define the amplitudes

$$A \equiv \langle f|H_w|D^0\rangle \quad ; \quad B \equiv \langle f|H_w|\bar{D}^0\rangle$$
$$\bar{A} \equiv \langle \bar{f}|H_w|\bar{D}^0\rangle \quad ; \quad \bar{B} \equiv \langle \bar{f}|H_w|D^0\rangle$$

and the ratio

$$\bar{\rho} \equiv \frac{A}{B} \qquad (28)$$

To simplify notation we consider the case when $f = CP$ eigenstate (e.g. $\pi\pi$, KK, etc). If mixing is a large effect, let us say right below the current upper limit, and therefore due to new physics contributions then it is very likely that $\Delta m_D/\Gamma \gg \Delta\Gamma_D/\Gamma$ is a very good approximation: non standard contributions to $\Delta\Gamma_D$ are constrained by actual branching ratios. Under these assumptions the asymmetry takes the form

$$a_{\rm CP} \simeq \frac{1 - \left|\frac{q}{p}\bar{\rho}\right|^2 - 2\frac{\Delta m_D}{\Gamma}Im\left[\frac{q}{p}\bar{\rho}\right]}{\left(1 + \left|\frac{q}{p}\bar{\rho}\right|^2\right)\left(1 + (\frac{\Delta m_D}{\Gamma})^2\right)} \qquad (29)$$

Small direct CP violation implies $|(q/p)\bar{\rho}|^2 \simeq 1$. This leads to

$$a_{\rm CP} \simeq \frac{\Delta m_D}{\Gamma} \times Im\left[\frac{q}{p}\bar{\rho}\right] \simeq \frac{\Delta m_D}{\Gamma} 2\eta A \qquad (30)$$

where the last step follows in the SM and from considering the contribution of the b quark to the imaginary part of the box diagram. The resulting asymmetry can be then of the order of $\Delta m_D/\Gamma$ even if only the SM phases intervene. However in models giving large mass differences it is also likely that there will be additional CP violating phases. These could be present in the *non* SM contributions to the mass difference and they mostly affect q/p and not $\bar{\rho}$. Several non standard scenarios for generating this additional phases are discussed in [20].

RARE AND RADIATIVE DECAYS

Let us first address the distinction made between rare and radiative decays. Radiative weak decays of charm mesons do no effectively test the SM. To see this let us take the transitions governed by the short distance flavor changing vertex $c \to u\gamma$. In the SM they occur only at one loop through the electromagnetic penguin, analogous to $b \to s\gamma$. However in this case, the inclusive branching ratio, even after very large QCD corrections, is very small: $B(c \to u\gamma) \simeq 10^{-12}$ [21]. This, however, does not constitute a window for new physics. There are more mundane contributions to the corresponding exclusive processes, like $D^0 \to \rho^0\gamma$, that do not arise from short distance physics. These "long" distance contributions can be thought of as coming from either pole diagrams (arising from quark exchange) and vector meson dominance diagrams, all of which are not calculable from first principles but can be estimated in models to give $B(D^0 \to \rho^0\gamma) \simeq (1-5) \times 10^{-6}$ [21]. This large rates preclude the use of these modes as SM tests. On the other hand, a better theoretical understanding of these modes is interesting in its own right as well as in order to understand possible long distance contamination in radiative B decays [22]. The availability of several decay modes -$D^0 \to \rho^0\gamma$, $D^0 \to \bar{K}^*\gamma$, $D_s \to \rho^+\gamma$, etc.- at branching fractions of $\mathcal{O}(10^{-6})$ or larger will improve our knowledge of strong dynamics at the charm scale [23].

Truly rare decays are those whose SM rates are extremely small or simply zero. Most of them proceed through FCNC induced at one loop in the SM. The simplest example is $D^0 \to \ell^+\ell^-$, with $\ell = e$ or μ. Their branching ratios are smaller than 10^{-15}, even after long distance contributions are taken into account. There are experimentally clean and any signal in any of these channels would imply new physics. However the helicity suppression is a factor of $\sim 10^{-3}$ in the case of μ, and 10^{-7} for the e. Modes without this suppression, like $D \to X\ell^+\ell^-$ and $D \to X\nu\bar{\nu}$, are more likely to show the first signals if, for instance, there is a new physics mechanism underlying the short distance transition $c \to u$. In the SM their branching ratios are expected to be of $\mathcal{O}(10^{-8})$ or smaller in the charged lepton cases and negligibly small in the neutrino modes. New upper limits in the $D^+ \to \pi^+\ell^+\ell^-$ channels have been recently reported by E791 [24]: $B_{\mu\mu} < 1.8 \times 10^{-5}$ and $B_{ee} < 6.6 \times 10^{-5}$. These are already constraining extensions of the SM in a way complementary with mixing. Predictions for new physics scenarios can be found in several places in the literature [25], but it is necessary to update them and most importantly to take into account other pieces of phenomenology now available.

CONCLUSIONS

We have seen that charm meson physics offers several opportunities to observe the effects of new physics. This is mainly due to the suppression of the signals in the SM. In the case of $D^0 - \bar{D}^0$ mixing, the estimate of the long distance SM contribution is very uncertain. However, the question of interest at the moment is whether

$$r_D^{SM} < 10^{-5} \tag{31}$$

is a correct upper limit. This is relevant because is the planned sensitivity of future high statistics charm experiments [12]. If r_D^{SM} is 10^{-10} or 10^{-7} is an issue theorists should worry about, but it will not affect the interpretation of the outcome of these experiments. The validity of (32) is not a settled question among theorists. We have seen two approaches to r_D^{SM} satisfying (32): a dispersive approach and the HQET approach. However, these are approximate calculations and there are those who point out that (6) with a unit coefficient is another way of estimating the effect [26]. The high end of this estimate violates (32). The theoretical community should make an effort to resolve this outstanding problem. More data in nonleptonic D decays is needed in order to see if there is a pattern of cancellations as suggested by the HQET. In the meantime, experiments might provide with an independent way of deciding on the origin of an observation of r_D: the direct measurement of the lifetime difference $\Delta\Gamma_D$. This can be done, assuming CP conservation, by looking at the decay time distribution of D decays to CP even and odd final states [27]. The difference of the slopes if proportional to $\Delta\Gamma_D$. This quantity is not prone to get contributions from new physics but rather to be entirely given by the SM: after all it is a sum over real intermediate states. Moreover, it should be of the same order of magnitude as the long distance contributions to Δm_D. Thus, not only would this allow the separation of Δm_D and $\Delta\Gamma_D$ but also, for instance, point at new physics if Δm_D is observed and $\Delta\Gamma_D$ is not seen at the corresponding level. Considering how many extensions of the SM saturate the experimental limit, this program makes $D^0 - \bar{D}^0$ mixing an important window for new physics in the future.

With respect to CP violation, we have seen that asymmetries at the level of one percent would signal new physics. We also pointed out that CP violation due to mixing will be enhanced by new physics only if both Δm_D and the entering phases, both from the SM and/or new physics, are large.

For rare decays a lot more work is needed in order to establish what level of branching ratios are allowed in each new physics scenario once all the constraints from low energy phenomenology are factored in.

AKNOWLEDGEMENTS

This presentation reflects work in collaboration with Eugene Golowich, JoAnne Hewett and Sandip Pakvasa. I would also like to thank Jose Repond and the rest of the organizers of the workshop for their excellent effort. This work was supported by the U.S. Department of Energy.

References

[1] J. C. Anjos et al., Phys. Rev. Lett. **60**, 1239 (1988).

[2] A. Datta and D. Kumbhakar, Z. Phys. **C27**, 515 (1985).

[3] J. F. Donoghue, E. Golowich and B. R. Holstein, Phys. Lett. **B135**, 481 (1984).

[4] J. F. Donoghue, E. Golowich, B. R. Holstein and J. Trampetić, Phys. Rev. **D33**, 179 (1986).

[5] L. Montanet et al., Phys. Rev. **D50**, 1173 (1994).

[6] L. Wolfenstein, Carnegie Mellon Univ. preprint CMU-HEP-95-04 (1995), hep-ph/9505285.

[7] T. E. Browder and S. Pakvasa, Univ. of Hawaii preprint UH 511-828-95 (1995).

[8] T. A. Kaeding, LBL preprint LBL-37224 (1995).

[9] G. Burdman, E. Golowich, J. Hewett and S. Pakvasa, in preparetion.

[10] H. Georgi, Phys. Lett. **B297**, 353 (1992).

[11] T. Ohl, G. Ricciardi and E. H. Simmons, Nucl. Phys. **B403**, 605 (1993).

[12] D. M. Kaplan, in these Proceedings. See also R. J. Morrison, in Proceedings of the Charm 2000 Workshop, Ed. D. M. Kaplan and S. Kwan, FERMILAB-Conf-94/190, (1994).

[13] S. L. Glashow and S. Weinberg, Phys. Rev. **D15**, 1958 (1977).

[14] S. Pakvasa and H. Sugawara, Phys. Lett **B73**, 61 (1978);
L. Hall and S. Weinberg, Phys. Rev. **D48**, 979 (1993).

[15] C. T. Hill, Phys. Lett. **B345**, 483 (1995).

[16] J. S. Hagelin, S. Kelley, T. Tanaka, Nucl. Phys. **B415**, 293 (1994); Mod. Phys. Lett. **A8**,2737 (1993).

[17] Y. Nir and N. Seiberg, Phys. Lett. **B309**, 337 (1993).

[18] M. Gronau and S. Wakaizumi, Phys. Rev. Lett. **68**, 1814 (1992);
A. Le Yaouanc, L. Oliver and J.-C. Raynal, Phys. Lett. **B292**, 353 (1992).

[19] P. L Frabetti *et al.*, Phys. Rev. **D50**, 2953 (1994);
J. Bartelt *et al.*, CLEO preprint CLNS 95/1333 (1995).

[20] G. Blaylock, A. Seiden and Y. Nir, Univ. of Sta. Cruz preprint SCIPP 95/16 (1995).

[21] G. Burdman, E. Golowich, J. Hewett and S. Pakvasa, FERMILAB-Pub-94/412-T (1995).

[22] E. Golowich and S. Pakvasa, Phys. Rev. **D51**,1215 (1995).

[23] I. I. Bigi, these Proceedings.

[24] E. M. Aitala *et al.*, Fermilab preprint FERMILAB-Pub-95/142-E (1995).

[25] For the effect of lepto-quarks see, W. Buchmuller and D. Wyler, Phys. Lett. **B177**, 377 (1986) and more recently M. Leurer, Phys. Rev. Lett. **71**, 1324 (1993).

[26] L. Wolfenstein, Phys. Lett. **B164**, 170 (1985).

[27] T. Liu, these proceedings and in Proceedings of the Charm 2000 Workshop, Ed. D. M. Kaplan and S. Kwan, FERMILAB-Conf-94/190, (1994).

High-Impact Charm Physics at the Turn of the Millennium

Daniel M. Kaplan
Illinois Institute of Technology, Chicago, IL 60616

Abstract

I review the sensitivities achieved by and projected for fixed-target charm experiments in CP violation, flavor-changing neutral-current and lepton-number-violating decays, and mixing, and I describe the Charm2000 experiment intended to run at Fermilab in the Year $\approx$2000. If approved, Charm2000 will in many of these areas exceed the sensitivities projected for a Tau/Charm Factory, but the Tau/Charm Factory retains certain qualitative advantages.

1 Introduction

A Tau/Charm Factory (τcF) may turn on early in the next millennium. At that time one can anticipate significant competition in charm physics from fixed-target experiments, as well as from e^+e^- colliders operating near $b\bar{b}$ threshold [1]. For many topics in charm physics (*e.g.* lifetimes and rare-decay searches), Fermilab fixed-target experiments now dominate the field. The progress of fixed-target charm experiments at Fermilab is sketched in Fig. 1, which shows roughly exponential growth in sensitivity since the late 1970s. While physics reach depends both on the number of signal events reconstructed and on the amount of background under the peaks, the former figure can still serve as a starting point for discussion. This number is expected to reach $\sim 10^6$ events during the next few years with the runs of Fermilab E781 (SELEX) and E831 (FOCUS) and the advent of CLEO III. In addition, a Letter of Intent has been submitted to CERN for an experiment (CHEOPS) aiming to reconstruct 10^7 charm [2], and one for a 10^8-charm experiment (Charm2000) at Fermilab in the Year $\approx$ 2000 is in progress [3, 4]. It is against this backdrop that the case for a Tau/Charm Factory must be evaluated.

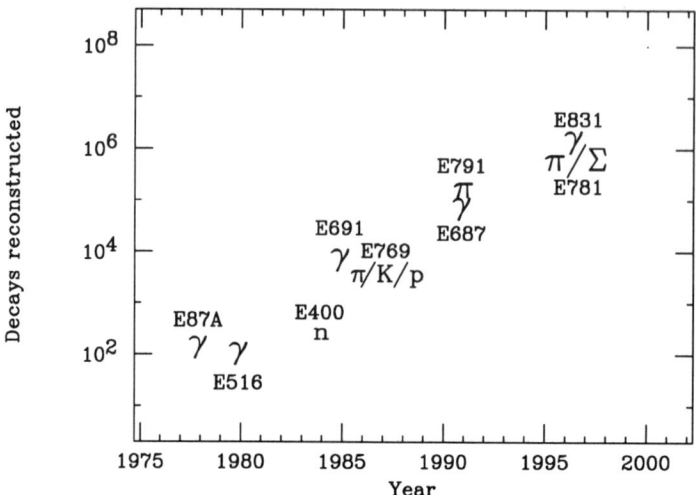

Figure 1: Yield of reconstructed charm *vs.* year of run for those completed and approved Fermilab fixed-target charm experiments with the highest statistics of their generation; symbols indicate type of beam employed.

2 High-Impact Charm Physics

"High-impact" denotes measurements which are particularly sensitive to new, non-Standard-Model physics [5]. The Standard Model (SM) contains two key mysteries: the origin of mass and the existence of multiple fermion generations [6, 7]. We seek to answer the first in experiments at the LHC, exploring the ≈ 1 TeV mass scale of electroweak symmetry breaking. The answer to the second appears to lie at higher mass scales, beyond what can be directly accessed at the LHC. But these scales can be probed in virtual loops in processes such as *CP* violation, mixing, and flavor-changing neutral or lepton-number-violating currents [7-9].

Such effects have been pursued with high sensitivity in the strange sector, and in the beauty sector they have become something of a holy grail, because of large SM contributions to mixing and *CP* violation in the decays of "down-type" quarks. These effects are enhanced for s and b quarks relative to those for "up-type" quarks by the pattern of the Cabibbo-Kobayashi-Maskawa (CKM) matrix and the large mass of the top quark, whose contribution in loops allows *CP* violation by virtue of the CKM phase [10]. It is precisely because these SM contributions are *small* in the charm sector that charm is a good place to look for new-physics contributions [8, 11-13]. Furthermore, charm is the only up-type quark for which these studies are possible, since the top quark is above $W+b$ threshold and decays too quickly to form bound states. The information available from charm studies is often complementary to that from strangeness and beauty [7, 14]. Finally, as we shall see, sensitivity to new physics at interesting levels is anticipated in upcoming charm experiments: levels at which even the failure to observe an effect imposes significant constraints on models.

Table 1 summarizes sensitivities in high-impact charm physics currently achieved and expected by the turn-on of the τcF, assuming approval of the Charm2000 project at Fermilab. Table 2 estimates yields of reconstructed events in various modes in Charm2000, some directly and some by extrapolation from E791; since these yields vary rapidly with vertex separation cuts, which are typically optimized differently for each physics analysis, they are necessarily ill-defined at the factor-of-2 level.[1] (To remind the reader of this effect, I have indicated in Table 2 the type of analysis for each E791 yield given.) I next discuss each physics topic in more detail,[2] following which I describe the salient aspects of the proposed Charm2000 experiment.

2.1 Direct CP violation

The Standard Model predicts direct CP violation in singly Cabibbo-suppressed decays (SCSD) of charm at the $\sim 10^{-3}$ level [5, 15 - 17], arising from interference between tree-level and penguin diagrams for the decay of the charm quark. CP violation in Cabibbo-favored (CFD) or doubly Cabibbo-suppressed (DCSD) modes would however be a clear signature for new physics [17, 6]. Asymmetries in all three categories could reach $\sim 10^{-2}$ in such scenarios as non-minimal supersymmetry [6] and left-right-symmetric models [8, 14]. There are also expected SM asymmetries of $\approx 3.3 \times 10^{-3}$ ($= 2\,Re(\epsilon_K)$) due to K^0 mixing in such modes as $D^+ \to K_S\pi^+$ and $K_S\ell\nu$ [18], which should be observed in Charm2000 (Table 1) or even in predecessor experiments. While observation of K^0-induced CP asymmetries might teach us little new about physics, they will at least constitute a calibration for the experimental systematics of asymmetries at the 10^{-3} level. However, Bigi has pointed out that a small new-physics contribution to the DCSD rate could amplify these asymmetries to $\mathcal{O}(10^{-2})$ [6].

Experimental limits at the 10% level have been set in SCSD modes; at present the most sensitive come from the photoproduction experiment Fermilab E687 [19] and from CLEO [20]. E687 has set limits in $D^0 \to K^+K^-$ and $D^+ \to K^-K^+\pi^+$, $\overline{K^{*0}}K^+$, and $\phi\pi^+$ as indicated in Table 1.[3] CLEO has

---

[1]Of course the statistical significance of signals, which directly determines physics sensitivities, goes as the square root of yield and is more stable with respect to cuts.

[2]The reach of Charm2000 in other physics areas such as charm spectroscopy, tests of QCD, lifetimes, form factors, and branching ratios will be discussed in a future publication.

[3]To avoid such cumbersome notations as $D^0(\overline{D^0}) \to K^\mp\pi^\pm$, here and elsewhere in this paper charge-conjugate states are generally implied even when not stated.

studied D^0 decays to the CP eigenstates K^+K^-, $K_S\phi$, and $K_S\pi^0$ as well as $K^\mp\pi^\pm$.

The signal for direct CP violation is an absolute rate difference between decays of particle and antiparticle to charge-conjugate final states f and $\bar{f}$:

$$A = \frac{\Gamma(D \to f) - \Gamma(\overline{D} \to \bar{f})}{\Gamma(D \to f) + \Gamma(\overline{D} \to \bar{f})}. \tag{1}$$

Since in photoproduction D and $\overline{D}$ are not produced equally, in the E687 analysis the signal is normalized relative to the production asymmetry observed in a CFD mode:

$$A = \frac{\eta(D \to f) - \eta(\overline{D} \to \bar{f})}{\eta(D \to f) + \eta(\overline{D} \to \bar{f})}, \tag{2}$$

where, for example,

$$\eta(D^0) = \frac{N(D^0 \to K^+K^-)}{N(D^0 \to K^-\pi^+)}, \tag{3}$$

and for the D^+ modes the normalization mode is $D^+ \to K^-\pi^+\pi^+$. (Thus in the unlikely event that there is a CP asymmetry from new physics in the CFD normalization mode which is equal to that in the corresponding SCSD mode, the signal would be masked.) A further complication is that to distinguish $D^0 \to K^+K^-$ from $\overline{D^0} \to K^+K^-$, D^* tagging must be employed; of course, no tagging is needed for charged-D decays. Typical E687 event yields are $\approx 10^2$ in signal modes and $\sim 10^3$ in normalization modes.

Given the sensitivity achieved in E687, one can extrapolate to that expected in Charm2000. E687 observed 4287 ± 78 (4666 ± 81) events in the normalization mode $D^+ \to K^-\pi^+\pi^+$ ($D^- \to K^+\pi^-\pi^-$). As an intermediate step in the extrapolation I use the event yield in E791, since that hadroproduction experiment is more similar to Charm2000 than is E687. Using relatively tight vertex cuts, E791 observed 37006 ± 204 events in $D^\pm \to K\pi\pi$ [21], and Charm2000 should increase this number by a factor ≈ 2000 (see Sec. 4). Thus relative to E687, the statistical uncertainty on A should be reduced by $\approx \sqrt{8000}$, implying sensitivities in various modes of 10^{-3} at 90% confidence. While the ratiometric nature of the measurement reduces sensitivity to systematic biases, at the 10^{-3} level these will need to be studied carefully.

For DCSD modes, I extrapolate from E791's observation of $D^+ \to K^+\pi^+\pi^-$ at 4.2σ based on 40% of their data sample [22]. The statistical significance in Charm2000 should be $\approx \sqrt{2000/0.4}$ better, implying few$\times 10^{-3}$ sensitivity for CP asymmetries. For $D^0 \to K^+\pi^-$, CLEO's observation [23] of $B(D^0 \to K^+\pi^-)/B(D^0 \to K^-\pi^+) \approx 0.8\%$ suggests $\approx 10^5$ D^*-tagged DCSD $K\pi$ events in Charm2000, giving few$\times 10^{-3}$ CP sensitivity. However, the need for greater background suppression for DCSD compared to CFD events is likely to reduce sensitivity. For example, preliminary E791 results show a $\approx 2\sigma$

signal in $D^0 \to K^+\pi^-$ [24], implying $\sim 10^{-2}$ sensitivity in Charm2000. These extrapolations are conservative and ignore expected improvements in vertex resolution and particle identification. Detailed simulations are underway to assess these effects.

Sensitivities at a τcF have been estimated at a few$\times 10^{-3}$ in SCSD modes [25], but clear qualitative advantages make a τcF complementary to fixed-target experiments [25 - 27]. For example, the equal production of D and $\overline{D}$ in e^+e^- annihilation allows study of CP violation at $< 10^{-3}$ sensitivity in CFD modes, a measurement which in a fixed-target experiment can be carried out to greater statistical precision (Table 1) but depends on effects differing in size among various CFD modes. A τcF also has a clear advantage in modes with final-state photons.

SM predictions for direct CP violation are rather uncertain, since they require assumptions for final-state phase shifts as well as CKM matrix elements [17, 6]; the predictions given in Table 1 are representative, but the theoretical uncertainties are probably larger than indicated there [28]. However, given the order of magnitude expected in charm decay, the Charm2000 experiment might make the first observation of direct CP violation outside the strange sector, or indeed the first observation anywhere if (as may well be the case [29, 30]) signals prove too small for detection in the next round of K^0 [31 - 33] and hyperon [34] experiments.

2.2 Flavor-Changing Neutral Currents

Charm-changing neutral currents are forbidden at tree level in the Standard Model due to the GIM mechanism [35]. They can proceed via loops at rates which are predicted to be unobservably small, e.g. for $D^0 \to \mu^+\mu^-$ (which suffers also from helicity suppression in the SM) the predicted branching ratio is $\sim 10^{-19}$ [36, 8, 7], and for $D^+ \to \pi^+\mu^+\mu^-$ it is $\sim 10^{-10}$ [12, 7]. Long-distance effects increase these predictions by some orders of magnitude, but they remain of order 10^{-15} to 10^{-8} [8, 37, 38]. Various extensions of the SM [12, 39] predict effects substantially larger than this, for example in models with a fourth generation, both $B(D^+ \to \pi^+\mu^+\mu^-)$ and $B(D^0 \to \mu^+\mu^-)$ can be as large as 10^{-9} [12]. Experimental sensitivities are now in the range $\sim 10^{-4}$ to 10^{-5} [21, 40 - 43] and are expected to reach $\sim 10^{-5}$ to 10^{-6} in E831 [44].

Limits on FCNC charm decays have recently improved considerably, with new results from Fermilab E653 and E791 and WA92 at CERN. E653 [43] studied charm decays to hadrons plus muon pairs in a variety of modes, E791 [21] studied charged-D decays to $\pi\mu^+\mu^-$ and πe^+e^-, and WA92 [41] searched for $D^0 \to \mu^+\mu^-$. Typically a normalization mode is used to determine the sensitivity, reducing systematic uncertainty. Thus E791 normalized to $K^-\pi^+\pi^+$ and WA92 to $K^\mp\pi^\pm$, eliminating normalization uncertainty due to the D production cross section. (Older limits [45, 46] on $D^0 \to \mu^+\mu^-$ used $J/\psi \to \mu^+\mu^-$ for normalization, reducing uncertainty due to muon identification and triggering efficiency.)

One can extrapolate from recent results to estimate sensitivities in Charm2000. While Charm2000 aims at a single-event branching-ratio sensitivity of $\approx 10^{-9}$, FCNC limits are typically background-limited, so sensitivites can be expected to improve as the square root of the number of events reconstructed. In some cases, however, more dramatic improvement may result from improved lepton identification. For $D^+ \to \pi^+\mu^+\mu^-$, scaling the E791 sensitivity by a factor $\sqrt{2000}$ as above gives $\approx$few$\times 10^{-7}$ 90%-confidence sensitivity in Charm2000. This estimate may be conservative, since the simple muon detection scheme employed by E791 (one layer of scintillation counters following 2.5 m of steel equivalent) resulted in a (momentum-dependent) π-μ misidentification probability ranging from 4.5 to 20% [21], and it should be possible to reduce this to $\approx 1\%$ in Charm2000. With modern calorimetry for electron identification one expects to do almost as well for πee as for $\pi\mu\mu$. For $D^0 \to \mu^+\mu^-$ and e^+e^-, extrapolating from WA92 implies sensitivity of 10^{-7} per mode.

Radiative charm decays present the opportunity to test models of nonperturbative long-distance effects, since short-distance (penguin) contributions are estimated to be negligible even in extensions of the SM such as models with a fourth generation [38]. Long-distance effects give branching ratios of order $10^{-5} - 10^{-6}$, whereas current experimental limits are $\sim 10^{-4}$ (see Table 1). It is important to test these calculations in the charm sector, where the predicted effects are large and not "contaminated" by short-distance physics, since small but non-negligible long-distance corrections are predicted in the b sector, where e.g. one would like to extract the CKM element V_{td} from $B(B \to \rho\gamma)$ [47, 7, 48]. In addition there may be a window for new physics, since e.g. non-minimal supersymmetry might make a substantial contribution to $D \to \rho\gamma$, and this may be distinguishable from a long-distance SM effect since the latter is Cabibbo-suppressed with respect to $D^0 \to K^*\gamma$ in the SM but not in SUSY [49, 47]. Observation of such modes as $D^0 \to \rho^0\gamma$ and $D^0 \to \phi\gamma$ may be within reach at a $\tau c F$ or B factory. It is not clear how well fixed-target experiments can do on these modes, since they must cope with large combinatoric photon backgrounds from π^0 decay.

2.3 Lepton-Number-Violating Decays

There are two lepton-number-violating effects which can be sought: decays violating conservation of lepton number (LNV) and decays violating conservation of lepton-family number (LFNV). LFNV decays (such as $D^0 \to \mu^\pm e^\mp$) are expected in theories with leptoquarks [39], heavy neutrinos [7], extended technicolor [50], etc.; LNV decays (such as $D^+ \to K^- e^+ e^+$) can arise in GUTs and have been postulated to play a role in the development of the baryon asymmetry of the Universe [51]. Since no fundamental principle forbids either type of decay, it is of interest to search for them as sensitively as possible.

Although much smaller decay widths can be probed in K decays, there are simple theoretical arguments why LFNV charm decays are nevertheless worth seeking. If such currents arise through Higgs exchange, whose couplings are proportional to mass, they will couple more strongly to charm than to strangeness [11]. Furthermore, LFNV currents may couple to up-type quarks more strongly than to down-type [39, 52].

As shown in Table 1, the best existing limits come in most cases from the e^+e^- experiments Mark II, ARGUS, and CLEO (although the hadroproduction experiment Fermilab E653 dominates in modes with same-sign dimuons) and are typically at the $10^{-3} - 10^{-4}$ level [42, 43]. E831 expects to lower these limits to $\sim 10^{-6}$ [44], and Charm2000 should reach $\sim 10^{-7}$.

2.4 Mixing and Indirect CP Violation

$D^0 \overline{D^0}$ mixing may be one of the more promising places to look for low-energy manifestations of physics beyond the Standard Model. For small mixing, the mixing rate is given to good approximation by [17]

$$r_{\text{mix}} \approx \frac{1}{2}\left[\left(\frac{\Delta M_D}{\Gamma_D}\right)^2 + \left(\frac{\Delta \Gamma_D}{2\Gamma_D}\right)^2\right]. \qquad (4)$$

In the SM the ΔM and $\Delta \Gamma$ contributions are expected [17] to be of the same order of magnitude and are estimated [17, 53] to give $r_{\text{mix}} < 10^{-8}$; any observation at a substantially higher level will be clear evidence of new physics.[4] Many nonstandard models predict much larger effects. An interesting example is the multiple-Higgs-doublet model lately expounded by Hall and Weinberg [55], in which $|\Delta M_D|$ can be as large as 10^{-4} eV, approaching the current experimental limit. In this model all CP violation arises from Higgs exchange and is intrinsically of order 10^{-3}, too small to be observed in the beauty sector and (except through mixing) in the kaon sector, but possibly observable in charm – another example of the importance of exploring rare phenomena in *all* quark sectors. The large mixing contribution arises from flavor-changing neutral-Higgs exchange (FCNE) [57], which can be constrained to satisfy the GIM mechanism for K^0 decay by assuming small phase factors ($\sim 10^{-3}$). (This is in distinction to the original "Weinberg model" of CP violation [58], in which FCNE was suppressed by assuming a discrete symmetry such that one Higgs gave mass to up-type quarks and another to down-type.) Multiple-Higgs models are one

---

[4]Earlier estimates [54] that long-distance effects can give $|\Delta M_D/\Gamma_D| \sim 10^{-2}$ are claimed to have been disproved [17, 55], but there remain skeptics [6, 56].

of the simplest extensions of the SM [8, 33, 57], and many other authors have also considered multiple-Higgs effects in charm mixing [49, 52, 59 - 61]. Large mixing in charm can also arise in theories with supersymmetry [49, 59, 62], technicolor [50], leptoquarks [39], left-right symmetry [63], or a fourth generation [8, 12].

The experimental situation regarding $D^0\overline{D^0}$ mixing is complicated by the presence of DCSD. Since both effects can lead to the same final states, one needs to distinguish them using time-resolved measurements [11]. In the notation of Refs. [64] and [65], the time dependence for wrong-sign decay is given by

$$\Gamma(D^0(t) \to K^+\pi^-) = \frac{e^{-\Gamma t}}{4}|B|^2|\frac{q}{p}|^2\{4|\lambda|^2 + (\Delta M^2 + \frac{\Delta\Gamma^2}{4})t^2 + 2Re(\lambda)\Delta\Gamma t + 4Im(\lambda)\Delta M t\}, \quad (5)$$

and there is a similar expression for $\overline{D^0} \to K^-\pi^+$ in which λ is replaced by $\bar{\lambda}$. In Eq. 5 the first term on the right-hand side is the DCSD contribution, which peaks at $t = 0$; the second is the mixing contribution, which peaks at 2 D^0 lifetimes because of the factor t^2; and the third and fourth terms reflect interference between mixing and DCSD and peak at 1 lifetime due to the factor t. λ and $\bar{\lambda}$ can acquire nonzero phases through indirect CP violation or through final-state interactions [65, 56]. While for sufficiently small $|\Delta M/\Gamma|$ experimental sensitivity to mixing is enhanced if there is interference [66], at present levels of sensitivity allowing an arbitrary interference phase when fitting decay-time distributions reduces the stringency of the resulting limit [67, 24].

The most sensitive limit on $D^0\overline{D^0}$ mixing (quoted in Table 1 and in the *Review of Particle Properties* [40]) comes from the Fermilab photoproduction experiment E691 [67]. The E691 analysis considered two modes, $D^0 \to K^\mp\pi^\pm$ and $K^\mp\pi^\pm\pi^+\pi^-$, and five possible values of the interference phase ϕ covering the range $-1 \leq \cos\phi \leq 1$. The limits in each mode were stable over most of the ϕ range, but worsened for $\cos\phi = -1$ by a factor 1.8 (3.3) for $K\pi$ ($K3\pi$). The final result was derived by combining the two modes neglecting interference.

Recently several authors have critiqued the E691 mixing analysis. Liu [66] has questioned the validity of the combined limit, suggesting that even if interference is negligible for one mode, it is less likely to be negligible for both. Blaylock, Seiden, and Nir [64] and Wolfenstein [56] suggest that whereas the E691 fit neglected the term in Eq. 5 proportional to ΔM but kept the term in $\Delta\Gamma$, the reverse should have been done. Browder and Pakvasa [65] have reconsidered the E691 analysis taking into account the role of final-state interactions; they conclude that even maximal destructive interference degrades the no-interference E691 limit only at the 10% level. However, the understanding of final-state phases is entirely phenominological, and more work and data are required to assess its reliability. Nevertheless it appears that the E691 limit is not "wrong" by much if at all, and interference does not appear to play a large role at present sensitivity.

While there is as yet no published mixing limit from E791, the preliminary indication is sensitivity to r at the $\approx 10^{-3}$ level if interference is neglected, ranging to perhaps a few times this if interference is allowed [24]. A simple extrapolation by $\sqrt{2000}$ suggests sensitivity of $\approx 2 \times 10^{-5}$ in Charm2000 neglecting interference, which with improvements in particle identification and resolution for the tagging pion might approach 10^{-5}. However, since the interference term is linear in ΔM_D while the mixing term is quadratic, the ratio of the interference and mixing contributions goes as $1/\Delta M_D$. Thus as experimental sensitivity improves and smaller and smaller values of $|\Delta M_D|$ are probed, interference becomes relatively more important. One therefore cannot extrapolate simply from the E691 or E791 sensitivity to that expected in Charm2000.

A first attempt to assess the impact of interference on mixing sensitivity in Charm2000 has been carried out by generating ten Monte Carlo samples of DCSD $D^0 \to K^+\pi^-$ events and fitting them allowing for interference or not. I conservatively assume 10^4 events observed after vertex cuts and fit each decay-time histogram only for $t > 0.88$ ps (2 D^0 lifetimes) as in the E691 analysis, following the prescription of Browder and Pakvasa [65] for the time dependence in the case of no CP violation (their Eq. 4). Within their suggested range of final-state phase ($5°$ to $13°$), the interference term improves sensitivity slightly, and 10^{-5} sensitivity is obtained. Since the interference contribution peaks at 1 lifetime it would be desirable to include shorter decay times in the fit, however more simulation studies are required to evaluate signal cleanliness in that region.

Semileptonic decays offer a way to study mixing free from the effects of DCSD. So far the only published limit on charm mixing from semileptonic decays (Table 1) is from the Fermilab dimuon hadroproduction experiment E615 [45], in which only the muons were detected and no vertex information was available. A preliminary result from E791 using D^*-tagged $D^0 \to Ke\nu$ events indicates sensitivity at the $\approx 0.5\%$ level [68]. Extrapolation by $\sqrt{2000}$ suggests 10^{-4} sensitivity in Charm2000, but use of muonic decays as well, plus improvements in lepton identification and resolution for the tagging pion, may give significantly better sensitivity. At the Charm2000 Workshop, Morrison suggested 10^{-5} sensitivity may be possible [69].

Liu has stressed the importance of setting limits on $\Delta\Gamma$ as well as on ΔM. Although typical extensions of the SM which predict large $|\Delta M|$ also predict $|\Delta M| \gg |\Delta\Gamma|$ [64, 65], from an experimentalist's viewpoint both should be measured if possible. $\Delta\Gamma$ can be studied quite straightforwardly by comparing the lifetime measured for CP-even modes such as $K^+K^-, \pi^+\pi^-$ with that for CP-odd modes or (more simply) with modes of mixed CP such as $K^-\pi^+$. No such result has yet been published, so it is difficult to extrapolate realistically to Charm2000 sensitivity. Liu [66] has estimated Charm2000 sensitivity (in an idealized case) at $\sim 10^{-5} - 10^{-6}$ in $(\Delta\Gamma/2\Gamma)^2$ (*i.e.* the contribution to r due to $\Delta\Gamma$).

The τcF can make a unique contribution to the study of mixing. DCSD are forbidden in decays such as $\psi'' \to D^0\overline{D^0} \to (K^-\pi^+)(K^-\pi^+)$ due to the $C = -1$ initial state and the Bose symmetry of the final state [60, 70, 26], allowing direct time-integrated observation of mixing in hadronic final states; sensitivity has been estimated at $\sim 10^{-4}$ per year of running [26].

2.4.1 Indirect CP violation

Since in the SM $D^0\overline{D^0}$ mixing is negligible, any indirect CP-violating asymmetries are expected to be less than 10^{-4} [6]. However, possible mixing signals at the $\approx 1\%$ level have been reported [23, 71]. While given the E691 limit these probably represent enhanced DCSD signals, if a significant portion of this rate is in fact mixing then new physics must be responsible [17, 56]. Then indirect CP violation at the $\lesssim 1\%$ level is possible [60, 47, 6, 56]. Some authors have suggested that the CP-violating signal, which arises from the interference term of Eq. 5, may be more easily detectable than the mixing itself [56, 64-66]. In particular, Browder and Pakvasa [65] point out that in the difference $\Gamma(D^0 \to K^+\pi^-) - \Gamma(\overline{D^0} \to K^-\pi^+)$, the DCSD and mixing components cancel, leaving only the fourth term of Eq. 5. Thus if indirect CP violation is appreciable this is a particularly clear way to isolate the interference term.

3 A Next-Generation Charm Spectrometer

A Letter of Intent is in progress for an experiment which can achieve the 10^8-reconstructed-charm sensitivity mentioned above. As we will see, the most demanding requirement is on the trigger. In particular, an on-line secondary-vertex trigger is needed if adequate trigger rejection is to be achieved without sacrificing sensitivity in hadronic decay modes. (More detailed discussions may be found in [3] and [4].)

3.1 Beam and target

To achieve the sensitivity discussed here in a fixed-target run of $\approx 10^5$ beam spills requires a primary proton beam [72]. Assuming 800-GeV beam energy the charmed-particle production rate is 7×10^{-3}/interaction if a high-A target (*e.g.* Au) is used, or 3×10^{-3} if diamond is used [73].

A target which is short compared to typical charm decay lengths is crucial for optimizing background suppression, both off-line and at trigger level. While multiple thin targets could be employed (as in E791 and E831), a single target facilitates fast vertex triggering. A ≈ 1 mm W, Pt, or Au target is one possibility, representing $\approx 1\%$ of an interaction length and on average $\approx 15\%$ of a radiation length for outgoing secondaries. A low-Z material such as diamond may be favored to minimize scattering of low-momentum pions from D^* decay [72]; then a ≈ 2 mm target is suitable, representing $\approx 1\%$ of an interaction length and $\approx 1\%$ of a radiation length. Given the mean Lorentz boost $\gamma \approx 35$, a 1–2 mm target is short enough that a substantial fraction even of charmed baryons will decay outside it.

For triggering purposes (see Sec. 3.3) and to optimize resolution in decay distance, it is desirable to minimize the rate of occurrence of multiple simultaneous interactions. We therefore assume a 5 MHz interaction rate, which given the Tevatron's 53 MHz bunch rate and the typical 50% spill duty factor implies a $\approx 20\%$ fraction of events with multiple interactions. The needed 0.5–1 GHz of primary proton beam is easily attainable. As shown in Sec. 4, this yields $\gtrsim 10^8$ reconstructed charm per few $\times 10^6$ s of beam ($\approx 10^5$ spills $\times$ 20 s/spill).

3.2 Spectrometer

We assume a highly rate-capable large-acceptance open-geometry spectrometer. A significant design challenge is posed by radiation damage to the vertex detectors. To configure detectors which can survive at the desired sensitivity, we choose suitable maximum and (in one view) minimum angles for the instrumented aperture, arranging the detectors along the beam axis with a small gap through which pass the uninteracted beam and secondaries below the minimum angle (Figs. 2, 3).[5] Thus the rate is spread approximately equally over several detector planes, with large-angle secondaries measured close to the target and small-angle secondaries farther downstream. Along the beam axis the spacing of detectors increases geometrically, making the lever arm for vertex reconstruction independent of production angle. Since small-angle secondaries tend to have high momentum, the multiple-scattering contribution to vertex resolution is also approximately independent of production angle. We have chosen an instrumented angular range $|\theta_x| \leq 200$ mr, $4 \leq \theta_y \leq 175$ mr, corresponding to the center-of-mass rapidity range $|y_{\rm CM}| \lesssim 1.9$ and containing over 90% of produced secondaries.

Assuming n charged particles per unit pseudorapidity, the rate per unit

---

[5]An alternative approach with no gap may also be workable if the beam is spread over sufficient area to satisfy rate and radiation-damage limits, however the approach described here probably allows smaller vertex detectors and is "cleaner" in that the beam passes through a minimum of material.

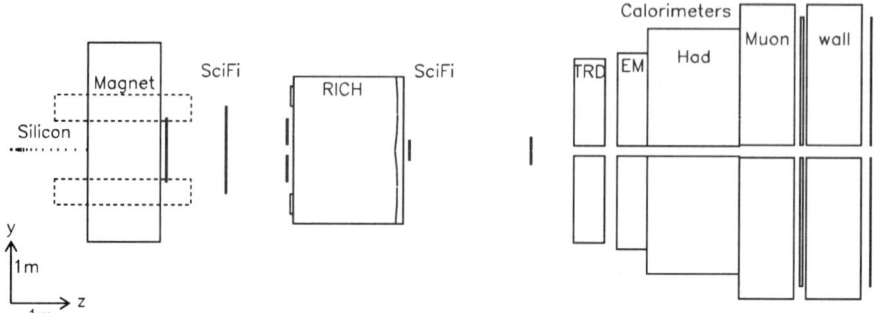

Figure 2: Spectrometer layout (bend view).

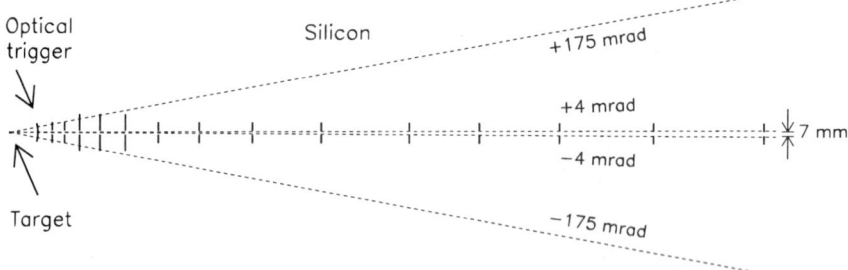

Figure 3: Detail of vertex region (showing optional optical impact-parameter trigger).

detector area at transverse distance r from the beam is given by $n/2\pi r^2$. Since in 800 GeV proton-nucleus collisions $n \approx 4$ for high-A targets [74] (less for C), in a run of n_{int} interactions, a detector which can withstand a maximum fluence of $R_{\max}$ particles/cm^2 has a "minimum survivable" inner detector radius

$$r_{\min} = \left(\frac{n}{2\pi} \frac{n_{\text{int}}}{R_{\max}}\right)^{\frac{1}{2}}. \qquad (6)$$

A typical run will yield fewer than 2×10^{13} interactions. If we assume currently-available silicon detectors ($R_{\max} \approx 10^{14}/\text{cm}^2$), we obtain conservatively $r_{\min} = 3.5$ mm. An order-of-magnitude improvement in radiation hardness would reduce $r_{\min}$ to ≈ 1 mm, which is close to the minimum half-gap through which the beam could be reliably steered. In Fig. 3 we have conservatively indicated 3.5 mm as the half-gap. Vertex resolution tends to improve as the half-gap is reduced, so the use of radiation-hard detectors (either diamond detectors [75] or improved silicon detectors) is highly desirable; such detectors are likely to be available by $\approx$Year 2000. The desired angular range can be covered with sufficient redundancy for pattern recognition using

14 double-sided vertex detectors above and 14 below the beam as shown in Fig. 3. These might be radiation-hard silicon-strip or -pixel or diamond-strip or -pixel detectors.

Downstream of the analyzing magnet we assume scintillating-fiber tracking using 3HF/PTP fibers with VLPC readout [76] as in the D0 [77] and CDF upgrades. The minimum half-gap for the fiber planes is determined by occupancy, which in the uniform-pseudorapidity approximation used above (and neglecting magnetic bending) is given by

$$\frac{n}{\pi} \frac{dy}{y} \arctan \frac{x_{\max}}{y} \qquad (7)$$

for a detector element of height dy a distance y from the beam which covers $-x_{\max} < x < x_{\max}$. For 800 µm fiber diameter, this implies ≈16% occupancy at $y = 1$ cm, ≈8% at 2 cm, and ≈4% at 4 cm. A full trackfinding simulation will be required to assess the maximum acceptable occupancy, but this suggests ≈1 cm as the minimum acceptable half-gap in the scintillating-fiber planes. The fibers near the gap could be split at $x = 0$ and read out at both ends, halving their occupancies. Since shorter fibers have less attenuation, a smaller diameter could be used near the gap, reducing occupancy still further. Since the fibers are more radiation-hard than silicon detectors and the fiber-plane beam gap is larger than that of the vertex detectors, radiation damage of the fibers will not be a problem.

The spectrometer sketched here accepts ≳ 50% of two-prong D^0 decays and ≈50% of three-prong decays, comparable to E687 and E791 acceptances. Assuming a 0.5 GeV analyzing-magnet p_t kick, the D mass resolution (≈5 MeV rms) is a factor ≈2 better than that of E687 or E791. With vertex detectors of 25 µm pitch read out digitally (i.e. no pulse-height information), vertex resolution is comparable to that of existing spectrometers; we are exploring the possible improvement from reduction of the half-gap to 1 mm and use of analog readout via flash ADCs as in E831. (Since the mass resolution is dominated by scattering, minimization of material is crucial, for example use of helium bags and avoidance of threshold Cherenkov counters employing heavy gas mixtures.)

3.3 Trigger

While previous Fermilab charm hadroproduction experiments E769 and E791 recorded and analyzed large charm samples using very loose triggers which accepted most inelastic interactions, this approach is unlikely to extrapolate successfully by three orders of magnitude! (Consider that E791 recorded 2×10^{10} events – 50 terabytes of data – on 24,000 8 mm tapes.) Thus our sensitivity goal requires a highly selective trigger. However, we wish to trigger on charm-event characteristics which bias the physics as little as possible. We therefore assume a first-level trigger requiring calorimetric E_t (as in E769 and E791) OR'ed with high-p_t-lepton and lepton-pair triggers. At second level, secondary-vertex requirements are imposed on the E_t-triggered events to achieve a rate ($\sim$100 kHz) which is practical to record.

3.3.1 E_t trigger

Based on experience in E791, and using PYTHIA to simulate the effect of pile-up in the calorimeter[6] [78], we expect a $\approx$10 GeV E_t threshold to give a minimum-bias rejection factor of 5 with $\approx$50% charm efficiency. (These are rough estimates based on a relatively crude calorimeter, and an optimized calorimeter may provide better rejection.) Such an E_t trigger yields a 1 MHz input rate to the next level; the leptonic trigger rates should be negligible by comparison.

3.3.2 Secondary-vertex trigger

An additional factor $\approx$10 in trigger rejection is desirable, and can be achieved by requiring evidence of secondary vertices. This might be accomplished using a hardware trigger processor, which would need to be an order of magnitude faster than existing vertex processors [79] to accept events at 1 MHz; fast readout and buffering of event information would also be required. Track-finding secondary-vertex triggers benefit from the use of focused beam and a single thin target, which allow simplification of the algorithm since the primary vertex location is known *a priori*.

Christian [80] has suggested a simple trigger-processor algorithm based on this idea. A PYTHIA-based simulation of this algorithm for the vertex-detector configuration of Fig. 3 shows good performance [4]. Assuming negligible spread in y of primary-interaction vertices,[7] requiring at least one track to miss the primary vertex by at least 200 μm in y rejects 95% of minimum-bias events while retaining 67% of all charm events. The simulation also tested the effects of making a preliminary pass through the data eliminating hits which lie on straight lines pointing to the primary vertex: rejection and efficiency were hardly affected. Since as the number of hits per detector plane (n) increases, the time to eliminate hits is linear in n, while the time to find tracks of finite impact parameter goes as n^2 (due to the required loops over hits in two seed planes), such a hit-elimination pass can reduce processing time substantially [80].

As alternatives to iterative trackfinding at a 1 MHz event rate, three other

---

[6]Given 20% probability for >1 interaction, pile-up degrades the rejection by a factor $\approx$2.
[7]achievable *e.g.* by use of a target of 100 μm height.

approaches also appear worth pursuing. The first is a secondary-vertex trigger implemented using fast parallel logic, *e.g.* PALs, neural networks, or pre-downloaded fast RAMs, to look quickly for patterns in the vertex detectors corresponding to tracks originating downstream of the target. The others are fast secondary-vertex trigger devices originally proposed for beauty: the optical impact-parameter [81] and Cherenkov multiplicity-jump [82] triggers; while results from prototype tests so far suggest lower than desired charm efficiency, these might with further development provide sufficient resolution to trigger efficiently on charm. For example, one simulation of an optical impact-parameter trigger [83] indicated 40% charm efficiency for a factor 5 minimum-bias rejection, which is good enough to be usable in Charm2000. In a very different regime of decay length and impact parameter, an optical trigger is in development for the hyperon *CP*-violation experiment Fermilab E871 [84]; experience gained from this effort should allow prediction of charm performance with good confidence. A charm multiplicity-jump trigger is under development for CHEOPS [2, 85].

4 Yield

The charm yield is straightforwardly estimated. Assuming a Au target and a typical fixed-target run of 3×10^6 live beam seconds, 10^{11} charmed particles are produced. The reconstructed-event yields in representative modes are estimated in Table 2 assuming (for the sake of illustration) that the optical trigger described in [83] is used for all-hadronic modes (but not for leptonic modes, for which the first-level trigger rate should be sufficiently low to be recorded directly) and performs as estimated above. Although due to off-line selection cuts not yet simulated, realistic yields could be a factor $\approx 2-3$ below those indicated, the total reconstructed sample is well in excess of 10^8 events. Given the factor ≈ 2 mass-resolution improvement compared to E791, one can infer a factor ≈ 50 improvement in statistical significance in typical decay modes.

5 Conclusions

A fixed-target hadroproduction experiment (Charm2000) capable of reconstructing in excess of 10^8 charm events is feasible using detector, trigger, and data acquisition technologies which exist or are under development. A typical factor ≈ 50 in statistical significance of signals may be expected compared to E791. We expect the spectrometer sketched here to cost substantially less than

HERA-B (whose cost was estimated at 33M DM in 1994 [74]).[8] Should such an experiment be carried out it will likely exceed the sensitivity of a τcF in the high-impact areas of charm CP violation, mixing, and flavor-changing neutral and lepton-number-violating currents. This conclusion might be questioned in light of recent scheduling experience at Fermilab. However, the typical ≈3-year interval between Fermilab fixed-target runs is offset by the need to divide τcF running time among various physics topics requiring differing beam energies.[9] Even without Charm2000, the CHEOPS experiment may come within an order of magnitude of Charm2000 sensitivity and rival that achievable in a τcF. Neverthelesss, the τcF complements charm hadroproduction experiments by its capability to make various unique measurements, not to mention its capabilities in τ physics [86]. Ideally, both projects will go forward.

Acknowledgements

I thank J. A. Appel, I. I. Bigi, C. N. Brown, G. Burdman, D. C. Christian, J. L. Hewett, S. Kwan, T. Liu, S. Pakvasa, and M. D. Sokoloff for useful discussions, and J. Repond for the invitation to participate in this Workshop. The organizers deserve particular thanks for a memorable workshop dinner.

References

[1] D. Z. Besson and A. P. Freyberger, "Future Experimental Charm Physics at $\sqrt{s} = 10$ GeV," in **The Future of High-Sensitivity Charm Experiments**, *Proc. CHARM2000 Workshop*, Fermilab, June 7–9, 1994, D. M. Kaplan and S. Kwan, *eds.*, FERMILAB-Conf-94/190 (1994), p. 35.

[2] Yu. Alexandrov *et al.*, "CHarm Experiment with Omni-Purpose Setup," Letter of Intent to CERN, CERN/SPSLC 95-22, March 28, 1995.

[3] D. M. Kaplan, "A High-Rate Fixed-Target Charm Experiment," in **The Future of High-Sensitivity Charm Experiments**, *op. cit.*, p. 229.

[4] D. M. Kaplan and V. Papavassiliou, "An Ultrahigh-Statistics Charm Experiment for the Year ~2000," IIT-HEP-95/2, hep-ex/9505002, to appear in *Proc. LISHEP95 Workshop*, Rio de Janeiro, Brazil, February 20–22, 1995.

[5] I.I. Bigi, "$D^0\overline{D^0}$ Mixing and CP Violation in D Decays: Can There Be High Impact Physics in Charm Decays?", *Proc. Tau Charm Factory Workshop*, Stanford, CA, May 23–27, 1989, SLAC-Report-343 (1989), p. 169.

---

[8] While HERA-B is potentially competitive with Charm2000 as a charm experiment, it lacks the capabilities to trigger efficiently on charm and to acquire the needed large data sample, and it probably has significantly poorer vertex resolution as well.

[9] The frequency of Fermilab fixed-target runs might also increase once Main Injector construction is completed.

Table 1: Sensitivity to high-impact charm physics.

Topic	Limit*	Charm2000 Reach*	SM prediction
Direct CP Viol.			
$D^0 \to K^-\pi^+$	$-0.009 < A < 0.027$ [20]		≈ 0 (CFD)
$D^0 \to K^-\pi^+\pi^+\pi^-$		few$\times 10^{-4}$	≈ 0 (CFD)
$D^0 \to K^+\pi^-$		$10^{-3} - 10^{-2}$	≈ 0 (DCSD)
$D^+ \to K^+\pi^+\pi^-$		few $\times 10^{-3}$	≈ 0 (DCSD)
$D^0 \to K^-K^+$	$-0.11 < A < 0.16$ [19]	10^{-3}	
	$-0.028 < A < 0.166$ [20]		
$D^+ \to K^-K^+\pi^+$	$-0.14 < A < 0.081$ [19]	10^{-3}	
$D^+ \to \overline{K}^{*0}K^+$	$-0.33 < A < 0.094$ [19]	10^{-3}	$(2.8 \pm 0.8) \times 10^{-3}$ [16]
$D^+ \to \phi\pi^+$	$-0.075 < A < 0.21$ [19]	10^{-3}	
$D^+ \to \eta\pi^+$			$(-1.5 \pm 0.4) \times 10^{-3}$ [16]
$D^+ \to K_S\pi^+$		few$\times 10^{-4}$	3.3×10^{-3} [18]
FCNC			
$D^0 \to \mu^+\mu^-$	7.6×10^{-6} [41]	10^{-7}	$< 3 \times 10^{-15}$ [7]
$D^0 \to \pi^0\mu^+\mu^-$	1.8×10^{-4} [43]	10^{-6}	
$D^0 \to \overline{K}^0 e^+e^-$	17.0×10^{-4} [42]	10^{-6}	$< 2 \times 10^{-15}$ [7]
$D^0 \to \overline{K}^0 \mu^+\mu^-$	2.6×10^{-4} [43]	10^{-6}	$< 2 \times 10^{-15}$ [7]
$D^+ \to \pi^+ e^+e^-$	6.6×10^{-5} [21]	few $\times 10^{-7}$	$< 10^{-8}$ [7]
$D^+ \to \pi^+\mu^+\mu^-$	1.8×10^{-5} [21]	few $\times 10^{-7}$	$< 10^{-8}$ [7]
$D^+ \to K^+ e^+e^-$	4.8×10^{-3} [42]	few $\times 10^{-7}$	$< 10^{-15}$ [7]
$D^+ \to K^+\mu^+\mu^-$	8.5×10^{-5} [40]	few $\times 10^{-7}$	$< 10^{-15}$ [7]
$D \to X_u + \gamma$			$\sim 10^{-5}$ [7]
$D^0 \to \rho^0\gamma$	1.4×10^{-4} [7]		$(1-5) \times 10^{-6}$ [7]
$D^0 \to \phi\gamma$	2×10^{-4} [7]		$(0.1 - 3.4) \times 10^{-5}$ [7]
LF or LN Viol.			
$D^0 \to \mu^\pm e^\mp$	1.0×10^{-4} [40]	10^{-7}	0
$D^+ \to \pi^+\mu^\pm e^\mp$	3.2×10^{-3} [42]	few$\times 10^{-7}$	0
$D^+ \to K^+\mu^\pm e^\mp$	3.3×10^{-3} [42]	few$\times 10^{-7}$	0
$D^+ \to \pi^-\mu^+\mu^+$	2.2×10^{-4} [43]	few$\times 10^{-7}$	0
$D^+ \to K^-\mu^+\mu^+$	3.4×10^{-4} [43]	few$\times 10^{-7}$	0
$D^+ \to \rho^-\mu^+\mu^+$	5.6×10^{-4} [43]	few$\times 10^{-7}$	0
Mixing			
$\stackrel{(-)}{D^0} \to K^\mp\pi^\pm$	$r < 0.37\%$ [67],	$r < 10^{-5}$,	
	$|\Delta M_D| < 1.3 \times 10^{-4}$ eV	$|\Delta M_D| < 10^{-5}$ eV	10^{-7} eV [17]
$\stackrel{(-)}{D^0} \to \ell\nu X$	$r < 0.56\%$ [45]		
$\stackrel{(-)}{D^0} \to K\ell\nu$		$r < 10^{-5}$	

*at 90% confidence level

Table 2: Estimated yields of reconstructed events (antiparticles included)
a) direct estimates

mode	charm frac.	BR (%)	acceptance	efficiency	yield
$D^0 \to K^-\pi^+$	0.5	4.0	0.6	0.1	1.3×10^8
$D^+ \to K^{*0}\mu\nu$	0.25	2.7	0.4	0.25	7×10^7
$\to K\pi\mu\nu$					
all	1	≈ 0.1	≈ 0.4	≈ 0.1	$\approx 4 \times 10^8$

b) extrapolations from E791

mode	BR (%)	E791 yield	Charm2000 yield	analysis
$D^+ \to K^-\pi^+\pi^+$	9.1	37000 ± 200	$(7 \pm 0.001) \times 10^7$	FCNC
$D^+ \to K_S\pi^+$	0.94		$(7 \pm 0.003) \times 10^6$	
$D^{*+} \to \pi^+ D^0 \to \pi^+ K^-\pi^+$	2.7	5000	10^7	mixing
$D^{*+} \to \pi^+ D^0 \to \pi^+ K^-\pi^+\pi^+\pi^-$	5.5	3200	0.6×10^7	mixing
$D^{*+} \to \pi^+ D^0 \to \pi^+ K^+\pi^-$	0.02?	45?	$10^4 - 10^5$	DCSD
$D^0 \to K^-\pi^+\pi^+\pi^-$	8.1		6×10^7	

[6] I. I. Bigi, "The Expected, the Promised and the Conceivable – On *CP* Violation in Beauty and Charm Decays," UND-HEP-94-BIG11, hep-ph/9412227, to appear in *Proc. HQ94 Workshop*, Charlottesville, VA, Oct. 7-10, 1994.

[7] J. L. Hewett, "Searching for New Physics with Charm," SLAC-PUB-95-6821, hep-ph/9505246, to appear in *Proc. LISHEP95 Workshop*, Rio de Janeiro, Brazil, Feb. 20-22, 1995.

[8] S. Pakvasa, "Charm as Probe of New Physics," in **The Future of High-Sensitivity Charm Experiments**, *op. cit.*, p. 85.

[9] M. D. Sokoloff and D. M. Kaplan, "Physics of an Ultrahigh-Statistics Charm Experiment," IIT-HEP-95/1, hep-ex/9508015, to appear in *Proc. HQ94 Workshop*, Charlottesville, VA, Oct. 7-10, 1994.

[10] For a recent review see J. L. Rosner, "Present and Future Aspects of CP Violation," hep-ph/9506364, to appear in *Proc. LISHEP95 Summer School*, Rio de Janeiro, Brazil, Feb. 7-11, 1995.

[11] I. I. Bigi, in **Charm Physics**, *Proc. Int. Symp. on Charm Physics*, Beijing, China, June 4-16, 1987, Gordon and Breach (1987), p. 339.

[12] K. S. Babu *et al.*, Phys. Lett. B **205**, 540 (1988); T. G. Rizzo, Int. J. Mod. Phys. **A4**, 5401 (1989).

[13] A. Le Yaouanc *et al.*, "Mixing and *CP* Violation in *D* Mesons," LPTHE-Orsay/95-15, hep-ph/9504270 (1995).

[14] A. Le Yaouanc, L. Oliver, and J.-C. Raynal, Phys. Lett. B **292**, 353 (1992).

[15] M. Golden and B. Grinstein, Phys. Lett. B **222**, 501 (1989).

[16] F. Buccella *et al.*, Phys. Lett. B **302**, 319 (1993); A. Pugliese and P. Santorelli, "Two Body Decays of *D* Mesons and *CP* Violating Asymmetries in Charged *D* Meson Decays," *Proc. Third Workshop on the Tau/Charm Factory*, Marbella, Spain, 1-6 June 1993, Edition Frontieres (1994), p. 387.

[17] G. Burdman, "Charm Mixing and CP Violation in the Standard Model," in **The Future of High-Sensitivity Charm Experiments**, *op cit.*, p. 75; G. Burdman, this Workshop.

[18] Z. Xing, Phys. Lett. B **353**, 313 (1995).

[19] P. L. Frabetti *et al.*, Phys. Rev. D **50**, R2953 (1994).

[20] J. Bartelt *et al.* (CLEO Collaboration), "Search for CP Violation in D^0 Decay," CLNS 95/1333, CLEO 95-7, submitted to Phys. Rev. (1995).

[21] E. M. Aitala *et al.*, "Search for the Flavor-Changing Neutral-Current Decays $D^+ \to \pi^+\mu^+\mu^-$ and $D^+ \to \pi^+e^+e^-$," FERMILAB-Pub-95/142-E, submitted to Phys. Rev. Lett. (1995).

[22] M. Purohit and J. Weiner, "Preliminary Results on the Decays $D^+ \to K^+\pi^+\pi^-$, $D^+ \to K^+K^+K^-$," FERMILAB-Conf-94/408-E, to appear in *Proc. DPF '94*, Albuquerque, NM, Aug. 1–8, 1994.

[23] D. Cinabro *et al.* (CLEO collaboration), Phys. Rev. Lett. **72**, 1406 (1994).

[24] M. Purohit, "$D^0 - \overline{D^0}$ Mixing Results from E791," to appear in *Proc. LISHEP95 Workshop*, Rio de Janeiro, Brazil, Feb. 20–22, 1995.

[25] J. R. Fry and T. Ruf, "CP Violation and Mixing in D Decays," *Proc. Third Workshop on the Tau/Charm Factory*, Marbella, Spain, 1–6 June 1993, Edition Frontieres (1994), p. 387.

[26] G. Gladding, "$D^0 - \overline{D^0}$ Mixing and CP Violation: Experimental Projections for a τ-Charm Factory," in *Proc. Tau-Charm Factory Workshop*, Stanford Linear Accelerator Center, Stanford, CA, May 23–27, 1989, SLAC-Report-343, p. 152.

[27] W. Toki, "BES Program and Tau/Charm Factory Prospects," in **The Future of High-Sensitivity Charm Experiments**, *op. cit.*, p. 57.

[28] F. Buccella *et al.*, Phys. Rev. D **51**, 3478 (1995).

[29] E. A. Paschos and Y. L. Wu, Mod. Phys. Lett. **A6**, 93 (1991).

[30] M. Lu, M. B. Wise, and M. J. Savage, Phys. Lett. B **337**, 133 (1994).

[31] K. Arisaka *et al.*, Fermilab Proposal 832 (1990).

[32] G. D. Barr *et al.*, NA48 Proposal, CERN/SPSC/90-22 (1990).

[33] B. Winstein and L. Wolfenstein, Rev. Mod. Phys. **65**, 1113 (1993).

[34] J. Antos *et al.*, Fermilab Proposal 871;
E. C. Dukes, "A New Fermilab Experiment to Search for Direct CP Violation in Hyperon Decays," to appear in *Proc. 11th Int. Symp. on High Energy Spin Physics*, Bloomington, IN, Sept. 15–22, 1994.

[35] S. L. Glashow, J. Iliopoulos, and L. Maiani, Phys. Rev. D **2**, 1285 (1970);
S. L. Glashow and S. Weinberg, Phys. Rev. D **15**, 1958 (1977);
E. A. Paschos, Phys. Rev. D **15**, 1966 (1977).

[36] M. Gorn, Phys. Rev. D **20**, 2380 (1979).

[37] A. J. Schwartz, Mod. Phys. Lett. **A8**, 967 (1993).

[38] G. Burdman, E. Golowich, J. L. Hewett, and S. Pakvasa, "Radiative Weak Decays of Charm Mesons," FERMILAB-Pub-94/412-T, hep-ph/9502329 (1995).

[39] W. Buchmuller and D. Wyler, Phys. Lett. **177B**, 377 (1986) and Nucl. Phys. **B268**, 621 (1986);
Miriam Leurer, Phys. Rev. Lett. **71**, 1324 (1993).

[40] L. Montanet *et al.* (Particle Data Group), Phys. Rev. D **50**, 1173 (1994).

[41] M. Adamovich *et al.*, "Search for the Decay $D^0 \to \mu^+\mu^-$," CERN-PPE/95-71, submitted to Phys. Lett. B (1995).

[42] P. D. Sheldon, "Searching for CP Violation, Flavor Changing Neutral Currents, and Lepton Number Violation in Charm Decay," in **The Future of High-Sensitivity Charm Experiments**, *op. cit.*, p. 25.

[43] K. Kodama *et al.* (E653 collaboration), Phys. Lett. B **345**, 85 (1995).

[44] J. Cumalat, "Fermilab Fixed-Target Charm Program," in **Proceedings: The Tau-Charm Factory in the Era of B-Factories and CESR**, Stanford Linear Accelerator Center, Stanford, CA, Aug. 15–16, 1994, L. V. Beers and M. L. Perl, *eds.*, SLAC-Report-451 (1994), p. 335.

[45] W. C. Louis *et al.*, Phys. Rev. Lett. **56**, 1027 (1986).

[46] C. S. Mishra *et al.*, Phys. Rev. D **50**, R9 (1994).

[47] I. I. Bigi, "Open Questions in Charm Decay Deserving an Answer," in **The Future of High-Sensitivity Charm Experiments**, *op cit.*, p. 323.

[48] A. Ali, V. M. Braun, and H. Simma, Z. Phys. **C63**, 437 (1994).

[49] I. I. Bigi, F. Gabbiani, and A. Masiero, Z. Phys. **C48**, 633 (1990).

[50] E. Eichten, I. Hinchliffe, K.D. Lane, and C. Quigg, Phys. Rev. D **34**, 1547 (1986).

[51] A. D. Sakharov, Pis'ma Zh. Eksp. Teor. Fiz. **5**, 32 (1967) [JETP Lett. **5**, 24 (1967)];
E. W. Kolb and M. S. Turner, "Baryogenesis," in **The Early Universe**, Addison-Wesley, 1990, p. 157.

[52] A. Hadeed and B. Holdom, Phys. Lett. **159B**, 379 (1985).

[53] See also Ref. [54]; J. F. Donoghue *et al.*, Phys. Rev. D **33**, 179 (1986); H. Georgi, Phys. Lett. B **297**, 353 (1992); and T. Ohl, G. Ricciardi, and E. H. Simmons, Nucl. Phys. **B403**, 605 (1993).

[54] L. Wolfenstein, Phys. Lett. **164B**, 170 (1985).

[55] L. Hall and S. Weinberg, Phys. Rev. D **48**, R979 (1993); such models were previously considered by S. Pakvasa and H. Sugawara, Phys. Lett. **73B**, 61 (1978); S. Pakvasa et al., Phys. Rev. D **25**, 1895 (1982); T. P. Cheng and M. Sher, Phys. Rev. D **35**, 3484 (1987); and M. Shin, M. Bander, and D. Silverman, in *Proc. Tau-Charm Factory Workshop*, Stanford Linear Accelerator Center, Stanford, CA, May 23–27, 1989, SLAC-Report-343, p. 686.

[56] L. Wolfenstein, "CP Violation in $D^0 - \overline{D^0}$ Mixing," CMU-HEP-95-04, hep-ph-9505285 (1995).

[57] Y. L. Wu and L. Wolfenstein, Phys. Rev. Lett. **73**, 1762 (1994) and references therein.

[58] S. Weinberg, Phys. Rev. Lett. **37**, 657 (1976).

[59] A. Datta, Phys. Lett. **154B**, 287 (1985).

[60] I. I. Bigi and A. F. Sanda, Phys. Lett. B **171**, 320 (1985).

[61] L. F. Abbott, P. Sikivie, and M. B. Wise, Phys. Rev. D **21**, 1393 (1980);
V. Barger, J. L. Hewett, and R. J. N. Phillips, Phys. Rev. D **41**, 3421 (1990);
J. L. Hewett, Phys. Rev. Lett. **70**, 1045 (1993).

[62] Y. Nir and N. Seiberg, Phys. Lett. B **309**, 337 (1993).

[63] A. S. Joshipura, Phys. Rev. D **39**, 878 (1989).

[64] G. Blaylock, A. Seiden, and Y. Nir, "The Role of CP Violation in $D^0\overline{D^0}$ Mixing," SCIPP 95/16, hep-ph-9504306 (1995).

[65] T. E. Browder and S. Pakvasa, "Experimental Implications of Large CP Violation and Final State Interactions in the Search for $D^0 - \overline{D^0}$ Mixing," UH-511-828-95 (1995).

[66] T. Liu, "The $D^0 - \overline{D^0}$ Mixing Search – Current Status and Future Prospects," in **The Future of High-Sensitivity Charm Experiments**, *op cit.*, p. 375, and
T. Liu, Ph.D. Thesis, Harvard University, HUHEPL-20 (1995).

[67] J. C. Anjos et al., Phys. Rev. Lett. **60**, 1239 (1988).

[68] A. Tripathi, "A Search for $D^0 - \overline{D^0}$ Mixing in the Mode $D^0 \to K^-e^+\nu_e$ ($\overline{D^0} \to K^+e^-\bar{\nu}_e$)," presented at the *General Meeting of the American Physical Society*, Washington, DC, April 18–21, 1995.

[69] R. Morrison, "Charm2000 Workshop Summary," in **The Future of High-Sensitivity Charm Experiments**, *op cit.*, p. 313; see also
T. Liu, "Summary Report of the Mixing Working Group," *ibid.*, p. 441.

[70] H. Yamamoto, Ph.D. Thesis, California Institute of Technology, CALT-68-1318 (1985).

[71] G. E. Gladding (Mark II collaboration), "Search for $\psi'' \to D^0\overline{D^0} \to$ Strangeness ± 2 Final States," in *Proc. 5th Int. Conf. on Physics in Collision*, Autun, France, Jul. 3–5, 1985, B. Aubert and L. Montanet, *eds.*, Editions Frontieres, 1985, p. 259.

[72] C. N. Brown, D. M. Kaplan, and D. J. Summers, "Report of the Working Group on Beams and Architectures," in **The Future of High-Sensitivity Charm Experiments**, *op cit.*, p. 459.

[73] We average together results on charged- and neutral-D production by 800 GeV proton beams from R. Ammar *et al.*, Phys. Rev. Lett. **61**, 2185 (1988); K. Kodama *et al.*, Phys. Lett. B **263**, 573 (1991); and M. J. Leitch *et al.*, Phys. Rev. Lett. **72**, 2542 (1994), assuming linear A dependence as observed by Leitch *et al.*

[74] T. Lohse *et al.*, "HERA-B: An Experiment to Study CP Violation in the B System Using an Internal Target at the HERA Proton Ring," Proposal to DESY, DESY-PRC 94/02, May 1994.

[75] R. Tesarek, "Diamond Detectors," in **The Future of High-Sensitivity Charm Experiments**, *op cit.*, p. 163.

[76] B. Baumbaugh *et al.*, Nucl. Instr. Meth. **A345**, 271 (1994).

[77] R. Ruchti, "Fiber Tracking," in **The Future of High-Sensitivity Charm Experiments**, *op cit.*, p. 173;
D. Adams *et al.*, "Cosmic Ray Test Results of the D0 Prototype Scintillating Fiber Tracker," FERMILAB-Conf-95-012-E, to appear in *Proc. 4th Int. Conference on Advanced Technology and Particle Physics*, Como, Italy, 3–7 Oct. 1994.

[78] C. J. Kennedy, R. F. Harr, and P. E. Karchin, "Simulation Study of a Transverse Energy Trigger for a Fixed Target Beauty and Charm Experiment at Fermilab," Sept. 9, 1993 (unpublished).

[79] C. Lee *et al.*, IEEE Trans. Nucl. Sci. **38**, 461 (1989);
M. Adamovich *et al.*, *ibid.* **37** No. 2, 236 (1990);
B. C. Knapp, Nucl. Instr. Meth. **A289**, 561 (1990).

[80] D. C. Christian, "Triggers for a High-Sensitivity Charm Experiment," in **The Future of High-Sensitivity Charm Experiments**, *op cit.*, p. 221.

[81] G. Charpak, L. M. Lederman, and Y. Giomataris, Nucl. Instr. Meth. **A306**, 439 (1991);
D. M. Kaplan *et al.*, *ibid.* **A330**, 33 (1993);
G. Charpak *et al.*, *ibid.* **A332**, 91 (1993).

[82] A. M. Halling and S. Kwan, Nucl. Instr. Meth. **A333**, 324 (1993).

[83] L. D. Isenhower *et al.*, "P865: Revised Letter of Intent for a High-Sensitivity Study of Charm and Beauty Decays," Letter of Intent to Fermilab, April 2, 1993.

[84] M. Atac *et al.*, "The Development of the Optical Discriminator," to appear in *Proc. 7th Vienna Wire Chamber Conference*, Vienna, Austria, 13–17 Feb. 1995;
G. Charpak *et al.*, "The Optical Trigger for the E871 Experiment," RD30 Note, March 16, 1995.

[85] S. Kwan, private communication.

[86] T. Huang, this Workshop.

An Overview of $D^0\bar{D}^0$ Mixing Search Techniques: Current Status and Future Prospects

Tiehui (Ted) Liu

Department of Physics, Princeton University, Princeton, NJ 08544

Abstract

The search for $D^0\bar{D}^0$ mixing may carry a large discovery potential for new physics since the $D^0\bar{D}^0$ mixing rate is expected to be small in the Standard Model. The past decade has seen significant experimental progress in sensitivity. This paper discusses the techniques, current experimental status, and future prospects for the mixing search. Some new ideas, applicable to future mixing searches, are introduced. In this paper, the importance of separately measuring the decay rate difference and the mass difference of the two CP eigenstates (in order to observe New Physics) has been emphasized, since the theoretical calculations for long distance effects are still plagued by large uncertainties.

1 Introduction

Particle-antiparticle mixing has always been of fundamental importance in testing the Standard Model and constraining new physics. This is because mixing is responsible for the small mass differences between the mass eigenstates of neutral mesons. Being a flavor changing neutral current (FCNC) process, it often involves heavy quarks in loops. Such higher order processes are of great interest since the amplitudes are sensitive to any weakly-coupling quark flavor running around the loop. Historically, $K^0\bar{K}^0$ mixing is the rare (FCNC) process that has been experimentally examined in the greatest detail. It has been amply demonstrated that in spite of many inherent uncertainties of strong interaction physics, the Standard Model predicts the correct phenomenology of the $K^0\bar{K}^0$ mixing. In fact, based on the calculation of the K_L - K_S mass difference, Gaillard and Lee [1] were able to estimate the value of the charm quark mass before the discovery of charm. Moreover, $B^0 \bar{B}^0$ mixing gave the first indication of a large top quark mass.

Although $D^0 \bar{D}^0$ mixing is very similar to $K^0 \bar{K}^0$ and $B^0 \bar{B}^0$ mixing, as all are FCNC processes, there are significant differences which make $D^0 \bar{D}^0$ mixing a possible unique place to explore new physics. Roughly speaking, in the case of K and B FCNC processes, the appearance of the top quark in the internal loop with $m_t > M_W >> m_c, m_u$ removes the GIM [2] suppression, making K and B decays a nice place to test FCNC transitions and to study the physics of the top. In the case of D FCNC processes, the FCNC are much stronger suppressed because the down-type quarks (d, s and b) with m_d, m_s, $m_b << M_W$ enter the internal loops and the GIM mechanism is much more effective [3]. Therefore the $D^0 \overline{D}^0$ mixing rate is expected to be small in the Standard Model, which means the mixing search may carry a large potential for discovery of new physics. There are many extensions of the Standard Model which allow $D^0 \overline{D}^0$ mixing (the mass difference between the two CP eigenstates) to be significantly larger than the Standard Model prediction (for example, see [4] to [15]). Recent reviews on FCNC processes in D decays can be found elsewhere [16, 17, 18, 19]. In general, there could be a large enhancement of the one-loop induced FCNC processes in D decays with no constraint from limits on FCNC processes in the K and B systems. Roughly speaking, this is because the couplings of FCNC to up-type quarks (u,c,t) could be completely different from those to down-type quarks (d,s,b). Thus one gains independent pieces of information when searching for FCNC in D decays, compared to what is learned searching for FCNC in K and B decays.

One can characterize $D^0 \bar{D}^0$ mixing in terms of two dimensionless variables: $x = \delta m / \gamma_+$ and $y = \gamma_- / \gamma_+$, where the quantities $\gamma_\pm$ and δm are defined by $\gamma_\pm = (\gamma_1 \pm \gamma_2)/2$ and $\delta m = m_2 - m_1$ with m_i, γ_i ($i = 1, 2$) being the masses and decay rates of the two CP (even and odd) eigenstates. Assuming a small mixing, namely, $\delta m, \gamma_- \ll \gamma_+$ or $x, y \ll 1$, we have $R_{\text{mixing}} = (x^2 + y^2)/2$. Mixing can be caused either by $x \neq 0$ (meaning that mixing is genuinely caused by the $D^0 - \bar{D}^0$ transition) or by $y \neq 0$ (meaning mixing is caused by the fact that the fast decaying component quickly disappears, leaving the slow decaying component which is a mixture of D^0 and $\bar{D}^0$). Theoretical calculations of $D^0 \bar{D}^0$ mixing in the Standard Model are plagued by large uncertainties. While short distance effects from box diagrams are known [1] to give a negligible contribution ($\sim 10^{-10}$), the long distance effects from second-order weak interactions with mesonic intermediate states may give a much larger contribution. Estimates of R_{mixing} from long distance effects range from 10^{-7} to 10^{-3} [20]. It has recently been argued by Georgi and others that the long distance contributions are smaller than previously estimated, implying that cancellations occur between contributions from different classes of intermediate mesonic states [22]. While many people now believe that within the Standard Model $R_{\text{mixing}} < 10^{-7}$ [16, 17, 18], others think R_{mixing} could be much larger [23, 24], say 10^{-4} [23] (meaning both x and y are above 10^{-3}). For example, Bigi [23] pointed out that observing a non-vanishing value for R_{mixing} between 10^{-4} and 10^{-3} would at present not constitute irrefutable evidence for New Physics, considering the large uncertainties in the long distance calculations. While there is some hope that the uncertainties can be reduced in the future, as pointed out by Bigi [23], partly through theoretical efforts and partly through more precise and comprehensive data (since a more reliable estimate can be obtained from a dispersion relation involving the measured branching ratios for the channels common to D^0 and $\bar{D}^0$ decays), one recent paper claims that the hope is rather remote [24]. Speculations abound, but (fortunately) physics is an experimental science, and only with solid experimental evidence will we be able to properly address these problems. As experimentalist, I think the best way is to measure x and y separately, as suggested in [26, 27]. As will be discussed, this is experimentally possible. If we can measure R_{mixing} as well as y, then we can in effect measure x. Within the Standard Model, x and y are expected to be at the same level, although we do not know exactly at what level as theoretical calculations for long distance effects (which contribute to both x and y) are still plagued by large uncertainties. We expect New Physics does not affect the decays in a significant way thus does not contribute to y, but only to x. The point I am trying to make here is that the long distance contribution can be measured, even if it cannot be calculated in a reliable way; that is, by measuring y directly. If we can experimentally confirm that indeed $x >> y$, then we can claim New Physics, regardless of what theoretical calculations for long distance effects are. Otherwise, if it turns out $x \sim y$, then mostly likely we are seeing the Standard Model Physics. Therefore, it is crucial to measure y in order to understand the size of x within the Standard Model. This is one of the major points I have been trying to make in the past [26, 27, 25] and in this paper.

Motivated by the experience with $K^0 \bar{K}^0$ system, experimenters have been searching for $D^0 \bar{D}^0$ mixing since shortly after the discovery of D^0 meson at SPEAR in 1976, in either hadronic decays $D^0 \to \bar{D}^0 \to K^+ \pi^-(X)$ [28], or semileptonic decays $D^0 \to \bar{D}^0 \to X^+ l^- \nu$. The past decade has seen significant experimental progress in sensitivity (from 20% to 0.37% [29] to [41]), as can be seen in Figure 1. The search for $D^0 \bar{D}^0$ mixing has a long and interesting history (see Figure 1). In the first few years, people searched for $D^0 \to K^+ \pi^-$ assuming that it would be due to mixing only. Normally, D^0 decays by Cabibbo favored decay $D^0 \to K^- \pi^+$ and $\bar{D}^0 \to K^+ \pi^-$. A signal for $D^0 \to K^+ \pi^-$

could indicate mixing of $D^0 \to \bar{D}^0$. But it could also indicate a different decay channel, namely, Doubly Cabibbo Suppressed Decay(DCSD) $D^0 \to K^+\pi^-$, which is suppressed with respect to the Cabibbo favored decay by a factor of $tan^4\theta_C \sim 0.3\%$ where θ_C is the Cabibbo angle. As will be discussed, around 1985 there were hints of $D^0 \to K^+\pi^-\pi^0$ observation, which could be due to DCSD or mixing. The popular interpretation neglected the possible DCSD contribution, giving the impression that $D^0\bar{D}^0$ mixing rate R_{mixing} could be of order $\mathcal{O}(1\%)$. This engendered much theoretical work to accommodate the possibly large mixing rate. At that time, the "theoretical prejudice" was that long-distance contributions dominated and would give a large mixing rate on the order of 1% level. Later on, fixed target experiments published limits which were not much larger than the naïve quark model DCSD rate. In light of these results, the commonly held impression was then that DCSD was much larger than mixing, and that exploring mixing by means of hadronic D^0 decays had been almost exhausted as a technique since the "annoying DCSD background" would inherently limit ones ability to observe the interesting physics - $D^0\bar{D}^0$ mixing. It was believed by many that the signature of mixing appears only at longer decay times; therefore, it will suffer from DCSD fluctuation, and destructive interference could wipe out the signature of mixing. Since semileptonic decays are not subject to this "annoying background", the general consensus was that semileptonic decays were a better avenue to explore $D^0\bar{D}^0$ mixing.

However, as will be discussed in more detail later, the commonly believed "annoying DCSD background" does not necessary inherently limit the hadronic method as the potentially small mixing signature could show up in the interference term [26]. Moreover, the possible differences between the resonant substructure in many DCSD and mixing decay modes could, in principle, be used to distinguish between DCSD and mixing candidates experimentally [26] (the importance of the mixing-DCSD interference effect will be more clear here). Our ability to observe the signature of a potentially small mixing signal depends on the number of $D^0 \to K^+\pi^-(X)$ events we will have. This means observing $D^0 \to K^+\pi^-(X)$ would be an important step on the way to observing mixing with this technique. Recently, CLEO has observed a signal for $D^0 \to K^+\pi^-$ (see Figure 2), and found R = $\mathcal{B}(D^0 \to K^+\pi^-)/\mathcal{B}(D^0 \to K^-\pi^+) \sim 0.8\%$ [42]. Unfortunately, without a precision vertex detector, CLEO is unable to distinguish a potential mixing signal from DCSD. If the number of reconstructed charm decays can reach 10^8 around the year 2000, that would allow one to reach a new threshold of sensitivity to $D^0\bar{D}^0$ mixing, and perhaps actually observe it. Therefore, it is time to take a detail look of all possible techniques for $D^0\bar{D}^0$ mixing search.

This paper [1] is organized as follows: in Section 2 there is a review of the experimental techniques which can be used to search for mixing, together with some thoughts on possible new techniques. In each case, the relevant phenomenology will be briefly presented. Section 3 discusses the history, present status and future prospects of searching for mixing at different experiments. In Section 4, a comparison of the future prospects of the different experiments with different techniques, in the light of the CLEO II signal for $D^0 \to K^+\pi^-$, will be given. A brief summary is given in Section 5. Some detailed formulae and discussions (including possible CP violation effect) are provided in the appendices.

2 The Techniques

The techniques which can be used to search for mixing can be roughly divided into two classes: hadronic and semi-leptonic. Each method has advantages and limitations, which are described below.

2.1 Hadronic method

The hadronic method is to search for the D^0 decays $D^0 \to K^+\pi^-(X)$. These decays can occur either through $D^0\bar{D}^0$ mixing followed by Cabibbo favored decay $D^0 \to \bar{D}^0 \to K^+\pi^-(X)$, or through DCSD $D^0 \to K^+\pi^-(X)$. This means that the major complication for this method is the need to distinguish between DCSD and mixing [48]. The hadronic method can therefore be classified according to how DCSD and mixing are distinguished. In principle, there are at least three different ways to distinguish between DCSD and mixing candidates experimentally: (A) use the difference in the decay time-dependence; (B) use the possible difference in the resonant substructure between DCSD and mixing events in $D^0 \to K^+\pi^-\pi^0, K^+\pi^-\pi^+\pi^-$, etc. modes; (C) use the quantum statistics of the production and the decay processes.

Method (A) requires that the D^0 be highly boosted and so that the decay time information can be measured. Method (B) requires knowledge of the resonant substructure of the DCSD decays, which is unfortunately something about which we have no idea at this time. Finally, method (C) requires that one use e^+e^- annihilation in the charm threshold region. In the following, we will discuss these three methods in some detail.

---

[1] This paper is essentially a revised version of Chapter 6 in [25] and [26].

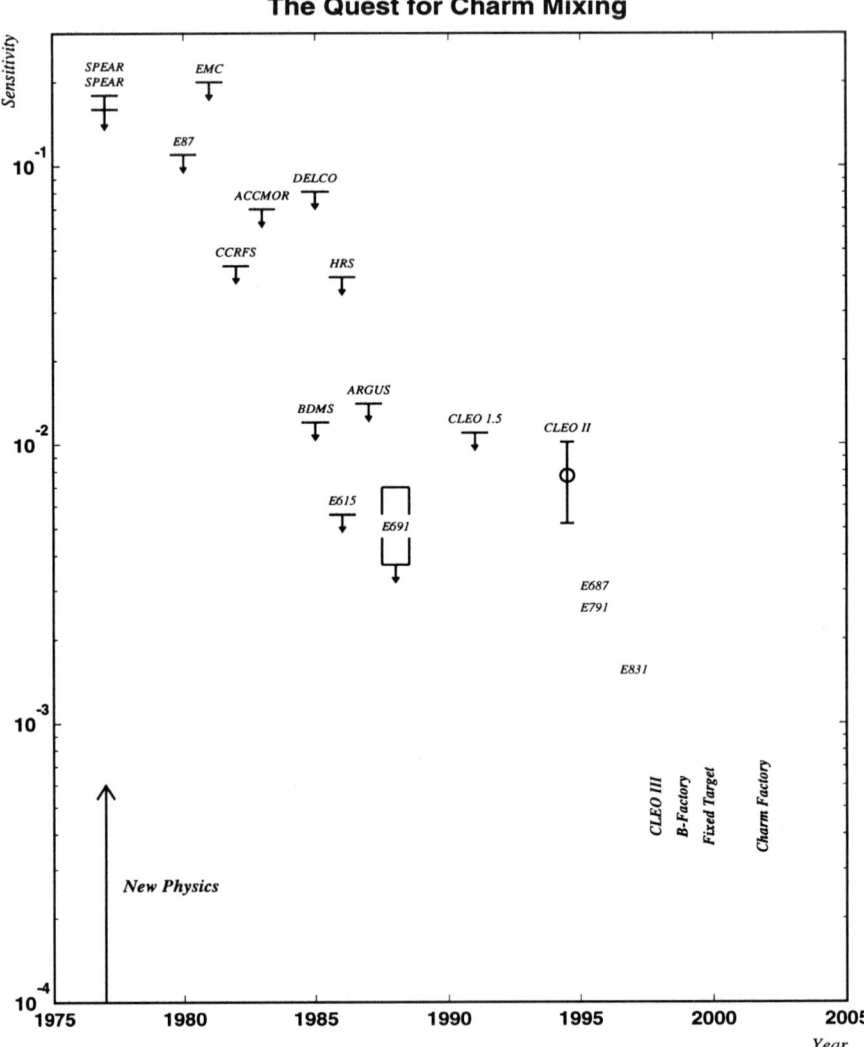

Figure 1: The history of the quest for $D^0\bar{D}^0$ mixing. Note that the range in E691 result reflects the possible effects of interference between DCSD and mixing, and the CLEO II signal could be due to either mixing or DCSD, or a combination of the two.

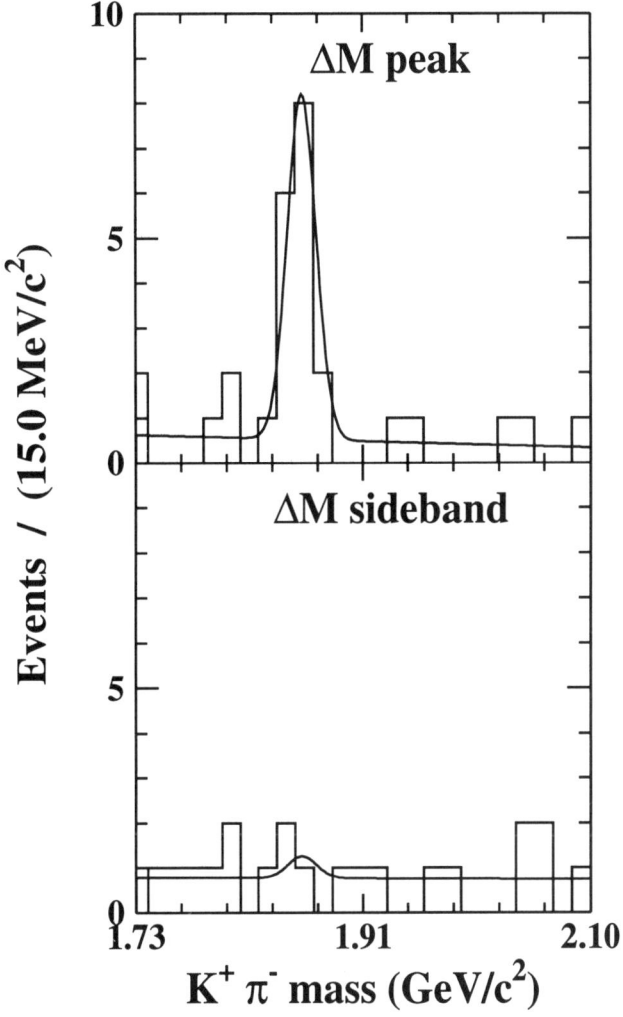

Figure 2: The CLEO II signal for $D^0 \to K^+\pi^-$. The D^0 mass for wrong sign events. (a) for events in the ΔM peak; (b) for events in the ΔM sidebands. The solid lines are the fits using the corresponding right sign mean and σ in data.

2.1.1 Method A –use the difference in the time-dependence of the decay

This method [49] is to measure the decay time of the $D^0 \to K^+\pi^-$ decay. Here the D^0 tagging is usually done by using the decay chain $D^{*+} \to D^0\pi_s^+$ followed by $D^0 \to K^+\pi^-$. The π_s^+ from D^{*+} has a soft momentum spectrum and is referred to as "the slow pion". The idea is to search for the wrong sign D^{*+} decays, where the slow pion has the same charge as the kaon arising from the D^0 decay. This technique utilizes the following facts: (1) DCSD and mixing have different decay time-dependence, which will be described below. (2) The charge of the slow pion is correlated with the charm quantum number of the D^0 meson and thus can be used to tag whether a D^0 or $\bar{D}^0$ meson was produced in the decay $D^{*+} \to D^0\pi_s^+$ or $D^{*-} \to \bar{D}^0\pi_s^-$. (3) The small Q value of the D^{*+} decay results in a very good mass resolution in the mass difference $\Delta M \equiv M(D^{*+}) - M(D^0) - M(\pi_s^+)$ and allows a D^{*+} signal to obtained with very low background. (4) The right sign signal $D^{*+} \to D^0\pi_s^+$ followed by $D^0 \to K^-\pi^+$ can be used to provide a model-independent normalization for the mixing measurement.

A pure D^0 state generated at $t = 0$ decays to the $K^+\pi^-$ state either by $D^0\bar{D}^0$ mixing or by DCSD, and the two amplitudes may interfere. The amplitude for a D^0 decays to $K^+\pi^-$ relative to the amplitude for a D^0 decays to $K^-\pi^+$ is given by (see appendix A)

$$A = \sqrt{R_{\text{mixing}}/2}\ t + \sqrt{R_{\text{DCSD}}}\ e^{i\phi} \qquad (1)$$

where ϕ is an unknown phase, t is measured in units of average D^0 lifetime. Detailed discusion on the interference phase ϕ can be found in Appendix A. Here $R_{DCSD} = |\rho|^2$ where ρ is defined as:

$$\rho = \frac{Amp(D^0 \to K^+\pi^-)}{Amp(\bar{D}^0 \to K^+\pi^-)} \qquad (2)$$

denoting the relative strength of DCSD. We have also assumed a small mixing; namely, $\delta m, \gamma_- \ll \gamma_+$ or $x, y \ll 1$, and CP conservation. Detailed formulae and discussions (including possible CP violation effect) can be found in Appendix A. In the following, we will simply discuss the basic idea of how to distinguish DCSD and mixing with this technique.

The first term, which is proportional to t, is due to mixing and the second term is due to DCSD. It is this unique attribute of the decay time-dependence of mixing which can be used to distinguish between DCSD and mixing. Now we have:

$$I(D^0 \to K^+\pi^-)(t) \propto (R_{\text{DCSD}} + \sqrt{2R_{\text{mixing}}R_{\text{DCSD}}}\ t\ \cos\phi + \frac{1}{2}R_{\text{mixing}}t^2)e^{-t} \qquad (3)$$

Note that this form is different from what people usually use (but equivalent), see Appendix A. I prefer this form since it is not only more convenient for discussion here, but also much easier to be used to fit data. Define $\alpha = R_{\text{mixing}}/R_{\text{DCSD}}$, which describes the strength of mixing relative to DCSD. Equation 3 can then be rewritten as:

$$I(D^0 \to K^+\pi^-)(t) \propto R_{\text{DCSD}}(1 + \sqrt{2\alpha}\, t\cos\phi + \frac{1}{2}\alpha t^2)e^{-t} \quad (4)$$

From this equation, one may read off the following properties [26]: (1) The mixing term peaks at $t = 2$. (2) The interference term peaks at $t = 1$. (3) A small mixing signature can be enhanced by DCSD through interference (with $cos\phi \neq 0$) at lower decay times, compared to the case without interference (with $cos\phi = 0$). The ratio between the interference term and the mixing term, denoted $\xi(t)$, is given by $\xi(t) = \sqrt{\frac{8}{\alpha}}\, cos\phi/t \propto \sqrt{\frac{1}{\alpha}}$. So when $\alpha \to 0$, $\xi \to \infty$. (4) Only for $t > \sqrt{\frac{8}{\alpha}}|cos\phi|$ does the interference term become smaller than the mixing term. (5) $I(t_0) = 0$ happens and only happens when $cos\phi = -1$, and only at location $t_0 = \sqrt{\frac{2}{\alpha}}$. (6) One can obtain a very pure DCSD sample by cutting at low decay time.

While Property (1) tells us that the mixing term does live at longer decay time, Property (3) tells us clearly that we should not ignore the interference term. In fact, that's the last thing one wants to ignore! (unless we know for sure $\cos\phi = 0$). The commonly believed "annoying background", namely DCSD, could actually enhance the chance of seeing a very small mixing signal through the interference, compared to the case without the interference. In other words, the "annoying DCSD background" does not necessary inherently limit the hadronic method since the potentially small mixing signature could show up in the interference term. For a very small mixing rate, almost all the mixing signature could show up in the interference term, not in the mixing term, as long as $\cos\phi \neq 0$. Property (2) tells us at which location one expect to find the richest signature of a potential small mixing, which is where the interference term peaks: $t \sim 1$ (why should one keep worrying about long lived DCSD tails? let's hope for $\cos\phi \neq 0$ first). Property (5) shows that destructive interference is not necessarily a bad thing. In fact, it could provide extra information. For example, if $\cos\phi = -1$, then one should find $I(t_0) = 0$ at $t_0 = \sqrt{\frac{2}{\alpha}}$, see Figure 5. Note this unique attribute will become more interesting in method B, see Appendix B. This tells us that the destructive interference does not necessarily wipe out the signature of mixing. For the general case, interference will lead to very characteristic time distribution, as can be clearly seen in Figure 6. Property (6) shows that we can study DCSD well without being confused by the possible mixing component. This will also become more important when we discuss method B.

Therefore the signature of mixing is a deviation from a perfect exponential time distribution with the slope of γ_+ [2]. Our ability to observe this signature depends on the number of $D^0 \to K^+\pi^-$ events we will have. Right now this is limited by the rather poor statistics. Figures 3 or Figure 4 shows each term with $\alpha = 10\%$ and $\cos\phi = \pm 1$ (with $R_{DCSD} = 1$).

It is worth to point out that the interference between mixing and DCSD also occurs in $B^0\bar{B}^0$ system. In this case, mixing is quite large and can be well measured while DCSD is small and unknown. The signature of the small DCSD would mostly show up in the interference term. But here we are not interested in measuring mixing nor measuring DCSD, what is interesting here is to measure CP violation. One can use $B_d^0 \to D^+\pi^-$ as an example to show the basic idea, see Appendix C.

It is interesting to point out here that there is also a possibility, previously unrecognized, of using the Singly Cabibbo Suppressed Decays (SCSD), such as $D^0 \to K^+K^-, \pi^+\pi^-$ to study mixing [26]. This is because (assuming CP conservation) those decays occur only through the CP even eigenstate, which means the decay time distribution is a perfect exponential with the slope of γ_1. Therefore, one can use those modes to measure γ_1. The mixing signature is not a deviation from a perfect exponential (again assuming CP conservation), but rather a deviation of the slope from $(\gamma_1+\gamma_2)/2$. Since $\gamma_+ = (\gamma_1+\gamma_2)/2$ can be measured by using the $D^0 \to K^-\pi^+$ decay time distribution, one can then derive $y = \gamma_-/\gamma_+ = (\gamma_2 - \gamma_1)/(\gamma_1 + \gamma_2)$. Observation of a non-zero y would demonstrate mixing caused by the decay rate difference ($R_{mixing} = (x^2+y^2)/2$). It is worth pointing out that in this case other CP even (odd) final states such as $D^0 \to K_S\rho^0$ can be also used to measure $\gamma_1(\gamma_2)$. In addition, there is no need to tag the D^0, since we only need to determine the slope. Note that this method is only sensitive to mixing caused by the decay rate difference between the two eigen states, not to mixing caused by the mass difference $x = \delta m/\gamma_+$ ($\delta m = m_2 - m_1$). Right after this technique was introduced [26] last summer, Fermilab fixed target experiments E791 started to apply this idea to their data [43]. The sensitivity of this method is discussed in Section 4.1.

2.1.2 Method B – use difference in resonance substructure

The idea of this new method [26] is to use the wrong sign decay $D^{*+} \to D^0\pi_s^+$ followed by $D^0 \to K^+\pi^-\pi^0$, $K^+\pi^-\pi^+\pi^-$, etc., and use the possible differences of the resonant substructure between mixing and DCSD to study

---

[2] One can use $D^0 \to K^-\pi^+$ to study the acceptance function versus decay time.

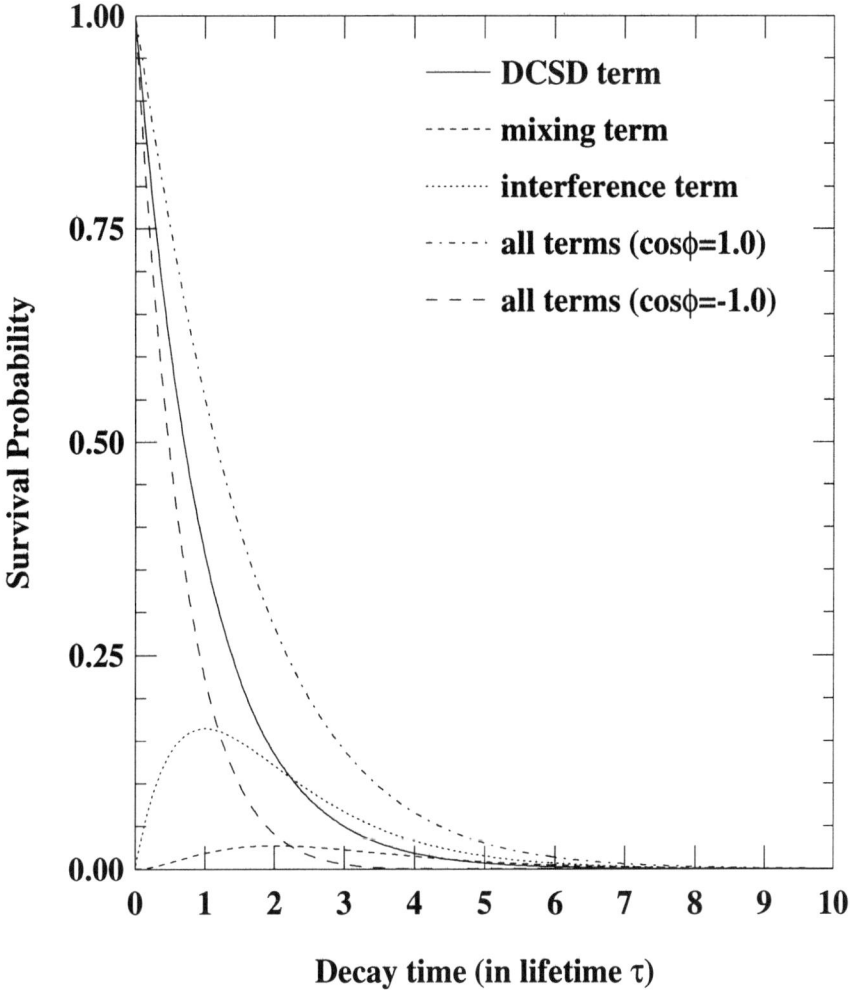

Figure 3: The decay time dependence of DCSD and mixing with $\alpha = R_{mixing}/R_{DCSD} = 10\%$.

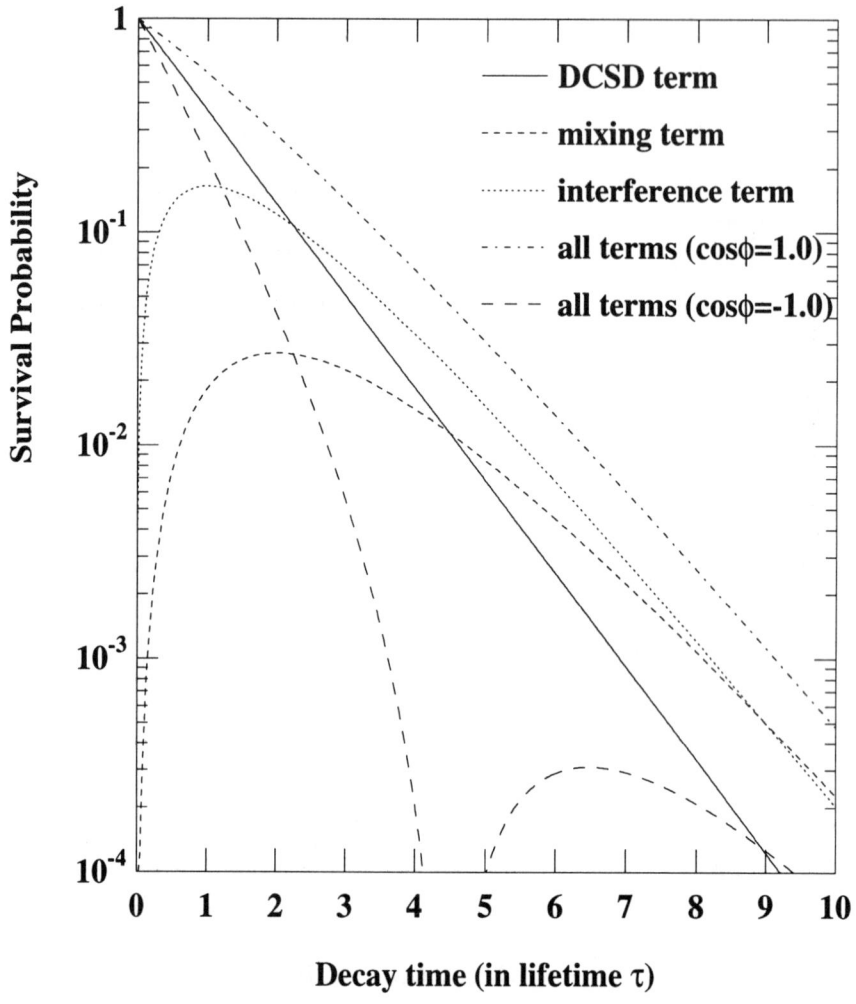

Figure 4: The decay time dependence of DCSD and mixing with $\alpha = R_{mixing}/R_{DCSD} = 10\%$, in log scale.

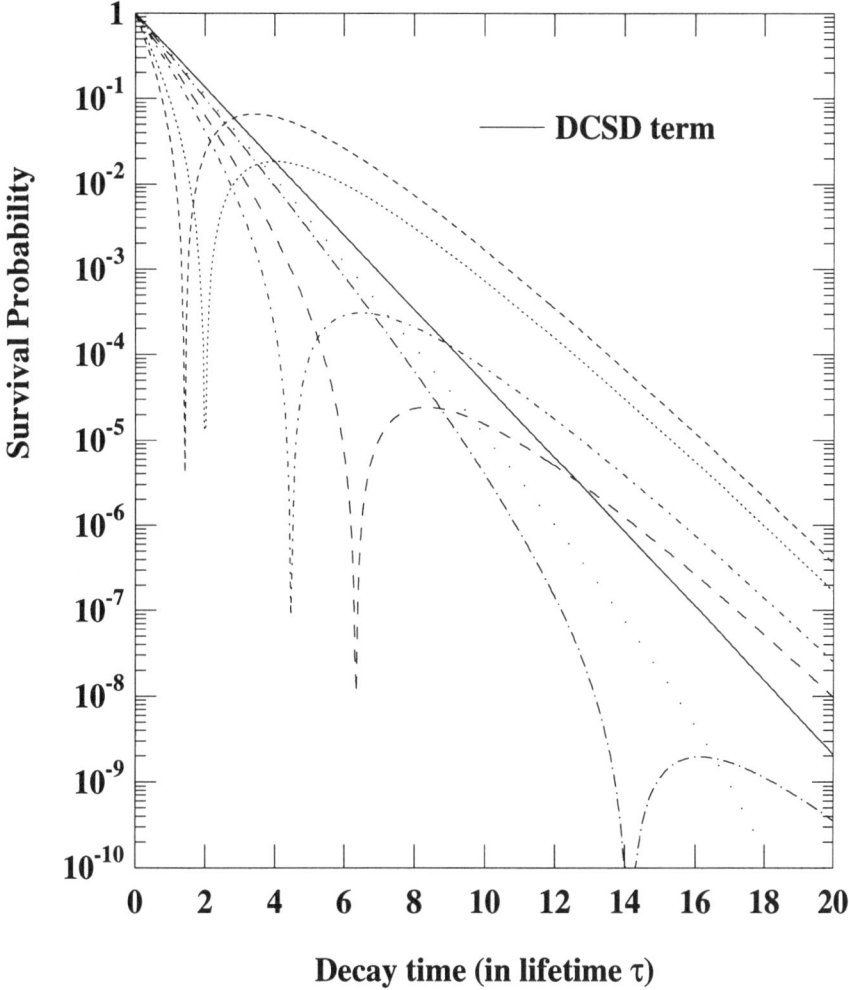

Figure 5: The decay time dependence of DCSD and mixing with maximal destructive interference $cos\phi = -1.0$. For different $\alpha = R_{mixing}/R_{DCSD}$ values: from left to right, $\alpha = 100\%, 50\%, 10\%, 5\%, 1\%, 0.5\%$ (with $R_{DCSD} = 10^{-2}$, this corresponds to $R_{mixing} = 10^{-2}, 5 \times 10^{-3}, 10^{-3}, 5 \times 10^{-4}, 10^{-5}, 5 \times 10^{-6}$).

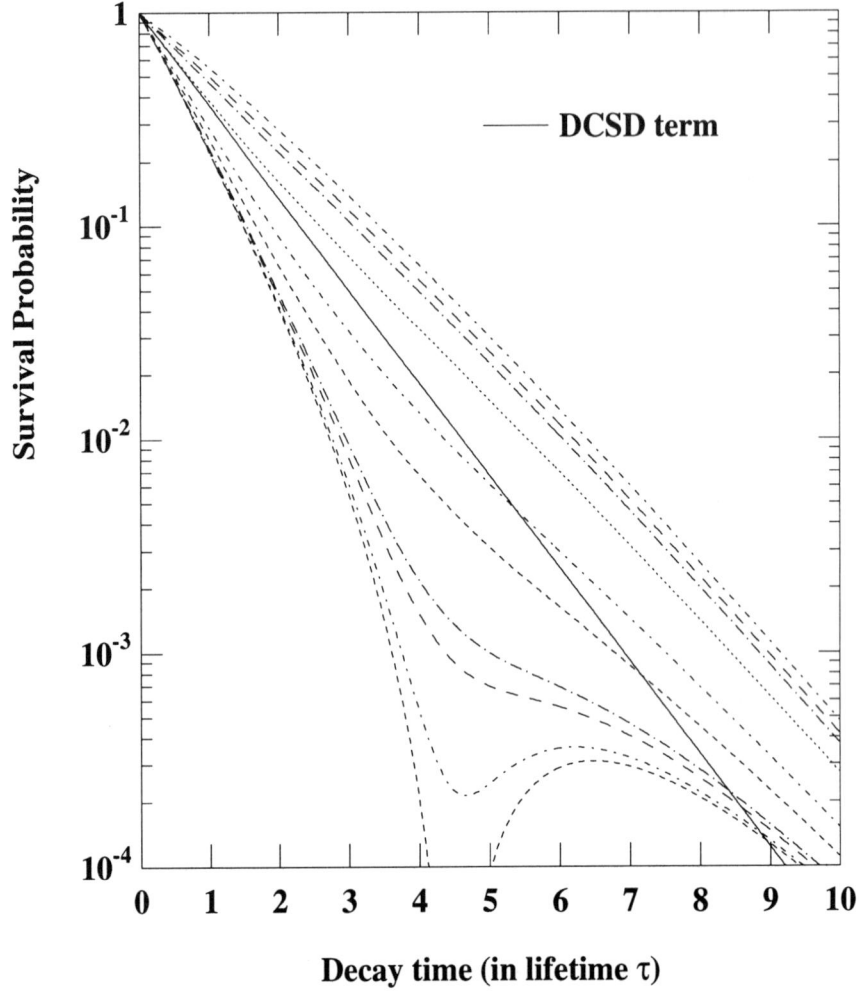

Figure 6: The decay time dependence of DCSD and mixing with $\alpha = R_{\text{mixing}}/R_{\text{DCSD}} = 10\%$. For different $cos\phi$ values: from bottom to top, $cos\phi = -1.0, -0.99, -0.96, -0.94, -0.80, -0.6, 0.0, 0.5, 0.7, 1.0$. The solid line is the DCSD term, as a reference line.

mixing. There are good reasons to believe that the resonant substructure of DCSD decay is different from that of mixing (Cabibbo favored decay, CFD). We can use the $D^0 \to K^+\pi^-\pi^0$ decay as an example. Detail discussion about this method can be found in appendix B (including possible CP violation effect), here we will just outline the basic idea.

For CFD and DCSD, the true yield density $n(p)$ at a point p in the Dalitz plot can be written as:

$$n(p) \propto |f_1 e^{i\phi_1} A_{3b} + f_2 e^{i\phi_2} BW_{\rho^+}(p) + f_3 e^{i\phi_3} BW_{K^{*-}}(p) + f_4 e^{i\phi_4} BW_{\bar{K}^{*0}}(p)|^2 \quad (5)$$

where f_i are the relative amplitudes for each component and ϕ_i are the interference phases between each submode. A_{3b} is the S-wave three-body decay amplitude, which is assumed to be flat across the Dalitz plot. The various terms BW are Breit-Wigner amplitudes for the $D^0 \to K^*\pi$ and $D^0 \to K\rho$ sub-reactions, which describe the strong resonances and decay angular momentum conservation: $BW_R \propto \frac{\cos\theta_R}{M_{ij} - M_R - i\Gamma_R/2}$ where M_R and Γ_R are the mass and width of the M_{ij} resonance (K^* or ρ), and θ_R is the helicity angle of the resonance. For CFD, f_i and ϕ_i have been measured by MARKIII [76], E691 [77] and are being measured by CLEO II. For DCSD, f_i and ϕ_i have not been measured. Note that in general

$$f_i^{DCSD}/f_i^{CFD} \neq f_j^{DCSD}/f_j^{CFD} \quad (i \neq j) \quad (6)$$

$$\phi_i^{DCSD} \neq \phi_i^{CFD} \quad (7)$$

This means that the resonant substructure (the true yield density $n(p)$) for DCSD is different from that of mixing. As both DCSD and mixing contribute to the wrong sign decay, the yield density for the wrong sign events $n_w(p)$ will have a complicated form. Just like in method A, for very small mixing, the interference term between DCSD and mixing could be the most important one.

Mathematically, the time-dependence of $D^0 \to K^+\pi^-\pi^0$ is the same as that of $D^0 \to K^+\pi^-$, the only difference is that now both R_{DCSD} and the interference phase ϕ (between DCSD and mixing) strongly depends on the location p on the Dalitz plot. As discussed in Appendix B, the time-dependence can be written in the form [25]:

$$I(\,|D^0_{\text{phys}}(t)> \to f\,)(p) =$$
$$\left[n_D(p) + \sqrt{2R_{\text{mixing}}\, n_D(p)\, n_C(p)}\, \cos\phi(p)\, t + \frac{1}{2} n_C(p)\, R_{\text{mixing}}\, t^2 \right] e^{-t} \quad (8)$$

where $n_D(p)$ and $n_C(p)$ are the true yield density for DCSD and CFD respectively. Detailed discussion on the interference phase ϕ can be found in Appendix B.

In principle, one can use the difference between the resonant substructure for DCSD and mixing events to distinguish mixing from DCSD. For instance, combined with method A, one can perform a multi-dimensional fit to the data by using the information on ΔM, $M(D^0)$, proper decay time t and the yield density on Dalitz plot $n_w(p,t)$. The extra information on the resonant substructure will, in principle, put a much better constraint on the amount of mixing. Of course, precise knowledge of the resonant substructure for DCSD is needed here and so far we do not know anything about it. Because of this, for current experiments this method is more likely to be a complication rather than a better method when one tries to apply method A to $D^0 \to K^+\pi^-\pi^0$ (see [27] and [25]) or $D^0 \to K^+\pi^-\pi^+\pi^-$. In principle, however, one can use wrong sign samples at low decay time (which is almost pure DCSD) to study the resonant substructure of the DCSD decays.

It is interesting to point out here (as discussed in detail in Appendix B) that the Dalitz plot changes its shape as the decay time "goes by" due to the interference effect. Note that the interference phase, unlike in the case of $D^0 \to K^+\pi^-$, strongly depends on the location on the Dalitz plot since there are contributions from the various Breit-Wigner amplitudes, which changes wildly across each resonance. One would expect that $\cos\phi(p)$ could have any value between [-1,1], depending on the location p. It is interesting to look at the locations on the Dalitz plot where $\cos\phi(p) = -1$ (maximal destructive mixing-DCSD interference). As pointed out in method A, Property (5) tells us that one should find $I(t_0) = 0$ at $t_0 = \sqrt{\frac{2}{\alpha}}$. This means that the mixing-DCSD interference would dig a "hole" on the Dalitz plot at time t_0 at that location. Since $t_0 \propto \sqrt{R_{\text{DCSD}}(p)} = \sqrt{n_D(p)/n_C(p)}$, the "holes" would show up earlier (in decay time) at locations where $R_{\text{DCSD}}(p)$ is smaller. Imagine that someone watches the Dalitz plot as the decay time "goes by", this person would expect to see "holes" moving from locations with $\cos\phi(p) = -1$ and smaller $R_{\text{DCSD}}(p)$ toward locations with $\cos\phi(p) = -1$ and larger $R_{\text{DCSD}}(p)$. The existence of the "moving holes" on the Dalitz plot would be clear evidence for mixing. Once again we see the importance of the mixing-DCSD interference effect.

In the near future, we should have a good understanding of DCSD decays and this method could become a feasible way to search for mixing (and CP violation).

2.1.3 Method C —use quantum statistics of the production and decay processes

This method is to search for dual identical two-body hadronic decays in $e^+e^- \to \Psi'' \to D^0\bar{D}^0$, such as $(K^-\pi^+)(K^-\pi^+)$, as was first suggested by Yamamoto in his Ph.D thesis [50]. The idea is that when $D^0\bar{D}^0$ pairs are generated in a state of odd orbital angular momentum (such as Ψ''), the DCSD contribution to identical two-body pseudo-scalar-vector ($D \to PV$) and pseudo-scalar-pseudo-scalar ($D \to PP$) hadronic decays (such as $(K^-\pi^+)(K^-\pi^+)$) cancels out, leaving only the contribution of mixing [50, 51, 52]. The essence of Yamamoto's original calculation for the $(K^-\pi^+)(K^-\pi^+)$ case is given below.

Let's define $e_i(t) = e^{-im_i t - \gamma_i t/2}$ ($i = 1, 2$) and $e_\pm(t) = (e_1(t) \pm e_2(t))/2$. A state that is purely $|D^0\rangle$ or $|\bar{D}^0\rangle$ at time $t = 0$ will evolve to $|D(t)\rangle$ or $|\bar{D}(t)\rangle$ at time t, with $|D(t)\rangle = e_+(t)|D^0\rangle + e_-(t)|\bar{D}^0\rangle$ and $|\bar{D}(t)\rangle = e_-(t)|D^0\rangle + e_+(t)|\bar{D}^0\rangle$. In $e^+e^- \to \Psi'' \to D^0\bar{D}^0$, the $D^0\bar{D}^0$ pair is generated in the state $D^0\bar{D}^0 - \bar{D}^0 D^0$ as the relative orbital angular momentum of the pair $\mathcal{L} = 1$. Therefore, the time evolution of this state is given by $|D(t)\bar{D}(t')\rangle - |\bar{D}(t)D(t')\rangle$, where t (t') is the time of decay of the D ($\bar{D}$). Now the double-time amplitude $\mathcal{A}_w(t,t')$ that the left side decays to $K^-\pi^+$ at t and the right side decays to $K^-\pi^+$ at t', giving a wrong sign event $(K^-\pi^+)(K^-\pi^+)$, is given by:

$$\mathcal{A}_w(t,t') = (e_+(t)e_-(t') - e_-(t)e_+(t'))(a^2 - b^2) \tag{9}$$

where $a = \langle K^-\pi^+|D^0\rangle$ is the amplitude of the Cabibbo favored decay $D^0 \to K^-\pi^+$, while $b = \langle K^-\pi^+|\bar{D}^0\rangle$ is the amplitude of DCSD $\bar{D}^0 \to K^-\pi^+$. Similarly, the double-time amplitude $\mathcal{A}_r(t,t')$ for the right sign event $(K^-\pi^+)(K^+\pi^-)$ is given by:

$$\mathcal{A}_r(t,t') = (e_+(t)e_+(t') - e_-(t)e_-(t'))(a^2 - b^2) \tag{10}$$

One measures the wrong sign versus right sign ratio R, which is:

$$R = \frac{N(K^-\pi^+, K^-\pi^+) + N(K^+\pi^-, K^+\pi^-)}{N(K^-\pi^+, K^+\pi^-) + N(K^+\pi^-, K^-\pi^+)} = \frac{\iint |\mathcal{A}_w(t,t')|^2 \, dt \, dt'}{\iint |\mathcal{A}_r(t,t')|^2 \, dt \, dt'} \tag{11}$$

Note in taking the ratio, the amplitude term $(a^2 - b^2)$ in Equations 9 and 10 drops out. Thus, clearly R does not depend on whether b is zero (no DCSD) or finite (with DCSD). Integrating over all times, one then obtains $R = (x^2 + y^2)/2 = R_{\text{mixing}}$, where x and y are defined as before.

This is probably the best way to separate DCSD and mixing. The exclusive nature of the production guarantees both low combinatoric backgrounds and production kinematics essential for background rejection. This method requires one use e^+e^- annihilation in the charm threshold region. Here the best final state is $(K^-\pi^+)(K^-\pi^+)$. In principle, one can also use final states like $(K^-\rho^+)(K^-\rho^+)$ or $(K^{*-}\pi^+)(K^{*-}\pi^+)$, etc., although again there are complications. For example, it is hard to differentiate experimentally $(K^-\rho^+)(K^-\rho^+)$ from $(K^-\rho^+)(K^-\pi^+\pi^0)$, where DCSD can contribute. With high statistics, in principle, this method could be combined with method B.

It has been pointed out that quantum statistics yield different correlations for the $D^0\bar{D}^0$ decays from $e^+e^- \to D^0\bar{D}^0, D^0\bar{D}^0\gamma, D^0\bar{D}^0\pi^0$ [53]. The well-defined coherent quantum states of the $D^0\bar{D}^0$ can be, in principle, used to provide valuable cross checks on systematic uncertainties, and to extract $x = \delta m/\gamma_+$ and $y = \gamma_-/\gamma_+$ (which requires running at different energies) if mixing is observed [53].

2.2 Semi-leptonic method

The semi-leptonic method is to search for $D^0 \to \bar{D}^0 \to Xl^-\nu$ decays, where there is no DCSD involved. However, it usually (not always!) suffers from a large background due to the missing neutrino. In addition, the need to understand the large background often introduces model dependence. In the early days, the small size of fully reconstructed samples of exclusive D^0 hadronic decays and the lack of the decay time information made it difficult to constrain the $D^0\bar{D}^0$ mixing rate using the hadronic method, many experiments used semi-leptonic decays. The techniques that were used were similar — searching for like-sign $\mu^+\mu^+$ or $\mu^-\mu^-$ pairs in $\mu^+N \to \mu^+(\mu^+\mu^+)X$ [32, 35] and $\pi^-Fe \to \mu^+\mu^+$ [33], $\pi^-W \to \mu^+\mu^+$ [38]. These techniques rely on the assumptions on production mechanisms, and the accuracy of Monte Carlo simulations to determine the large conventional sources of background.

There are other ways of using the semi-leptonic method. The best place to use the semi-leptonic method is probably in e^+e^- annihilation near the charm threshold region. The idea is to search for $e^+e^- \to \Psi'' \to D^0\bar{D}^0 \to (K^-l^+\nu)(K^-l^+\nu)$ or $e^+e^- \to D^-D^{*+} \to (K^+\pi^-\pi^-)(K^+l^-\nu)\pi_s^+$ [57, 58]. The latter is probably the only place where the semi-leptonic method does not suffer from a large background. It should have a low background, as there is only one neutrino missing in the entire event, threshold kinematics constraints should provide clean signal.

It has been pointed out that one can not claim a $D^0\bar{D}^0$ mixing signal based on the semi-leptonic method alone (unless with the information on decay time of D^0). Bigi [53] has pointed out that an observation of a signal on $D^0 \to l^-X$ establishes only that a certain selection rule is violated in processes where the charm quantum number is changed, namely the rule $\Delta \text{Charm} = -\Delta Q_l$ where

Q_l denotes leptonic charge. This violation can occur either through $D^0\bar{D}^0$ mixing (with the unique attribute of the decay time-dependence of mixing), or through new physics beyond the Standard Model (which could be independent of time). Nevertheless, one can always use this method to set upper limit for mixing.

3 Mixing Searches at Different Experiments

3.1 e^+e^- running on $\Psi''(3770)$ –MARK III, BES, Tau-charm factory

The MARK III collaboration was the first (though hopefully not the last) to use the $e^+e^- \to \Psi'' \to D^0\bar{D}^0$ technique. They reported three events consistent with $|\Delta S = 2|$ transitions [55]. One event is observed in the final state $K^+\pi^-$ vs $K^+\pi^-\pi^0$. The other two are reconstructed in the final states $K^+\pi^-\pi^0$ vs $K^+\pi^-\pi^0$, and a Dalitz plot analysis finds one to be consistent with $K^-\rho^+$ versus $K^-\rho^+$ and the other consistent with $K^{*0}\pi^0$ versus $K^{*0}\pi^0$ (note the $D^0 \to K^-\pi^+\pi^0$ decays are dominated by $D^0 \to K^-\rho^+$ and $D^0 \to \bar{K}^{*0}\pi^0$ channels). Using a maximum likelihood analysis, they interpreted the results for two limiting cases: a). if there is no DCSD in $D^0 \to K^+\pi^-$ and $D^0 \to K^+\pi^-\pi^0$; then the events imply $R_{mixing} = (1.2\pm 0.6)\%$ or $R_{mixing} > 0.4\%$ at 90% C.L.; b). if there is no DCSD for $D^0 \to K^+\pi^-$ and no $D^0\bar{D}^0$ mixing, and also at least one of the $K^+\pi^-\pi^0$'s in each of those two events are non-resonant, then the results imply $R_{DCSD} = \Gamma(D^0 \to K^+\pi^-\pi^0)/\Gamma(D^0 \to K^-\pi^+\pi^0) = (7\pm 4)\tan^4\theta_C$ or $R_{DCSD} > 1.9\tan^4\theta_C$ at 90% C.L.. This was a interesting result at that time, and had a strong influence on the subject. However, one cannot draw a firm conclusion about the existence of $D^0\bar{D}^0$ mixing based on these events. There are at least two reasons: (1) The background study has to rely on Monte Carlo simulation of the PID (particle identification – Time-of-Flight) [3]. As Gladding has pointed out: "These results must be considered preliminary because the calculation of the confidence level is sensitive to the tails of PID distribution for the background" [56]; (2) Assuming that the Monte Carlo background study is correct, and that the events are real, one still cannot claim the two events are due to mixing, for example, the non-resonant decays $D^0 \to K\pi\pi^0$ may contribute to one side of the pair in each of the events, in which DCSD can contribute.

---

[3] In principle, one can use kinematics to check whether the events are due to doubly misidentified $D^0 \to K^-\pi^+\pi^0$: if one inverts the K^+, π^- assignments and recalculates the D^0 mass (let's call this M_{flip}), the M_{flip} will not be within D^0 mass peak if it is real, unless the K and π momentum is close. Unfortunately, M_{flip} for all the three events are within D^0 mass peak. That's why one needs to totally rely on PID.

The MARK III puzzle can be solved at a τ-charm factory, which is a high luminosity ($10^{33} cm^{-2} s^{-1}$) e^+e^- storage ring operating at center-of-mass energies in the range 3-5 GeV. The perspectives for a $D^0 \bar{D}^0$ mixing search at a τ-charm factory have been studied in some detail [57, 58]. I will outline here the most important parts. The best way to search for mixing at τ-charm factory is probably to use $e^+e^- \to \Psi'' \to D^0\bar{D}^0 \to (K^-\pi^+)(K^-\pi^+)$. The sensitivity is not hard to estimate. Assuming a one year run with a luminosity of $10^{33} cm^{-2} s^{-1}$, 5.8×10^7 D^0s would be produced from Ψ''. Therefore about 9×10^4 $(K^-\pi^+)(K^+\pi^-)$ events would be produced. About 40% of them (3.6×10^4) could be fully reconstructed. A study [57] has shown that the potential dominant background comes from doubly misidentified $(K^-\pi^+)(K^+\pi^-)$, and if TOF resolution is 120 ps, this background could be kept to the level of one event or less. This means one could, in principle, set an upper limit at the 10^{-4} level.

As mentioned Section 2.2, the best place to use the semi-leptonic method is probably at a τ-charm factory. One good example is to search for $e^+e^- \to D^-D^{*+} \to (K^+\pi^-\pi^+)(K^+l^-\nu)\pi_s^+$. It is expected that this method can also have a sensitivity at the 10^{-4} level. There are many other independent techniques that one can use for a mixing search at a τ-charm factory. By combining several independent techniques (which require running at different energies), it was claimed that $D^0\bar{D}^0$ mixing at the 10^{-5} level could be observable [58].

There have been several schemes around the world for building a τ-charm factory. If such a machine is built, it could be a good place to study mixing. The history of the τ-charm factory can be found in reference [59]: one was proposed at SLAC in 1989 and one at Spain in 1993; one was discussed at Dubna in 1991, at IHEP (China), and at Argonne [60] in 1994 and at this workshop. It will be discussed again at IHEP (China) soon. Let us hope that we will have one in the not-too-distant future.

3.2 e^+e^- running near $\Upsilon(4S)$ –ARGUS, CLEOII, CLEO III, B factory

Without a precision vertex detector, CLEO II can only in effect measure the rate $\mathcal{B}(D^0 \to K\pi)$ integrated over all times of a pure D^0 decaying to a final state $K\pi$. The ratio R=$\mathcal{B}(D^0 \to K^+\pi^-)$ /$\mathcal{B}(D^0 \to K^-\pi^+)$ is given by integrating equation 3 over all times (see appendix A)

$$R = R_{\text{mixing}} + R_{\text{DCSD}} + \sqrt{2R_{\text{mixing}}R_{\text{DCSD}}} \cos\phi. \qquad (12)$$

CLEO finds [42] R = $(0.77 \pm 0.25\,(\text{stat.}) \pm 0.25\,(\text{sys.}))\%$. This signal could mean one of two things: (1) mixing could be quite large, which would imply that mixing can be observed in the near future; (2) the signal is dominated by DCSD. The theoretical prediction for R_{DCSD} is about $(2-3)tan^4\theta_C \sim (0.6-0.9)\%$ [51, 61, 62], which is quite consistent with the measured value. It is, therefore, believed by many that the signal is due to DCSD, although it remains consistent with the current best experimental upper limits on mixing, which are $(0.37-0.7)\%$ [40] and 0.56% [38].

CLEO has also tried to use hadronic method B, by searching for $D^0 \to K^+\pi^-\pi^0$. The excellent photon detection at CLEO II allows one to study this mode with a sensitivity close to $D^0 \to K^+\pi^-$ mode. The main complication faced here is that (as discussed in method B) the resonant substructure is not necessarily the same for wrong sign and right sign decays. Because of this, as discussed in appendix B, the interpretation of R as R_{mixing} or R_{DCSD} will be complicated by the lack of knowledge of the details of the interference between submodes (and also the decay time information). Moreover, one has to worry about the detection efficiency across the Dalitz plot. Setting an upper limit for each submode is clearly very difficult. CLEO has set an upper limit [27, 25] on the inclusive rate for $D^0 \to K^+\pi^-\pi^0$ as $R = \mathcal{B}(D^0 \to K^+\pi^-\pi^0) / \mathcal{B}(D^0 \to K^-\pi^+\pi^0) < 0.68\%$. Note this upper limit includes the possible effects of the interference between the DCSD and mixing for each submode as well as the interference between submodes.

This summer, CLEO will install a silicon vertex detector (SVX) with a longitudinal resolution on vertex separation around 75 μm. This will enable CLEO to measure the decay time of the D^0, and reduce the random slow pion background (the resolution of the D^{*+} - D^0 mass difference is dominated by the angular resolution on the slow pion, which should be greatly improved by the use of the SVX). By the year 2000, with CLEO III (a symmetric B factory) and asymmetric B factories at SLAC and KEK, each should have thousands of $D^0 \to K^+K^-(X), \pi^+\pi^-(X)$ and a few hundred $D^0 \to K^+\pi^-$ (and perhaps $D^0 \to K^+\pi^-\pi^0, K^+\pi^-\pi^+\pi^-$ too) signal events with decay time information for one year of running. The typical decay length of D^0 ($\mathcal{L}$) is about a few hundred μm, and the resolution of the decay length ($\sigma_\mathcal{L}$) is about 80 μm ($\mathcal{L}/\sigma_\mathcal{L} \sim 3$). The sensitivity to mixing at CLEO III and asymmetric B factories has not been carefully studied yet. A reasonable guess is that it could be as low as 10^{-4}. If mixing rate is indeed as large as DCSD rate, it should be observed by then.

3.3 Fixed target experiments

A significant amount of our knowledge has been gained from Fermilab fixed target experiments, and in fact the current best upper limits on mixing have emerged from these experiments (E615, E691), and will come from their successors E687, E791 and E831 soon.

The best upper limit using the semi-leptonic method comes from the Fermilab experiment E615, which used a 255 GeV pion beam on a tungsten target. The technique is to search for the reaction $\pi N \to D^0 \bar{D}^0 \to (K^- \mu^+ \nu) D^0 \to (K^- \mu^+ \nu)(K^- \mu^+ \nu)$, where only the final state muons are detected (i.e. the signature is like-sign $\mu^+ \mu^+$ or $\mu^- \mu^-$ pairs). Assuming $\sigma(c\bar{c}) \sim A^1$ nuclear dependence, they obtained $R_{\text{mixing}} < 0.56\%$ [38].

The best upper limit using the hadronic method by measuring the decay time information comes from E691, which is the first high statistics fixed target (photoproduction) experiment. In fact, E691 was the first experiment which used the decay time information (obtained from the excellent decay time resolution of their silicon detectors) to distinguish DCSD and mixing. The decay chains $D^{*+} \to D^0 \pi_s^+$ followed by $D^0 \to K^+\pi^-$, $K^+\pi^-\pi^+\pi^-$ were used. Their upper limits from the $D^0 \to K^+\pi^-$ mode are $R_{\text{mixing}} < (0.5-0.9)\%$ and $R_{DCSD} < (1.5-4.9)\%$, while the upper limits from $D^0 \to K^+\pi^-\pi^+\pi^-$ are $R_{\text{mixing}} < (0.4-0.7)\%$ and $R_{DCSD} < (1.8-3.3)\%$. The ranges above reflect the possible effects of interference between DCSD and mixing with an unknown phase (ϕ). Although the combined result gives $R_{\text{mixing}} < (0.37-0.7)\%$, in principle, one cannot combine the results from the two modes since the interference phases are totally different, as discussed in appendix B. Note that in their analysis for $D^0 \to K^+\pi^-\pi^+\pi^-$, the resonant substructure in the Cabibbo favored and DCSD decays was ignored. As discussed in appendix B, in general, one cannot treat $D^0 \to K^+\pi^-\pi^+\pi^-$ exactly the same way as $D^0 \to K^+\pi^-$ when one ignores the resonant substructure. Moreover, in principle, the detection efficiency vs decay time may not be studied reliably by using $D^0 \to K^-\pi^+\pi^-\pi^+$ as the resonant substructure could be different for CFD and DCSD.

At the Charm 2000 workshop [63], both E687 and E791 reported their preliminary result from part of their data. The best upper limits on mixing should come from these two experiments soon. Some most recent preliminary results can be found in [46], progress has been made [44, 45] on the measurement of the lifetime difference between $D^0 \to K^-\pi^+$ and $D^0 \to K^+K^-(\pi^+\pi^-)$.

4 Comparison of Different Experiments

4.1 Hadronic method A

This measurement requires: (1) excellent vertexing capabilities, at least good enough to see the interference structure; (2) low background around the primary vertex. The background level around the primary vertex could be an important issue as the interference term in Equation 3 does peak at $t = 1$. In addition, low background around primary vertex means that one does not suffer much from random slow pion background and also one can measure the DCSD component at short decay times well. This also means that good acceptance at short decay times are very important. These are also important for understanding DCSDs at large decay times. The vertexing capabilities at e^+e^- experiments ($\mathcal{L}/\sigma \sim 3$) for CLEO III and asymmetric B factories at SLAC and KEK may be sufficient for a mixing search. The extra path length due to the Lorentz boost, together with the use of silicon detectors for high resolution position measurements, have given the fixed target experiments an advantage in vertex resolution (typically $\mathcal{L}/\sigma \sim 8-10$) over e^+e^- experiments. One major disadvantage at fixed target experiments is the poor acceptance at short decay times. The low background around the primary vertex at e^+e^- experiments is a certain advantage. It is worth pointing out here that at the e^+e^- experiments (especially at an asymmetric B factory or Z factory) it maybe possible to use $\bar{B}^0 \to D^{*+}l^-\nu$, where the primary ($D^{*+}$ decay) vertex can be determined by the l^- together with the slow pion coming from the D^{*+}. In this case, the background level around the primary vertex is intrinsically very low [26].

However, in the case of $D^0 \to K^+K^-, \pi^+\pi^-$, etc., the requirement on the background level around the primary vertex is not so important. In this case, the mixing signature is not a deviation from a perfect exponential (again assuming CP conservation), but rather a deviation of the slope from $(\gamma_1+\gamma_2)/2$. It is worth pointing out that there are many advantages with this method. For example, one can use Cabibbo favored decay modes, such as $D^0 \to K^-\pi^+$, to measure the average D^0 decay rate $(\gamma_1 + \gamma_2)/2$ (which is almost a pure exponential, mixing followed by DCSD effect should be tiny, see Appendix A). This, along with other SCSD CP even (or odd) final states, would allow for valuable cross checks on systematics uncertainties. In addition, since we only need to determine the slope here, we do not need to tag the D^0 and do not have to use the events close to the primary vertex. The sensitivity of this method depends on how well we can determine the slope difference, which needs to be carefully studied. This is currently under study [45, 44]. Roughly speaking, in the ideal case, the sensitivity to y would be $\sim 1/\sqrt{N}$, where N is the number of $D^0 \to K^+K^-, \pi^+\pi^-$, etc. events, which means that the sensitivity to mixing caused by the decay rate difference ($\sim y^2/2$) would be close to $\sim 1/N$. For example, a fixed-target experiment capable of producing $\sim 10^8$ reconstructed charm events could, in principle, lower the sensitivity to $\sim 10^{-5} - 10^{-6}$ level for the y^2 term in $R_{\text{mixing}} = (x^2 + y^2)/2$. In reality, the sensitivity depends on many things and should be carefully studied.

It is worth to point out that the current PDG experimental upper limit (90% CL) on the life time difference is only [64]

$$\frac{|\tau_{D_1^0} - \tau_{D_2^0}|}{\tau_{D^0}} = 2\,y < 17\% \tag{13}$$

This is based on the upper limit $R_{\text{mixing}} = (x^2 + y^2)/2 < 0.37\%$, which is the E691 [40] combined results on the $D^0 \to K^+\pi^-$ and $D^0 \to K^+\pi^-\pi^+\pi^-$ modes by assuming no interference between DCSD and mixing for both modes at the same time.

4.2 Hadronic method B

In the near future, we should be able to have a good understanding of DCSD [4] in $D^0 \to K^+\pi^-\pi^0$, $D^0 \to K^+\pi^-\pi^+\pi^-$, etc. modes, then method B will become a feasible way to study mixing and the sensitivity should be improved. Just like method A, this method requires very good vertexing capabilities and very low background around the primary vertex (this is even more important than in method A, since precise knowledge of DCSD is very important here). In addition, this method requires that the detection efficiency (for the mode being searched) across Dalitz plot be quite uniform (at least the detector should have good acceptance on the Dalitz plot at locations where DCSD and mixing resonant substructure are different). This is necessary so that detailed information on the resonant substructure can be obtained in every corner on the Dalitz plot.

The excellent photon detection capabilities will allow e^+e^- experiments to study the $D^0 \to K^+\pi^-\pi^0$ mode with very low background. From the CLEO II $D^0 \to K^+\pi^-\pi^0$ analysis [27, 25], the detection efficiency across the Dalitz plot will have some variations due to cuts needed to reduce background, however, it is still good enough to obtain detailed information on the resonant

---

[4]It may be possible that good understanding of DCSD can be reached by measuring the pattern of D^+ DCSD decays where the signature is not confused by a mixing component. It is worth pointing out that the D^+ DCSD decays can be studied very well at future fixed target experiments and B factories.

substructure. Future fixed target experiments may have a good chance to study $D^0 \to K^+\pi^-\pi^+\pi^-$ mode, since the detection efficiency across Dalitz plot should be quite flat. The sensitivity that each experiment can reach by using this method depends on many things and need to be carefully studied in the future.

4.3 Hadronic method C

The sensitivity of this method depends crucially on the particle identification capabilities. Since the D^0 is at rest, the K and π mesons will have the same momentum, so a doubly misidentified $D^0 \to K^-\pi^+$ ($K^- \to \pi^-, \pi^+ \to K^+$) mimics a $D^0 \to K^+\pi^-$ with almost the same D^0 mass. It is worth pointing out here that particle identification is not as crucial to method A as it is to this method (C), as far as this particular background is concerned. This is because in method A, the D^0 is highly boosted, and doubly misidentified $D^0 \to K^+\pi^-$ decays will have a broad distribution in the D^0 mass spectrum around the D^0 mass peak; this background can be kinematically rejected with only a small reduction of the efficiency for the signal events.

Once the sensitivity reaches $\mathcal{O}(10^{-5})$, one may have to worry about other contributions, such as contributions from continuum background, contributions from $e^+e^- \to 2\gamma \to D^0\bar{D}^0$ which can produce C-even states where DCSD can contribute [52].

4.4 Semi-leptonic method

The semi-leptonic method usually suffers from large background (except at a τ charm factory), the traditional method of looking for like sign $\mu^+\mu^+$ or $\mu^-\mu^-$ pairs is an example. New ideas are needed in order to improve the sensitivity significantly. Some promising techniques have been suggested by Morrison and others at the Charm 2000 workshop [63] and have been discussed in the working group [65].

The technique suggested by Morrison is very similar to that of the hadronic method: one uses the decay chain $D^{*+} \to D^0\pi^+$, instead of looking for $D^0 \to K^+\pi^-$, one can search for $D^0 \to K^+l^-\nu$ where there is no DCSD involved. Of course, due to the missing neutrino, this mode usually suffers from large background. However, for events in which the neutrino is very soft in D^0 rest frame, $D^0 \to K^+l^-\nu$ is quite similar to $D^0 \to K^+\pi^-$ kinematically. In this case, one has the same advantages as $D^{*+} \to D^0\pi^+$ followed by $D^0 \to K^+\pi^-$ has. In addition, as the neutrino is soft, the proper decay time of the D^0 can be reasonably estimated from K^+l^-. The potential mixing signal therefore should show up as a t^2 term in the proper decay time distribution. To select the events with soft neutrino, one can require the K^+l^- mass above 1.4 GeV. This requirement will keep about 50% of the total signal. One major background here is the random slow pion background, as the effective mass difference width is still much larger (a factor of 10) than $D^{*+} \to D^0\pi^+$ followed by $D^0 \to K^+\pi^-$. In order to reduce this background, Morrison has suggested to look for a lepton with the correct charge sign in the other side of the charm decay. Another background is DCSD decay $D^0 \to K^+\pi^-$ when the π^- fakes a l^-, however, this background will only populates at the higher end of the K^+l^- mass spectrum where the neutrino energy is almost zero. This can be eliminated by cutting off that high end of the K^+l^- mass. In principle, this idea can be used in a fixed target experiment as well as in a e^+e^- experiment. The sensitivity of this method depends on the lepton fake rate (meson fakes as a lepton). One can find some detail discussions in Morrison's Charm2000 workshop summary paper [66].

Another technique, suggested by Freyberger at CLEO, is based on the technique which has been used by ALEPH, HRS and CLEO to extract the number of $D^{*+} \to D^0\pi_s^+$ events. The technique utilizes the following facts: (1) Continuum production of $c\bar{c}$ events are jet like. (2) The jet axis, calculated by maximizing the observed momentum projected onto an axis, approximates the D^{*+} direction. (3) The $D^{*+} \to D^0\pi_s^+$ decay is a two-body process, and the small amount of energy available means that the π_s^+ is very soft, having a transverse momentum $p_\perp$ relative to the D^{*+} direction which cannot exceed 40 MeV/c. This low transverse momentum provides the $D^{*+} \to D^0\pi_s^+$ signature.

The facts are used in the following way. The maximum momentum in the lab that the π_s^+ can have perpendicular to the line of flight of the D^{*+} is 40MeV. One can define this quantity as $p_\perp = |p_\pi|sin\theta_\pi$, where $sin\theta_\pi$ is the angle between the D^{*+} and the π_s^+ in the lab frame, and p_π is the magnitude of the π_s^+ momentum. Hence, the π_s^+ from D^{*+} will populate the low $p_\perp$ (or $sin\theta_\pi$) region. The signal is enhanced if one plots $p_\perp^2$ (or $sin^2\theta_\pi$) instead of $p_\perp$. One then looks for an lepton in the jet with the correct sign, namely, $\pi_s^+l^+$ right sign combination and $\pi_s^+l^-$ wrong sign combination. The signal $D^{*+} \to D^0\pi_s^+$ followed by $D^0 \to K^-l^+\nu$ will peak in the low $p_\perp^2$ (or $sin^2\theta_\pi$) region for the wrong sign events. It is worth pointing out that one can look for a lepton in the other side of the event to reduce background.

There are many kinds of background to this method one has to worry about. One of the major backgrounds is fake lepton background. For example, the decay chain $D^{*+} \to D^0\pi_s^+ \to (K^-X)\pi_s^+$ will also peak at the low $p_\perp^2$ (or $sin^2\theta_\pi$) if the K^- is misidentified as a l^-. Another major background is probably the π^0 dalitz and γ conversions in $D^0 \to X\pi^0$ followed by $\pi^0 \to \gamma e^+ e^-$ or $D^0 \to X\pi^0$ followed by $\pi^0 \to \gamma\gamma$ and then $\gamma \to e^+ e^-$. These two major backgrounds are at about 0.3% level in the current CLEO II data. Understanding these backgrounds is the major difficulty faced by this method. Although for CLEO III, things should improve, it is not clear what kind of sensitivity one can expect from this method for future experiments. Nevertheless, it is an interesting idea and worth investigating. In fact, this technique is currently under study at CLEO [67].

5 Summary

The search for $D^0\bar{D}^0$ mixing carries a large discovery potential for new physics since the $D^0\bar{D}^0$ mixing rate is expected to be very small in the Standard Model. The past decade has seen significant experimental progress in sensitivity (from 20% down to 0.4%). Despite these 18 years of effort there is still much left to be done.

As was discussed in the introduction, the observation of $D^0 \to K^+\pi^-$ is an important step on the way to observing a potentially small mixing signal by using this technique. With the observation of $D^0 \to K^+\pi^-$ signal at CLEO at the level R = 0.0077 ± 0.0025 (stat.) ± 0.0025 (sys.), any experimenter can estimate the number of reconstructed $D^0 \to K^+\pi^-$ events their data sample will have in the future and what kind of sensitivity their experiment could have. Eighteen years after the search for $D^0 \to K^+\pi^-$ started, we have finally arrived at the point where we could take advantage of possible DCSD-mixing interference to make the mixing search easier.

In light of the CLEO's $D^0 \to K^+\pi^-$ signal, if the mixing rate is close to that of DCSD (above 10^{-4}), then it might be observed by the year 2000 with either the hadronic or the semi-leptonic method, either at fixed target experiments, CLEO III, asymmetric B factories (at SLAC and KEK), or at a τ-charm factory. If the mixing rate is indeed much smaller than DCSD, then the hadronic method may have a better chance over the semi-leptonic method. This is because the semi-leptonic method usually suffers from a large background due to the missing neutrino, while the hadronic method does not. Moreover, the commonly believed "annoying DCSD background" does not necessary inherently limit the hadronic method as the potentially small mixing signature could show up in the interference term. The design of future experiments should focus on improving the vertexing capabilities and reducing the background level around the primary vertex, in order to fully take advantage of having the possible DCSD and mixing interference. In addition, we have learned that the very complication due to the possible differences between the resonant substructure in many DCSD and mixing decay modes $D^0 \to K^+\pi^-(X)$ could, in principle, be turned to advantage by providing additional information once the substructure in DCSD is understood (the method B) and the sensitivity could be improved significantly this way. This means that understanding DCSD in D decays could be a very important step on the way to observe mixing. Experimenters and theorists should work hard on this.

In the case of $D^0 \to K^+\pi^-(X)$ and $D^0 \to X^+l^-$, we are only measuring $R_{\text{mixing}} = (x^2 + y^2)/2$. Since many extensions of the Standard Model predict large $x = \delta m/\gamma_+$, it is very important to measure x and y separately. Fortunately, SCSD can provide us information on y. This is due to the fact that decays such as $D^0 \to K^+K^-, \pi^+\pi^-$, occur only through definite CP eigenstate, and this fact can be used to measure the decay rate difference $y = \gamma_-/\gamma_+ = (\gamma_2 - \gamma_1)/(\gamma_1 + \gamma_2)$ alone. Observation of a non-zero y would demonstrate mixing caused by the decay rate difference. This, together with the information on R_{mixing} obtained from other methods, we can in effect measure x. I should point out here that x and y are expected to be at the same level within the Standard Model, however we do not know for sure exactly at what level since theoretical calculations for the long distance contribution are still plagued by large uncertainties. Therefore, it is very important to measure y in order to understand the size of x within the Standard Model, so that when $D^0\bar{D}^0$ mixing is finally observed experimentally, we will know whether we are seeing the Standard Model physics or new physics beyond the Standard Model.

In this sense, it is best to think of the quest to observe mixing (new physics) as a program rather than a single effort.

Acknowledgement

Much of my knowledge on this subject has been gained by having worked closely with Hitoshi Yamamoto over the past four years, I would like to express my sincere gratitude to him. Besides, I would like to thank the organizers of the workshop, especially Jose Repond, for his patient and encouragement. This work is supported by Robert H. Dicke Fellowship Foundation at Princeton University.

A Appendix A — The Time Dependence of $D^0 \to K^+\pi^-$

Note that due to the excessive length of the paper, appendices A, B and C are omitted. But they can be found in references [84] or [25].

A.1 The time-dependent effect

A.2 The time-integrated effect

B Appendix B — The Time Dependence of $D^0 \to K^+\pi^-\pi^0$

B.1 The time-dependent Dalitz analysis

B.2 The time-integrated Dalitz analysis

B.3 The time-dependent Non-Dalitz analysis

B.4 The time-integrated Non-Dalitz analysis

B.5 Summary of Appendix B

C Appendix C –The Interference Between Mixing and DCSD in B^0 Case

References

[1] M. K. Gaillard and B. W. Lee, Phys. Rev. D **10**, 897 (1974);
A. Datta, Phys. Lett. **154B**, 287 (1985).

[2] S. L. Glashow, J. Iliopoulos, and L. Maiani, Phys. Rev. D **2**, 1285 (1970).

[3] See A. J. Buras and M. K. Harlander, "A Top Quark Story: Quark Mixing, CP violation and Rare Decays in the Standard Model" and references therein, in Advanced Series on Directions in High Energy Physics - Vol. 10, "Heavy Flavours", editors: A. J. Buras and M. Lindner, World Scientific, 1992.

[4] K. S. Babu *et al.*, Phys. Lett. **205B**, 540 (1988).

[5] S. L. Glashow and S. Weinberg, Phys. Rev. D **15**, 1958 (1977).

[6] J. L. Hewett, "Probing New Physics in Rare Charm Processes", to appear in the Proceedings of the Eighth Meeting of the Division of Particles and Fields of the American Physics Society (DPF'94), Albuquerque, New Mexico. (World Scientific).

[7] A. Antaramian, L. J. Hall and A. Rasin, Phys. Rev. Lett. **69**, 1871 (1992).

[8] S. Pakvasa and H. Sugawara, Phys. Lett. **73B**, 61 (1978).

[9] T. P. Cheng and M. Sher, Phys. Rev. D **35**, 3484 (1987).

[10] L. Hall and S. Weinberg, Phys. Rev. D **48**, R979 (1993).

[11] S. King, Phys. Lett. **195B**, 66 (1987).

[12] A. Hadeed and B. Holdom, Phys. Lett. **159B**, 379 (1985).

[13] R. N. Cahn and H. Harari, Nucl. Phys. B **176**, 135 (1980) and references therein;
G. L. Kane and R. L. Thun, Phys. Lett. **94B**, 513 (1980).

[14] Y. L. Wu and L. Wolfenstein, Phys. Rev. Lett. **73**, 1762 (1994).

[15] Y. Nir and N. Seiberg, Phys. Lett. **309B**, 337 (1993)

[16] G. Burdman, "Charm mixing and CP violation in the Standard Model", Proceedings of the Charm 2000 Workshop, FERMILAB-Conf-94/190, Fermilab, June 7-9, 1994.

[17] S. Pakvasa, "Charm as Probe of New Physics", Proceedings of the Charm 2000 Workshop, FERMILAB-Conf-94/190, Fermilab, June 7-9, 1994.

[18] J. L. Hewett, "Searching for New Physics with Charm", SLAC-PUB-95-6821, presented at the Lafex International School on High Energy Physics (LISHEP95), Rio de Janeiro, Brazil, February 6-22, 1995.

[19] A. J. Schwartz, Modern Physics Letters A, Vol.8, No. 11 (1993) 967-977.

[20] J. Donoghue, E. Golowich, B. R. Holstein and J. Trampetic, Phys. Rev. D **33**, 179 (1986).

[21] X. G. He and S. Pakvasa, Phys. Lett. **156B**, 34 (1985);
T. G. Rizzo, *Int. J. Mod. Phys.* A **4**, 5401 (1989).

[22] H. Georgi, Phys. Lett. **297B**, 353 (1992);
T. Ohl, G. Ricciardi and E. H. Simmons, Nucl. Phys. B **403**, 603 (1993).

[23] I. Bigi, "Open Questions in Charm Decays Deserving An Answer". Proceedings of the Charm 2000 Workshop, FERMILAB-Conf-94/190, Fermilab, June 7-9, 1994.

[24] T. A. Kaeding, "D Meson Mixing in Broken SU(3)", LBL-37224, UCB-PTH-95/14, HEP-PH-9505393.

[25] T. Liu, "$D^0\bar{D}^0$ mixing and Doubly Cabibbo Suppressed Decays", Ph.D thesis, HUHEPL-20, May 1995.

[26] T. Liu, "The $D^0\bar{D}^0$ Mixing Search - Current Status and Future Prospects", Proceedings of the Charm 2000 Workshop, FERMILAB-Conf-94/190, Fermilab, June 7-9, 1994. Harvard preprint HUTP-94/E021, HEP-PH 9408330.

[27] T. Liu, "$D^0\bar{D}^0$ mixing and DCSD - Search for $D^0 \to K^+\pi^-(\pi^0)$", to appear in the Proceedings of the Eighth Meeting of the Division of Particles and Fields of the American Physics Society (DPF'94), Albuquerque, New Mexico. (World Scientific).

[28] We discuss D^0 decays explicitly in the text, its charge conjugate decays are also implied throughout the text unless otherwise stated.

[29] G. J. Feldman *et al.*, Phys. Rev. Lett. **38**, 1313 (1977).

[30] G. Goldhaber *et al.*, Phys. Lett. **69B**, 503 (1977).

[31] P. Avery *et al.*, Phys. Rev. Lett. **44**, 1309 (1980).

[32] J. J. Aubert *et al.*, Phys. Lett. **106B**, 419 (1981).

[33] A. Bodek *et al.*, Phys. Lett. **113B**, 82 (1982).

[34] R. Bailey *et al.*, Phys. Lett. **132B**, 237 (1983).

[35] A. Benvenuti *et al.*, Phys. Lett. **158B**, 531 (1985).

[36] H. Yamamoto *et al.*, Phys. Rev. Lett. **54**, 522 (1985).

[37] S. Abachi *et al.*, Phys. Lett. **182B**, 101 (1986).

[38] W. C. Louis *et al.*, Phys. Rev. Lett. **56**, 1027 (1986).

[39] H. Albrecht *et al.*, Phys. Lett. **199B**, 447 (1987).

[40] J. C. Anjos *et al.*, Phys. Rev. Lett. **60**, 1239 (1988);
For details of this study, see T. Browder, Ph.D. Thesis, UCSB-HEP-88-4, (1988).

[41] R. Ammar *et al.*, Phys. Rev. D **44**, 3383 (1991).

[42] D. Cinabro *et al.*, Phys. Rev. Lett. **72**, 1406 (1994). The detials of this study can be found in [25].

[43] S. Kwan and K. C. Peng, private communication last summer.

[44] K. C. Peng, H. Rubin and S. Kwan, "A study of Singly Cabibbo Suppressed Decay (SCSD) of $D^0 \to K^+K^-$ and $D^0 \to \pi^+\pi^-$", E791 OFFLINE DOCUMENT 207, Feb. 1995.

[45] M. V. Purohit, "Measuring $\Delta\tau$ in the $D^0 \to K^-\pi^+$ and $D^0 \to K^-K^+$ modes", E791 internal memo, March, 1995.

[46] R. Zaliznyak, G. Blaylock and M. Purohit, "E791 Study of $D^0\bar{D}^0$ Mixing in Hadronic Modes", APS talk, 1995. ALso see M. Purohit, "$D^0\bar{D}^0$ Mixing Results from E791", to appear in Proc. LISHEP95 Workshop, Rio de Janeiro, Brazil, Feb. 20-22, 1995.

[47] M. Purohit, "A $D^0\bar{D}^0$ mixing and DCSD analysis of E791 data", E791 internal memo. March, 1994.

[48] In the early days when people searched for $D^0\bar{D}^0$ mixing using the hadronic method, the lack of knowledge on DCSD decays and the lack of statistics and decay time information have forced one to assume that DCSD is negligible in order to set upper limit for mixing [29, 30, 31, 34, 36, 37, 39, 41].

[49] This idea was first suggested by J.D. Bjorken in 1985, private communication with Tom Browder and Mike Witherell.

[50] H. Yamamoto, Ph.D Thesis, CALT-68-1318 (1985).

[51] I. Bigi and A. I. Sanda, *Phys. Lett.* B **171**, 320 (1986).

[52] D. Du and D. Wu, *Chinese Phys. Lett.* **3**, No. 9(1986).

[53] I. Bigi, in Proceedings of the τ-charm factory workshop, SLAC-Report-343, June 1989.

[54] I. Bigi, in SLAC summer institute on Particle Physics, Stanford, 1987; edited by E. C. Brennan (SLAC, 1988); Other predictions can be found in R. C. Verma and A. N. Kamal, Phys. Rev. D **43**, 829 (1991).

[55] G. E. Gladding, in Proceedings of the Fifth International Conference on Physics in Collision, Autun, France, 1985 (World Scientific).

[56] G. E. Gladding, "$D^0\bar{D}^0$ Mixing, The Experimental Situation" in Proceedings of International Symposium on Production and Decay of Heavy Flavors, Stanford, 1988.

[57] G. E. Gladding, in Proceedings of the τ-charm factory workshop, SLAC-Report-343, June 1989.

[58] R. H. Schindler, in Proceedings of the τ-charm factory workshop, SLAC-Report-343, June 1989.

[59] W. Toki, "BES Program and Tau/Charm Factory Physics". Proceedings of the Charm 2000 Workshop, FERMILAB-Conf-94/190, Fermilab, June 7-9, 1994.

[60] J. Norem and J. Repond, "A Tau-Charm-Factory at Argonne". Proceedings of the Charm 2000 Workshop, FERMILAB-Conf-94/190, Fermilab, June 7-9, 1994.

[61] L. L. Chau and H. Y. Cheng, ITP-SB-93-49, UCD-93-31 (HEP-PH-9404207).

[62] F. Buccella *et al.*, Phys. Rev. D **51**, 3478 (1995).

[63] "The Future of High-Sensitivity Charm Experiments", Charm 2000 Workshop, sponsored by Fermilab and Northern Illinois University, FERMILAB-Conf-94/190, Fermilab, June 7-9, 1994.

[64] Particle Data Group, L. Montanet *et al.*, Review of Particle Properties, Phys. Rev. D **50**, 1 (1994).

[65] T. Liu, "Charm mixing working group summary report". Proceedings of the Charm 2000 Workshop, FERMILAB-Conf-94/190, Fermilab, June 7-9, 1994.

[66] R. J. Morrison, " Charm 2000 Workshop Summary". Proceedings of the Charm 2000 Workshop, FERMILAB-Conf-94/190, Fermilab, June 7-9, 1994.

[67] D. Gibaut, A. Freyberger and R. Morrison, "Using Inclusive Semileptonic D^0 Decays With D^{*+} Tag to Study $D^0\bar{D}^0$ Mixing", APS talk, 1995.

[68] G. Blaylock, A. Seiden and Y. Nir, preprint SCIPP 95/16, WIS 95/16/Apr-PH, April 1995;

[69] L. Wolfenstein, report-no: CMU-HEP95-04, DOE-ER/40682-93.

[70] T. E. Browder and S. Pakvasa, University of Hawaii preprint UH 511-828-95.

[71] R. Briere, Private communication. Also see K, Arisaka *et al.*, "KTeV Design Report - Physics Goals, Technical Components, and Detector Costs", FN-580, Jan. 1992.

[72] L. Wolfenstein, Phys. Lett. **164B**, 170 (1985).

[73] A. Pais and S. Treiman, Phys. Rev. D **12**, 2744 (1975),
L. Okun and V. Zakharov, and B. Pontecorvo, *Lett. Nuovo Cimento* **13**, 218 (1975).

[74] H. Cheng, Phys. Rev. D **26**, 143 (1982),
A. Datta and D. Kumbhakar, Z. Phys. C **27**, 515 (1985).

[75] S. Weinberg, *Physica* 96A, 327 (1979);
A. Manohar and H. Georgi, Nucl. Phys. B **234**, 189 (1984);
H. Georgi and L. randall, Nucl. Phys. B **276**, 241 (1986).

[76] A. Weinstein, "Final results from Dalitz plot analysis of $D^0 \to K^-\pi^+\pi^0$", SLAC Memorandum, March 21,1987. Also see J. Adler *et al.*, (MARKIII Collaboration), J. Adler *et al.*, Phys. Lett. **196B**, 107 (1987).

[77] J. C. Anjos *et al.*, Phys. Rev. D **48**, 56 (1993).

[78] R. G. Sachs, Physics of Time Reversal (University of Chicago Press, Chicago). R. G. Sachs, Report No. EFI-85-22.

[79] I. Dunietz and J. L. Rosner, Phys. Rev. D **34**, 1404 (1986).

[80] D. Du, I. Dunietz and D. Wu, Phys. Rev. D **34**, 3414 (1986).

[81] D. Du, X. Li and Z. Xiao, "Reliability of the Estimation of CP Asymmetries for Nonleptonic $B^0\bar{B}^0$ Decays into Non-CP-Eigen states", BIHEP-TH-94-32.

[82] H. Fritzsch, D. Wu and Z. Xing, "On Measuring CP Violation in Neutral B meson Decays at the $\Upsilon(4S)$ Resonance", CERN-TH.7194/94, PVAMU-HEP-94-2.

[83] Z. Xing, "On Test of CPT Symmetry in CP-violating B Decays", LMU-2/94.

[84] T. Liu, "An Overview of $D^0\bar{D}^0$ Mixing Search Techniques: Current Status and Future Prospects", PRINCETON/HEP/95-6, HEP-PH-9508415.

Rare D Decays, $D^0 \overline{D^0}$ Mixing, and CP Violation at a Tau Charm Factory

Gary Gladding

Department of Physics
University of Illinois at Urbana-Champaign
Urbana, Illinois 61801

Abstract. Results of experimental projections for measurements of rare D Decays, $D^0\overline{D^0}$ Mixing and CP Violation at a Tau-Charm Factory are presented. Emphasis is given to those measurements which are unique to a Tau-Charm Factory. Sensitivities for some Lepton Family Number Violating Decays and Flavor Changing Neutral Current Decays can reach the level of 10^{-7} - 10^{-8}, while the sensitivity for the unique decay $D^0 \rightarrow \nu\overline{\nu}$ should reach the 10^{-5} level. Several different kinds of measurements are presented which should be sensitive to $D^0\overline{D^0}$ mixing (without DCSD complications!) at the level of 10^{-4}. Finally, examples are given of both asymmetry and rate experiments which are sensitive to direct and indirect CP violation.

CAVEAT

No new experimental studies have been performed for this paper. Rather, this paper is meant to be a brief summary of previous results. In particular, heavy use is made of results from the original Tau-Charm Factory Workshop held at SLAC in 1989 and the Marbella Workshop held in Marbella, Spain in 1993.

RARE D DECAYS

The pursuit of very rare D Decays at a Tau-Charm Factory should be seen as largely a search for new physics. In particular, searches for Lepton Family Number Violating Decays (which are forbidden in the Standard Model) and Flavor Changing Neutral Current Decays (which are strongly suppressed in the Standard Model) should be vigorously pursued at a Tau-Charm Factory. The initial experimental projections for these searches were carried out by Ian Stockdale (1).

The main advantages of a Tau-Charm Factory for these decays lies in the extremely clean environment in which only D pairs are produced at the Ψ'' resonance. In addition, an excellent detector (99.5% solid angle coverage, e/π rejection $\approx 5\times 10^{-4}$ and μ/π rejection $\approx 10^{-2}/p(GeV)$) was found to be necessary in order to establish the existence of a signal in these rare decays at the 5σ level. The

© 1996 American Institute of Physics

detector solid angle coverage was found to be particularly important for the search for the decay $D^0 \to \nu\bar{\nu}$, which probably could only be done at a Tau-Charm Factory. This signal would be seen in events in which all observable particles reconstruct to a single D^0 mass. The high solid angle coverage is necessary to eliminate backgrounds from the decay $D^0 \to K^0_L \pi^0$. In particular, Stockdale estimates a branching ratio of 10^{-5} could be seen (at the 5σ level) with 10^8 produced D^0's.

With the exception of $D^0 \to \nu\bar{\nu}$, the rare D decays that can be pursued involve all charged particles in the final state. For these decays, the Tau Charm Factory's advantage over B factories or fixed target experiments are significantly reduced. Estimates for the branching ratio limit which can be achieved from a sample of 10^8 produced D^0's for the Lepton Family Number Violating Decay $D^0 \to \mu^+ e^-$ are at the 5×10^{-8} level. For Flavor Changing Neutral Current Decays ($D^0 \to \mu^+ \mu^-$, $D^0 \to e^+ e^-$, $D^0 \to \rho^0 e^+ e^-$), estimates of the branching ratio limits range from $2 - 4 \times 10^{-8}$.

$D^0 \overline{D^0}$ MIXING

$D^0 \overline{D^0}$ Mixing will occur if the mass eigenstates are not the flavor eigenstates. The mixing is usually discussed in terms of the parameters $x \equiv \Delta M/\Gamma$ and $y \equiv \Delta\Gamma/\Gamma$. where ΔM and $\Delta\Gamma$ refer to the difference in mass and width of the two states. Experimental limits on mixing are usually given in terms of the parameter $r_D \equiv \Gamma(D^0 \to e^- X) / \Gamma(D^0 \to e^+ X) \approx 1/2 (x^2 + y^2)$. In the Standard Model, short distance calculations are quite small ($r_D \approx 10^{-7}$) due to a GIM suppression. However, long distance effects can increase the prediction for r_D to be in the range $10^{-5} - 10^{-4}$.

The experimental signature for mixing is the observation of the flavor at decay being different from the flavor at birth. The tag of the flavor at birth is usually accomplished in one of two ways: (i) the "D* trick" in which the charge of the pion from the decay $D^{*+} \to D^0 \pi^+$ tags the charm of the D, or (ii) correlated production in which the charm of one D tags the charm of the other as in $e^+ e^- \to D^0 \overline{D^0}$. The tag of the flavor at decay is also determined in one of two ways: (i) the charge of the kaon in a hadronic decay of the D^0, or (ii) the charge of the lepton in a semileptonic decay of the D^0. Both methods for determining the flavor at decay have difficulties for experiments at a B factory or at fixed targets. Namely, if hadronic decays are used, one has to account for, in some way, Doubly Cabibbo Suppressed Decays (DCSD) which can also lead to the same final state. Usually, this separation has to be made by measuring the time evolution, a tricky business, especially if all the interference terms are properly treated. On the other hand, if semileptonic decays are used, then there is a problem with the D* tag since there is no mass peak which results in larger backgrounds. All these difficulties can be removed at a Tau-Charm Factory!

Semileptonic decays can be used at a Tau-Charm Factory to tag the flavor at decay since the correlated production of D pairs can be exploited to produce clean signals. In particular, two different final states can be used in this way. The process $e^+ e^- \to D^{*+}(\to \pi^+ D^0 (\to K e \nu)) D^- (\to K^+ \pi^- \pi^-)$ was initially studied by Constantine Simopolous (2). The mixing signal in this process is very clean since the final state is fully reconstructed and the flavor is tagged by the charge of the D. Simopolous estimates that 19000 right sign events would be observed in a year's run with luminosity = 10^{33} cm^{-2} s^{-1} while backgrounds should be at the < 1 event level. Presumably, the sensitivity could be increased by adding other semileptonic decays and other D^- tags.

Semileptonic decays can also be used to search for mising in the process: $e^+ e^- \to D^0 (\to K e \nu) \overline{D^0}(\to K e \nu)$. This measurement is more challenging in that good particle identification is needed for both kaons and electrons. In addition, good calorimetric coverage is needed to identify π^0's and K^0_L's which may be produced. I have reported elsewhere (3) my estimates that $\approx$ 22000 right sign events would be observed in a year's run with luminosity = 10^{33} cm^{-2} s^{-1} while backgrounds should be at the < 5 event level. This sensitivity can be increased by adding other semileptonic decays involving muons or K^*'s.

Hadronic decays can also be used at a Tau-Charm Factory to unambiguously search for $D^0 \overline{D^0}$ mixing. The complications arising from DCSD can be completely eliminated at a Tau-Charm Factory, due to a quantum statistics argument (4). In particular, if the $D^0 \overline{D^0}$ pair is produced in an odd charge conjugation state (as it is in $e^+ e^- \to D^0 \overline{D^0}$ or $e^+ e^- \to D^0 \overline{D^0} \pi^0$), then Bose statistics precludes, in the absence of mixing, the decay of both D's to the same final state! For example, the observation of the reactions (i) $e^+ e^- \to D^0 (\to K^- \pi^+) \overline{D^0}(\to K^- \pi^+)$ or (ii) $e^+ e^- \to \pi^0 D^0(\to K^- \pi^+) \overline{D^0}(\to K^- \pi^+)$ would be unambiguous evidence for the existence of $D^0 \overline{D^0}$ mixing. These final states are particularly clean since all particles are observed and D mass peaks must be seen. In particular, kaons and pions must be identified (a time-of-flight system with resolution $\sigma \approx$ 120ps will suffice) and cuts must be made on the beam-constrained and invariant masses. I have previously reported estimates (3) of $\approx$ 38000 right sign events would be observed in a year's run with luminosity = 10^{33} cm^{-2} s^{-1} while backgrounds should be at the < 1 event level. The background can be easily predicted since it is dominated by the double misidentification of the kaon and pion. This rate is small and calculable since it is the product of two larger and easily measured numbers.

In conclusion, there are several methods available at a Tau-Charm Factory to detect mixing at the few $\times 10^{-4}$ level. Table 1 gives estimates of the sensitivities to $D^0 \overline{D^0}$ mixing which can be achieved for many final states in a year's run with luminosity = 10^{33} cm^{-2} s^{-1}.

Table I
$D^0\bar{D}^0$ Mixing
Rate Summary (1 year's run at L = 10^{33})

$e^+e^- \to$	Final State	Rate	Events (right sign)	r_D 90% CL	r_D 5σ signal (8 evts)
$D^0\bar{D}^0$	$(K^-\pi^+)(K^-\pi^+)$	r_D	37500	6.0 ×10^{-5}	2.1 ×10^{-4}
$D^0\bar{D}^0\ \gamma$	$(K^-\pi^+)(K^-\pi^+)\ \gamma$	$3r_D + 8(\frac{\Delta\Gamma}{2\Gamma})tan^2\theta_c\hat{\rho}$ $+4tan^4\theta_c\hat{\rho}^2$	3300*	--	--
$D^0\bar{D}^0\ \pi^0$	$(K^-\pi^+)(K^-\pi^+)\ \pi^0$	r_D	2750*	8.3 ×10^{-4}	2.9 ×10^{-3}
$D^0\bar{D}^0$	$(K^-e^+\nu)(K^-e^+\nu)$	r_D	21600		
	$(K^-e^+\nu)(K^-\mu^+\nu)$	r_D	40000*	2.8 ×10^{-5}	9.8 ×10^{-5}
	$(K^-\mu^+\nu)(K^-\mu^+\nu)$	r_D	20000*		
$D^0\bar{D}^0\ \gamma$	$(K^-l^+\nu)(K^-l^+\nu)\ \gamma$	$3r_D$	11600*	5.1×10^{-5}	1.8×10^{-4}
$D^0\bar{D}^0\ \pi^0$	$(K^-l^+\nu)(K^-l^+\nu)\ \pi^0$	r_D	9600*		
$(D^0\pi^+)D^-$	$[\pi^+(K^+e^-\nu)(K^+\pi^-\pi^-)]$	r_D	19000		
	$[\pi^+(K^+\mu^-\nu)(K^+\pi^-\pi^-)]$	r_D	15000*		
	$[\pi^+(K^+e^-\nu)(\text{other } D^-\text{ tag})]$	r_D	15000*	3.6 ×10^{-5}	1.2 ×10^{-4}
	$[\pi^+(K^+\mu^-\nu)(\text{other } D^-\text{ tag})]$	r_D	15000*		

* estimates based on scaling acceptances of similar processes

CP VIOLATION

Both direct and indirect (*ie* induced through $D^0\bar{D^0}$ mixing) can be searched for at a Tau-Charm Factory. If $D^0\bar{D^0}$ mixing is really small, then there is not much hope of seeing indirect CP violation. However, if mixing is not really small, then once again the quantum coherence of the initial state can be used to advantage. For example, consider the following asymmetry

$$A \equiv \frac{R(D^0 \to KK) - R(\bar{D^0} \to KK)}{R(D^0 \to KK) + R(\bar{D^0} \to KK)}$$

If the $D^0\bar{D^0}$ state has odd C (*eg* $D^0\bar{D^0}$ or $\pi^0\ D^0\bar{D^0}$), then A = 0 in the limit of no direct CP violation! If the $D^0\bar{D^0}$ state has even C (*eg* $\gamma^0 D^0\bar{D^0}$), then A ≈ x, assuming y <<x<<1. The reaction $e^+ e^- \to \gamma\ D^0\bar{D^0}$, where one D decays semileptonically and the the other D decays to a "shared state", *ie* one which can be reached from either a D^0 or a $\overline{D^0}$, was initially studied by Uri Karshon (5). This study was extended by Joe Izen (6) to include hadronic decays of the D. It was estimated that ≈13000 semileptonic events and ≈27000 hadronic events would be observed in a year's run with luminosity = 10^{33} cm^{-2} s^{-1}, leading to an asymmetry sensitivity ≈5×10^{-3}. Fry and Ruf (7) propose to use inclusive tagging while ignoring the π^0 / γ distinction to obtain 100000 - 400000 events with a somewhat diluted asymmetry. They estimate this technique would yield an asymmetry sensitivity ≈$2\text{-}3\times10^{-3}$.

Direct CP violation can be searched for at a Tau-Charm Factory through either asymmetry or rate experiments. The general asymmetry here is defined as:

$$A = \frac{R(D \to f) - R(\overline{D} \to \overline{f})}{R(D \to f) + R(\overline{D} \to \overline{f})}$$

This asymmetry can be measured using either charged or neutral D decays. Fry and Ruf (7) investigated the charged D modes, $D^+ \to K^- K^+ \pi^+$ and $D^+ \to \eta \pi^+$. The main difficulty with this approach is that the need to measure absolute rates requires careful corrections for detector asymmetries using Cabibbo favored decays in which no effect is expected. They estimate ≈200000 events in each mode, leading to an asymmetry sensitivity ≈2×10^{-3}. This asymmetry can also be measured using the neutral D decays, $D^0 \to K^- K^+$ and $\overline{D^0} \to K^- K^+$. Fry and Ruf (7) considered D^0's from the decays of D^*'s, while I studied D^0's produced at the Ψ'' (3). It is difficult to make firm estimates here due to the uncertainites in how inclusive a flavor tag can be used as well as possible problems understanding the systematics involved. Probably asymmetry sensitivities in the range of $2\text{-}5\times10^{-3}$ can be reached.

Direct CP violation can also be probed by searching for Ψ'' decays into two CP eigenstates since Ψ'' has CP = +1, but $D^0\overline{D^0}$ has C=-1 and L=1. Consequently, D^0 and $\overline{D^0}$ cannot decay to CP eigenstates of the same value unless CP is violated! Once again, the quantum coherence available at a Tau-Charm Factory allows a unique measurement to be made. I have reported elsewhere (3) my estimates for the rates for decays into the CP eigenstates $K^- K^+$, $\pi^- \pi^+$, and $K^0_L \pi^0$ for 100% CP violation. These estimates (a few thousand events) are not particularly encouraging since the rates are expected to vary as the square of the CP-violating amplitude and hence should be quite small. However, the simple experimental technique of measuring a rate, rather than an asymmetry offers a useful complementary approach.

CONCLUSIONS

A Tau-Charm Factory is well-suited for searches for new physics with unique methods in the areas of rare D decays, $D^0\overline{D^0}$ mixing and CP violation. Lepton Family Violating Decays and Flavor Changing Neutral Current Decays can be detected at the level of 10^{-7} - 10^{-8}. Several (≈5) different kinds of measurements can be made to search unambiguously (*ie* without DCSD complications) for mixing at the level of $\approx 10^{-4}$. Searches for indirect CP violation can be made using asymmetries which are only possible due to the quantum coherence of the initial state. Searches for direct CP violation can be made both in asymmetry measurements and in rate measurements.

REFERENCES

1. Stockdale, I., "A Study of Rare D Decays for the Tau-Charm Factory," in *Proceedings of the Tau-Charm Workshop.* 1989, SLAC-Report-343, pp. 724-732.
2. Simopolous, C., talk presented at the Tau-Charm Workshop, 1989.
3. Gladding, G.," $D^0\overline{D^0}$ Mixing and CP Violation: Experimental Projections for a τ-Charm Factory," in *Proceedings of the Tau-Charm Workshop.* 1989, SLAC-Report-343, pp. 152-168.
4. Bigi, I. & Sanda, A.I., Phys Lett **171B**, 320 (1986).
5. Karshon, U., "Monte Carlo Study of CP Asymmetry Measurement at a Tau-Charm Factory," in *Proceedings of the Tau-Charm Workshop.* 1989, SLAC-Report-343, pp. 706-723.
6. Izen, J., talk presented at the Tau-Charm Workshop, 1994.
7. Fry, J.R. and Ruf,T., "CP Violation and Mixing in D Decays," in *Proceedings of the Marbella Workshop on the Tau-Charm Factory,* 1993, pp. 387-405.

CONCLUSION

Tau/charm Workshop Summary

Frederick J. Gilman

Department of Physics
Carnegie Mellon University
Pittsburgh, PA 15213

I. Introduction

The concept of a Tau/charm Factory is centered on the construction of an electron-positron collider with a center-of-mass energy of approximately 3 to 6 GeV/c^2. This permits probing the physics of the τ lepton starting from the threshold for its pair production and of the charm quark through the regions of narrow charm-anticharm resonances and of production of pairs of charmed mesons and baryons. A detector with calorimetry, tracking, and particle identification capabilities that are optimized to the physics is an essential part of this concept.

Such a machine explores the high intensity/high precision frontier rather than the high energy frontier. The task of building the collider and doing physics at a Tau/charm Factory is not easy, and it is complicated by the fact that we have a broader range of energies and a more diverse set of physics topics to which we wish to optimize both the machine and the detector than in many other particle "factories".

The original proposal [1] in 1987 has been followed by a series of workshops [2] to develop the machine and detector designs and clarify the physics objectives. Attention focused in the early 1990s on the possibility of building a Tau/charm Factory in Spain. More recently, sites where such a machine has been actively considered include Argonne National Laboratory in the USA, Dubna and Novosibirsk in Russia, and Beijing in China.

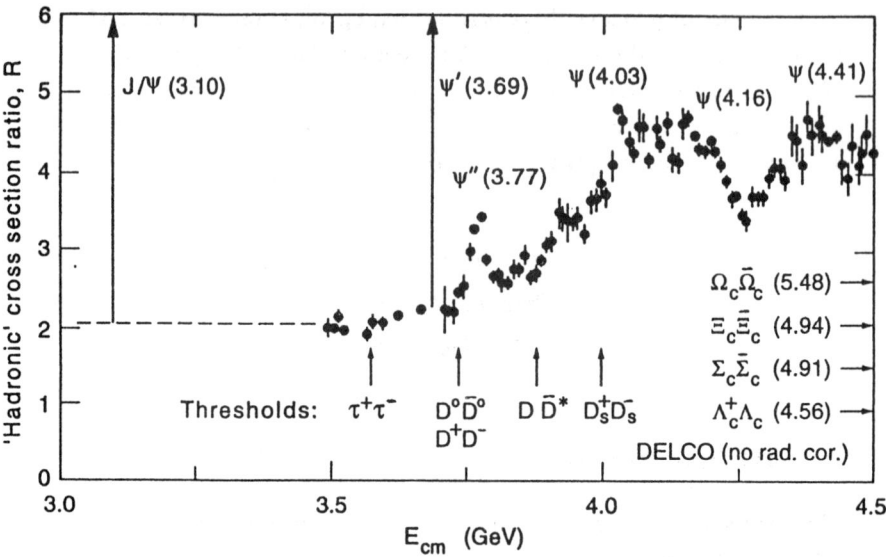

Figure 1. The hadronic total cross section in the center-of-mass energy range of the Tau/charm Factory.

II. The Collider

Current designs at all these laboratories involve two rings of magnets carrying electron and positron beams, respectively, of equal energies. A high luminosity requirement, $10^{33}/\text{cm}^2/\text{sec}$ at around 4 GeV/c^2 in the center-of-mass, leads to high currents, of order an ampere in each ring. In addition, some of the physics discussions have led to requiring, or at least not precluding, the options of extension of the collider's design to having:

- Monochromator optics – corresponding to an order of magnitude smaller spread in the center-of-mass energy and an order of magnitude greater peak cross section on narrow resonances

- Higher luminosity

- Beam polarization

- A larger center-of-mass energy range.

Examples of machine designs from Beijing [3], Dubna [4], and ANL [5] have been presented and discussed at this meeting. These designs have many state-of-the-art features (see Table 1. and Fig. 2.); in fact they are faced with many of the challenges in RF, vacuum, feedback, interaction region design, etc., that are not unlike those that face the B-factories [6].

Scheme	Standard	Monochr.	Crossing Angle
Nominal energy E(GeV)	2.0	1.5	2.0
Ring circumference C(m)	367.5	367.5	367.5
Crossing angle at IP θ (mrad)	0.0	0.0	3.0
β-function at IP β_x^*/β_y^* (m)	0.2/0.01	0.01/0.15	0.8/0.01
Dispersion at IP D_y^* (m)	0.00	0.35	0.00
Momentum compaction α_p	0.022	0.008	0.022
Natural emittance ϵ_x (nm rad)	251	20	251
Emittance with wiggler		10 (J_x=2)	
Vertical emittance ϵ_{yc}	12	2	4.8
Energy spread σ_E	5.4×10^{-4}	8×10^{-4}	5.4×10^{-4}
Energy loss per turn U_0 (keV)	142.6	45.0	427.8
Damping time $\tau_x/\tau_y/\tau_e$ (ms)	34/34/17	41/80/80	34/34/17
RF frequency (MHz)	500	500	500
RF voltage (MV)	9.00	9.00	9.00
Numbers of bunches K_B	32	32	32×3
Bunch spacing S_B	11.48	11.48	11.48/3
Total current per beam (A)	0.55	0.215	1.65
Particles per bunch	1.32×10^{11}	5.14×10^{10}	1.32×10^{11}
Natural bunch length (cm)	1.0	0.78	1.0
Impedance $\mid Z/n \mid_{\parallel}$ (Ω)	0.32	>0.32	0.32
Beam-Beam effect ξ_x/ξ_y	.04/.04	.031/.015	.04/.04
Beam life time τ (hours)	4.8	1.5	
Transverse tune Q_x/	11.192/	13.18/	11.15/
Q_y	10.192	9.24	10.18
Synchrotron tune Q_s	0.098	0.068	0.098
Natural chromaticity Q_x' /	-26.6/	-35.9/	-20.0/
Q_y'	-32.0	-44.5	-35.09
Luminosity $\mathcal{L}$ (cm^{-2}s^{-1})	1×10^{33}	2.2×10^{32}	3×10^{33}
CM energy spread σ_W (MeV)	1.53	0.105	1.53

Table 1. Main parameters of the τcF from Ref. 3.

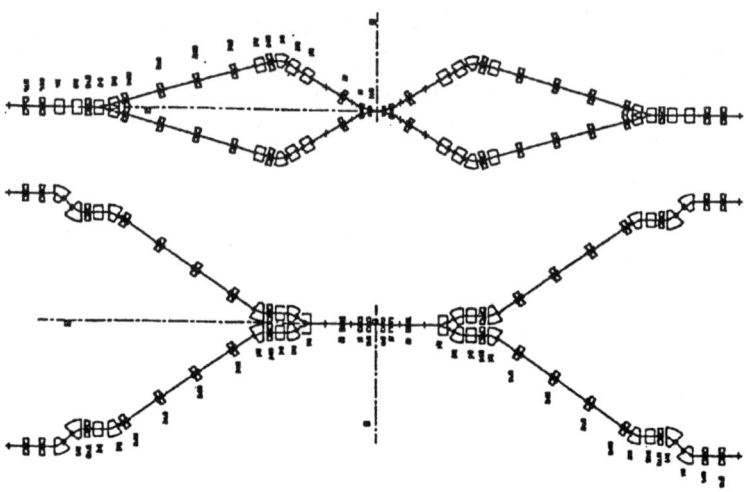

Figure 2. Horizontal and vertical interaction region views from Ref. 4.

Thus the design of a Tau/charm Factory should greatly benefit from the solution of the technical challenges now facing the B-factories. Nevertheless, the Tau/charm Factory has its own challenges and must operate over a wider range of energy where the beams are less 'stiff' than is the case for the B-factories [7].

A number of important questions need to be addressed soon, such as, what are the tradeoffs (e.g., in luminosity or beam lifetime) of exercising one or more of the options listed above. The machine/detector interface needs to be explored in detail. For these and other issues an extended, real workshop is required [7].

III. The Detector

The detector will need to be optimized to the physics and to the machine in order to reduce the systematics and backgrounds. While in general such a detector will have a resemblance to the Mark III and BES, it is clear that in some regards, such as calorimetry, much better performance will be required to explore tau/charm physics at the requisite sensitivity. Some of the considerations developed in earlier meetings were summarized by Kirkby [8], and a comparison of detector performance is shown in Table 2.

	Mark III (SPEAR) / BES (BEPC)	EXACT (τcF)
Charged particles:		
Momentum res.: σ_p/p(GeV/c)	1.5%$p \oplus$ 1.5%/β [MkIII] 0.7%$p \oplus$ 1.3%/β [BES]	0.4%$p \oplus$ 0.4%/β
Angular resolution: σ_ϕ (mr)	$2 \oplus 2/p\beta$	$0.5 \oplus 1.1/p\beta$
p^π_{min}(MeV/c) for efficient tracking	80	50
Ω(barrel) ($\times 4\pi$ sr)	70%	90%
Photons:		
Energy resolution: σ_E/E(GeV)	17%/$\sqrt{E}$	2%/$E^{\frac{1}{4}} \oplus$ 1%
Angular resolution: $\sigma_{\theta,\phi}$ (mr)	10 [MkIII]; 5 [BES]	$1.7 + 2/\sqrt{E}$ ($\theta=90°$)
2γ angular separation: $\Delta\theta_{2\gamma}$ (mr)	20	50
E^γ_{min}(MeV) for efficient detection	100	10
Particle identification:		
$\pi \to K$ separation	3σ at 0.7 GeV/c	3σ at 1.0 GeV/c (10^{-4} inc. Č)
$\pi/K \to e$ separation	4% at 0.5 GeV/c	0.1% (10^{-5} inc. Č)
$\pi/K \to \mu$ separation	5% at 1.0 GeV/c	1.5%/p + (1-4)%
K^0_L detection efficiency	60%	95%
n mean detection efficiency	-	50%
ν 'detection': $p^\perp_{min}$(MeV/c)	-	100

Table 2. Comparison from Ref. 8 of the performances of the τcF detector concept, EXACT, with previous detectors at these energies. The symbol '$\oplus$' denotes addition in quadrature.

As with the machine, the construction of the B-Factory detectors will be of great benefit in dealing with some, but not all, of the issues that confront the Tau/charm Factory detector. We need much more work along the lines already started [9, 10, 11] on physics/detector simulation, subsystem development, integration of the pieces into an overall detector concept, etc. One significant advantage in understanding the performance of the detector once it is built is the presence of the J/ψ and ψ' resonances for detector calibration.

Computing and data acquisition requirements when running off the J/ψ and ψ' are not rigorous by contemporary or future standards, for example the LHC detectors. However, at the J/ψ and ψ' one will likely find oneself in the position, unknown at previous electron-positron colliders, of not being able to take every annihilation event and a selective trigger being required! In any case, this requires serious attention soon, along with the rest of the detector design.

IV. Tau Physics Issues

There is a long list of physics issues related to the τ that can be addressed at a Tau/charm Factory: the mass of the τ and τ neutrino; the structure of the the weak current in τ decay; measurement of branching ratios with high precision; measurement of rare decays; limits on (or discovery!) of forbidden decays (such as $\tau \to 3$ leptons, $\tau \to \gamma e$, $\tau \to \gamma \mu$, . . .); and CP violation in τ couplings. To carry out these measurements, the Tau/charm Factory offers: small backgrounds (often directly measurable by running below τ pair production threshold); high statistics; single-tagged samples; a small Lorentz boost; and possible longitudinal polarization, especially for studies of the structure of the weak current and CP violation.

The status of measurements of properties of the τ from CLEO and from LEP were presented by White [12] and by Wasserbaech [13], respectively. Among the important goals in τ physics is to lower the present limit on the mass of the τ neutrino of 24 MeV/c^2 to a few MeV, whence cosmological arguments [14] can be used to restrict τ neutrino masses to less than about 100 eV/c^2. Such a limit can be established at a Tau/charm Factory, and it seems difficult to do at any other facility. The present data on the structure of the weak current was reviewed by Wasserbaech [13], especially the global fits to τ decays done by ALEPH that give results consistent with what is expected from V - A charged current couplings. Pich [15] summarized the measurements from a theoretical point of view, and indicated the substantial improvements that could come from a Tau/charm Factory, where, for example, one can expect a precision for the Michel parameters comparable to that existing in muon decay (see Table 3).

Parameter	1990	1995	τcF sensitivity
m_τ (MeV)	$1784.1^{+2.7}_{-3.6}$	1777.0 ± 0.3	0.1
m_{ν_τ} (MeV)	< 35 (a)	< 24 (a)	1-2
τ_τ (fs)	303 ± 8	291.6 ± 1.6	-
B_e (%)	17.7 ± 0.4	17.79 ± 0.09	0.02
B_μ (%)	17.8 ± 0.4	17.33 ± 0.09	0.02
$B(\pi^-\nu_\tau)$ (%)	11.0 ± 0.5	11.09 ± 0.15	0.01
$B(K^-\nu_\tau)$ (%)	0.68 ± 0.19	0.68 ± 0.04	0.003
$B(\pi^-\eta\nu_\tau)$	$< 9 \times 10^{-3}$ (a)	$< 3.4 \times 10^{-4}$ (a)	10^{-6}
$B(l^- G)$	$< 10^{-2}$	$< 2.7 \times 10^{-3}$ (a)	10^{-5}
$B(\mu^-\gamma)$	$< 5.5 \times 10^{-4}$ (b)	$< 4.2 \times 10^{-6}$ (b)	10^{-7}
$B(e^-e^+e^-)$	$< 3.8 \times 10^{-5}$ (b)	$< 3.3 \times 10^{-6}$ (b)	10^{-7}
$\rho_{\tau \to \mu}$	0.84 ± 0.11	0.738 ± 0.038	0.002
$\eta_{\tau \to \mu}$	-	-0.14 ± 0.23	0.003
$\xi_{\tau \to \mu}$	-	1.23 ± 0.24	0.02
$(\xi\delta)_{\tau \to \mu}$	-	0.71 ± 0.15	0.02
$\xi'_{\tau \to \mu}$	-	-	0.15
h_{ν_τ}	-	-1.014 ± 0.027	0.003
a_τ^γ	< 0.1 (b)	< 0.01 (a)	0.001
d_τ^γ (e cm)	$< 6 \times 10^{-16}$ (b)	$< 5 \times 10^{-17}$ (a)	10^{-17}

(a) 95% CL ; (b) 90% CL

Table 3. Recent improvements in τ physics and expected precision at the Tau/charm Factory. [15]

Especially interesting is the exploration of the possibility of observing CP violation in τ couplings. The τ provides a unique opportunity to search for CP violation in the lepton sector: the τ is massive; it decays to a τ neutrino plus both hadrons and leptons; and these same decays provide a polarization analyzer. Further, CP could be violated in both τ production and decay. When produced in e^+e^- collisions, for example, the τ could have an electric dipole moment in addition to its charge and magnetic moment couplings to the photon and Z^0. The limits [16] coming from LEP of order 10^{-17} e-cm for such an electric dipole moment are impressive, and it will take polarized beams to get to levels of sensitivity that are substantially lower at a Tau/charm factory. The importance of beam polarization for studying CP violation in τ decay has been emphasized by Tsai [17]; near threshold for τ pair production the final state is in an s-wave, with the simple consequence that the electron polarization ($\vec{w}_{initial}$) in the initial state is transferred to the τ in the final state. One then looks for non-zero values of terms like $(\vec{w}_{initial} \times \vec{p}_\mu) \cdot \vec{w}_\mu$ in the case where the final τ decays into a neutrino, antineutrino, and muon with momentum and polarization $\vec{p}_\mu$ and $\vec{w}_\mu$, respectively. The full theoretical apparatus exists to look in hadronic decays as well [18]. In the Standard Model we do not expect to see such effects, but the uniqueness of being able to probe the lepton sector, the lack of definitive knowledge of the origin of CP violation, and the spectacular consequences of seeing a CP-violating effect make this an important aspect of building a Tau/charm Factory. In particular, we would strongly urge that the possibility of having beam polarization not be precluded.

V. Charmonium Physics Issues

While a great amount of data has been accumulated on charmonium over the past two decades and we think that we understand the basic physics that underlies the spectroscopy and decays of $c\bar{c}$ states, interesting questions remain to be answered and puzzles/discrepancies resolved. Included in these categories are confirmation/establishment of the η'_c state, comparisons of the ψ' to J/ψ decays to the same final state, discovery of narrow d-wave states above $D\overline{D}$ (but not $D\overline{D}^*$) threshold, branching ratios for many decay modes of states other than the J/ψ and ψ', the observation of many rare hadronic and electromagnetic decays, etc. For masses and widths we are moving on the theoretical side from a QCD inspired phenomenology to first-principles calculations [19]. While important information is coming from $p\bar{p}$ annihilation [20], many of the open questions can only be settled or the precision vastly improved at a Tau/charm Factory, as shown in Table 4 from the talk of Toki [21].

The potential samples of J/ψ and ψ' decays, ranging from 10^{10} to 10^{11} per year are incredible, and represent three to four orders of magnitude over present experiments, as shown in Table 5. Especially with monochromator optics, event rates like these should permit the study of the weak decays of a heavy $Q\overline{Q}$ system, namely $c\bar{c}$. An example of such a decay is $c\bar{c} \to c\bar{s}e\bar{\nu}$ at the quark level or $\psi \to D_s e\bar{\nu}, D_s^* e\bar{\nu}, \ldots$ at the hadron level. A rough calculation gives branching ratios for weak J/ψ decays in the range 10^{-8} to 10^{-9}, with the higher end of the range pertaining to inclusive weak decays and the lower end relevant to some of the major semileptonic decay modes [22]. Aspects of these decays lend themselves to application of Heavy Quark Effective Theory

(HQET), including calculation in cases like $J/\psi \to D_s e\bar{\nu}$ of absolute rates in terms of $|V_{cs}|$ and f_{D_s}, in analogy to work that was done for decays of the B_c meson [23].

State	Mass [MeV]	Width	Spin	BR [%] (ee or $\gamma\gamma$)	hadronic decays
η_c	2974.4± 1.9	> 15 MeV*	0^{-+}	7± 3 keV*	few decays*
J/ψ	3096.9± .09	85.5± 6 keV*	1^{--}	5.92±.15 ± .2*	many decays*
ξ_0	3417± .8	13.5± 3	consistent*	0.04±0.02	few decays*
ξ_1	3510± .04	0.88± 0.11	consistent*		few decays*
ξ_2	3556± .07	1.98± 0.17	consistent*		few decays*
$^1P_1$	3526.2± .2	.9± 0.44	?*	?	$p\bar{p}, J/\psi + \pi^*$
$\eta_c(2s)$	no signal E760	?*	?*	?*	?*
ψ'	3686±.1	0.308± 0.036*	1^{--}	0.88±0.13*	many decays*
$\psi(3.77)$	3764±5	24±5*	1^{--}		DD^*
$\psi(4.03)$	4040±10	52±10*	1^{--}		$D^*D^*, D^*D,^*$ $DD, D_s D_s$
$\psi(4.14)$	4159±20	78±20*	1^{--}		$D^*D^*, D^*D,^*$ $DD, D_s D_s^*$

Table 4. Status of Charmonium Measurements. Areas where a Tau/charm factory would make a significant improvement are shown with an asterisk (*).

Resonance	Peak σ [nb]	Instantaneous Rate at 10^{33}	#events/day at 50 % Efficiency
J/ψ	~2600	2.6 kHz	112×10^6
ψ'	~800	0.8 kHz	34×10^6
J/ψ monochrometer	~16000	16 kHz	688×10^6
ψ' monochrometer	~5000	5 kHz	215×10^6

Table 5. J/ψ and ψ' Event Rates [21].

With these same enormous J/ψ and ψ' data samples comes the possibility of exploring light-quark/gluon spectroscopy in great depth. It is especially interesting to look for hadrons that involve gluonic degrees of freedom such as glueballs (gg) or hybrid states ($q\bar{q}g$). Such states surely exist, often mixed with ordinary ($q\bar{q}$) mesons, making finding them problematic. One way to establish their existence would be to find a "smoking gun" – a meson resonance that has "exotic" quantum numbers that cannot be formed from a quark and antiquark (such as $J^{PC} = 1^{-+}$). Even with the luck of finding such a state, a real understanding of this spectroscopy will require intensive experiments coupled with exhaustive theoretical analysis. The situation until now has been murky and inconclusive; the status of theory and experiment was reviewed for us by Barnes [24], while Meyer [25] gave us an example of the massive data sets and isobar analysis being done with LEAR data in the channels $p\bar{p} \to 3\pi\eta$, and $\pi\eta\eta$. Since the J/ψ is predicted to have appreciable decays to γgg, radiative J/ψ and ψ' decays have long been realized to be a natural place to look for glueballs. New data from BES discussed by Huang [26] are especially intriguing in this regard, as they show not only confirmation of the $\xi(2230)$ state first observed at SPEAR, but ratios of decay rates into $\pi\pi$ and $K\bar{K}$ that are consistent with those of an SU(3) singlet. This, together with the quite narrow width (for a state with a mass above 2 GeV/c^2), is suggestive of a glueball interpretation for this state [26]. While an enormous amount remains to be done to unscramble glueballs and hybrid states from the hadronic spectrum, this recent work highlights the more general point that a Tau/charm Factory would be a unique facility for the study of QCD, and especially for investigating spectroscopy, both old and new.

This is an appropriate place to note that the electron-positron total cross section for annihilation into hadrons is itself an important input into other calculations. A case in point is the running of the electromagnetic coupling from $q^2 = 0$ to $q^2 = M_Z^2$, involving vacuum polarization graphs, which has been re-examined by Swartz [27] in the last year in connection with understanding the consistency and implications of the LEP and SLC data. Aside from correcting previous analyses, the upshot from this work is that the dominant uncertainty in $\alpha(M_Z^2) - \alpha(0)$ arises from the portion of an integral over the weighted hadronic total cross section in the 1 to 5 GeV region. Since the experimental uncertainty in the cross section translates into a "theoretical" uncertainty in $\sin^2\theta_W$ that is only somewhat smaller than its present experimental error bars, making progress in the longer term in pinning down the electroweak parameters is dependent on improving e^+e^- total cross section measurements accessible at a Tau/charm Factory.

VI. Charm Physics Issues

The "laundry list" of physics issues in the charm sector is a long one as well. Semileptonic decays of charmed hadrons were discussed here for fixed target experiments by Wiss [28] and for CLEO by Fujino [29]. Many observed decays still lack precisely measured branching ratios and form factors, while there are semileptonic decays of baryons and Cabibbo-suppressed decays of both mesons and baryons that are unmeasured and are likely to remain largely as the domain of a Tau/charm Factory [30]. By working with tagged samples, accurate measurement of the leptonic decays of charmed mesons and the couplings f_D and f_{D_s} will be possible. Non-leptonic charm decays remain puzzling, but with interesting lessons [31] to teach us in the domain between light quarks and truly heavy quarks through study of the systematics of non-leptonic decays.

As is often the case, things that are difficult experimentally, such as rare decays, $D\overline{D}$ mixing and CP violation, are the most interesting theoretically and have the biggest payoff if found. The prospects for fixed target experiments were reviewed by Kaplan [32] and for B-Factories by Liu [33], while Gladding [34] covered some of the possibilities at a Tau/charm Factory. Flavor-changing neutral currents induced at the one-loop level can drive processes like $D \to \gamma\rho$, $D \to \gamma\omega$, and $D \to \ell^+\ell^- X$. They are predicted to be very small in the Standard Model, and are likely obscured by much larger long-distance contributions, although these are interesting to understand for their own sake as they relate to similar contributions to rare B decays. Rare decays, and forbidden decays such as $D \to \mu e$, could be pushed to the level of 10^{-7} to 10^{-8} in branching ratio in the very clean environment of a Tau/charm Factory [34].

One of the more interesting measurements at a Tau/charm Factory should be that of $D\overline{D}$ mixing. [35] The present upper limit on $x_D = \Delta m_D/\Gamma < 0.083$ is far from theoretical expectations of a number of order 10^{-4} or so, based on long-distance, Cabibbo-suppressed contributions to Δm_D. We therefore have a place where there is lots of room for flavor-changing effects to show up from physics beyond the Standard Model. Such is the case in some versions of the Two-Higgs Doublet Model (see Fig. 3) and in other multi-Higgs models with flavor-changing couplings, [35, 36] where it is possible to get contributions to Δm_D that are orders of magnitude larger than that from the Standard Model.

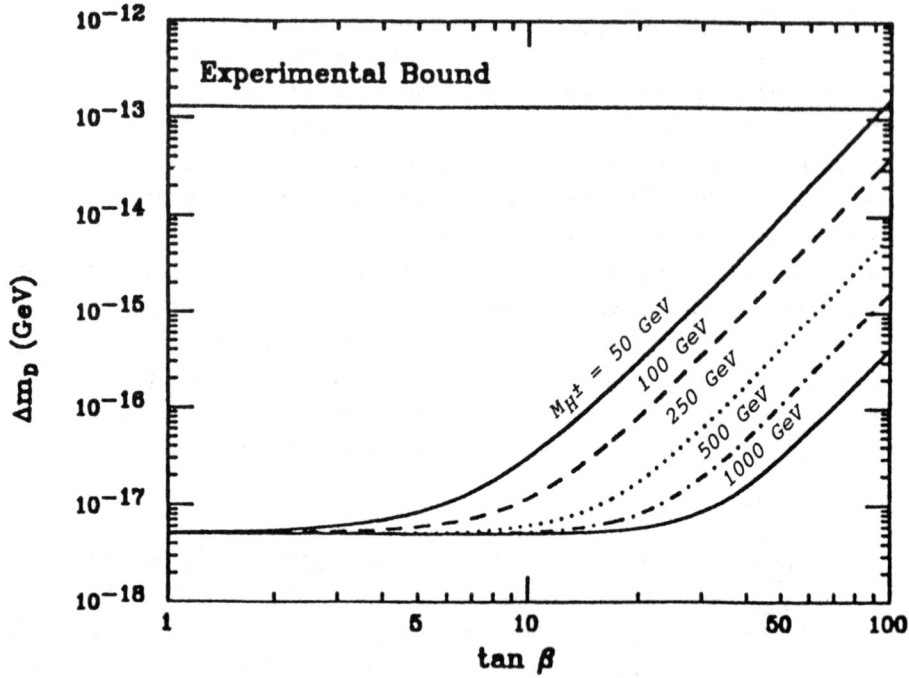

Figure 3. Δm_D in a Two Higgs Doublet Model [36].

Within the Standard Model, CP-violating effects in the charm sector are also small. One way observable effects could arise is from the interference of two weak decay amplitudes with different Cabibbo-Kobayashi-Maskawa (CKM) factors, such as those from tree and penguin diagrams in CKM suppressed modes. Estimates put possible asymmetries at the 10^{-3} level, [35] and a Tau/charm Factory should be able to probe to roughly this level of sensitivity in several years of running [1]. The asymmetries produced through $D\overline{D}$ mixing and decay into a CP eigenstate, whose analogues in the B system produce large asymmetries, are small here from the combined effects of small $D\overline{D}$ mixing and of small phase differences or interfering amplitudes in the Standard Model. This is not necessarily the case if the source of large $D\overline{D}$ mixing is also associated with a significant CP violating phase, as happens in a number of extensions of the Standard Model [37]. In particular, one can then envisage large asymmetries when looking at the interference of doubly suppressed amplitudes with those that arise from relatively modest ($x_D \sim 10^{-2}$, still much larger than expected in the Standard Model) $D\overline{D}$ mixing.

VII. Conclusion

The Tau/charm Factory has a broad and exciting physics program. There are both programatic, bread-and-butter physics issues and more narrowly focussed issues where there is a large window for physics from beyond the Standard Model. While there are overlaps with the physics of fixed target and B-Factory experiments, much is complementary to what will be done at other facilities. Futhermore, even looking ahead years from now, a number of important physics topics are unique and/or probed with higher precision.

Both the machine and the detector must separately push the state-of-the-art if we are to aim for the physics goals we have been discussing, and they must be carefully integrated if we are to realize the physics potential. Considerable effort must still be expended to arrive at final, detailed designs.

On both a technical and a physics basis, a Tau/charm Factory merits an international effort to build it. The time has come to choose a site, develop the final machine and detector designs, and start construction.

References

[1] See the "Tau/charm Factory Overview" by J. Kirkby, these proceedings and references therein.

[2] The proceedings of the most recent such meeting is found in "The Tau-Charm Factory in the Era of B-Factories and CESR", Stanford, California, August 15 - 16, 1994, edited by Lydia V. Beers and Martin L. Perl (Stanford Linear Accelerator Center, Stanford, 1994), SLAC Report No. 451.

[3] Y.-Z. Wu, these proceedings.

[4] B. Perelstein, these proceedings.

[5] L. Teng, these proceedings.

[6] M. Zisman, these proceedings.

[7] M. Tigner, talk and roundtable discussion, these proceedings.

[8] J. Kirkby, "Tau/charm Factory Detector Overview" in these proceedings and references therein.

[9] Y. Huang, these proceedings.

[10] T.-Y. Chen, these proceedings.

[11] S.-H. Wang, these proceedings.

[12] C. White, these proceedings.

[13] S. Wasserbaech, these proceedings.

[14] H. Harari, invited talk at the Third Workshop on Tau Lepton Physics, Montreux, Switzerland, September 19-22, 1994, unpublished.

[15] A. Pich, these proceedings.

[16] D. Buskulic et al., Phys. Lett. **346B**, 371 (1995); R. Akers et al., Z. Phys. **C66**, 31 (1995).

[17] Y. S. Tsai, these proceedings; Phys. Rev. **D51**, 3172 (1995); and SLAC preprint, SLAC-PUB-6916, 1995 (unpublished).

[18] E. Mirkes, these proceedings.

[19] A. X. El'Khadra, these proceedings.

[20] C. M. Ginsburg, these proceedings.

[21] W. Toki, these proceedings.

[22] F. Gilman, unpublished

[23] E. Jenkins et al., Nucl. Phys. **B390**, 463 (1993).

[24] T. Barnes, these proceedings.

[25] C. A. Meyer, these proceedings.

[26] T. Huang, these proceedings, and T. Huang et al., Beijing IHEP preprint, BIHEP-TH-95-11, 1995 (unpublished).

[27] M. Swartz, these proceedings and SLAC preprint, SLAC-PUB-6710, 1994 (unpublished), submitted to Phys. Rev. D.

[28] J. Wiss, these proceedings.

[29] D. Fujino, these proceedings.

[30] P. Roudeau, these proceedings.

[31] I. Bigi, "Quo Vadis, Fascinum?" in these proceedings and references therein.

[32] D. Kaplan, these proceedings.

[33] T. Liu, these proceedings.

[34] G. Gladding, these proceedings.

[35] G. Burdman, these proceedings.

[36] J. Hewett, in " The Tau-Charm Factory in the Era of B-Factories and CESR", Stanford, California, August 15-16, 1994, edited by Lydia V. Beers and Martin L. Perl (Stanford Linear Accelerator Center, Stanford, 1994), SLAC Report No. 451, p. 206 and references therein.

[37] G. Blaylock, A. Seiden, and Y. Nir, Phys. Lett **B355**, 555 (1995); L. Wolfenstein, Carnegie Mellon University preprint, CMU-HEP95-04, 1995 (unpublished).

APPENDICES

Program Advisory Committee

Peter Blüm	University of Karlsruhe, Karlsruhe, Germany
Michel Davier	LAL, Orsay, France
Shou Xian Fang	IHEP, Beijing, China
Frederick J. Gilman	Carnegie-Mellon, Pittsburgh, PA, U.S.A.
Nathan Isgur	CEBAF, Newport News, VA, U.S.A.
John Jowett	CERN, Geneva, Switzerland
Vladimir Kadishevsky	JINR, Dubna, Russia
Thomas, B. W. Kirk	BNL, Upton, NY, U.S.A.
Jasper Kirkby	CERN, Geneva, Switzerland
T.D. Lee	Columbia University, New York, NY, U.S.A.
Wolfgang K. H. Panofsky	SLAC, Stanford, CA, U.S.A.
Lawrence E. Price	ANL, Argonne, IL, U.S.A.
Zhipeng Zheng	IHEP, Beijing, China

Local Organizing Committee

Jeffrey A. Appel	FNAL, Batavia, IL, U.S.A.
Edmond L. Berger	ANL, Argonne, IL, U.S.A.
Ikaros I. Bigi	University of Notre-Dame, Notre-Dame, IN, U.S.A.
Thomas H. Fields	ANL, Argonne, IL, U.S.A.
Gary Gladding	University of Illinois, Urbana, IL, U.S.A.
Sandra A. Klepec (secretariat)	ANL, Argonne, IL, U.S.A.
Lawrence E. Price	ANL, Argonne , IL, U.S.A.
José Repond (chairman)	ANL, Argonne, IL, U.S.A.
Kamal K. Seth	Northwestern University, Evanston, IL, U.S.A.
Lee C. Teng	ANL, Argonne, IL, U.S.A.
Walter Toki	Colorado State University, Fort Collins, CO, U.S.A.
A. Barry Wicklund	ANL, Argonne, IL, U.S.A.

SCIENTIFIC PROGRAM

Wednesday, 21 June, 1995

Introduction   chairman: L.E. Price

8:30 - 8:45	Welcome to the Workshop on the Tau/charm Factory	A.Schriesheim
8:45 - 9:00	Welcome to the Workshop on the Tau/charm Factory	Z.Zheng
9:00 - 9:30	Tau/charm Factory Overview	J.Kirkby

Tau Lepton Physics   chairman: P.Roudeau

9:30 - 10:00	Importance of Precision Measurements of the τ Lepton	A.Pich
10:00 - 10:30	Tau Physics at CLEO II with Prospects for the B-Factories	C.White
11:00 - 11:30	Measurements of Properties of the τ Lepton at LEP	S.Wasserbaech
11:30 - 12:00	Prospects for High Precision Measurements at the τcF Including Current Achievements of BES	T.Huang
12:00 - 12:30	Test of CP, T and CVC Violation in τ Decay using the Tau/charm Factory	Y.S.Tsai

Accelerator Designs   chairman: J. H. Norem

14:00 - 14:25	Tau/charm Factory Collider Design at IHEP	Y.Z.Wu

14:25 - 14:50	JINR τcF: Status and Perspectives	E.Perelstein
14:50 - 15:15	Argonne Tau/charm Factory Collider Design Study	L.C.Teng
15:15 - 15:40	B Factory Collider Designs and Future Plans	M.S.Zisman
16:10 - 17:00	Round table Discussion of Collider Designs	M.Tigner (chair)

Detector Studies

chairman: **M.Tigner**

17:00 - 17:25	Tau/charm Factory Detector Overview	J.Kirkby
17:25 - 17:45	A Fast Time-of-Flight Detector also Used as Tracking Detector	T.Y.Chen
17:45 - 18:00	Monte Carlo Simulation of the Tau/charm Factory at IHEP	Y.Z.Huang
18:00 - 18:15	Program of the Beijing τcF Feasibility Study	S.H.Wang

Thursday, 22 June, 1995

Charmonium Physics and Hadronic Spectroscopy

chairman: **T.Huang**

8:30 - 9:05	Status of the Theory of the Charmonium System and Future Outlook	A.El'Khadra
9:05 - 9:40	Fermilab E760 and E835: Charmonium Formation in $\bar{p}p$ Annihilation	C.Ginsberg
9:40 - 10:00	Electroweak Radiative Corrections and Measurements of R_{had}	M.Swartz
10:30 - 11:05	Current Theoretical Expectations for (Light) "Exotic" Hadrons	T.Barnes
11:05 - 11:40	Status of Hadron Spectroscopy at LEAR	C.A.Meyer

11:40 - 12:05   Prospects for Precision Charmonium . . . . W.Toki
                Measurements

Colloquium   chairman: E.L.Berger

14:00 - 15:15   Symmetries and Asymmetries . . . . T.D.Lee

Charm Physics: Tests of the Standard Model

                                     chairman: J.Kirkby

15:30 - 16:00   Quo Vadis, Fascinum? . . . . I.I.Bigi
16:30 - 17:10   Charm Decay in Fixed Target: Present . . . . J.Wiss
                and Future
17:10 - 17:45   Charm Physics at CLEO and Future Prospects . . . . . . D.Fujino
                at a B Factory
17:45 - 18:20   Prospects for Charm Physics at the $\tau c F$ . . . . P.Roudeau
18:20 - 18:40   Decay Rates, Structure Functions and . . . . E.Mirkes
                New Physics Effects in Hadronic Tau Decays

Friday, 23 June, 1995

Charm Physics: Beyond the Standard Model

                                       chairman: W.Toki

8:30 - 9:00   Potential for Discoveries in Charm . . . . G.Burdman
                Meson Physics

9:00 - 9:30	High-Impact Charm Physics at the Turn of the Millenium	D.M.Kaplan
9:30 - 10:00	$D^0\overline{D^0}$ Mixing, CP Violation and Rare Charm Decays	T.Liu
10:00 - 10:30	Rare D Decays, $D^0\overline{D^0}$ Mixing, and CP Violation at the Tau/charm Factory	G.Gladding

Conclusion chairman: **Z.Zheng**

11:00 - 12:15 Summary of the Tau/charm Factory Workshop ...... F.Gilman

Optional Tours

14:00 - 14:45 Argonne Wakefield Accelerator
14:45 - 16:00 Advanced Photon Source

List of Participants

Ted Barnes	Oak Ridge National Laboratory
Edmond L. Berger	Argonne National Laboratory
Ikaros I. Bigi	University of Notre Dame
Geoffrey T. Bodwin	Argonne National Laboratory
Eric A. Braaten	Northwestern University
Gustavo Burdman	Fermi National Accelerator Laboratory
Ting-Yang Chen	Nanjing University
Claudio Coriano	Argonne National Laboratory
Daniel A. Crane	Argonne National Laboratory
James W. Cronin	University of Chicago
Edwin A. Crosbie	Argonne National Laboratory
Malcolm Derrick	Argonne National Laboratory
Aida El-Khadra	Ohio State University
Xiaoling Fan	Northwestern University
Thomas H. Fields	Argonne National Laboratory
Don H. Fujino	Ohio State University
Wei Gai	Argonne National Laboratory
Frederick J. Gilman	Carnegie-Mellon University
Camille M. Ginsburg	Northwestern University
Gary Gladding	University of Illinois
Lionel Gordon	Argonne National Laboratory
Tao Huang	Institute of High Energy Physics
Yinzhi Huang	Institute of High Energy Physics
Lihui Jin	Institute of High Energy Physics
Daniel M. Kaplan	Illinois Institute of Technology
Jasper Kirkby	CERN
Eve Kovacs	Argonne National Laboratory
Shinichi Kurokawa	KEK
T. D. Lee	Columbia University
Tiehui Liu	Princeton University
Curtis A. Meyer	Carnegie Mellon University
Daiva Mikunas	Argonne National Laboratory
Frederick E. Mills	Argonne National Laboratory
Erwin Mirkes	University of Wisconsin
James H. Norem	Argonne National Laboratory
Todd K. Pedlar	Northwestern University
Elcouno Perelstein	Joint Institute for Nuclear Research
Antonio Pich Zardoya	IFIC-Valencia
Lawrence E. Price	Argonne National Laboratory
James Proudfoot	Argonne National Laboratory
José Repond	Argonne National Laboratory

Patrick Roudeau	LAL, Orsay
Morris Swartz	Stanford Linear Accelerator Center
Lee C. Teng	Argonne National Laboratory
Maury Tigner	Cornell University
Walter Toki	Colorado State University
Yung-Su Tsai	Stanford Linear Accelerator Center
Yau W. Wah	University of Chicago
Shu-Hong Wang	Institute of High Energy Physics
Steven R. Wasserbaech	University of Washington
Christopher White	Ohio State University
A. Barry Wicklund	Argonne National Laboratory
Jim Wiss	University of Illinois
Yin-Zhi Wu	Institute of High Energy Physics
Bing-Lin Young	Iowa State University
Zhi-Peng Zheng	Institute of High Energy Physics
Michael S. Zisman	Lawrence Berkeley Laboratory

AUTHOR INDEX

A

ALEPH Collaboration, 72

B

Barnes, T., 285
Bigi, I. I., 331
Burdman, G., 409

C

Chen, S. M., 203
Chen, T.-Y., 212
Cheng, B. S., 203
CLEO Collaboration, 62
Crosbie, E. A., 160

E

El-Khadra, A. X., 233
E760/E835 Collaboration, 251

F

Fan, X. L., 203
Finkemeier, M., 119
Fujino, D., 375

G

Gilman, F. J., 491
Ginsburg, C. M., 251
Gladding, G., 480

H

He, M., 212
Hu, J. L., 203
Huang, N., 139
Huang, T., 89
Huang, Y. Z., 203

J

Jin, L. H., 139
Jin, S., 203

K

Kaplan, D. M., 425
Kirkby, J., 11
Kurokawa, S., 196

L

Liu, H. M., 203
Liu, T., 447

M

Ma, A. M., 203
Meyer, C. A., 307
Mirkes, E., 119

N

Norem, J., 160

P

Perelstein, E., 152, 196
Pich, A., 45

R

Repond, J., ix, 160
Roudeau, P., 391

S

Schrieschiem, A., 5
Swartz, M. L., 270

T

Teng, L. C., 160, 196
Tigner, M., 196
Toki, W., 319
Tsai, Y. S., 104, 113

W

Wang, D., 139
Wang, S.-H., 222
Wasserbaech, S. R., 72
White, C. G., 62
Wiss, J., 345
Wu, Y. Z., 139, 196

X

Xiong, W. J., 203

Y

Ye, S. Z., 203

Z

Zhang, D. H., 203
Zhang, N.-J., 212
Zhang, X.-Y., 212
Zheng, Z., 7
Zisman, M. S., 172, 196